FOUR CENTURIES OF SPECIAL GEOGRAPHY

Geography as a university subject dates from the late nineteenth century. However, in the preceding 400 years, more than 900 English-language books devoted to describing all the countries in the world, known collectively as special geography, were published. This book is the first comprehensive guide to this genre, which then formed the mainstream of geography. It lists, in a series of main entries, all known works published before 1888, providing extensive bibliographical information about them, together with critical notes about most of the books intended for an adult audience, as well as about many school books.

The main entries are preceded by an introduction in which the author evaluates special geography as a genre and discusses its relation to the evolving ideas of the period. A detailed guide to the organization of the main entries is provided, as are chronological and short-title indexes.

This book will interest scholars examining the development of geography in the sixteenth to the nineteenth centuries, as well as those working in the history of education and the evolution of interests among the reading public.

O.F.G. Sitwell is an associate professor in the Department of Geography at the University of Alberta.

Geography Anatomiz'd:
OR, THE
Geographical Grammar.
Being a Short and Exact
ANALYSIS
Of the whole Body of
Modern Geography,
After a New and Curious Method.
COMPREHENDING,

I. A General View of the Terraqueous Globe. Being a Compendious *System* of the true Fundamentals of *Geography*; Digested into various Definitions, *Problems*, Theorems, and *Paradoxes*: With a Transient Survey of the *Surface* of the *Earthly Ball*, as it consists of Land and Water.

II. A Particular View of the Terraqueous Globe. Being a clear and pleasant Prospect of all Remarkable Countries upon the Face of the whole Earth; shewing their Situation, *Extent*, Division, *Subdivision*, Cities, *Chief Towns*, Name, *Air*, Soil, *Commodities*, Rarities, *Archbishopricks*, Bishopricks, *Universities*, Manners, *Languages*, Government, *Arms*, Religion.

Collected from the Best Authors, and Illustrated with divers Maps.

The Sixth Edition, Corrected, and somewhat Enlarg'd.

By PAT. GORDON, M. A. F. R. S.

Omne tulit punctum qui miscuit utile dulci. Hor.

LONDON,
Printed for J. Nicholson, J. *and* B. Sprint, *and* S. Burroughs, in *Little Britain*; Andr. Bell, at the *Cross-Keys* and *Bible* in *Cornhil*, and R. Smith under the *Royal-Exchange*, 1711.

Title page, *Geography anatomized: or, a compleat geographical grammer*, Pat Gordon, 1711 (6th ed.)

O.F.G. Sitwell

Four Centuries of Special Geography

An annotated guide to books that purport to describe all the countries in the world published in English before 1888, with a critical introduction

UBC PRESS / VANCOUVER

Printed in Canada on acid-free paper ∞

ISBN 0-7748-0444-0

Canadian Cataloguing in Publication Data

Sitwell, O.F.G., 1933-
Four centuries of special geography

Includes bibliographical references and index.
ISBN 0-7748-0444-0

1. Geography – Bibliography. 2. Geography – Early works to 1800 – Bibliography. I. Title.

Z6002.S58 1993 016.91 C93-091687-5

This book has been published with the help of a grant from the Social Science Federation of Canada, using funds provided by the Social Sciences and Humanities Research Council of Canada.

UBC Press gratefully acknowledges the ongoing support to its publishing program from the Canada Council, the Province of British Columbia Cultural Services Branch, and the Department of Communications of the Government of Canada.

UBC Press
University of British Columbia
6344 Memorial Road
Vancouver, BC V6T 1Z2
(604) 822-3259
Fax: (604) 822-6083

To archivists and librarians

Contents

Tables and Figures

Preface

The ambition to identify every book whose author claimed to have described every country in the world took hold of me gradually. One root lay in a sense of dissatisfaction with my doctoral thesis. My graduate training coincided roughly with an episode known to geographers as the Quantitative Revolution. When it began I was a regional geographer; when it ended I was a spatial scientist. I learned much in the process, but nobody would call the thesis that was the proof of my newly acquired intellectual rigour a polished piece of work. Then there was my interest in the history of geography. The groundwork there was laid by George Tatham. His specialty was the eighteenth century, then a neglected wasteland so far as those interested in our traditions were concerned. When I set out to follow in his footsteps, I decided that I would survey that *terra incognita* and establish its boundaries. In less metaphoric speech, I decided that, since in those days geographers were people who described strange lands and distant places, I would compile a catalogue of them. Because I wanted the project to be manageable, I decided to limit the list to those authors who had sought to describe every country, not just some of them. Not long after I had made that decision, the then Keeper of Older Printed Books and Special Collections in the Library of Trinity College Dublin asked me, 'Do you know Cox?' As I did not, she produced the three volumes in which Edward Godfrey Cox had sought to list every book of travel, exploration, or discovery published in English before the beginning of the nineteenth century.

As I reconnoitred the field, it dawned on me that the men who had dominated my graduate program had been right to insist that their students cite their sources. Everyone makes mistakes, but those who show where they got their facts make it relatively easy for those who come later to correct an error. Thus, gradually, my ambition took its final shape. I would list every book, and as far as possible every edition of every book, whose author had sought to describe every country in the world, provided that the books had been written in English in the days before geography was taken over and made into an academic discipline by the university reforms of the late nineteenth century. In addition, because I intended to compile the guide that, had it been available, would have made it possible for me to write about the geography of times past rather than merely lay the foundation for later workers to do just that, I decided to provide information on where copies of the books could be found, and to suggest whether a particular book would or would not interest those who drew on my research.

What I had not expected, and what turned the task from the work of a few summer months, as I had supposed it would be, into something resembling the fate of Sisyphus, was the conviction that took root in the eighteenth century that children should be taught 'geography,' that is, that they should learn by heart the names, locations, and some further information about the countries of the world. If, at the moment when I realized that the scholarly requirement to be consistent would entail my adding books of that type to the four dozen or so titles that Cox had already noted, plus a few others that lay outside his scope, I had had a realistic idea of how many authors had dedicated their lives to putting into books a set of facts that children 'ought' to know, I would have abandoned the project. On the other hand, even as vistas still to be travelled unfolded before me, the conviction grew that perhaps if people came to reflect on the enormous effort that our predecessors had devoted to compelling their children to learn this particular body of information (which, after all, had not been thought necessary in other civilizations, either those ancestral to our own, or elsewhere in the world), we might learn some lessons from the experience. To understand history, it is necessary to know what happened. The belief that others would find my guide helpful in establishing what had happened in one particular branch of the intellectual history of our people kept me plodding on.

To an extent that I could not have imagined when I began, this work is the product of countless heads and hands, many of them unknown to me. If there is a group of people who, besides being absolutely indispensable to the scholar, are more courteous, helpful, and self-effacing than librarians, they must live in some world where I can only hope to go when I die. In the thirteen years since working on this guide became a major project, I have exchanged letters with more than a hundred librarians as I have sought information about books I believed to be in their particular collection. They will find that the information they contributed is acknowledged individually in the notes accompanying the entries that form the body of the guide. I seize this opportunity to thank them collectively, and not just the ones I have dealt with, either by mail or in person when I was lucky enough to get to their libraries, but also the many others who for generations have catalogued collections and worked to compile general or union catalogues. The genius in the laboratory or at the podium may be the star in the academic universe, but it is the librarians, or the information-retrieval specialists, as we are beginning to call them, who make the university possible.

To this collective thank you I want to add some particular debts of gratitude. Though not all whose names follow are librarians, they all helped me by going further than their professional training required in directing me to, or helping me gather or verify, some of the information that I have compiled. In the order in which I called on them, they are: Gordon Davies, Mary (Paul) Pollard, Elspeth Buxton, Gene Olson, Norma Gutteridge, Richard Kuhn, Anne Yandle, and Stephen Siegnoski. In addition I want to thank the staff of the libraries of the University of Alberta.

There are debts of other sorts. First, to Fran Metcalfe, who typed the main entries into the computer; this made it possible for me to edit and re-edit them to satisfy the spirits who guard the standards of bibliography and scholarly publication. Brian Robinson, Ken Fairbairn, and Wilf Hamley provided help or advice at critical moments. Then there are computer people: first, the people who designed the hardware and the software I used; though they will never know it, without their many different contributions, I would never have been able to come as close as I have to mastering the skills needed by those who compile bibliographies. Then there are three people whom I know: first, Dave Halliwell, who designed programs that changed the text from a word-processing system that, though it works quickly on the monitor, prompts spleen when it comes to getting text onto the page, into another that performs marvels of 'typesetting,' but only after plodding like an ageing carthorse on the screen; then John Honsaker and Dan Hemenway, both ready with advice on how to translate my vague desires into specific requests that machines could carry out.

Last in time, though far from last in terms of the debt I owe them, are the team who work at the University of British Columbia Press. Once the guide was in their hands, the speed at which work went forward changed from the plod of a tortoise to the sprint of a hare almost overnight. In particular, I want to thank Frank Chow, who undertook to copy-edit this guide. What is good about its appearance I owe to him; its flaws are my responsibility.

I thank the University of Alberta for the time and facilities it has provided. Without them I could not have compiled the work that follows. I am also happy to acknowledge that this book has been published with the help of a grant from the Aid to Scholarly Publications Programme of the Social Science Federation of Canada, using funds provided by the Social Sciences and Humanities Research Council of Canada.

Finally, and put last rhetorically because they should be first, I want to thank the Historical Geography Research Group of the Institute of British Geographers. At one time it seemed that they would publish this guide as part of their Research Series. That ultimately proved to be impracticable, but not before they, and in particular Charles Withers, their editor, had provided me with support, advice, and encouragement at a time when it seemed quite possible that, for want of such elsewhere, the guide might end up as a typescript in a filing cabinet.

Introduction

As its title implies, this book is intended to provide a guide to a particular body of writings, namely, those known as *special geography*. The Latin version of this term was coined in 1611 by Bartholomäus Keckermann (Büttner 1978).[1] In the decades following the publication, in 1650, of the *Geographia generalis* by Bernard Varen (usually referred to by the Latinized version of his name as Varenius), the term diffused into English with the meaning of books that purport to describe all the countries in the world. It is with this meaning that it is used in this guide.

It is anachronistic to apply the term *special geography* to books written in the sixteenth century. But as Taylor (1930) showed empirically, the question of which writings in that century can or cannot be called geography at all, far less be assigned to some subfield or other, is not easily answered. The outstanding problem for those with literal minds is that the word *geography* itself was scarcely used at that time. This prevented neither Taylor nor the scholars who came after her from asserting that some of the things written at the time were geography. It is sufficient for the purposes of this guide that all the books identified in it as special geography or as closely related, and that were published before Varen, were included by Taylor in the lengthy bibliographies that form the core of her monographs (Taylor 1930, 1934). Thus justified, I have taken the beginning of printing as my starting point.

The starting point thus anchored, what justifies making 1887 the terminus? As we shall see below, so far as books written for the general public, as distinct from textbooks intended for students, are concerned, the 1880s saw the end of the genre. There are thus empirical grounds for choosing that decade. It is a fact, moreover, that the last quarter of the nineteenth century is seen by many authorities as a critical period in the history of geography. Capel (1981a), for example, has argued that, in the words of Glick: '[The] academic geography which emerged in the 1890s had very little to do with what had been called geography previously, no institutional connections with it and few personal links' (Glick 1984:278). The reference to institutional connections points in particular to the establishment of departments of geography in universities. So far as the English-speaking world is concerned, the event that marks more clearly than any other the beginning of the institutionalization of geography at the university level was the appointment of Halford Mackinder as Reader in Geography at Oxford. This happened in 1887 (Firth 1918).

Having established the temporal frame occupied by special geography, this Introduction has three further functions. One is to provide readers with a key to the contents of the guide. Those who already know that they want to learn whether a particular book was or was not special geography, or to learn basic bibliographic information about some instance of the genre, or in what library a copy of a particular book can be found (and also, in many though not all cases, some critical information about it), may turn to the section 'A Guide for Users,' on page 28. The two sections of the Introduction that follow immediately are intended for those who want to know either more about the historical context of special geography, and of the way in which public interest in it waxed and waned, or more about the character of the genre. The latter topic is dealt with first, followed by the review of the historical context. This broad survey of the genre ends with a preliminary assessment of its place in the history of ideas.

Special Geography as a Genre

Although special geography is a convenient term in the context of this guide, it was less well known, even at the time of its widest acceptance, than three others: *cosmography*, *chorography*, and *topography*. The use of these terms, as a set, is attributed to Ptolemy (Broc 1980:66-7). He used them to refer to descriptions of the earth considered in its relations to the rest of the universe, of regions, and of small places, respectively. The differentiating criterion is thus one of scale. Geography itself belongs in the same series, being distinguished from cosmography by being 'la description "des principaux lieux et parties connues de la terre ... Elle diffère de la cosmographie en ce qu'elle distingue la terre par montagnes, fleuves, rivières, mers ... sans avoir regard aux cercles de la sphère"' (Broc 1980:67). Thus geography differs from chorography and topography in terms of scale, but from cosmography because the two are concerned with different sets of facts about the earth. Moreover, cosmography involves the application, if not the understanding of, mathematical principles, while geography proceeds by way of verbal description.

The use of the differentiating criterion of scale leads to questions about the nature of the 'graphos' to be found at each of the scales. The nub of these questions is seen most easily if we turn from written descriptions to their cartographic counterparts. Do maps at small, intermediate, and large scales differ in more than their scale? The empirical answer is *yes*; they differ not only in the amount of information but also in the classes of information that are provided on them. Consider, for example, maps of North America at scales of $^{1}/_{50,000}$, and $^{1}/_{12,000,000}$, respectively. The latter cannot show the relatively small features that not only appear on the former but are the very reason why we make 'topographic' maps. The converse is probably, though not necessarily, true. Maps of continents are generally used to show highly generalized information, such as the density of population or the distribution of climates or of ethnic groups. Such information could be shown on large-scale maps, but if it was, there might easily be no variation in the information presented on many individual sheets from one end to the other. We thus reach the question: Does the same state of affairs exist in the case of written descriptions?

A response, and one that would be satisfying from a narrowly personal point of view, would be to say that I compiled this guide so that a valid answer to that question, among others, could be sought. Until someone has examined either the whole body of the genre or at least a representative sample of it, we cannot say with confidence how the authors of special geography responded to this problem. The first objective of my research has been to establish the limits of the genre.

If I cannot give a definitive answer, I can, however, offer a provisional one. As a first step, I shall reword the question in terms that are appropriate to written descriptions. Is it either necessary or possible for a description of the whole earth to be anything other than the aggregation of the separate descriptions of every individual part of it? In other words, is (special) geography anything other than a compilation of individual topographies (with chorography being a resting point along the way)?

Ultimately, the answer to this question rests on the nature of generalization, and, thanks to recent research on the neurological processes at work in the brain, one can now be given with a fair degree of confidence. In the words of Laughlin and d'Aquili: 'All sentient beings exist in an "over-rich data environment" – that is, no organism is equipped with memory capable of storing all sensory input. More importantly, all such beings ... must for survival, be capable of discriminating between relevant and irrelevant input. [They do so] by constructing models of their environment against which they match sensory input for relevance' (Laughlin and d'Aquili 1974:78).

The point is clear: even the relatively elementary mental skills needed at the level of basic animal survival require the capacity to generalize. In other words, most of the things that we tend to think of as being individual facts either are or involve generalizations. But if generalization is necessary if we are to survive even at the animal level, and if virtually everything that we are used to calling facts are themselves generalizations, it is hard to see why we cannot have high-level generalizations that are appropriate to written descriptions of large parts of the earth's surface, or even to the earth as a whole, just as we have cartographic counterparts to such generalizations in the case of small-scale maps.

We thus see that, if geography is taken to be a body of information, as distinct from the 'facts' out there in the material universe, then it began (or so it seems safe to presume) when travellers returned and told the people back home what they had seen on their travels. Such stories necessarily included only those things that the travellers thought worth reporting. It also seems reasonable to believe that the things that caught the travellers' eyes were the 'unfamiliar.' However, there is nothing about that categorization that provides any clues as to the actual content of the information that was recorded. In other words, there is no way of telling a priori the form such tales took, or the classes of information they contained.

At some point, however, someone somewhere grew dissatisfied with the disorderly nature of the information that had accumulated about strange lands and distant places. From such a sense of dissatisfaction emerged geography in a narrower sense, as tidy-minded people undertook the task of gathering the miscellaneous information available in travellers' tales, sorting it, and putting it in some order. We know neither the first order nor the original reasoning that justified it. Then, or perhaps later, the need to include equivalent information about 'home' also became apparent.

We can sum up what happened up to that point by saying that a body of 'facts' had been related together, or, in more abstract terms, that a system of relationships had been established. The key concepts that governed those relationships were, in the terminology of today, those of *place* and *country*. Because we have used these terms since childhood, we take for granted that their meanings are clear, that everyone knows what a place is. But to make the assertion is to raise a doubt. Of the definitions of place that I know, the most satisfactory is that provided by Robinson. He argued (1973) that a place is an area on the surface of the earth whose approximate location and extent is known, as shown by the fact that *people have a name for it*. The emphasis is added to underline the fact that the criteria for establishing the existence of a place are subjective; given the arbitrary nature of language, there is no guarantee that the criteria will be understood and so used in the same way by different people.

Places thus defined vary greatly in size. They may be as large as a continent or an ocean. At the lower end of the range, the limit is also set empirically. Places are larger than a human being by at least an order of magnitude. Examples are provided by a cluster of houses, a lake, a wood, a hill. What these have in common is the fact that they are so large that we cannot take them in at a glance unless we are removed from them to a considerable distance – which is to say, not until they have been, in terms of our literal perception of them, reduced in scale.

Within this range of scales a special place is reserved for *countries*. As used by the authors of special geography, a country is the largest type of place that is subject to central political control; in other words, the term is all but synonymous with *state*. That basic concept is commonly modified in either of two related ways. On the one hand, political control over some unit of the earth's surface may be fragmented among a number of authorities, and yet that unit may still be recognized as a country (such as Italy and Germany for most of the period under consideration). Alternatively, units within a single state may be recognized as countries even though they lack political autonomy (such as Wales and, after 1707, Scotland). In both cases the justification is provided by reference to past conditions. That Wales had been politically independent in the recorded past was a fact known to all educated people in the British Isles. Germany and Italy were considered to be countries because they appear as such in the Classical literature that was still taken to be authoritative by those who wrote special geography.

Judging from the information provided by the great majority of the special geographies identified in this guide, one of the characteristics that led those who wrote them to take *country* as their basic organizing concept was the fact that political units have boundaries that are well defined, at least in the minds of the people who talk about them. The importance of the boundaries is shown by the fact that, with only a small number of exceptions, the first class of information that the author of a special geography gave about a country was its boundaries.

Beyond this, we can note that the information provided by special geography with respect to any particular country – the *things* that lie within its boundaries, its *content* as it were – was, initially at least, the information that would be available to someone looking at a map of the country. First and foremost were the subunits of the political/administrative hierarchy into which the country was divided, such as the counties of England and the regions of France. In some of the shorter special geographies, a large part of the text is devoted simply to naming these. Other features whose presence can be attributed to their prior inclusion on maps, and which are also almost invariably present in the text, are rivers and cities.

Counties, to use the English term, are themselves made up of parishes (or villages), and some authors carried the disaggregational approach to the level of describing sub-sub-units of the spatial-governmental hierarchy. Still lower in the hierarchy come the properties of individuals. No author went so far as to try to list, still less describe, them. Reflection on this act of restraint leads to some observations on the historical context within which the authors of special geography worked.

In those days, and this is especially true if we consider the state of affairs that prevailed when special geography was taking root in the world of letters, the state was the estate of the monarch. The control exercised by monarchs over their people was perceived to be not merely analogous to that of the father (as it would have been in those days) over his family, but the same in every way save for scale – a difference that was clearly perceived to be without significance. Of course, in the case of large kingdoms, monarchs could not administer their estates in person; instead, they delegated the administration of regional units to members of their courts – hence the derivation of *county* from *count*. Evidence of this interpretation is provided by the title of AVITY (1615), which begins: 'The *estates*, empires, and principallities [sic] of the world' (emphasis added). Furthermore, in the body of the book, the description of America, which is extremely brief, does not stand on its own, but is included as part of the description of that part of the earth ruled by the king of Spain.

Monarchs live in palaces; in the more detailed special geographies, it is quite common to find a description of the palace of a country. In others, yet more detailed, some of the great houses of the landed aristocracy are also described. In none, however, is there any discussion of the individual house of a single 'ordinary' family. The world of the special geographer is a hierarchy; the higher individuals stand in that hierarchy, the more important they are. And attention is granted by the author in proportion to importance. In no case do those who form the base of the pyramid qualify for individual attention.

Without this limitation it seems unlikely that any author would have attempted a complete special geography. The amount of information needed would have been self-evidently overwhelming. As it is, some of them do seem to have tried to describe every named settlement in their own country. The author who came closest to succeeding was probably BÜSCHING (1762); in English translation he needed three large volumes amounting to a total of almost 2,000 pages to describe Germany alone.

Reverting now to the problem of generalization and the issue of whether there are classes of information that are appropriate to high levels of generalization, one possibility that needs only to be named to seem entirely appropriate is national character. In Glacken's opinion, 'there was a conspicuous interest in the subject in the seventeenth century in England' (Glacken 1967:451). Yet neither he nor Taylor (1934), on whom he draws, mentions any author of a special geography who made systematic use of the concept. This is not to say that it is never found in the genre, but the only work in which the idea is discussed in detail lies at its margin (BARCLAY 1631). When it is used, as it was quite often in the schoolbooks of the nineteenth century, it is taken for granted; in other words, a set of characteristics is attributed to the people of some country, but without discussion, still less justification.

There is no sign that the author of any special geography recognized explicitly the methodological advantage that would accrue to them from the possession of such a device.

An alternative approach, and one that recent controversy has made familiar to geographers, does appear from time to time. In some cases an item of information is provided because it is unique to the country being described: 'There is nothing more famous in this kingdome then [sic] the Salike lawe: whereby it is prouided: that no woman, nor the heire of her, as in right, shall enioy the crowne of France: but it goeth alwayes to the heire male' (ABBOT 1599:A2 verso). Such an approach is uninfluenced by scale, and its use shows that some special geographers were aware, even if unconsciously, of a need to have some criterion by means of which to decide when, once they had gone beyond identifying those features visible on a map, they ought to stop adding more facts. This problem was never solved except in the empirical sense that every author made the necessary decisions in the course of compilation.

This is the appropriate point at which to address the question of exactly what information about countries was provided in special geographies. A complete inventory awaits further research, but a provisional answer is offered by one of the rare programmatic statements to appear within the genre. It was the work of Bernard Varen. As Baker (1955) notes, though Varen never wrote a description of all the countries of the world, he did, in his *Geographia generalis*, list the topics that a special geography should include. He also wrote a description of Japan in which he exemplified, though not perfectly, his own precepts.

Here, as Baker (1955:53-4) records them, are the three classes of 'particulars' that Varen recommended. The first class comprises the 'terrestrial':

1. The limits and bounds of the country
2. The longitude and situation of places
3. The figure of the country (i.e., its shape on a map)
4. Its magnitude
5. Its mountains; their names, situations, altitudes, properties, and things contained in them
6. Its mines
7. Its woods and deserts
8. Its waters; as seas, rivers, lakes, marshes, springs; their rise, origin, and breadth; the quantity, quality, and celerity of their waters, with their cataracts
9. The fertility, barrenness, and fruits of the country
10. The living creatures

The second class of particulars are 'celestial':

1. The distance of the place from the equator and pole
2. The obliquity of the motion of the stars above the horizon
3. The length of days and nights
4. The climate and zone
5. The heat and seasons, wind, rain, and other meteors
6. The rising and continuance of the stars above the horizon
7. The stars that pass through the zenith of the place
8. The celerity with which each place revolves, according to the Copernican system

Last come the 'human' particulars:

1. The stature of the inhabitants; their meat, drink, and origin
2. Their arts, profits, commodities, and trade
3. Their virtues and vices; their capacity and learning
4. Their ceremonies at birth, marriages, and funerals
5. Their speech and language
6. Their political government
7. Their religion and church government
8. Their cities
9. Their memorable histories
10. Their famous men and women, artificers, and inventions

To some eyes the human facts seem arbitrary, and Varen himself attributed them to 'convention and usefulness' (Baker 1955:58). By convention he was presumably referring to his predecessors, who stretch in a line that extends through Sebastian Münster to Strabo (Broc 1980, ch. 6). The topic of 'usefulness' is considered in the survey of the historical context of special geography that follows this section.

Those with much time at their disposal can examine the books listed in this guide, and so establish the extent to which the authors who followed Varen chronologically also followed his precepts. I can say only that many of the longer versions of special geography did so quite closely, though usually with a higher degree of fidelity to the terrestrial and human classes than to the celestial.

The Literary Form of the Genre

Another problem that deserves to be considered in this review of the genre of special geography is that of the literary form within which information is presented. The existence of the problem is revealed by the fact that, during the period covered in this guide, there were, besides special geographies as such, books generally referred to as geographical dictionaries. The difference between them is that special geographies present information in the form of continuous prose, and the countries are described in an order that approaches the contiguous. In the dictionaries the same information is divided into small units. Each unit pertains to an individual place, and these places are listed alphabetically.

The question of which form to use arose because geographers had difficulty with the order in which to present their information. Choosing the alphabetical sequence of a dictionary was one solution. The problem with a dictionary, which is of course a problem only for an author with a particular frame of mind, is that they are read only for reference. Authors who envisaged their readers taking up their books for the pleasure of reading set themselves to writing continuous prose. They were then faced with a pair of problems: the spatial framework within which to enclose their facts, and the order of their presentation. As a rule, their response was to proceed at the level of continents where facts were sparse, at the level of countries normally, and at the level of still smaller units where the facts were superabundant. All three levels could be used in a single book: for 'remote' parts of the world, for Europe, and for the author's native land, respectively. As for the order of treatment, as noted above, something resembling a contiguous sequence would be attempted. I have discussed the issues involved in such cases (Sitwell 1972, 1980), and we need not linger over them here.

Talk of dictionaries raises the question of gazetteers. According to the definition provided earlier, a gazetteer is a geographical dictionary. It is, however, one in which the information about a place is usually limited to its location on the surface of the earth, generally, though not invariably, identified in terms of its coordinates of latitude and longitude. Few gazetteers in the modern sense were compiled during the period covered by this guide, though the word itself was used as the title of a geographical dictionary (EACHARD 1692), in the process acquiring the meaning that gradually evolved into the one that it has today. No gazetteers in the modern sense are included in this guide.

The Audience to Which Special Geography Was Addressed

Although what I have just written implies that the geographers who wrote continuous prose did so in the hope that their readers would read for pleasure, or at least for voluntary self-enlightenment, I must now acknowledge that this can be asserted confidently of only some of them. Though well over a hundred books were clearly written with an adult audience in mind, far more were written to be read by 'youth,' and to be read not for pleasure but so that those who read would be educated by the experience. The conceptual difference between the two classes of book is straightforward. But although the conceptual distinction is clear, the issue of whether it resulted in distinctions of style, content, or organization between the two classes of book is still open. It is certainly a commonplace among commentators that the schoolbooks are dull reading. Though many of the adult works might also appear to deserve that judgment when read today, the fact that so many of them were published, and also that most of them reached a second edition, and some many more than that, suggests that they appealed to a sizable audience.

Special Geography in Its Historical Context and the Chronology of Publication

No serious scholar today will disagree with Livingstone when he argues that those who study the evolution of geography should relate 'the history of their subject to [the] broader social and intellectual currents' of its time (Livingstone 1988:269). Special geography does, however, offer to the historian the challenge of its manifest continuity of form and content throughout a period when the world of European ideas was changing greatly. The key to resolving this seemingly paradoxical independence of one 'unit' of the history of ideas from its changing context lies in the ambiguity of language. Those elements of the world of ideas that remained constant must be separated from those where the vocabulary is the same but the words have changed their reference.

The Historical Constants

As we saw in the previous section, two constants belong to what might be called the internal history of special geography: the objective of the authors who worked in the genre, and the types of information they included as content. In addition, three elements belonging to the general world of ideas changed little during the time covered by this guide: first, a belief in the primacy of particulars; second, an enthusiasm for useful knowledge; third, voyages of exploration that radiated from northwestern Europe throughout the period, and so made known to its people the extent and nature of the rest of the world. In other words, the 'discovery' of new lands was an intellectual constant during the period.

The Primacy of Particulars: The Influence of Francis Bacon

Though the influence of the ideas of Francis Bacon on the history of science has long been acknowledged (Singer 1959), it is only recently that, thanks to the work of Bowen (1981), the particular way in which they shaped the geographical writings of the time has become apparent.

Consider the following words of Bacon that Bowen quotes: 'I ... dwelling purely and constantly among the facts of nature, withdraw my intellect from them no further than may suffice to let the images and rays of natural objects meet in a point, *as they do in the sense of vision*; whence it follows that the strength and excellency of the wit has but little to do in the matter' (Bowen 1981:46; emphasis added). Do any scholars dwell more obviously among the 'facts of nature,' and, relying solely on their 'sense of vision,' seek to record those facts, than those who set out to identify places and describe them?

Bacon's influence, however, goes beyond this well-known 'privileging' of fact. Bowen makes this point by first quoting the following passage and then elaborating on Bacon's intention: 'The men of experiment are like the ant; they only collect and use; the reasoners resemble spiders, who make cobwebs out of their own substance. But the bee takes the middle course; it gathers its material from the flowers of the garden and of the field, but transforms and digests it by a power of its own' (Bowen 1981:46).

Then, in her own words: 'The particulars of natural history, collected by means of the senses, were to be *recorded and arranged in tables* as an aid to the memory, so that the intellect could then begin the work of extracting axioms from experience, as the basis for new experiments' (Bowen 1981:55; emphasis added).

All that the geographers of the time had to do to be able to adapt Bacon's method to their own interests was substitute the particulars of 'place' for those of 'natural history,' and replace the 'extraction of axioms' with the 'acquisition of the information needed to attain utilitarian ends' as being the objective of intellectual activity.

In 1693 Patrick GORDON published his *Geography Anatomiz'd: Or, the Geographical Grammer* [sic]. In the unpaginated preface, he claimed to have 'reduc'd the whole body of modern geography to a true grammatical method; this science being as capable of being taught by grammar as any tongue whatever.' Because Gordon himself did not justify his claim by an explicit exposition of his method, we are left to infer the latter from what he did. And what he did was provide table after table containing information of some of the types that Varen had declared were appropriate to special geography.

The seventeenth century has often been 'cited for its fascination with language' (Cohen 1977:1), so, although the particular source of Gordon's idea that the pedagogy associated with language could be a model for geographers is not known, that he should have had the idea at all is not surprising. Furthermore, whatever question marks may hang over this issue, there can be no doubt as to Gordon's dual acceptance of facts as being the heart of knowledge, and of the presentation of particular sets of facts in the form of tables as being the true and proper method for a scientist to use.[2]

While there can be little doubt that Bacon's influence on special geography was great, it is important to recognize that the earliest special geographies are either earlier than or contemporaneous with Bacon's major publications. But Bacon did not conceive of his ideas *de novo*. He too was aware of the currents of his time, and the sixteenth century provided, in Tycho Brahe and Johann Kepler, two astronomers who, as a pair, seemed to personify the inductive method. Brahe, who never abandoned a belief in a terracentric universe, devoted many years to recording the apparent position of the planets in the sky as they move along their orbits. Equipped with this greatly enlarged body of data, Kepler, driven by the conviction that the orbits of the heavenly bodies must be related according to some principle of harmony, searched until he found those famous 'laws' of his demonstrating that this was indeed the case (Butterfield 1957). It was easy for near contemporaries of Kepler, who knew only the results of his work and not the man himself, to misunderstand the relative priority of fact and 'philosophy' in what he had done.

Utility

That special geography had a utilitarian aspect we have already established in our consideration of the work of Varen. As further evidence we can note the opening lines of the title of Joseph BENTLEY (1839): *Modern geography, for the student, the man of business, and all classes who wish to know something of the world we live in; containing all the topographical, physical, historical, commercial, and political facts worthy of notice, relating to every empire, kingdom, republic, state, and country in the world ...* Though references to businessmen are not common in the titles of special geographies, they occur quite often in the prefatory statements of objective that their authors usually provide.

The root of that utilitarian spirit was well expressed, if it did not have its absolute source, in the writings of Descartes and Francis Bacon. Glacken (1967:426) refers to a 'famous passage of [Descartes's] *Discourse on Method* (1637)' that set the attaining of control over nature as a, if not the, goal of 'mankind.' In the case of Bacon, as Singer pointed out: '[The] grand motive in his attempt to found the sciences anew was the conviction that the knowledge man [then] possessed was of little service to him. Sovereignty over nature, which can be founded on knowledge alone, had been lost, and instead we [had] nothing but vain notions and blind experimentations. To restore the original commerce between man and nature, and to re-establish the *imperium hominis*, is [for Bacon] the object of all science' (Singer 1929, 2:885).

Though the ideas of Descartes and Bacon may have fertilized the utilitarian spirit of special geography, that they cannot be its root follows from the fact that was established above, namely, that a number of

English examples of the genre, as well as the famous Continental work of Sebastian Münster, preceded the major publications of both Descartes and Bacon.

Recently Büttner and Burmeister have argued that the Protestant Reformation played a critical role in turning the attention of many scholars in northern Europe towards the discovery of both the everyday world and distant countries (Büttner 1973, 1978, 1979a, 1979b; Büttner and Burmeister 1979). Among the points of contention between the Reformers and the Roman authorities was the degree of involvement shown by God in what might be called the day-to-day operation of the world. In the Catholic scheme, God created the world; this was a clear-cut act, located at a specific time in the past; that world had then been spoiled by human sin, and, as a result, was no longer an object of special interest. The Protestants concurred as far as the premises were concerned, but differed as to the conclusion, holding that God continued to act in the world by means of an agency they referred to as *providentia*; thus it was that God 'takes care that for mankind's sake everything in this world functions properly' (Büttner 1979a:156). As a result, a number of major scholars among the Reformers (Melanchthon, Peucer, Neander, Münster) turned their attention, in varying degrees, to the study of the earth, its form, and both the spatial organization of its external features and the processes visibly at work in them. Following on their heels came Keckermann, the man who, as we have already seen, coined the term *special geography*.

In conclusion, we can see that the reason Protestant scholars were interested in geography may have been their conviction that the knowledge gained from their empirical research was evidence of the beneficence of a divine architect. It is also true, however, that the facts they assembled in their books were profoundly useful to people engaged in commerce and navigation, and also to those who proselytized on behalf of voyages of exploration for non-theological reasons, such as the hope of gain from trade with previously unknown lands (Büttner 1979a, Hodgen 1954, Parks 1961, Taylor 1930).

Thus special geography appears as a symptom of a *Weltanschauung* that valued the collection of information in a systematic way, and did so not only because the facts were useful per se but also because, following Descartes and Bacon, it was believed that their mere accumulation laid the foundation for science.

The Discovery of New Lands

In the course of his magisterial survey of European attitudes towards, and relations with, the physical environment, Glacken provides evidence for the contention that the 'discoveries' of the explorers played a role in the world of ideas from early in the sixteenth century until well into the nineteenth. For the beginning of the period, he cites the case of Sebastian Münster, whose 'fame rests on the *Cosmography*, the most impressive of the early compilations of geography published after the voyages of discovery' (Glacken 1967:363). The first edition of that book appeared in 1544. An even earlier example of the new ideas at work is provided by Claval, who cites the example of Thomas More, whose *Utopia*, published in 1516, reveals a knowledge of the Inca empire (Claval 1972:16).

At the other end of the period, Glacken notes that 'the closing decades of the [eighteenth] century mark the fresh stimulus to natural history and ethnography coming from Cook and the Forsters ... The son's book ... not only charmed and inspired Alexander von Humboldt, but was the harbinger of the coming era of scientific travel' (Glacken 1967:502).

In that era of scientific travel the Royal Geographical Society, established in 1830, played an important role. Furthermore, we should also note that, in the words of its official historian, the society was just one of three that were founded 'in response to the stimulus given to geography by Cook's great voyages' (Mill 1930:9).

That the knowledge of new-found lands exerted a great influence on European thought is a truism. What is less well known is the great length of time it took for that influence to spread widely through society in general. We move here into the margins of myth. For at least half a dozen generations it was widely believed that the early sixteenth century marked a major watershed in British, as well as, though less generally, European history. Once upon a time, which is to say in the Middle Ages, superstition and ignorance reigned hand in hand; then, in quick succession, the events of the Renaissance led simultaneously to the Reformation and the Age of Discovery; suddenly, where there had been darkness, now there was light! This point of view is implicit in the words of Evans: 'the immortal Galileo lay rotting in the dungeons of the Inquisition for having dared to assert *the rotundity of the world*' (EVANS 1809:x; emphasis added).

Historians have been chipping at that simple, dramatic picture of history since at least the time of Whitehead (1927). Signs of the shift can be found in Taylor: 'In England ... notwithstanding the work of the Cabots ... medieval conceptions remained but little disturbed for half a century after [1497]' (Taylor 1930:1). Later, Glacken went much further. Writing about Jean Bodin, whose major works of geographical interest were published in 1566 (*Methodus*) and 1576 (*Republic*), he states:

> What is striking about both works is their reliance on classical and medieval ideas, contemporary European travel and opinion within Europe, and the slight consideration given to ... the New World ... It is revealing that an important thinker, writing almost seven-

> ty-five years after the discovery of America, still bases his arguments on classical and contemporary European evidence ... [It] is a sobering reminder that the full fruits of the age of discovery – the new ideas of humanity, human culture, the revelation of new environments – were not harvested, except for a few atypical individuals, until the seventeenth and eighteenth centuries. (Glacken 1967:434-5)

Education and Special Geography

Another factor, constant at first sight, but on closer examination found to be in flux during this period, was the conviction that geography should be taught to youth. The flux lies in the different assumptions made as to the age of those to be taught, and thus, by implication, the level of intellectual development they were assumed to have reached. The latter was, though again by implication, a corollary of some characteristic of geography that was to be taught. We may here take that characteristic to be the degree of difficulty entailed in learning either the subject matter or the ideas that gave it form.

In general terms, there is a clear development in the intentions of those who wrote special geographies for use in education. Initially the intended readers were students attending university; there follows a gradual lowering of expectations, a decline that reaches the point where, by the middle of the nineteenth century, it is likely to be children *entering school* who would be taught about the sphericity of the earth and its location in the universe as a background to learning enormous quantities of information about the countries of the world.

Though some authors continued to write for adults until the 1880s, it seems that roughly a hundred years before that, the reading of special geography ceased to be part of the studies at British universities. According to Warntz (1964), a similar change took place in the United States, though it happened there one to two generations later. The change is clear from the titles of the books, and further evidence is to be found in the statements of intent provided by the authors in their prefaces, advertisements, and introductions. The question of why or how change in the assessment of the 'difficulty' of geography came about has not attracted scholarly attention.

The Chronology of Publication

There is no space here for a detailed examination of the evolution of the genre whose members are listed in the pages that form the body of this guide. What follows is little more than a statistical summary of the chronology of publication, with a few remarks added to begin the task of relating the evolution of the genre to its historical context.

Adult Special Geography

We can begin by noting that the pattern of publication shown by the special geographies intended for an adult audience corresponds to what would be expected on the basis of Glacken's observation. The frequency of publication per decade is shown in Figure 1.

FIGURE 1

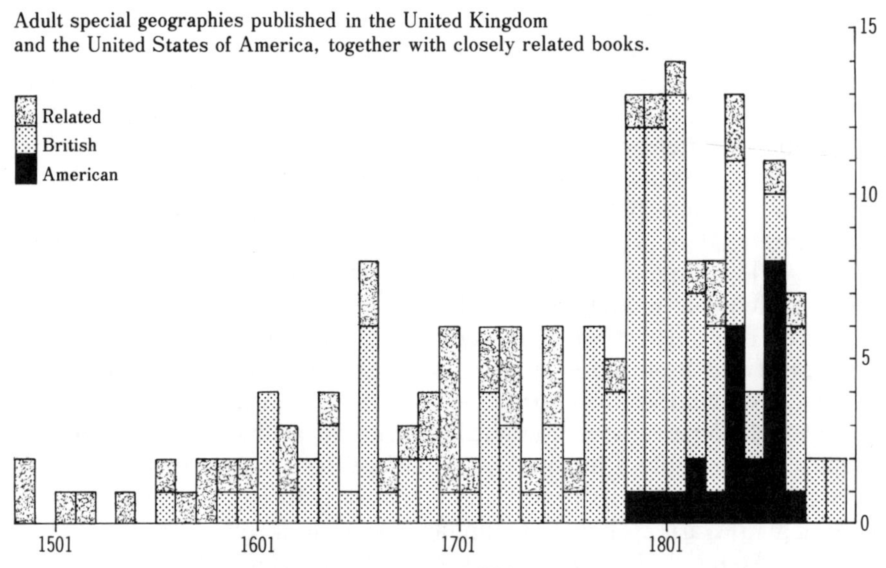

Notes: 1. In 1806 one title was published simultaneously in the U.K. and the U.S.A.
2. In 1511, 1636, 1768, and 1848, titles were published in other countries.

Beginning with an almost negligible interest in the sixteenth century, the pace of publication picked up in the seventeenth, though in an erratic fashion. There was a peak of six books in the decade following the almost complete dearth of works in the 1640s. The latter was the decade of the Civil War in England, and it is easy to believe that the book trade was adversely affected. The burst of activity in the 1650s then appears to have been a response to pent-up demand. If this is what happened, it is also clear that the demand was easily satisfied, for the remaining four decades of the century saw the publication of only six more special geographies. Interest then strengthened a little, but it was only in the last forty years of the next century that the rate of publication picked up strongly, reaching a plateau that extended into the nineteenth century.

The growth of interest identified by Glacken was followed by a decline. Though the rate of publication was relatively brisk in both the 1830s and 1850s, the trend was downward. Only two titles were published in the last full decade dealt with in this guide, and there were two more in the last eight years.

In the context of this discussion, we can also note that the plateau was aided by the publication of only three special geographies in America, whereas nineteen of the fifty-one books published after 1810 were American.

The Supply of Books for Youth

Books intended for young readers followed a pattern of publication similar to that for the adult audience in its general structure, although it was more compressed in time and contained a larger number of titles (Figure 2). In considering Figure 2, it must be noted that, as Table 1 shows, I felt more confident about the classification of books intended for youth published in America than I did in the case of their British counterparts. In preparing the figure, I included the books that I merely felt probably described all the countries in the world, as well as those where I knew this to be the case.

Excluding translations of Classical texts, the first special geography written explicitly for young people was published in 1671 (MERITON 1671). A full hundred years then passed before five or more were published in a single decade. After that output grew almost decade by decade to a peak of more than seventy in the 1850s (giving exact numbers is misleading because some books were published in both Britain and the United States, but see Table 1 for the numbers as I classified them). Once the decline set in, it was rapid, with some forty-five special geographies being published in the last full decade and around twenty in the next eight years. Unlike books for adults, some writing for youth and students has continued to the present.

As with books for an adult audience, the United States emerged as an important market in the last quarter of the eighteenth century. The market grew to the point that in each of the four decades from 1811 to 1850, British and American publications were equal or almost equal in number. The peak in the 1850s and 1860s was primarily a consequence of a surge in British publishing, but when the tide turned it ebbed more rapidly in that country than in America. The 1840s also saw a further diffusion of activity, with schoolbooks being published in the British colonies that were to become Canada and Australia.

FIGURE 2

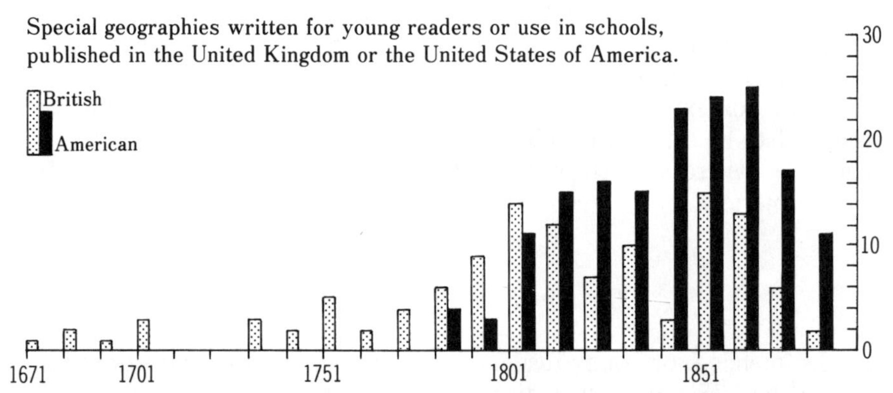

Notes: 1. In the U.K. in both the 18th. and 19th. centuries an additional title is known to have been published.
2. One title published in the U.K. in the 1570s.

A Contextual Interpretation of the Chronology

As a preliminary to a discussion of some specific points in the story of the slow growth and subsequent rapid decline in the books intended for adults, I shall summarize a thesis that explains the slowness of growth by seeing it as a small-scale counterpart to a general process in the larger world of ideas as a whole. The thesis is essentially that of Lovejoy. He maintained that 'in bringing about the change from the medieval to the modern conception of the scale of magnitude and the general arrangement of the physical world in space, it was not the Copernican hypothesis, nor even the splendid achievements of scientific astronomy during the two following centuries, that played the most significant and decisive part' (Lovejoy 1936:99).

There then follow a dozen closely written pages that lead to the conclusion that 'the more important features of the new conception of the world, then, owed little to any new hypotheses based upon the sort of observational grounds which we should nowadays call "scientific." They were [instead] chiefly derivative from philosophical and theological premises' (Lovejoy 1936:111).

TABLE 1
Special geography and its subclasses

Countries of Publication	UK	USA	Other	Total
Books for Adult Readers				
Special geography	111[1]	23[2]	3	137
Marginal special geography	47	5	0	52
Dictionaries	39	6	0	45
Other geography	36	4	0	40
Atlases	5	0	0	5
Total	238	38	3	279
Books for Youth				
Categories Certain				
Special geography	128	167	21	316
Marginal special geography	29	12	1	42
Other geography	8	4	1	13
Subtotal	165	183	23	371
Categories Uncertain				
Special geography	138	25	3	166
Marginal special geography	12	7	0	19
Subtotal	150	32	3	185
Total Geography, Including Unclassified				
Adult	238	38	3	279
Youth	315	215	26	556
Unclassified	91[2]	2	7[3]	100
Total	644	255	36	935
Not geography				22
Ghosts				29
Total Titles				**986**

1 One book was published simultaneously in the UK and the USA; it contributes to the count of both. All other books published in both countries contribute only to the count of the country where they were published first.
2 A pair of books that between them deal with the whole world are treated as a single work, although they appear separately in the main entries.
3 These are books whose country of publication is uncertain.

There is no space in this Introduction to summarize Lovejoy's full thesis in a way that would do it justice. Instead I shall condense it brutally as follows. There were present, from the time of Plato until at least the nineteenth century, in the realm of ideas that formed the Western *Weltanschauung*, two strands that played out a slow but implacable Hegelian gavotte. Lovejoy called them 'otherworldliness and this-worldliness' (Lovejoy 1936:24). He also argued that both strands were present, even though logically incompatible, in Plato's conception of God. Once these ideas were in place in the Western heritage of thought, they continued there throughout all the periods before the present, though the influence of one or the other usually dominated the ideas of any given period. Thus it was that the this-worldly Romans, who built a politico-military empire, gave way to the fantastical people of Christendom, whose eyes were obsessed with religious visions. Then, following the Renaissance, came men such as Galileo and Newton who laid the foundation of this-worldly science, and so made possible our own modern, practical era.

If we now focus our attention on the Middle Ages, argues Lovejoy, we can see that there was present in the God of the scholastics a built-in element of this-worldliness that, in a sense, 'demanded,' by the logic of its nature, to be let loose.[3] The logical working out of this idea did not take place then; the Middle Ages would not have had their characteristic otherworldliness if it had. It took place instead in early modern times, with its climax coming in the eighteenth century.

While some people might be willing to accept Lovejoy's thesis, others would reject it as being the product of an idealist outlook. I would like to argue, however, that it is possible to follow Lovejoy without retreating into idealism by seeing in his thesis an anticipation of a new materialism. The source of this interpretation lies in the findings of recent research in neuropsychology that were referred to earlier in the discussion of the human capacity to form 'mental' concepts such as generalizations.

It is only a matter of time before it is realized that once a theory of ideas that sees them as literal biochemical/electronic complexes of synaptic relationships extending over wide cerebral areas and operating in four dimensions (to paraphrase Laughlin and d'Aquili 1974:101-2) is accepted, then it will also be possible to acknowledge that ideas can act as agents in history 'without invoking some sort of Neo-Platonic idealism' (Laughlin and d'Aquili 1974:124).[4]

The consequences for us are that, just as the Middle Ages were filled with individuals who devoted themselves with the utmost practicality to everyday tasks, so the sixteenth, seventeenth, and eighteenth centuries were strongly influenced by individuals of pious outlook and a religious cast of mind. Even E.G.R. Taylor, who would have much preferred the evidence to have been otherwise, was compelled to admit, when speaking of the mid-sixteenth century, that 'Protestant theology and the pronunciation of Greek were the subjects that filled the minds of learned men in England' (Taylor 1930:74). To people of such an outlook, new lands in general, and most certainly those inhabited by savages, were of little interest.

It is thus not surprising that it was not until sixty years after Columbus had discovered America, and thirty years after some of Magellan's sailors had crossed the ocean that lay beyond it, that the small part of Münster's *Cosmography* (MÜNSTER 1553) that is devoted to the New World should have been published in England; nor is it surprising that of the eleven special geographies or related works published in England between 1553 and 1599, four were translations of Greek or Roman authors, while five more were based largely if indirectly on Ptolemy. The England of the time may have been home to Shakespeare, but, in the general world of European civilization, the country and its scholars occupied a peripheral, dependent, and largely conservative position throughout the eight generations following the invention of printing. The point has been made before (Taylor 1930), but it is worth repeating: it is only in the second half of the seventeenth century that original works by English authors became more common than translations either of Classical and medieval texts or of contemporary works by Continental authors.

Before turning to other observations on the historical context of special geography, we can note that the tradition of translation never ceased altogether. To cite only the largest debts: BÜSCHING (1762), MALTE-BRUN (1822-33), and STANFORD (1878-85) are all either direct translations from French or German, or followed such originals very closely.

It was not necessary for people to lose all interest in theology, however, for special geography to flourish. As Glacken (1967, ch. 8) has shown, not only was the eighteenth century the period when physico-theology flourished, but it is also a fact that that very branch of learning was itself extremely hospitable to cosmology, geology, and the life sciences (Glacken 1967:406-26). Though special geography as such falls outside Glacken's field of interest, much of what he has to say about the relationship between physico-theology and the emerging earth and life sciences is relevant to the genre. More particularly, it can be argued that there were two characteristics of physico-theology that favoured an interest in special geography: its utilitarian bias and its near obsession with detail. To take the latter point first: 'If one granted the initial assumption [of physico-theology] that every natural phenomenon was the product of [God's] design, then the justification for observation and the gathering of detail was that each bit became additional proof of the design' (Glacken 1967:406).

That an interest in detail was also a hallmark of the Baconian approach to learning was a point made above. So we are not surprised to find Lovejoy observing: 'In the eighteenth century ... the Baconian temper (if not precisely the Baconian procedure), the spirit of patient empirical inquiry, continued its triumphant march in science, and was an object of fervent enthusiasm among a large part of the general educated public' (Lovejoy 1936:183).

It was that same public that provided the readership of special geography, and it was from its ranks that its authors came. It was in this spirit that one of the wordier of the special geographers wrote that 'plain facts are worth a thousand theories' (PINKERTON 1802, 2:296).

Having thus offered a historical context for the observed record of publication up to the period when it peaked, it is worth noting that an examination of the full titles of the books published in the two decades that began in 1780 suggests strongly that the discoveries of Captain Cook in the Pacific were instrumental in making these books popular with the public, for references to his voyages figure prominently in them.

Before moving on to consider the period of declining frequency of publication, we can note that Taylor provides an explanation of the intimate association between special geography and maps. It is simple. In the evolution of the Middle Ages, when a growing number of learned scholars began to take an interest in the secular aspects of this world, the Latin phrase *mappa mundi* was applied both to maps of the world and to written descriptions of the same (Taylor 1930:4-6 and 1956:115-16, taken together). On this basis, Robinson (1971) made a critical observation that helps us understand why places and countries came to the fore as an organizing concept in the sixteenth century. Places are what maps show, and one of the changes that accompanied the replacement of the medieval world with that of the Renaissance was 'the [new] sense of objectivity which medieval topography seldom attained' (Robinson 1971:209).

Once the association was in place, it took until the nineteenth century for two trends working in parallel to reach the point where cartography and geography were perceived to be distinct from one another.[5] The first trend was the ever-increasing specialization of scholarship that has continued to the present. The other was the supplanting of those gentlemen whom we call amateurs, and who pursued their studies as a hobby, by professionals (Levine 1986). It is only to be expected that, during the period when the two activities overlapped in the perception of the people, those who practised them tended also to perceive them as having a common form and structure.

When we turn to the question of why special geography declined in popularity, a factor that may well have contributed greatly was the widespread adoption of special geography as a class of writing particularly well suited to the education of young children. It is my opinion that few things discredit a subject, and devalue the quality of the ideas that it has to offer, more thoroughly than obliging people to learn it by heart when they are young. Even leaving that issue to one side, however, it seems reasonable to believe that those who had 'learned' a subject in school would reach the conclusion that they now knew what it had to offer, and would in consequence see no reason to study it again in their adult years.

As for the question of why special geography should have been adopted as a part of school curricula to the degree that it was, Capel argues that 'geography fulfilled a role which ... was absolutely essential in the epoch of the appearance of ... European nationalism' (Capel 1981b:52). It combined knowledge about the students' own country with, in the case of the imperial powers, and thus very much so in the case of the United Kingdom, a knowledge of their colonies, the whole being set in the context of a sometimes-impartial survey of all the countries of the world. Given another point to which Capel also draws attention, namely, the vast expansion in primary and secondary education that took place in the nineteenth century, the explosive growth in the publication of books devoted to that end is fully understandable.

A Preliminary Assessment

As a genre, special geography has received some attention, much of it hostile, from historians of geography. Warntz, in particular, advanced a thesis that sprang from two premises: first, in the eighteenth century special geography came to be accepted as constituting the whole of the subject; second, geography so conceived was 'devoid of intellectual appeal' (Warntz 1964:36). The inference that Warntz drew, and one that he presented as being valid in logical as well as in historical terms, is that, if geography is nothing other than the description of the countries of the world, then it is not a subject worthy of a place in the curriculum of a university. Recently, Capel (1981b) has argued in support of this thesis, at least in general terms.

It can be argued, however, that the genre deserves a sympathetic re-examination. A case can be made for the view that it suffers, along with the languages that we speak, from 'the fact that we ... lose sight of the need for explanation when phenomena are too familiar and "obvious"' (Chomsky 1968:22). As noted above, in the discussion about place and country, we all have a tendency to assume that terms we have used since childhood are free of ambiguity. Since the work of Ogden and Richards (1923) and the identification of what has subsequently become known as the *semiotic triangle*, however, the academic world should have been aware that very few words, and certainly none that contain more than a very limited degree of generalization, can be regarded as being free of ambiguity. Where ambiguity is present, critical analysis is called for if it is to be reduced, let alone removed.

Following on the work of Jean Piaget (1959/1923), there is a growing realization that the acquisition of the ability to take a critical view of the world is a process that cannot begin until the secondary level of education, and is only developed fully at levels beyond that. That is why we find academic departments in our universities devoting their energy to literary criticism. If few contemporary professional geographers see their task as being equivalent to that of a scholarly critic, that should be seen as a historical problem, not a consequence of the logic of ideas.

If, then, we accept the contention that at least the more ambitious authors of special geography were attempting, according to the beliefs that prevailed in their day, to account for the character of the countries that they were describing, their place in the history of geography becomes worth investigating.

With that point established, the corpus of texts identified in this guide prove to be interesting in at least four ways. Two have a special appeal for geographers, while the other two are of wider interest.

Because of the modest degree of attention that has been paid to this period, many geographers may not be aware of the fact, noted above, that in the centuries that separate the Middle Ages from the twentieth century, the word *geographer* had two primary meanings. It identified someone who drew maps of the world, and by extension, maps of continents and countries, or it referred to such a person's counterpart who described the places that appeared on those maps. Only after the emergence of geography as a university discipline in the late nineteenth century did the word *geographer* begin to acquire other meanings, and even these are largely confined to the members of the academic community. Thus it can be said that the list of books I have compiled represents the mainstream of geography for the period it covers. As we saw above, Warntz concurs in this view.

These books are also interesting to historians of geography in that they represent what might be called, by analogy, the 'vernacular' tradition. Given the recent historiographical shift in attention from the lives and actions of 'great' and important individuals to concern with the doings and being of ordinary people, I have borrowed the term 'vernacular' as a way of showing that the focus of interest is not the achievements of giants but rather the doings of lesser mortals.

This point was made with great vigour more than sixty years ago by Lovejoy:

> Another characteristic of the study of the history of ideas, as I should wish to define it, is that it is especially concerned with the manifestations of specific unit-ideas in the collective thought of large groups of persons, not merely in the doctrines of opinions of a small number of profound thinkers ... It seeks to investigate the effects of the sort of factors which it has ... isolated, in the beliefs, prejudices, pieties, tastes, aspirations, current among the educated classes through, it may be ... many generations. It is, in

> short, most interested in ideas which attain a wide diffusion, which become a part of the stock of many minds. (Lovejoy 1936:19)

Shortly thereafter he quotes Palmer (*The English Works of George Herbert*, 1905:xii): 'The tendencies of an age appear more distinctly in its writers of inferior rank than in those of commanding genius. These latter tell of past and future as well as of the age in which they live. They are for all time. But on the sensitive responsive souls, of less creative power, current ideas record themselves with clearness' (Lovejoy 1936:20).

He continues: 'And it is, of course, in any case true that a historical understanding even of the few great writers of an age is impossible without an acquaintance with their general background in the intellectual life and common moral and aesthetic valuations of that age; and that the character of this background has to be ascertained by actual historical inquiry into the nature and interrelations of the ideas then generally present' (Lovejoy 1936:20).

The only possible question that might seem to rise from Lovejoy's thesis is whether special geography is built around a unit-idea of the type that he specifies as properly forming a focus of attention within a general history of ideas. I maintain that the *nature of a place* and the *character of a country* both fall in the same class as Lovejoy's *great chain of being*.

These observations provide a bridge to the first of the points of interest that transcend geography, namely, the interaction of power and the hierarchical form of society. Since the work of Foucault, there has been a spreading realization that the interrelations between power and knowledge need to be explored in all their ramifications (Gordon 1980). As we saw earlier, in the case of special geography, interest was focused on the top of the socio-political hierarchy. Thus, when children read their geographies, they learned, among other things, that society had such-and-such a form. This was as much a fact as the location on the surface of the globe of the countries where such societies had their being. In other words, an acceptance of the status quo was being instilled in the rising generation.

The next question follows easily: What is it that youth needs to know? Geographers have long contended that everybody should know something of their subject. In this it seems they have the support of the general public, at least in the United States.

> [The National Geographic Society] has commissioned two Gallup surveys – the International Gallup Survey of Geography in 1988, and a follow-up ... in 1989. The 1988 survey sought to gain information primarily about the knowledge and attitudes of U.S. adults in geography. In the U.S. sample, 1,600 interviews were made

> among adults, ages 15-55+. The results of the survey showed ...
> • 9 in 10 American adults think it is important to know something about geography in order to be considered a well-rounded person.
> • 7 in 10 believe it is absolutely necessary to be able to read a map. (Bockenhauer 1990, n.p.)

It was not always thus. Everybody with the slightest knowledge of the intellectual history of the Western world knows that formal education in the Middle Ages was so different from that of today that it is not surprising that geography had no part in it. But what about the Classical period? Another fact universally known is that in the centuries before the reform of education in the nineteenth century, education focused on the teaching and learning of Latin and Greek, the explicit purpose being to make the classics of the Classics accessible to those who would be thought educated. But geography did not play a central role in the education of young Romans or their Greek predecessors. One special geography is known from the Classical period, but Strabo was a classic neither in his own time, nor, except of a most minor kind, in the centuries that followed. So the expectation that the educated people of the eighteenth and nineteenth centuries would possess a certain class of knowledge may have been a consequence of the fact that 'the Greeks had a word for it,' but it was not a consequence of what happened to Greek and Roman youth.

While a study of the geography taught in the schools of the Age of Reason is unlikely, by itself, to show us why it was widely held at that time that geography should be taught, the isolation of a particular 'unit-idea' that persisted over some three centuries provides a potential focus for the contextualistic approach to history that all authorities now agree is the one to take.

The Contribution of This Guide

Although it has been implicit in much of the foregoing, it is now time to state explicitly that I have tried to identify in this guide the complete body of the genre known as special geography. While what I have said already might seem sufficient justification, I can also refer to implicit calls for such a work made by two among the few recent authorities in the field. When David Stoddart prepared his assessment of the growth and structure of geography he wrote: 'It is not at present possible to give any precise estimate of the growth of geographical literature in terms of the number of books ... published' (Stoddart 1967:1). Then, in the course of introducing her thesis with respect to the relationship of empiricism and geographical thought from Bacon to von Humboldt,

Margarita Bowen reported sadly that 'a major obstacle in carrying out this project has been the lack of research in the history of geography, as reflected in ... the absence of even a catalogue of geographical publications from 1650 to 1860' (Bowen 1981:13).

I believe that in the pages that follow I have been able to provide a record of publication that would have satisfied Stoddart. I admit that, because Bowen's enterprise called for a survey of the Continental literature as well as that of the English-speaking world, this guide is only a partial contribution to the 'data base' whose absence she lamented. But even she would have found it useful.

Besides specialists in geography, historians of education and, indeed, of middle-class culture in general will find much useful information in the guide. A sign of this is provided by the work of Ruth Elson. Her ambition was to identify the ideas held by ordinary Americans in the nineteenth century. She argued that the books most widely read at that time were 'schoolbooks written by printers, journalists, teachers, ministers, and future lawyers earning their way through college.' She continued: 'However ill qualified to do so, the authors of schoolbooks both created and solidified American traditions' (Elson 1964:vii). She then took advantage of the existence of the Plimpton Collection of schoolbooks held by the Columbia University Library to document the 'information and standards of behavior and belief that the adult world expected the child[ren] to make [their] own' (1964:11). She based her conclusions on the examination of 'more than one thousand of the most popular schoolbooks used in the first eight years of schooling' (1964: viii). In her list they are divided into the following categories, arranged in descending order of the number listed: readers, spellers, geographies, histories, and arithmetics.

We can obtain a rough idea of how representative of the books used were the ones she examined by comparing the geographies she listed with those in this guide. Of the 192 books she classified as geography, 172 were published before 1888. Of these, 158 were either special geographies or books closely related to the genre. Focusing attention on books of that type, we find that she lists 95 titles by 69 authors (although it should be noted that some of her entries are ambiguous because in some cases she identifies a book by its series (or head) title rather than its individual title). Table 1 shows that, in round numbers, Elson examined about half the minimum number of books that were used. Thus we can say that, although the Plimpton Collection has a bias towards books published before the middle of the nineteenth century, it would be unreasonable to say that the selection of schoolbooks she examined was obviously unrepresentative of the whole genre. However, both a more truly representative selection, or specialist selections by period, author, or regional source, can now be identified.

A GUIDE FOR USERS

In terms of content, the objective of this guide is to list every book, and every edition of every book, that purports to provide a description of all the countries of the world, provided that they were published in English before 1888. For the purposes of this list, the books are divided into seven major classes.

1. Books intended to be read, as distinct from being consulted, by adults; they are referred to hereafter as adult special geographies.
2. Schoolbooks and other books intended for young readers.
3. Geographical dictionaries.
4. Books that are closely related in objective or content to special geography, but which were not explicitly intended to provide information in a systematic way about all the countries of the world; if they provide information about all countries, the information is not of the type associated with the genre. These are classified as 'marginal' to special geography or as 'other geography,' respectively. The occasional 'atlas' also comes under this heading. As a rule atlases are included only if they contain text providing information about countries in addition to maps of them; in such cases, they would be included under either (1) or (2) above. Where the full title does not explain the classification, justification is provided in a note at the end of the entry.
5. Books whose short titles showed that they were certainly geography, but about which I had too little information to be able to assign them to a particular subcategory.
6. A few books with promising titles that turned out, on examination, to be unconcerned with geography at all; they are classified as 'not geography.'
7. Ghosts: titles appear in some library catalogue or other publication, but the books that should correspond to the titles do not exist.

Except for those books that I examined myself, sources are cited for all information contained in the main entries.

A claim to have compiled an exhaustive list of works in any field should be justified, or, if that is not possible, qualified. In this particular case I make the claim with respect only to adult special geographies. Moreover, even in this case I am not categorical, for, had I been able to examine all the books I included because of their titles, I might have judged that some of those I did not see were also instances of the genre.

The claim is based on a thorough search of two kinds of sources: first, partial listings of special geographies compiled by historians of geography in the course of pursuing some other objective within the field; second, bibliographies that purport to list every book published in

English during specified periods of time, provided that they had a subject index containing 'Geography' as a heading.

As far as historians of geography are concerned, the most important sources are the following. Bowen (1981:312-19) is placed first because she alone is completely reliable with respect to the bibliographic information she gives. She lists 28 special geographies among her primary sources. Wright (1959:144-8), who limits himself to 'folios and multivolume works of smaller format' and to the period 1649-1810, lists 17 such works, of which 9 are not recorded by Bowen. Inspired by Wright, Downes (1971) surveyed the publications of the period 1714-1830; he added 7 to Wright's 17, though only 3 of his additions are not in Bowen.

In separate categories are the compilations of Cox (1935:69-87, 1938:332-57) and Taylor (1930, 1934). In the two chapters Cox devotes largely to special geography will be found more than forty examples of the genre published before 1801, together with some useful annotations.[6] His listing is thus more extensive than those compiled by specialists in the field, but it is also the most variable in terms of the amount and type of information provided about individual titles. Taylor's ambition was to list every item of 'English geographical literature' written before 1650. Because she provides only author, short title, and year of publication, her lists are less useful as guides to special geography than might be thought at first glance.

Besides these published works, a number of graduate theses have been devoted to various aspects of the evolution of geography as an academic or didactic discipline in the period between the Middle Ages and the twentieth century. One of these, Allford (1964), was an excellent source of British special geographies, adding nineteen to those in the authors already cited, though in fairness to them it should be noted that only five of these books were published in the period they were concerned with, namely, roughly 1600-1840. The other theses - Culler (1945), Lochhead (1980), Sahli (1941), and Wise (1969) - yielded few special geographies, and only two that were not recorded by other authors.[7] However, they were still useful because in many cases they made it possible to distinguish books intended for adults from those addressed to youth; in addition, Culler and Wise are important sources for American schoolbooks, and are even more useful in confirming that the books they survey did take the form of descriptions of all the countries in the world.

The sources discussed so far have in common the following facts. First, none of them can be relied on either to have identified the first edition of every special geography that they list, or to show that the edition that they list is not the first when that negative fact is the case; nor, with the exception of Bowen, is the bibliographic information they provide completely reliable. In addition, none of them identifies the sources used in searching for the books that are included.

Regarding general bibliographies that identified geography as a subject heading, I used the following: Allibone (1854), Averley and others (1979), General Catalogue (1786), London Catalogue (1811), Low (1976, 1979a, 1979b, 1979c), Peddie and Waddington (1914), Watt (1824), and Phase I of the Nineteenth Century Short Title Catalogue (1984). Almost needless to say, the appearance of a book in one of these sources under the heading of *Geography* is no guarantee that the book in question is a special geography. In some cases the short title by itself is enough to show that it is not. In other cases the question can be settled only by an examination of the book itself. Where that was not possible (and this body of writings is extremely scattered as far as library holdings are concerned), secondary sources were used. These ranged from the contents of the full title, which could run to more than 200 words in the eighteenth century, to a letter from a library with a particular book in its collection.

Four other general sources of information were drawn on. First, the printed catalogues of the British Library (British Museum) (1881-1900, 1959-66, 1968, 1971-2) and the American *National Union Catalog* (1968-81). Using these, it was possible to compile complete lists of all the books written by the authors found in other sources, and also extensive information on the dates of all editions. NUC also provides information with respect to library holdings, making possible subsequent visits and correspondence. Use was also made of the comprehensive listings by Pollard and Redgrave (1946, 1976, 1986) and Wing (1945, 1972, 1982) of all books published in the British Isles before 1701. These confirmed and supplemented the information derived from the two major sources as far as British library holdings were concerned.

Through the use of these sources, I was able to identify all but a handful of the 137 books shown in Table 1 as forming the genre of adult special geography. It should be borne in mind, however, when studying Table 1, and also Figures 1 and 2, that it is sometimes difficult to decide whether two particular works that have much in common are yet so different that they should be considered as completely distinct, or whether they are better seen as different editions of the same book; see, for example, BLOME (1670, 1680-2), the various versions of BROOKES, C. BUTLER (1846, 1848), and CHISHOLM (1882, 1884).

In addition, I looked for titles far and wide elsewhere, often prompted by references made in the sources already cited, or made by the authors of special geographies. These are too numerous and too diverse to discuss in any detail, but they will all be found in the lists of secondary sources or of abbreviations located elsewhere in the guide. It was in them that I found many of the books written for young people that make up the bulk of the titles listed hereafter.

In summary, I am satisfied that, with respect to the subcategories that cluster around the core of adult special geography, the 45 geographical dictionaries and 52 'marginal' geographies account for all but a small fraction of the total originally published. In the case of books for young readers, I have identified more than 550 of them. It is probable that there were others, especially in British colonies, but at the very least I have found all those that sold well, as well as a large proportion of those that appeared in only one edition, or came from a press located far from London or New York. Of that large number, I am able to classify with confidence only 316 as presenting descriptions of all the countries in the world, although I think it likely in the case of 166 other titles. Table 1 presents this information in further detail, including a cross-classification by country of publication.

As far as separate editions are concerned, it is only with respect to the adult books that I hope to have identified every one. In the case of schoolbooks, where some titles were reissued as separate 'editions' on an annual basis, the information gathered in the guide falls well short of a literally comprehensive listing. As for dictionaries, the listing is probably not far short of the level of completeness achieved for the adult special geographies. Because inclusion in the 'marginal' and 'other' classes is a matter of subjective judgment, no precise limit can be set to their size. The number of ghosts, too, is indefinitely large, because, at least in the case of editions, they are always liable to be created by the slip of a finger.

When I began work on this guide, I thought of it as a bibliography. During the course of compilation, however, I was referred to Bowers's *Principles of Bibliographical Description* (New York 1949). As I learned the extent and type of information said to be necessary for a list of books to be worthy of being called a bibliography, I was reminded of the cry of anguish that Cox gave at a similar stage of his labours: 'No man can ever be a bibliographer!' (Cox 1935:v). That conviction did not prevent Cox from going on to compile and publish three volumes of bibliographic-like information concerned with the literature of travel. Nor did it stop me from trying to compile this guide. Moreover, I learned to follow Bowers in terms of the basic principles to be used when supplying information about the books listed in it. I have, however, refrained from trying to scale the higher slopes of bibliographical presentation. In particular, I have not gathered the fullest possible range of information about the collation of the books. For more on this topic, see 'The Presentation of Information in This Guide,' which follows. I have also not tried to distinguish various states from separate issues, or, in an exhaustive way, issues from distinct editions.

On the other hand, because it was my hope to produce a guide for people who would want to examine the books listed in it, I tried to identify libraries where copies of the books, all editions included, could be located. In the course of time, I learned, however, that even major research libraries dispose of or lose books, so that the presence of a code indicating that such and such a library holds a copy of a particular book should be treated as an allegation, not a statement of absolute fact.[8]

Finally, in cases where I examined a book in addition to gathering the information already reviewed, I tried to make a quick assessment of its 'quality,' paying attention to such features as the amount of information about individual countries, the apparent audience for which the book was intended, whether the book was scholarly or popular in style, and so on. Such information, whether obtained personally or from a secondary source, is presented as an annotation.

The Presentation of Information in This Guide

Most of the information in this guide is contained in the main entries. For them, as far as possible, nine classes of information are provided: author; class of book; edition, date, and number; title; publication statement; physical description; location of copies; source of information and annotations; and published studies. All but the last of these classes of information are given for all adult special geographies, and for all books that I examined myself. In other cases I recorded either all the information that was provided in the source or all that was compatible with the constraints on space that came with preparing this work for publication. To save space, codes are used to provide some of the information.

Three examples of fictitious main entries follow in Table 2 in the section 'Codes and Abbreviations.' Referring to them in association with the introductory information provided below will make it possible to extract the information contained in the main entries.

Author

Today many of the people who prepared the books listed in this guide would be referred to as editors or compilers rather than as authors. However, to decide where to draw the line between original composition and the compilation of the work of others in an age when an 'author' could openly state, 'I reckon myself no plagiary, to grant, that I have taken the assistance of others; esteeming it needless sometimes to alter the character of a people or country, when I found it succinctly worded by a credible person' (GORDON 1693; quoted from unpaginated Preface of 15th ed. of 1737) would call for the judgment of Solomon. To then go on and establish which of the writers should be called an editor and which a compiler was a task far beyond the scope intended for this guide. Where the modern practice is followed on the title page, the fact is available here; where a secondary source makes the judgment, I have noted the fact.

Class of Book

All the books that I examined, and all the others where the necessary information was available, have been identified as adult special geography, geographical dictionary, and so on, as the case may be. This information is given in code on the same line as the author's name.

Date, Edition, and Number

The information for edition and date is what would be expected. The 'number' of an edition is that of the edition as recorded in this guide. It is shown in parentheses, and is used for cross references. In the case of first editions, this class of information follows the line with the author's name. In the case of all subentries from (02) on, the edition, date, and number mark the beginning of a set of information about a new edition. The author's name appears only at the beginning of a main entry.

Title

For every main entry where I examined the books themselves, the title appears in full at least once; in other words, all the information contained on the title page is reproduced here except (1) the author's name; (2) information about the publisher and, if on the title page, the bookseller(s); such information is provided next, in the publication statement; (3) mottoes, verses, and Latin tags; and (4) date of publication.

If the information about the edition was obtained from a secondary source, the title is given as it is recorded there. This information is given for the (01) entry where possible. If I could not find the full title for a (01) entry, it is usually given in the subentry corresponding to the earliest edition of the book that I examined. In the case of some schoolbooks, however, where the information is given in a compressed form, information from later editions is combined with that from the first.

In all cases the original spelling is given. Only occasionally, though more often when modern spelling would be expected, are variant spellings identified by the use of '[sic].'

Publication Statement

The place of publication is given first, followed by the publisher, and, if present on the title page, information about the booksellers where the book could be purchased. If the information about the book is derived from a secondary source, it is shown as in the source.

Physical Description

This refers to information about the dimensions of the book and the number of pages. As noted earlier, the information given under this heading is limited relative to the highest bibliographic standards.

Location

Libraries that are known to have a copy of the book are listed after the word ***Libraries***. British libraries are listed first, then Irish, American, and Canadian, in that order. It may be assumed that in all cases both BL and NUC were searched for the title in question. Unless otherwise specified, the source of information on British holdings may be assumed to be Pollard and Redgrave, Wing, or ESTC (and of course BL in the case of books in the British Library), and NUC for their American counterparts. More on this point follows in Table 2.

Source and Annotations

Two sets of information are given under this heading. First, if all or some of the information about the main entry is drawn from secondary sources, those sources are, with one exception, identified. The exception involves cases where only one library is known to have a copy of a particular book. Because the location is given, the source is implied. It is NUC if the book is held by a North American library. If a book is held by the British Library, the catalogue of that library is the source; if it is held by other British or Irish libraries, the source is Pollard and Redgrave, Wing, or ESTC, depending upon the period when the book was published. Where secondary sources provide information about the book beyond that already referred to under this or the preceding headings, I have included that information in the entry if it helps in distinguishing one edition of the book from another.

Second, comments about the book are provided. These are my own opinions based on examination of the book, or they are derived from the secondary sources. Such comments are provided in notes.

Published Studies

If the book to which the main entry is devoted has been discussed in a published study, the source where that study can be found is given. Such sources include graduate theses, even if they are technically unpublished; they do not include brief references made in standard histories of geography.

Cross References

Two kinds of cross references are used: those within a main entry, and others. Those within a main entry follow a form similar to: 'See (03).' Other cross references follow either of two forms. If the cross reference is to a work that appears in the main entries, the author's surname appears in upper case; if the reference is to secondary literature, the author's name appears in lower case. A list of the secondary literature cited in this guide is given in the section 'Sources.'

Final Comments

For the first recorded edition of a main entry, the foregoing information is laid out in such a way that each class begins on a new line. For most of the special geographies, this format is also used for most subsequent editions. However, where little or no fresh information was available about a subsequent edition, and also in the case of most of the schoolbooks, in particular those of the nineteenth century, when annual 'editions' came from some publishers, the information about subsequent editions is commonly both abbreviated in content, and also laid out in a compressed form.

Besides the main entries, the guide contains two indexes. The first contains an alphabetical list of the short titles that appear in the main entries, together with their authors and the dates of the earliest known editions. The titles of subsequent editions, where they vary from those of the first edition, also appear together with a cross reference. The second appendix contains a chronologically ordered list showing the earliest known edition of each main entry.

NOTES

1 Where the name of an author is given in lower case, as here, it is a reference to a work listed in the 'Sources' section at the end of the guide. Where the name is given in upper case, the reference is to a work listed in the main entries that form the body of the guide.

2 Gordon's initiative was taken up enthusiastically by both the public and, consequently no doubt, other geographers. His own *Grammar* appeared in twenty editions, the last of which was published in 1760. At least eighteen other 'grammars' of geography published during the next two hundred years are listed in this guide. The most frequently republished were those of SALMON (1749), GUTHRIE (1770), and PHILLIPS, R. (1803, 1804). It was 1872 before the type came to an end.

3 'A timeless and incorporeal One became the logical ground as well as the dynamic source of the existence of a temporal and material and extremely multiple and variegated universe' (Lovejoy 1936:49).
4 As Marxists of a traditional outlook would be quick to point out, however, even if ideas are material, it does not necessarily follow that they will play an independent role of the type apparently envisaged by Lovejoy.
5 The word *cartographer* is first recorded in 1863 (OED).
6 As was noted in the Preface, it was Cox's work that inspired this guide.
7 There are also Hart (1957) and Mayo (1964); I did not come across their theses until a very late stage in compiling this work.
8 The problem of books disappearing from libraries can be expected to quickly grow worse with respect to those published since the use of paper made from softwood pulp began rather more than a hundred years ago. In the words of one archivist I met, 'they are biodegrading on our shelves.' Unless steps are taken soon, there will be little material evidence of the nature of the schoolbooks used from the middle of the nineteenth century onward.

Codes and Abbreviations

Three sets of codes and abbreviations are used in this guide:

1. Codes and abbreviations used only in the main entries. These are shown for three sample main entries that are presented in Table 2.
2. Those for libraries where books listed in the guide can be found. These are listed under 'Library Codes,' on page 47.
3. Other abbreviations. For the most part these are conventional and are taken for granted. Only those specific to this guide are listed in the section immediately following the sample entries.

TABLE 2
Sample entries

Parts of Entry	Code	Interpretation
Sample Entry No. 1		
[AUTHOR, Andrew N.] **A**	[…]	Book published anonymously, though author now known
	A	Adult special geography; for other category codes, see p. 45.
1788 (01)	1788	No parentheses. I have examined a copy of this edition of this book. Compare with […] and {…} below.
	(01)	The subentry number for this edition of the book
A description of all the countries in the world, as located in Europe, Asia, Africa, and America. More particularly, providing information with respect to their location	A description …	The wording of the title on the title page, except author, place, and publisher, as well as mottoes, Latin tags, and the like. The short title is set in italics.
LONDON; I. Swell, U. Puffit, and J. Overwork, in Paternoster Row; and sold also by booksellers in Scotland and Ireland	LONDON	Place of publication

	I. Swell ...	Publisher, plus any information about booksellers provided on the title page
4to, 25 cm.; pp. ix, 155, [2]	4to	Number of times a full sheet of paper is folded to make the leaves: folio, folded once to form two leaves; 4to, folded twice to form four leaves; etc.
	25 cm	Length of a page
	pp. ix, 155, [2]	Collation. In this case: 9 numbered preliminary pages, 155 pages of text, and 2 unnumbered final pages. This is the arrangement I used. Where I did not examine the book but took the information from a secondary source, the collation is provided as in the source. Examples of other uses occur below.
Libraries **L**, OSg, [IDT] : CtY, DLC, MH : CaBVaU	***Libraries*** **L** ...	For the codes used to identify libraries where copies of the book are located, see 'Library Codes' on page 47. Libraries where I have examined a copy of a book, or obtained information about one in person, are shown in boldface.
	OSg, [IDT] : CtY, DLC, MH : CaBVaU	In this subentry, copies of the book are held in two British libraries (given first), in an Irish library (code in [ ]), three American libraries (following the first :), and one Canadian library (following the second :).

Notes 1. Provides a greater range of information than its title suggests, but not consistently; far more is provided about the countries of the British Isles than any others.	***Notes***	Notes contain two general types of information: first, observations I made upon examining the book; second, information obtained from secondary sources. In many cases, the latter contains condensed information with respect to later editions.
1795 (02)	1795	The date of the next edition or issue
	(02)	The second subentry with respect to this book
A description of all the countries in the world. Whether located … America, noting especially …	A description of all the countries in the world …	The title of this edition. Because I have examined the book, the short title is given. Thereafter, I include only wording, including punctuation, that differs from the previous edition whose complete title is known.
DUBLIN; Merryweather and Careless 4to, 24 cm; pp. ix, 155, [5] ***Libraries*** **L** *Notes* 1. Spot-checking revealed no changes in the text besides minor changes in the punctuation of the tp. An index has been added.		
{1801} (03)	{ … }	All the information about this subentry has been obtained from secondary sources.

A description of all the countries in the world PHILADELPHIA; Harder, Tuff, and co.		
22 cm; 1 p. l., [vii], 10-164 ***Libraries*** : NN, ViU	1 p. l., [vii], 10-164	The collation begins with one unnumbered leaf, followed by 7 pages of preliminary matter, followed by the text, which contains 155 numbered pages of material.
Source NUC (NA 9876543)	***Source*** NUC (NA 9876543)	The source of the information contained in the subentry is the (American) National Union Catalog. The NUC entry number is shown in parentheses.
Insight (1957:112-13) and Drudgery (1982) discuss (01).	Insight (1957: 112-13)	Citation of a secondary source. A list of such sources is provided at the end of the guide.

Sample Entry No. 2

WORTHY, Uriah Norman (Peter Pilgrim) **Y SG? Am**	WORTHY ...	The author of the book
	(Peter Pil-grim)	The pseudonym used by the author at the time of publication

{1831}* (01)	{ … }*	There is evidence to show that this edition is not, or may not be, the first.
An introduction to geography, by a method totally new, for students in primary schools … A new ed. BOSTON; Noble and Son		
{1855 [c1850]} (02)	[…]	The date of publication is provided by a source other than the title page (the normal source during the period covered by this guide).
	c …	For books published in America, the date when copyright was granted is sometimes shown thus in library catalogues.
An introduction to geography … NEW YORK; Piratt, Steel, and Prophet 72 pp. ***Libraries*** : PPGeo ***Source*** NUC (NW 8765432), which notes: 'c1850.'		
{1857} (03) *An introduction to geography* … CHICAGO; Robber, Baron & Co.		

Entry	Element	Explanation
18 cm.; iii, 72, 8, [2] p.	iii, 72, 8, [2] p.	In this collation, there are two separate series of numbered pages. The explanation is provided in the ***Source*** of the subentry; see there, and also below.
Libraries : PPGeo ***Source*** NUC (NW 8765433), which notes that the second series of numbered pages contains 'A Geography of Illinois.'		
Notes 1. NUC records further eds.: 1858 (NN); 1858, 72, 10 pp. (PPL) {NW 8765434 notes: '1. A Geography of Pennsylvania, 10 p. at the end. 2. [c1857]'}; 1859 (**)	1. NUC ...	Compressed information drawn from NUC with respect to two further editions. The first is held by the New York Public Library, whose catalogue provides only the author, title, and year of publication. The second is held by Library Company of Philadelphia; its catalogue notes that, although the title page carries the date 1858, copyright was granted in 1857. In addition, information about the existence and content of a second series of numbered pages is provided.
	{... [...]'}	Different bracket styles are used thus in notes containing condensed information, to help in deciphering the information.
	1859(**)	In notes where condensed information is being given, a double asterisk (**) following the year of an edition indicates that the collection to which the book belonged is no longer available.

{1858} (04) *An introduction to geography.*		
? Copy & Sons	?	Information required at this point is lacking.
Libraries n.r.	n.r.	(Libraries) not recorded
Source Advertisement in end pages of SCHOLAR (1860)	Upper case, followed by date in parentheses	A cross reference to another author listed in the body of this guide

Sample Entry No. 3

DESCRIPTION ... (1777) **Y** {1777} (01) *A description of all the countries in the world*, composed for the benefit of young ladies and gentlemen ...		The author being unknown, the opening words of the title are used here. The date occurs if there is more than one title with the same opening words.
[LONDON]; A. Publisher *Libraries* : NBuHi *Source* NUC (NS 7654321)	[...]	Place of publication provided by a secondary source

Book Categories

As noted in the Introduction, the books listed in this guide are divided into different categories. Information on membership in the categories is provided in the form of an alphabetic code at the end of the line carrying the author's name.

1. Adult special geography: **A** (note that **SG?** = may be adult special geography)
2. Books for youth: **Y** (special geography unless otherwise noted)
3. Books on the margin of the genre: **mSG**
4. Geographical dictionaries: **Dy**
5. Atlases, provided that they contain only maps, with no text describing the countries that appear on the maps: **At**
6. Books that would probably be classified as geography today, but which do not belong to any of the above classes: **OG**
7. Books that are not geography at all: **NG**
8. Ghosts: **Gt**
9. Books about which I did not have sufficient information to make a judgment: **?** (where it is only probable, not certain, that a book belongs in a particular category, this is indicated by the use of a '?' as in 'Dy?')

Some supplementary information is also provided:

10. Books published outside the United Kingdom are identified by the following codes:

    Published in Australia: **Aus**
    Published in Canada: **Can**
    Published in the United States: **Am**
    Published simultaneously in Britain and the United States: **UK+US**
    Published elsewhere: **els**

11. Books translated from foreign languages: **tr**

General Abbreviations

anr	Another
BCB	Index to the British Catalogue of Books (see British ... in 'Sources, Part 1')

BL	*The General Catalogue of Printed Books* published by the British Museum or the British Library in 1959 or thereafter
BoC	*A Bibliography of Canadiana*; see (1) Staton and Tremaine (1934), (2) Boyle and Colbeck (1959), and (3) Alston and Evans (1985, 1986, 1989) in 'Sources, Part 1.'
ca.	About (N.B.: distinguish 'ca.' from 'c[date],' the date when copyright was granted)
CGBN	*Catalogue générale des livres imprimes de la Bibliothèque Nationale*, Paris (see 'Sources, Part 1')
CIHM	Canadian Institute for Historical Microreproductions
DNB	[British] *Dictionary of National Biography*
ECB	Index to the English Catalogue of Books (see English Catalogue ... in 'Sources, Part 1')
ed./eds.	Edition/editions, and editor/editors
ESTC	*Eighteenth Century Short Title Catalogue* (this refers to the manuscript version I consulted in the British Library, not the current computer version)
l.	Leaf (an unnumbered sheet of paper bound as part of a book)
Johnson, or TOT	W.J. Johnson, Norwood, NJ, and Theatrum Orbis Terrarum, Amsterdam (publishers of *The Early English Experience*: a series of facsimile reprints of books originally published in England between 1475 and 1640). TOT also published facsimile editions of several early atlases.
n.c.	Not complete
n.d.	No date
n.r.	Not recorded
NSTC	*Nineteenth Century Short Title Catalogue*
NUC	[American] National Union Catalog (see 'Sources, Part 1')
OED	*Oxford English Dictionary*

STC	*Short-title catalogue*; for books published before 1641, Pollard and Redgrave (1946, 1976, 1986); for books published from 1641 to 1700, Wing (1945, 1972, 1982)
tp./t.-p.	Title page
v./vol./vols.	Volume, volumes

Library Codes

Where I examined a copy of a book at a library, the code of that library is shown in boldface in the main entries. Where I obtained information in person about a book at a library, the code of the library is also shown in boldface, even if I did not see the book itself.

Departmental and other specialist libraries at large institutions are, with occasional exceptions, not listed separately.

British Libraries

AbU	Aberdeen University
AbrwthU	Aberystwyth, University College of Wales
AbrwthWN	–, National Library of Wales
ArmP	Armagh Public Library
BatP	Bath Public Library
BdmnLk	Bodmin, Cornwall, Lanhydrock Library
BfQ	Belfast, the Queen's University
BidP	Bideford Public Library
BikP	Birkenhead Public (Wirral Metropolitan) Library
BlaR	Blackburn Reference Library (Lancashire County Library)
BmP	Birmingham (Central Reference) Public Library (West Midlands regional catalogue)
BmU	Birmingham University
BmUE	Birmingham University, Institute of Education
Brf	Bradford Metropolitan District Library
BrP	Bristol Public Library (South Western regional catalogue)
BrU	Bristol University
ByseC	Bury St Edmunds Cathedral
C	Cambridge University
CGc	–, Gonville and Caius College
CCl	–, Clare College

CEm	–, Emmanuel College
CJ	–, Jesus College
CMp	–, Magdalene College, Pepysian Library
CStJ	–, St John's College
CSyS	–, Sydney Sussex College
CT	–, Trinity College
CaKU	Kent University, Canterbury
CfU	Cardiff University
ChC	Chichester Cathedral
CheR	Chester Reference Library
CoP	Colchester Public Library
CrlC	Carlisle Cathedral
Crf	Crief, Perthshire; The Schoolhouse, Innerpeffray (by Crief)
DC	Dulwich College
DU	Durham University (includes Cosin Library)
E	Edinburgh, National Library of Scotland
ECP	Edinburgh, Royal College of Physicians
ES	Edinburgh, Signet Library
EU	Edinburgh University
ExC	Exeter Cathedral
ExDeI	Exeter, Devon and Exeter Institution
GlP	Gloucester Public Library
GP	Glasgow Public (the Mitchell) Library
GU	Glasgow University
HeP	Hereford Public Library (Hereford Reference; Hereford & Worcester County Library)
HlU	Hull University
HuC	Humberside County Library, Hull
Lam	Lampeter, St David's University College
LcM	Leicester (Art Gallery) and Museum
LcP	Leicester Public Library (East Midlands regional catalogue)
LcU	Leicester University
LeP	Leeds Public (Metropolitan District/Reference) Library
LeU	Leeds University
LinC	Lincoln Cathedral
LinL	Lincoln, Lincolnshire Library Service
L	British Library, London
LA	London, Admiralty
LC	London, Congregational Library
LCP	London, Royal College of Physicians
LDW	London, Dr Williams' Library
LG	London, Guildhall Library
LILS	London, Incorporated Law Society

LIT	London, Inner Temple
LLam	London, Lambeth Palace
LLI	London, Lincoln's Inn
LMT	London, Middle Temple
LNC	London, National Central Library
LPO	London, Patent Office
LRAg	London, Royal Agricultural Society
LRGS	London, Royal Geographical Society
LRIS	London, Royal Institute of Chartered Surveyors
LRS	London, the Royal Society
LRUS	London, Royal United Service Institution library
LSC	London, Sion College
LSe	London School of Economics
LSM	London, Science Museum library
LU	London University, the university library (Senate House)
LUBd	London University, Bedford College
LUU	London University, University College
LV	London, Victoria and Albert Museum
LWI	London, Wellcome Institute for the History of Medicine
LWS	London, Westminster School
LonU	Londonderry, University of Ulster, Magee College
LvU	Liverpool University
MaPL	Maldon, Essex, the Plume Library
MC	Manchester, Chetham's Library
MP	Manchester Public Library (North Western regional catalogue)
MRu	Manchester University, John Ryland's Library
NcP	Newcastle (Central) Public Library
NcU	Newcastle University
NwP	Newport (Gwent) Public Library
NC	Norwich Cathedral
NP	Norwich Public Library (Norwich Central Library; Norfolk County Library)
NoP	Nottingham (Reference) Public Library
NoU	Nottingham University
O	Oxford University, the Bodleian Library
OB	–, Balliol College
OChch	–, Christ Church College
OE	–, Exeter College
OH	–, Hertford College
OJ	–, Jesus College
OMa	–, Magdalene College
OMan	–, Manchester College
OMe	–, Merton College

OSg	–, School of Geography
OStJ	–, St. John's College
OT	–, Trinity
OW	–, Worcester College
PetC	Peterborough Cathedral
PrR	Preston Reference Library
RdU	Reading University
RgtP	Reigate Public Library
RchP	Rochester Public Library
SanU	Saint Andrews University
	Scotland, National Library; see E
ShP	Sheffield Public City Libraries
SkPt	Skipton, North Yorks, Petyt Library
SoM	Southwell Minster, the parish library
SothnU	Southampton University
SsaC	St Asaph's Cathedral
StkP	Stoke (Staffordshire County) Library
SwnU	Swansea, University College
SwnR	Swansea Reference (West Glamorgan County) Library
	Wales, National Library; see AbrwthWN
WarSE	Ware, Herts., St Edmund's College
WiC	Winchester Cathedral
WigMD	Wigan Metropolitan District Library
WinStG	Windsor Castle, St. George's Chapter Library
WnC	Winchester Cathedral
WorC	Worcester Cathedral
YM	York Minster

Irish Libraries

[ICD]	Cashel Diocesan Library
[IDKI]	Honourable Society of King's Inn, Dublin
[IDM]	Dublin, Marsh's Library
[IDT]	Trinity College, University of Dublin

American Libraries

AAP	Auburn University, Auburn, AL
AkU	University of Alaska, Fairbanks
AU	University of Alabama, Tuscaloosa
AzTeS	Arizona State University, Tempe
AzU	University of Arizona, Tucson
CBPac	Pacific School of Religion, Berkeley, CA

CCC	Honnold Library, Claremont College, CA
CLSU	University of Southern California, Los Angeles
CLU	University of California at Los Angeles
CoCC	Colorado College Library, Colorado Springs
CoDU	University of Denver, Denver, CO
CoFS	Colorado State University, Fort Collins
CoGrS	Colorado State University, Greeley
CoMC	Mills College, Oakland, CA (not CO)
CP	Pasadena Public Library, Pasadena, CA
CPa	Palo Alto City Library, Palo Alto, CA
CoU	University of Colorado, Boulder
CSmH	Henry E. Huntington Library, San Marino, CA
CSt	Stanford University Libraries, Stanford, CA
CT	Torrance Public Library, Torrance, CA
CtH	Hartford Public Library, Hartford, CT
CtHT	Trinity College, Hartford, CT
CtW	Wesleyan University, Middletown, CT
CtY	Yale University, New Haven, CT
CU	University of California, Berkeley
CU-A	University of California, Davis
CU-I	University of California, Irvine
CU-S	University of California, San Diego, La Jolla
DAU	American University Library, Washington, DC
DCU	Catholic University of America, Washington, DC
DeGE	Eleutherian Mills Historical Library, Greenville, DE
DeU	University of Delaware, Newark
DeWI	Wilmington Institute and the New Castle County Free Library, DE
DFo	Folger Shakespeare Library, Washington, DC
DGU	Georgetown University Library, Washington, DC
DHEW	U.S. Department of Education Research Library, Washington, DC
DI	U.S. Department of the Interior, Washington, DC
DI-G	–, Geological Survey Library, Washington, DC
DLC	Library of Congress, Washington, DC
DN	U.S. Department of the Navy Library, Washington, DC
DN-Ob	–, Naval Observatory Library
DNAL	U.S. National Agricultural Library, Washington, DC
DNLM	U.S. National Library of Medicine, Washington, DC
DNW	U.S. National War College, Fort McNair, Washington, DC
DP	U.S. Patent Office Library, Washington, DC

DS	U.S. Department of State Library, Washington, DC
DSI	Smithsonian Institution Library, Washington, DC
F	Florida State Library, Tallahassee
FJ	Jacksonville Public Library System, Jacksonville, FL
FMU	University of Miami, Coral Gables, FL
FSaHi	Saint Augustine Historical Society, St. Augustine, FL
FTaSU	Florida State University, Tallahassee
FU	University of Florida, Gainesville
FWpR	Rollins College, Winter Park, FL
GA	Atlanta Public Library, Atlanta, GA
GASC	Georgia State College, Atlanta
GASU	Georgia State University, Atlanta, GA
GEU	Emory University, Atlanta, GA
GHi	Georgia Historical Society, Savannah
GMM	Mercer University, Macon, GA
GMW	Wesleyan College, Macon, GA
GU-De	DeRenne Georgia Library, University of Georgia, Athens
HH	Hawaii State Library System, Honolulu
I	Illinois State Library, Springfield
Ia	Iowa State Library Commission, Des Moines
IaAS	Iowa State University of Science and Technology, Ames
IaSlB	Buena Vista College, Storm Lake, IA
IaU	University of Iowa, Iowa City
ICA	Art Institute of Chicago
ICarbS	Southern Illinois University, Carbondale
ICJ	John Crerar Library, Chicago
ICN	Newberry Library, Chicago
ICRL	Center for Research Libraries, Chicago
ICU	University of Chicago, Chicago
IdPI	Idaho State University, Pocatello
IdU	University of Idaho, Moscow
IEdS	Southern Illinois University, Edwardsville
IEG	Garrett Theological Seminary, Edwardsville, IL
IHi	Illinois Sate Historical Library, Springfield
INS	Illinois State University, Normal
InSNHi	Northern Indiana State Historical Society, South Bend
InStme	St. Meinrad's College and Seminary, St. Meinrad, IN
InU	Indiana University, Bloomington
IObT	Bethany and Northern Baptist Theological Seminaries Library, Oak Brook, IL

IRoC	Rockford College, Rockford, IL
IU	University of Illinois, Urbana
KAS	St. Benedict's College, Atchison, KS
KBB	Baker University, Baldwin City, KS
KEmT	Kansas State Teachers College, Emporia
KMK	Kansas State University, Manhattan
KU	University of Kansas, Lawrence
KyBgW	Western Kentucky State College, Bowling Green
KyHi	Kentucky Historical Society, Frankfort
KyLoF	Filson Club, Louisville, KY
KyLoU	University of Louisville, Louisville, KY
KyLx	Lexington Public Library, Lexington, KY
KyLxT	Transylvania College, Lexington, KY
KyU	University of Kentucky, Lexington
LNT	Tulane University, New Orleans
LU	Louisiana State University, Baton Rouge
M	Massachusetts State Library, Boston
MA	Amherst College, Amherst, MA
MB	Boston Public Library
MBAt	Boston Athenaeum, Boston
MBBL	Massachusetts Bureau of Library Extension, Boston
MBCo	Countway Library of Medicine (Harvard-Boston Medical Libraries)
MBrZ	Zion Research Library, Brookline, MA
MChB	Boston College, Chestnut Hill, MA
MCE	Episcopal Divinity School, Cambridge, MA
MdAN	U.S. Naval Academy, Annapolis, MD
MdBE	Enoch Pratt Free Library, Baltimore
MdBG	Goucher College, Baltimore
MdBJ	Johns Hopkins University, Baltimore
MdBP	Peabody Institute, Baltimore
MdBS	Saint Mary's Seminary and University, Roland Park, Baltimore
MdE	Mount St. Mary's College, Emmitsburg, MD
MdHi	Maryland Historical Society, Baltimore
MdU	University of Maryland, College Park
MdW	Woodstock College, Woodstock, MD
MeAu	University of Maine, Augusta
MeB	Bowdoin College, Brunswick, ME
MeWC	Colby College, Waterville, ME
MF	Fall River Public Library, Fall River, MA
MFi	Fichtburg Public Library and Regional Center for Central Massachusetts Regional Library System, Fichtburg
MH	Harvard University, Cambridge, MA
MHi	Massachusetts Historical Society, Boston

Mi	Michigan State Library, Lansing
MiAlbC	Albion College, Albion, MI
MiD	Detroit Public Library
MiDSH	Sacred Heart Seminary, Detroit
MiDW	Wayne State University, Detroit
MiEM	Michigan State University, East Lansing
MiKC	Kalamazoo College, Kalamazoo, MI
MiPon	Pontiac Public Libraries, Pontiac, MI
MiU	University of Michigan, Ann Arbor
MMeT	Tufts University, Medford, MA
MNBedf	New Bedford Free Public Library, New Bedford, MA
MnHi	Minnesota Historical Society, St. Paul
MnU	University of Minnesota, Minneapolis
MoSU	St. Louis University, St. Louis
MoSW	Washington University, St. Louis
MoU	University of Missouri, Columbus
MPB	Berkshire Athenaeum, Pittsfield, MA
MSaE	Essex Institute, Salem, MA
MShM	Mount Holyoke College, South Hadley, MA
MsSM	Mississippi State University, State College
MsSU	Sunflower County Library, Sunflower, MS
MtBC	Montana State University, Bozeman
MtBuM	Montana College of Mineral Science and Technology, Butte
MtHi	Montana Historical Society, Helena
MtU	University of Montana, Missoula
MU	University of Massachusetts, Amherst
MWA	American Antiquarian Society, Worcester, MA
MWC	Clark University, Worcester, MA
MWelC	Wellesley College, Wellesley, MA
MWH	College of the Holy Cross, Worcester, MA
MWiW	Williams College, Williamstown, MA
N	New York State Library, Albany
NB	Brooklyn Public Library, Brooklyn, NY
NbHi	Nebraska State Historical Society, Lincoln, NE
NBu	Buffalo and Erie County Public Library, Buffalo
NbU	University of Nebraska, Lincoln
NBuC	State University of New York, College at Buffalo
NBuG	Grosvenor Reference Division, Buffalo and Erie County Library, Buffalo
NBuHi	Buffalo and Erie County Historical Society, Buffalo
NBuU	State University of New York at Buffalo
NcA	Pack Memorial Public Library, Asheville, NC
NCanHi	Ontario County Historical Society, Canandaigua,

	NY
NcCU	University of North Carolina at Charlotte
NcD	Duke University, Durham, NC
NcGN	University of North Carolina at Greensboro
NCH	Hamilton College, Clinton, NY
NcRS	North Carolina State University at Raleigh
NcRSh	Shaw University, Raleigh, NC
NcU	University of North Carolina, Chapel Hill
NcWsW	Wake Forest University, Winston-Salem, NC
NEC	Explorers' Club, New York
NFQC	Queens College, Flushing, NY
NGlc	Glen Cove Public Library, Glen Cove, NY
Nh	New Hampshire State Library, Concord
NH	Hamilton Public Library, Hamilton, NY
NhD	Dartmouth College, Hanover, NH
NHi	New York Historical Society, New York
NhM	Manchester City Library, Manchester, NH
NIC	Cornell University, Ithaca, NY
NjN	Newark Public Library, NJ
NjNbS	New Brunswick Theological Seminary, New Brunswick, NJ
NjP	Princeton University, Princeton, NJ
NjPT	Princeton Theological Seminary, Princeton, NJ
NjR	Rutgers - The State University, NJ
NmU	University of New Mexico, Albuquerque
NN	New York Public Library
NNBG	New York Botanical Garden, Bronx, NY
NNC	Columbia University, New York
NNE	Engineering Societies Library, New York
NNG	General Theological Seminary of the Protestant Episcopal Church, New York
NNH	Hispanic Society of America, New York
NNHi	New York Historical Society
NNNAM	New York Academy of Medicine, New York
NNPM	Pierpont Morgan Library, New York
NNQ	Queens Borough Public Library, New York
NNR	City College of the City University of New York, New York
NNU	New York University Libraries
NNU-W	–, Washington Square Library
NNUT	Union Theological Seminary, New York
NPV	Vassar College, Poughkeepsie, NY
NRAB	Samuel Colgate Baptist Historical Library of the American Baptist Historical Society, Rochester, NY
NRCR	Colgate-Rochester Divinity School, Rochester, NY
NRU	University of Rochester, Rochester, NY

NSyU	Syracuse University, Syracuse NY
NUt	Utica Public Library, Utica, NY
NWh	West Hampstead Public Library, West Hampstead, NY
NWM	U.S. Military Academy, West Point, NY
OAk	Akron Public Library, Akron, OH
OC	Public Library of Cincinnati and Hamilton County, Cincinnati
OCH	Hebrew Union College, Cincinnati
OCl	Cleveland Public Library
OClJC	John Carroll University, Cleveland
OClND	Notre Dame College, Cleveland
OClStM	Saint Mary's Seminary, Cleveland
OClW	Case Western Reserve University, Cleveland
OClWHi	Western Reserve Historical Society, Cleveland
OCU	University of Cincinnati, Cincinnati
OCX	Xavier University, Cincinnati
ODW	Ohio Weslyan University, Delaware
OFH	Rutherford B. Hayes Library, Fremont, OH
OGK	Kenyon College, Gambier, OH
OHi	Ohio State Historical Society, Columbus
OKentU	Kent State University, Kent, OH
OkU	University of Oklahoma, Norman
OO	Oberlin College, Oberlin, OH
OOxM	Miami University, Oxford, OH
OrCS	Oregon State University, Corvallis
OrHi	Oregon Historical Society, Portland
OrP	Library Association of Portland, Portland, OR
OrPR	Reed College, Portland, OR
OrPS	Portland State College, Portland, OR
OrU	University of Oregon, Eugene
OSW	Wittenberg University, Springfield, OH
OU	Ohio State University, Columbus
OW	Warren Public Library, Warren, OH
OWoC	College of Wooster, Wooster, OH
OWorP	Pontifical College Josephinum, Worthington, OH
PBa	Academy of the New Church, Bryn Athyn, PA
PBL	Lehigh University, Bethlehem, PA
PBm	Bryn Mawr College, Bryn Mawr, PA.
PCarlD	Dickinson College, Carlisle, PA
PCC	Crozer Theological Seminary, Chester, PA
PHC	Haverford College, Haverford, PA
PHi	Historical Society of Pennsylvania, Philadelphia
PLatS	Saint Vincent College and Archabbey, Latrobe, PA
PMA	Allegheny College, Meadville, PA
PNt	Newtown Library Company, Newtown, PA

PP	Free Library of Philadelphia
PPA	Athenaeum of Philadelphia
PPAmP	American Philosophical Society, Philadelphia
PPAN	Academy of Natural Sciences, Philadelphia
PPB	Philadelphia Bar Association
PPC	College of Physicians of Philadelphia
PPCC	Carpenters' Company, Philadelphia
PPCCH	Chestnut Hill College, Philadelphia
PPD	Drexel Institute of Technology, Philadelphia
PPDrop	Dropsie University, Philadelphia
PPeSchw	Schwenkfelder Historical Library, Pennsburg, PA
PPF	Franklin Institute, Philadelphia
PPFr	Friends' Free Library of Germantown, Philadelphia
PPG	German Society of Pennsylvania, Philadelphia
PPiPT	Pittsburgh Theological Seminary
PPiU	University of Pittsburgh
PPL	Library Company of Philadelphia
PPLas	La Salle College, Philadelphia
PPLT	Lutheran Theological Seminary, Philadelphia
PPPCPh	Philadelphia College of Pharmacy and Science, Philadelphia
PPPD	Philadelphia Divinity School
PPPH	Pennsylvania Hospital, Philadelphia
PPPrHi	Presbyterian Historical Society, Philadelphia
PPRF	Rosenbach Foundation, Philadelphia
PPStCh	Saint Charles Borromeo Seminary, Overbrook, Philadelphia
PPT	Temple University, Philadelphia
PPULC	Union Library Catalogue of Pennsylvania, Philadelphia
PPWa	Wagner Free Institute of Science, Philadelphia
PPWI	Wistar Institute of Anatomy and Biology, Philadelphia
PSC	Swarthmore College, Swarthmore, PA
PSt	Pennsylvania State University, University Park, PA
PU	University of Pennsylvania, Philadelphia
PV	Villanova College, Villanova, PA
PWcS	West Chester State College, West Chester, PA
RNHi	Newport Historical Society, Newport, RI
RNR	Redwood Library and Athenaeum, Newport, RI
RP	Providence Public Library, Providence, RI
RPB	Brown University, Providence, RI
RPJCB	John Carter Brown Library, Providence, RI
ScCleU	Clemson University, Clemson, SC
ScNC	Newberry College, Newberry, SC
ScU	University of South Carolina, Columbia

T	Tennessee State Library and Archives, Nashville
TKL	Knoxville Public Library System, Knoxville, TN
TNF	Fisk University, Nashville, TN
TNJ	Joint University Libraries, Nashville, TN
TU	University of Tennessee, Knoxville
TxDaM	Southern Methodist University, Dallas
TxHU	University of Houston, Houston
TxU	University of Texas, Austin
UU	University of Utah, Salt Lake City
Vi	Virginia State Library, Richmond
ViFreM	Mary Washington College of the University of Virginia, Fredericksburg
ViHo	Hollins College, Hollins College, VA
ViL	Jones Memorial Library, Lynchburg, VA
ViLxW	Washington and Lee University, Lexington, VA
ViNeM	Mariners' Museum, Newport News, VA
ViU	University of Virginia, Charlottesville
ViW	College of William and Mary, Williamsburg, VA
VtMiM	Middlebury College, Middlebury, VT
VtMiS	Sheldon Art Museum, Middlebury, VT
VtU	University of Vermont and State Agricultural College, Burlington
W	Wisconsin State Law Library, Madison
Wa	Washington State Library, Olympia
WaPS	Washington State University, Pullman
WaS	Seattle Public Library, Seattle, WA
WaSpG	Gonzaga University, Spokane, WA
WaT	Tacoma Public Library, Tacoma, WA
WaTC	University of Puget Sound, Tacoma, WA
WaU	University of Washington, Seattle
WaWW	Whitman College, Walla Walla, WA
WGr	Brown County Library, Greenbay, WI
WHi	State Historical Society of Wisconsin, Madison
WU	University of Wisconsin, Madison
Wy-Ar	Wyoming State Archives and Historical Department, Cheyenne

Canadian Libraries

CaAEU	University of Alberta, Edmonton
CaBVaU	University of British Columbia, Vancouver
CaBViP	Provincial Library, Victoria, BC
CaBViPA	Provincial Archives, Victoria, BC
CaNSPH	Hector Centre Trust, Pictou, NS
CaNSWA	Acadia University, Wolfville, NS

CaNBFU	University of New Brunswick, Frederiction
CaNBSaM	Mount Allison University, Sackville, NB
CaNBSM	New Brunswick Museum, Saint John
CaOLU	University of Western Ontario, London
CaOKU	Queen's University, Kingston, ON
CaOOA	The Library of the Public Archives, Canada, Ottawa
CaOOU	University of Ottawa
CaOONL	The National Library of Canada, Ottawa
CaOTER	The Ontario Institute for Studies in Education, Toronto
CaOTP	Toronto Public Library
CaOTU	University of Toronto
CaOWaU	Waterloo University, Waterloo, ON
Capc	(Belongs to a private collection)
CaQMM	McGill University, Montréal
CaQMNB	Bibliothèque Nationale du Québec, Montréal
CaQQS	Le Séminaire de Québec, PQ
CaQSherU	Université de Sherbrooke, PQ
CaSSU	University of Saskatchewan, Saskatoon

Main Entries

In the following list, the books are arranged alphabetically by author. For a given author, the order in which books are listed is determined by the wording of their titles. Where the information comes from secondary sources, the wording of the title as it is given in the source is used to determine the order in which the books are listed below. For example, *Cornell's primary geography*, by Sarah Cornell, is listed before the same author's *Intermediate geography*. This is done because other procedures lead to wording based on guesses.

A

[ABBOT, George] **A**

- **1599 (01)**
- *A briefe description of the whole worlde.* Wherein is particularly described, all the monarchies, empires, and kingdomes of the same: with their seuerall titles and scituations thereunto adioyning.
- LONDON; printed by T. Iudson, for Iohn Browne, and are to be sould at the signe of the Bible in Fleete-streete
- 4to in 8's, 18 cm.; collates A-D^8 E^2, pp. [67] of which 65 are text
- ***Libraries*** DU, **L** : CSmH, MiU, NjP, NN
- ***Notes***

1. The first special geography in English to describe at least some countries in every continent. Identifies the location of the country by noting, except in remote places, what seas or countries border it on the north, east, south, and west. Also gives, at the very least, a few words about its political condition. The author is nowhere identified, and this title was, in 1955, catalogued by BL under 'Description ...' Compared with BOTERO (1601), the information provided about the different parts of the world is more evenly balanced.
2. Facsimile ed., 1970 (Johnson/TOT).

- **{1599} (02)**
- *A brief description of the whole worlde* ... Wherein are particularly described all ... situations ...
- LONDON; printed by ...
- 4to, 16 cm.; [collates A-H^4 (Pollard and Redgrave)] {[pp. [63]. [NUC (NA 0010211)]}
- ***Libraries*** CEm : MiU

- **1600 (03)**
- *A briefe description of the whole worlde* ... al the ...
- LONDON; printed by R.B[radock] for Iohn Browne, and are to be sould at the signe of the Bible ...
- 4to, 19 cm.; collates A-H^4, pp. [63] of which 61 are text
- ***Libraries*** **L, O** : CoDU, DFo, ICN, MH, NN
- ***Notes***

1. The text differs from that of (01) in the spelling of many words. It also corrects at least one error in the earlier version, supplying three words missing from the second paragraph on China.
2. Pollard and Redgrave (1986) supply the printer.

- **{1605} (04)**
- *A briefe description of the whole worlde* ... Wherein is ... all ... of the same: newly augmented and enlarged; with their ... scituations ...
- LONDON; printed [W. White] for Iohn Browne, and are to be sold at his shoppe in S. Dunstans churchyard in Fleet-streete
- 4to, 17.5 cm.; collates A-I^4 K-T^4 V-V^4 X-X^2, pp. [164]
- ***Libraries*** C, CCl, E, **L** : CoDU, CSmH, NN, NNH
- ***Notes***

1. The new material is fairly well distributed, though the countries of western Europe and the Americas receive the greatest proportional increase since 1600. The descriptions of a few countries in Asia are unaltered from 1600. The final leaf carries a list of European universities, though no reference is made to this on the tp. until (05).
2. Pollard and Redgrave (1986) supply the printer.

- **1608, 3rd (05)**
- *A briefe description of the whole worlde* ... *of the same, with their academies*. Newly augmented ... adioyning. The third edition
- LONDON; printed [R. Bradock] for Iohn Browne ... Saint Dunstans churchyard in Fleetstreet
- 4to, 18 cm.; collates A-I^2 J^3-N M^2-M^4 O-T^4 V-V^4, pp. [160] of which 156 are text
- ***Libraries*** C, **L**, LLam, O, OT : CoDU, DFo, MH, MnU, NHi, NjP, NN, NNH, RPJCB
- ***Notes***

1. The type has been reset, with each page carrying a little more text than in (04); this accounts for the reduction in the number of pages. The Latin names of countries have been replaced by their English counterparts, and the spelling has been changed in some instances. There is an English translation of the Latin epigram that sums up the national character of the Eng-

lish (M verso), and a trivial change in the wording devoted to America (Q^2 verso). Otherwise no changes were noted in the text.
2. Pollard and Redgrave (1986) supply the printer.

- **1617, 4th (06)**
- *A briefe description of the whole worlde* ... fourth ...
- LONDON; printed [T. Snodham] for Iohn Browne ...
- 4to, 20 cm.; pp. [170] of which 166 are text
- ***Libraries*** CEm, CT, EU, **L**, O : CoDu, CSmH, CtY, DFo, DLC, MBAt, MHi, MiU
- ***Notes***

1. Modest amounts of additional material fairly well distributed, with North America, Peru, and Brazil receiving the most, proportionately, and Spain and France the next.
2. Pollard and Redgrave (1986) supply the printer.

- **1620, 5th (07)**
- *A briefe description of the whole worlde* ... fift ...
- LONDON; printed for J. Marriott
- 4to, 18 cm.; pp. [170] of which 166 are text
- ***Libraries*** C, CGc, CStJ, **L**, LU : CoDU, CSmH, DCU, DFo, MH, NN, NSyU
- ***Notes***

1. Seems identical to (06).
2. Pollard and Redgrave (1986) supply the printer.

- **1624, 6th (08)**
- *A briefe description of the whole worlde* ... sixt ...
- LONDON; printed [A. Mathewes] for Iohn Marriott, and are to bee sold at his shop in Saint Dunstans Churchyard in Fleet-Street
- 4to, 19 cm.; pp. [170] of which 166 are text
- ***Libraries*** **L**, LcM, LinC, **O** : DFo, DLC, IU, MBAt, NN
- ***Notes***

1. Although the type has been reset since (06), the text seems unaltered. The margins contain numerous headings summarizing the contents of the accompanying text.
2. Pollard and Redgrave (1986) supply the printer.

- **1634 (09)**
- *A briefe description of the whole world* ... academies. As also their severall ... situations ... Written by the most reverend father in God, George, late Archbishop of Canterburie
- LONDON; printed [T. Harper] for William Sheares, at the signe of the Harrow in Britaines Burse
- 12mo, 16 cm.; pp. [4], 329, [3]
- ***Libraries*** ArmP, C, EU, **L**, LWI, O : CoDU, CSmH, DFo, DLC, MB, MnU, MWiW, NN, RPJCB
- ***Notes***

1. 'The engr[aved] tp. has the same imprint but "George Abbott ..." in full' (Pollard and Redgrave 1986, who also supply the printer).

2. Contains the full text of (06) to (08). Can it be a coincidence that a new publisher brought out a 'pocket' ed., and also identified Abbot as the author, the year after he died?

- **{1635} (10)**
- [*A brief description of the whole world ...*]
- LONDON; printed [T. Harper] for W. Sheares
- 12mo, 15 cm.; 2 p. l., 350, [5] p.
- ***Libraries*** AbrwthWN, BdmnLk, DU [lacks engr. tp.] (Pollard and Redgrave 1986), YM : CtY, DFo, DLC, MH, MWA, NjP
- ***Notes***

1. Pollard and Redgrave (1986) show a copy at L, adding: '[letterpress tp. only, Ames III.1291,' and then: 'Engr. tp. still dated 1634.' Pollard and Redgrave seem to have been led into error. The collation of the L copy with the tp. dated 1634 is that shown in (09). The BL catalogue, as of April 1989, shows no entry for 1635.
2. Pagination from NUC (NA 0010228).

- **1636 (11)**
- *A briefe description of the whole world* ... empires, and kingdomes ... archbishop of Canterbury
- LONDON; printed by T. H[arper] and are to be sold by Wil. Sheares at the signe of the Harrow in Brittains Burse
- 12mo, 15 cm.; pp. [2], 350, [5]
- ***Libraries*** AbrwthWN, **O** : CtHT, CLU, DFo, HH, IU, NjP, NNC, NNH, ViW
- ***Notes***

1. Text reset but apparently unchanged since (09).

- **1636 (12)**
- *A briefe description of the whole world* ... empires, and kingdomes ... archbishop of Canterbury
- LONDON; T. H[arper] f. W. Sheares
- 12mo, 14 cm.; pp. [4], 350, [5]
- ***Libraries*** **L** : MnU
- ***Notes***

1. Pollard and Redgrave (1986) record this as an issue with a variant tp., and add: 'The L copy has engr. tp. dated 1642.' On this point they are correct.

- **{1642} (13)**
- *A briefe description of the whole world ...*
- LONDON; printed for Will: Sheares at the Bible in Couen Garden
- 15 cm.; 2 p. l., 329, [3] p.
- ***Libraries*** CT, EU : CtY, IU
- ***Source*** NUC (NA 0010232)

- **{1656} (14)**
- *A briefe description of the whole world* ... empires and kingdomes ... situations ... George Abbot, late archbishop ...
- LONDON; printed for W. Sheares

- 12mo, 15 cm.; 2 p. l., 329 [l] (i.e., 337 [l]) p., numbers 189-96 repeated in paging
- ***Libraries*** : CLU, CtY, ICN, NIC, NN, OCl, PBL, PPULC
- ***Source*** NUC (NA 0010233)

- **1664, 5th (15)**
- *A briefe description of the whole world* ... Canterbury. The 5th edition
- LONDON; printed for Margaret Sheares, at the Blew Bible in Bedford-Street in Coven Garden, and John Playfere at the White-Beare in the upper Walk in the New Exchange
- 12mo, 15 cm.; pp. [2], 340, of which 337 are text
- ***Libraries*** L, LV, **O** : DLC
- ***Notes***

1. Text reset but apparently unchanged since (09).
2. Discussed in Gilbert (1972:44-5) and Bowen (1981:68-9).

ABRIDGEMENT OF GEOGRAPHY **Y**

- **{[1800]} (01)**
- *An abridgement of geography.* Adorned with cuts representing the dress of each country. For the juvenile; or child's library
- LONDON; printed and sold by John Marshall
- 9.5 cm.; pp. v, 68
- ***Libraries*** : : CaOTP
- ***Source*** St. John (1975, 1:177), which notes that the date is taken from the frontispiece

ADAM, Alexander **A**

- **1794 (01)**
- *A summary of geography and history, both ancient and modern;* containing an account of the political state, and principal revolutions of the most illustrious nations in ancient and modern times; especially of such as have been distinguished by memorable event: with an abridgement of the fabulous history or mythology of the Greeks. To which is prefixed, an historical account of the progress and improvements of astronomy and geography, from the earliest periods of time to the time of Sir Isaac Newton: Also, a brief account of the principles of Newtonian philosophy, occasionally compared with the opinions of the ancients, concerning the general and particular properties of matter; the air, heat and cold, light, and its effects; the laws of motion; the planetary system, etc. – With a short description of the component parts of the terraqueous globe, according to the notions of the ancients, and the more accurate discoveries of modern chemists. Designed chiefly to connect the study of Classical learning with that of general knowledge
- EDINBURGH; printed for A. Strahan and T. Cadell, London; and Bell & Bradfute and W. Creech, Edinburgh
- 8vo, 22 cm.; pp. xii, 720, [8]
- ***Libraries*** GU, **L** : ICN, MH, NcU, NIC, NjPT, OC, PPL, PPULC, PU

▸ *Notes*
1. It is a special geography although it contains much historical material, especially in the section on Greece, where there are 95 pp. of mythology. For much of the world the chorographical information is very condensed. The BL copy has bound with it 'A geographical index containing the Latin names of the principal countries, cities ... in the Greek and Roman Classics; with modern names subjoined; also, the Latin names of the inhabitants ... and an explanation of difficult words and phrases, being a supplement to the summary of ancient and modern geography' (Edinburgh 1795:145).

▸ **1797, 2nd (02)**
▸ *A summary of geography* ... knowledge. The second edition, corrected. To which is added, a geographical index, containing the Latin names of the principal countries, cities, rivers, and mountains mentioned in the Greek and Roman Classics; with the modern names subjoined. Illustrated with Maps
▸ EDINBURGH; printed for A. Strahan, and T. Cadell jun. and W. Davies (sucessors to Mr. Cadell) in the Strand; and W. Creech, at Edinburgh
▸ 8vo, 22 cm.; pp. xii, 858, [1]
▸ ***Libraries*** **L** : MiD, OrU
▸ *Notes*
1. Spot-checking reveals no changes in the text since (01).
2. NUC records further eds. (all London): 1802, 3rd, 858 pp. (CtY, MH, NCH); 1809, 4th (PP); 1816, 5th, 858 pp. (DLC, MB, NcU, NjR : CaBViP); and 1824, 6th (CtY).
3. Discussed in Downes (1971:383-4).

ADAMS, Daniel **Y Am**
▸ **1814 (01)**
▸ *Geography; or, a description of the world.* In three parts. Part I. Geographical orthography, divided and accented. Part II. A grammar of geography to be committed to memory. Part III. A description of the earth, manners and customs of the inhabitants, manufactures, commerce, government, natural and artificial curiosities, etc. To be read in classes. Accompanied with an atlas. To which is added an easy method of constructing maps, illustrated by plates. For the use of schools and academies
▸ BOSTON; West and Blake
▸ 12mo, 17 cm.; pp. 359, [1]
▸ ***Libraries*** **L** : CtY, MB, MH, Nh, NNC
▸ *Notes*
1. Pt. I, which is a guide to the pronunciation of place names, occupies pp. 13-22. Pt. II (pp. 23-76) is an abstract of Pt. III, which is special geography. Pp. 77-99 carry questions that can be used to test a student's knowledge of Pt. II.
2. NUC records further eds. (all Boston and 323 pp., except as noted): 1816, 2nd, 359 pp. (DLC, MH, MHi, NH, OClWHi, ViU) {NA 0059093 records '3rd ed., [c1816]' (ICU); 1818 is a more probable date for the 3rd ed.}; 1818, 3rd, 335 pp. (CtY, MB, MH, MiU); 1819, 4th, 335 pp. (CtY, NN); 1820, 3rd, 335 pp. (PMA, ViU); 1820, 5th, 335 pp. (CoCC, DLC, DSI,

OHi, NSyU); 1821, 6th (MH); 1823, 7th (CtY, MH, MnU, PSt); 1823, 316 pp. (KyLx); 1825, 8th (DLC, NIC, OClW, PHi, PPULC, ViU); 1826, 9th (AU, MH, PPAN); 1827, 10th (CtY, DLC, MnHi, NjR, PPULC, ViU); 1828, 11th (DLC, IdU, PP); 1830, 12th (NNC, ViU); 1831, 13th, 335 pp. (CtY, MH); 1832, 14th (NNC); 1834, 15th (CtY, MH, ViU); 1834, 15th, '[i.e., 1835]' (ViU); and 1891, 44th (OClWHi).

3. Discussed in Brigham and Dodge (1933:13-14), cited by Elson (1964:396-400), and drawn on by Hauptman (1978:431). Nietz (1961:201, 208, 213, 214, 217-20) and Sahli (1941, *passim*) discuss the 3rd ed. of 1818.

ADAMS, Daniel **Y Am**

- **{1838} (01)**
- *Modern geography [in three parts.* Pt. 1, A grammar of geography concisely arranged; to be committed to memory; with practical questions on the maps.] ... To which is added a brief sketch of ancient geography; a plain method of constructing maps; and an introduction to the use of globes. Illustrated by numerous engravings. Accompanied by an improved atlas
- BOSTON; R.S. Davis
- 19 cm.; 2 p. l., [vii]-viii, 316 pp.
- ***Libraries*** : DLC, N, NcD
- ***Source*** NUC (NA 0059132)
- ***Notes***

1. NUC records a further ed.: 1839, Boston, Davis, 315 pp. (PPL, PPWI).

ADAMS, Dudley

- **{1791} (01)**
- *Complete system of geography* ... (see PICTET and ST. QUENTIN 1791)

ADAMS, Edwin **Y**

- **1863 (01)**
- *Geography classified:* a systematic manual of mathematical, physical, and political geography; with geographical, etymological, and historical notes. For the use of teachers and upper forms in schools. By ... F.R.G.S., Junior Master, Lower School, Dulwich College; Master of the Central Committee of Educational Unions in connexion with the Society of Arts; author of 'The Geographical Word-Expositor and Dictionary;' 'Notes on the Geology, Mineralogy, and Springs of England and Wales;' 'Etymological Geography;' etc.
- LONDON; Chapman and Hall, 193 Piccadilly
- 8vo, 19 cm.; pp. [5], vi-viii, 357, [3]
- ***Libraries*** **L** : MB, MF, NN
- ***Notes***

1. Special geography occupies pp. 43-357. It seeks to steer a middle course between books that are too bulky to be practical as textbooks and ones that are too dry to be interesting. However, the bulk of the information is presented in the form of tables, with many statistics. There is also a marked ethnocentric bias in which 'British' equals 'English.' The author ac-

knowledges (pp. v-vi) his debt to his predecessors, notably: BOHN (1861), HUGHES, W. (1856), MacKAY (1861), and MILNER (1850). A. expresses surprise at the inconsistency of the statistics they provide, especially with respect to populations. Etymological information is limited to European languages.

ADAMS, John **Y mSG**

- **1818 (01)**
- *The young lady's and gentleman's atlas*, for assisting them in the knowledge of geography. By ... teacher of the mathematics, Edmonton
- LONDON; printed for Darton, Harvey, and Darton, No. 55, Gracechurch-Street
- 18.5 cm.; pp. [3], 4-40
- *Libraries* : : **CaVBaU**
- *Notes*

1. An atlas with text, devoted almost exclusively to Europe, rather than a special geography. Odds and ends of information are provided about the countries for which there are maps. Begins with astronomy (pp. [3]-11) and then definitions of geographical terms (pp. 11-14).

ADAMS, Michael **A**

- **1794? (01)**
- *The new royal system of universal geography*, containing complete, full, particular and accurate histories and descriptions of Europe, Asia, Africa & America, as divided into empires, kingdoms, states, provinces, republics, governments, continents, islands, oceans, seas, rivers, gulphs, lakes, etc. etc. Including all the new discoveries. Written and compiled by ... Assisted by many gentlemen, eminent in the science of geography.

  [Thus the first tp.; there is a second, which reads]: The new ... geography. Containing a complete ... particular, and ... history and description of all the several parts of the whole world; as divided into ... lakes, etc. Together with new accounts of their soil, situation, extent, and boundaries, in Europe ... Africa, and America: with a very particular account of their subdivisions and dependencies; their cities, chief towns, universities, and harbours; with their commerce, trade, learning, policy, arts, manufactures, genius, manners, customs, etc. as well as their revenues, forces, revolutions, curiosities, buildings, antiquities, ruins, mountains, mines, animals, vegetables, and minerals; and whatever is found curious, useful, and entertaining, at home and abroad. To which will be added, a new and easy guide to geography and astronomy, the use of globes, maps, etc. and the doctrine of the sphere; with an account of the rise and progress of navigation, its improvement and utility to mankind; together with chronological tables of the sovereigns of the whole world. Including every interesting discovery and circumstance in the narratives of Captain Cook's voyages round the world, together with all the recent discoveries made in the Pelew islands, New Holland, New South-Wales, Botany-Bay, Port-Jackson, Norfolk-Island, north and west coasts of America, the interior parts of America, Africa, China, Caffraria, India east and west, Arabia, Madagascar, Russia, etc.

Carefully written and compiled from the late journals of the voyages and travels of Captains Phillips, King, Ball, Hunter, White, Dixon, Portlock, Morse, Blyth, Brissot, Hodges, etc. The whole forming a complete, authentic, copious, and real new geographical library. By ... esq. of Lincoln's-Inn, London; author of the new, complete, and general Biographical Dictionary, now publishing, in ninety-six periodical numbers, with universal applause. Assisted by a Society of gentlemen who have respectable correspondents in the various parts of the world. Illustrated with a beautiful set of engravings, consisting of maps, charts, plans, harbours, views of cities, towns, etc.

- ▸ LONDON; printed for the proprietors: published by Alex. Hogg, at the King's-Arms, No 16, Paternoster-Row; and sold by all the booksellers and news-carriers in England, Wales, Scotland, and Ireland
- ▸ 4to, 26.5 cm.; pp. [7], iv-viii, 9-960
- ▸ ***Libraries*** : DLC, PPL : **CaEAU**
- ▸ ***Notes***

1. This is one of five large special geographies that were published in a span of six years (BALDWYN 1794?, BANKES and others 1788?, COOKE, G.A. 1790?, PAYNE, J. 1791). Some may see them as predecessors of the modern coffee-table book. As with three of the others, this one failed to carry its date of publication on the tp. It is inferred to be 1794 on the basis of the dates either attached to some of the illustrations or presented in the 'chronological list of remarkable events' (pp. 944-51).

   Alex. Hogg was responsible for the publication of this work and BALDWYN (1794?). It seems odd that he should have published two such similar works in the same year. In both cases they seem to have been issued in installments (weekly? monthly?), and each carries a list of subscribers at the end. In this work, as also in BALDWYN, BANKES, and COOKE, the text starts with more than 100 pp. devoted to the description of the new discoveries in the Pacific, then recently made by the voyages of Captain Cook, as well as other, less-known explorers. In this respect they are signs of the great interest shown by educated Europeans in the 'noble savages' who played such a large role in the imagination of the period.

   Two other special geographies equivalent in size to these five (MIDDLETON 1777-8, MILLAR 1782) share in this same enthusiasm, to a lesser degree; but then, they were published before the third of Cook's voyages. PINKERTON (1802) might also be classified with this group, but it is, both by intent and achievement, more scholarly than the others. It also differs from them in not devoting special attention to the discoveries in the Pacific.

   This work (i.e., ADAMS) and BANKES follow PAYNE in presenting the continents in the unusual order of Asia, Africa, America, and Europe. In terms of the distribution of attention devoted to the various parts of the world, the general range of topics dealt with, and the balance of attention paid to them, this work seems to resemble BANKES more closely than the others. It differs strikingly from BALDWYN in that ADAMS reports the execution of the king and queen of France (which had taken place in 1793), while the former makes no reference to the French Revolution and still identifies Louis XVI as the ruler of his country. They are similar, among many other ways, in that, while both acknowledge that France was now divided into *departments* for administrative purposes, their description of that country is organized in terms of the old provinces.

- **{1796} (02)**
- *A new system of universal geography ...*
- LONDON
- ***Libraries*** : NJR

ADAMS, W. **Y SG?**

- **{1858} (01)**
- *First lessons in geography*
- ? Mozley
- ***Libraries*** n.r.
- ***Source*** ECB, Index (1856-75:130)

ADDINGTON, STEPHEN **Y**

- **{1770} (01)**
- *The youth's geographical grammar;* containing geographical definitions ... to which is added I. An alphabetical index of kingdoms, states ... II. An alphabetical index of cities ...
- LONDON; printed and sold by the author and J. Buckland
- 18 cm.; pp. iv, 364
- ***Libraries*** : MnHi, PPL
- ***Source*** NUC (NA 0068483)

ADDINGTON, Stephen
See BUTLER, S. (1809).

ADDINGTON, Stephen and T. Watson **Y mSG? Am**

- **{1806} (01)**
- *Questions relating to geography;* with particular reference to Workman's system. To which is added promiscuous questions, on the study of geography in general
- PHILADELPHIA; printed and sold by John McCulloch, No. 1 North Third Street
- 15 cm.; pp. iv, [5]-69, [2]
- ***Libraries*** : PHi, PPL
- ***Source*** NUC (NA 0068472). Not likely to be special geography. See WORKMAN (1790).

AIKIN, John **A UK+US**

- **1806 (01)**
- *Geographical delineations*: or a compendious view of the natural and political state of all parts of the globe
- LONDON; printed for J. Johnson, No. 72, St. Paul's Church-Yard

- 2 vols., 8vo, 20 cm.; Vol. I, pp. viii, [2], 374; Vol. II, pp. iv, 408
- ***Libraries*** L, **OSg** : MBAt, MiD, PP, PPL, ViU : CaNSWA
- ***Notes***

1. 'In the prosecution of this design I have been guided by two leading considerations respecting each country – what nature has made of it, and what man has made of it' (p. iv). Europe occupies Vol. I, pp. 6-320, Asia the balance of Vol. I and also Vol. II, pp. 1-184. There are no sections devoted exclusively to history.

- **{1806} (02)**
- *Geographical delineations ...*
- PHILADELPHIA; Kimber
- Pp. 416
- ***Libraries*** : PP, PPULC
- ***Source*** NUC (NA 0108858)

- **{1806} (03)**
- *Geographical delineations ...*
- PHILADELPHIA; Nichols
- ***Libraries*** : PPULC, RNHi
- ***Source*** NUC (NA 0108859)

- **{1807} [anr ed., corrected and enlarged] (04)**
- *Geographical delineations ...*
- PHILADELPHIA; printed for F. Nichols, by Kimber, Conrad & Co., No. 93, Market Street
- 23 cm.; 4 p. l., 416 p.
- ***Libraries*** : CtY, InU, MH, MiU, MMeT, MSaE, MWA, MWH, PHi, PMA, PP, PU, PPULC, ViU
- ***Source*** NUC (NA 0108860)
- ***Notes***

1. Whatever may have been the relationship of Kimber to Nichols in 1806, it seems that they combined to publish this edition. Judging by the number of libraries who hold copies of the three eds., the alliance was productive.

AINSWORTH, William Francis, ed. **Dy**

- **1863 (01)**
- *The illustrated universal gazetteer*
- LONDON; John Maxwell and company, 122, Fleet Street; Houlston and Wright, 65, Paternoster Row
- 24.5 cm.; pp. [3], iv-vi, [3], 2-1040
- ***Libraries*** **L**, [IDT] : OCX
- ***Notes***

1. The letters A to G occupy pp. 2-639, H to Z pp. 640-1040. Recognizes the independence of the Confederate States of America.

ALDIS, William **Y SG?**

- **{[1858]}, 2nd (01)**

- *The first book of geography*
- LONDON; Jarrold & Sons
- 16mo; pp. iv, 194
- ***Libraries*** L

ALEXANDER, A. **Y SG?**

- **{1797} (01)**
- *Summary of geography and history, both ancient and modern*
- LONDON
- ***Libraries*** n.r.
- ***Source*** Allford (1964:241)

[ALINGHAM, William] **mSG**

- **1698 (01)**
- *A short account, of the nature and use of maps.* As also some short discourses of the properties of the earth, and of the several inhabitants thereof. To which is subjoin'd a catalogue of the factories and places now in possession of the English, French, Dutch, Spanish, Portuguese and Danes, both in the East and West-Indies
- LONDON; printed, and are to be sold by Mr. Mount, at the Postern on Tower-Hill; Mr. Lea at the Atlas and Hercules in Cheapside; Mr. Worgan mathematical instrument-maker under St. Dunstan's Church, Fleetstreet; and by William Alingham, mathematick-teacher, in Channel-Row, Westminster
- 8vo, 16 cm.; pp. [6], 56
- ***Libraries*** L, LPO
- ***Notes***

1. The author is not identified until (02). Contains some systematic human geography following the discussion of climates (in the Classical sense). Comes to the edge of special geography with a 'catalogue of some of the chiefest places in the world, with their latitudes and longitudes' (p. 57).

- **1703 (02)**
- *A short account of* ... discourses of the division of the earth into zones, climes and parallels; with the properties of the several ... To which is subjoined a ... West Indies. With several tables very useful in geography and navigation. By ... teacher of the mathematicks in Channel-Row, Westminster
- LONDON; printed by R. Janeway, for Benj. Barker at the White-Hart in Westminster-Hall
- 8vo, 16 cm.; pp. [8], 84
- ***Libraries*** L : CtY, DLC, MiU, MWA, NN
- ***Notes***

1. The opening sections on the globe and its properties, and on maps, have been enlarged considerably.

ALLEN, Fordyce **Y Am**
- **{1862} (01)**
- *Primary geography*
- PHILADELPHIA; J.B. Lippincott
- ***Libraries*** n.r.
- ***Notes***

1. Discussed in Culler (1945, *passim*). See (02) of Shaw (1854).

ALLEN, Joseph **Y mSG Am**
- **{1825} (01)**
- *Easy lessons in geography and history*, by question and answer ...
- BOSTON; Cummings, Hilliard and Co.
- ***Libraries*** : MH
- ***Source*** NUC (NA 0184511)

- **{1827}, 2nd ... (see below) (02)**
- *Easy lessons in geography and history*, by question and answer. Designed for the use of the younger classes in the New England schools. 2d ed., rev. and improved. To which are prefixed the elements of linear drawing
- BOSTON; Hilliard, Gray, Little and Wilkins
- 15.5 cm.; v, [1], 7-54
- ***Libraries*** : DLC, MR
- ***Source*** NUC (NA 0184512)
- ***Notes***

1. NUC notes: 'Part II. Relating to history, and particularly to the history of America': pp. [37]-54.
2. NUC records further eds. (both Boston): 1829, 3rd, 96 pp. (DLC, MH); 1832, enl., 116 pp. (DLC, MB).
3. Elson (1964:400) cites the ed. of 1832.

ALLISON, M.A. **Y SG?**
- **{1828} (01)**
- *First lessons in geography*, for the use of the nursery, and the junior classes in schools
- LONDON; Baldwin and Cradoc
- 14 cm.; pp. ii, 74
- ***Libraries*** : DLC
- ***Notes***

1. BL records further eds. (all London): 1835?, 7th; 1857, 25th; and 1860, 27th, 114 pp. BCB, Index (1837-57) records ed. of 1849.
2. BL notes that the ed. of 1857 was 'improved and corrected' by A.B. Power.

ALLISON, M.A. **?**
- **{1852} (01)**

- *Geography*
- [LONDON]; Simpkin
- 18mo
- ***Libraries*** n.r.
- ***Source*** BCB, Index (1837-57:109), which notes, after the author's name: 'by Power.'

ALLOTT, Robert
See WITS ... (1599).

ANDERSON, John **Y Am**
- **{1851} (01)**
- *A practical system of modern geography for exercises on maps*
- NEW YORK; J.S. Redfield
- 15.5 cm.; pp. 108
- ***Libraries*** : DLC
- ***Notes***

1. Discussed in Culler (1945, *passim*).

ANDERSON, Robert **Y SG?**
- **{1864} (01)**
- *Class-book of geography: physical and descriptive*
- LONDON; [T. Nelson]
- 8vo; pp. l. 12-224
- ***Libraries*** L.
- ***Source*** BL, which notes: 'Nelson's School Series'

ANDERSON, Robert **Y SG?**
- **{1858} (01)**
- *Geography for junior classes*
- [LONDON]; Nelson
- 18mo
- ***Libraries*** n.r.
- ***Source*** ECB, Index (1856-75:130), which records further ed. of 1859
- ***Notes***

1. BL and NUC record further eds. (both London): 1863 (L : MoU), 1868 (L). According to an unidentified source, the last had 109 pp.

ANDERSON, Robert **Y SG?**
- **{1856}, 4th (01)**
- *Modern geography for the use of schools*
- LONDON; Nelson
- 8vo; pp. viii, 221 (223 according to NUC)
- ***Libraries*** L : PPULC, PU
- ***Source*** BL and NUC (NA 0304446), which notes: 'Nelson's school series'

▸ *Notes*

1. NUC records further eds. (both London): 1861, 6th (OkU, N); 1866, 6th (MB, MH, PPULC, PU). CaBVaU has 1863, 6th. ECB, Index (1856-75: 131) records ed. of 1872.

ANSTED, David Thomas **Y**

▸ **1871 (01)**

▸ *Elementary geography adapted for teaching in primary schools*

▸ LONDON, Paris, and New York; Cassell, Petter, and Galpin

▸ 8vo, 16cm; pp. v, [1], 160, [16]

▸ ***Libraries*** **L**

▸ *Notes*

1. A special geography intended as a textbook for primary schools.

ARNOLD, Richard **mSG**

▸ **{1502 (NUC), 1503 (BL)} (01)**

▸ *The copy of a carete cumposynge the circuit of the worlde* and the compace of every land

▸ [ANTWERP? (NUC)]; [Adrian van Berghen (BL)]

▸ Folio, 26 cm.; 11 p. l., cxviii (i.e., cxix) numb. l. (NUC)

▸ ***Libraries*** **L**, LG, LU, O : CtY, DFo, DLC, ICN, MH, MWiW, WU

▸ *Notes*

1. 'This work consists of a small section of *Arnold's Chronicle*. The *Chronicle* is an untitled work that begins: In this booke is conteyned ...' (NUC). The section opens with a reference to the sources of the information to be presented; all are Classical authors. The world is then described as consisting of four parts: east, west, north, and south. These are all divisions of the Old World, the existence of America not being acknowledged. The description of the four parts is little more than a list of places, chiefly countries and provinces, with some indication as to their relative location. The section ends with some information on the sailing routes from Calais to Venice and Joppa, and on the lengths of coastlines.

   The complete *Chronicle* was reprinted in 1811; see below. In that edition, this section occupies pp. 140-4.
2. Discussed in E.G.R. Taylor (1930:6, 13).

▸ **{1521}, anr ed. (02)**

▸ *Chronicle or customs of London* (untitled work containing 'The copy ...' as before)

▸ SOUTHWARK; Peter Treveris

▸ Folio, 28 cm.; pp. [267]

▸ ***Libraries*** C, L, LLam, O, [IDT] : CtY, DLC, MH, MWiW, PU

▸ *Notes*

1. Holdings from STC and NUC (NA 0427218).

▸ **1535, anr ed. (03)**

▸ *Mappa mundi, otherwyse called the compasse and cyrcuet of the worlde,* and also the compasse of every ilande, comprehendyd in the same

- LONDON; Robert Wyer [on the last page]
- 12mo, 13 cm.; pp. [20], of which 16 are text
- *Libraries* **L** : DFo
- *Notes*

1. Despite the change in title, this is Arnold's *Copy of a Carete.* According to E.G.R. Taylor (1930:166-7), Wyer also published this section of *Arnold's Chronicle* at about the same time (ca. 1535) under the title of 'The Rutter of the Distances from one Porte or Countrie to another,' as an addendum to the 'Compost of Ptholomaeus' (see PTOLEMY 1530?). According to Park, this title (*Mappa Mundi*) was published in 1536 'with the "Rutter" of the preceding item' (Park 1928:270); the reference is to the 3rd ed. of the *Compost of Ptholomeus* (see PTOLEMY 1530?). This implies, in contrast to Taylor, that the *Mappa Mundi* and the Rutter are different works; in the 1961 ed. of Park (1928), this statement is not to be found.
2. British Museum (1959-66) does not recognize this work as an ed. of *Arnold's Chronicle*, but catalogues it under *Mappa Mundi*.

- **1811 (04)**
- *The Customes of London, otherwise called Arnold's Chronicle ...*
- LONDON; F.C. and J. Rivington
- 4to; pp. lii, 300
- *Libraries* **L** : CtY, DLC
- *Notes*

1. This is the edition of 1502/1503 in modern type.

ARROWSMITH, Aaron (Sr) **?**

- **{1794} (01)**
- *A companion to a map of the world*
- LONDON
- 4to; pp. 25
- *Libraries* L

ARROWSMITH, Aaron (Jr) **Y**

- **1831 (01)**
- *A compendium of ancient and modern geography* for the use of Eton school
- LONDON; published for the author, by E. Williams, Eton, and at the Eton warehouse, Red Lion Court, Fleet-Street; Whittaker, Treacher & Co. Ave Maria Lane; J. & J.I. Deighton, Cambridge; J. Parker, Oxford; Waugh & Innes, Edinburgh; and Milliken & Son, Dublin
- 8vo, 22 cm.; pp. [1], viii, [2], 906
- *Libraries* **L** : DLC, PP, PPL, PPULC
- *Notes*

1. The Preface suggests that it was a commissioned work. The bias towards information about the Classical world is very marked. It lists the principal ancient authors that were consulted.
2. BL and NUC record further eds. (both London): 1839, 847 pp. (L : CtY, DI-GS, DLC, MH, MiD); 1856, 795 pp. (L : DLC). [A new Preface summarizes the changes that have been made to Arrowsmith's original work.]

ARROWSMITH, Aaron (Jr) **Y**

- **1832a (01)**
- *A grammar of modern geography*, with an introduction to astronomy and the use of globes, compiled for the use of King's College school
- LONDON; S. Arrowsmith, Soho Square; and B. Fellowes, bookseller and publisher to the College, Ludgate Street
- 12mo, 18 cm.; pp. [5], 461, [2]
- ***Libraries*** **L**
- ***Notes***

1. A school textbook. 34 pp. at the beginning and 14 at the end are devoted to the world in general and to the use of globes, respectively. The balance is special geography, with the continents receiving descending amounts of attention in the order of presentation: Europe, Asia (including Australasia), Africa, and America.
2. BL records a further ed.: 1846, 464 pp.

ARROWSMITH, Aaron (Jr) **Y mSG**

- **1832b (01)**
- *A praxis on the grammar of modern geography and astronomy*, compiled for the use of King's College school
- LONDON; S. Arrowsmith, Soho Square; and B. Fellowes, bookseller and publisher to the College, Ludgate Street
- 12mo, 18 cm.; pp. [4], 60
- ***Libraries*** **L**
- ***Notes***

1. Entirely devoted to questions testing a student's knowledge of ARROWSMITH (1832a).

ASPIN, J[ehoshaphat] **Y mSG**

- **1827 (01)**
- *Cosmorama*; a view of the costumes and peculiarities of all nations. By ..., esq. author of 'A picture of the manners, customs, sports, and pastimes of the people of England'
- LONDON; John Harris, St. Paul's Church-Yard
- 12mo, 16.5 cm.; pp. [5], vi-viii, 404
- ***Libraries*** L : **MH**, MnU, NB, NN, PP, PPULC : CaOTP
- ***Notes***

1. The date of publication does not appear on the tp.; the end of the text is dated 'Christmas 1826,' and the plates 'Jan. 1, 1827.'

   The book is addressed to 'my young friends ... desirous of extending your information from a knowledge of the land in which you were born, and the people among whom you live, to an acquaintance with the whole world' (p. 1). It is described as following after 'Abbé Gaultier's excellent ... geography' (p. 2). As a result it does not mention 'boundaries, divisions, and other geographical particulars, of the countries we must visit.' Aspin was responsible for the 1821 ed. of GAULTIER (1792; see (07)), so it is presumably to this work that he is referring here.

   Fairly extensive use is made of direct quotations; the authors of the quo-

tations are identified only occasionally. See note to (02) of ASPIN (1839), below.
2. BL and NUC record further eds. (both London): 1827? (: MH, OClWHi); and 1834 (L : ICN, MH, MtU, NB, NN, TxU).

ASPIN, Jehoshaphat **Y mSG Am**

- **{1839}, New ... (as below) (01)**
- *Picture of the world*: or, a description of the manners, customs and costumes of all nations ... Illustrated by engravings. New and enlarged edition
- HARTFORD; Philemon Canfield
- 16 cm.; pp. viii, 256
- ***Libraries*** : OKentU
- ***Notes***

1. Although the tp. asserts that this is a new ed., it is treated here as a new work; see note to (02), below.

- **{1840}, New ... (as below) (02)**
- *Picture of the world*; or ... nations. A new and enlarged edition. Illustrated by engravings
- HARTFORD; Philemon Canfield
- 15.5 cm.; pp. 256
- ***Libraries*** : DSI, MWA, NcD, OClWHi
- ***Notes***

1. OClWHi supplied a photocopy of the tps., tables of contents, and selected pp. of text of both this work and ASPIN (1826), above. They also compared the copies of the eds. of the two titles that they hold and reached the conclusion that: 'While there are distinct differences in size, illustrations, and publication details, the 1840 work appears to be a shortened version of the earlier work. Much of the phraseology is exactly the same.' The two tables of contents show that the American publisher reordered the sequence in which countries were described, placing North America at the beginning, in place of Europe.

ASPIN, Jehoshaphat **Y mSG Am**

- **{1834(?)} (01)**
- *Sequel to geography*, being a view of the world as distinguished by manners, costumes and characteristics of all nations. Originally written by J. Aspin and now improved and adapted to the use of American schools
- PHILADELPHIA; Desilver, Thomas
- Pp. 365
- ***Libraries*** : PP
- ***Source*** NUC (NSA 0090612) notes: 'On spine: "A view of the world"'
- ***Notes***

1. NUC records further eds. (Cooperstown, NY, 365 pp., except as noted): 1835, Philadelphia (ViU, imperfect); 1836, New York (IaAS); 1840, 3rd (NCH); 1841, 3rd (IU, MH, ODW); 1842, 5th (N); 1844, 5th (MF, OClW); and 185-?, 5th, New York, 408 pp. (DLC, ICRL)

ATLAS GEOGRAPHUS **A**

- **1711-17 (01)**
- *Atlas geographus*: or, a compleat system of geography, ancient and modern. Containing what is of most use in Blaeu, Varenius, Cellarius, Cluverius, Baudrand, Brietius, Sanson, etc. With the discoveries and improvements of the best modern authors to this time. Illustrated with about 100 new maps, done from the latest observations, by Herman Moll, geographer; and many other cuts by the best artists. Europe is two volumes, with sixty eight maps, Sanson's tables, etc. Vol. I. [Thus the title of the first of the five vols. published (a sixth was planned); the titles of the succeeding vols are listed below.]
- LONDON; in the Savoy: printed by John Nutt; and sold by Benjamin Barker and Charles King in Westminster-Hall; Benjamin Tooke at the Middle-Temple Gate; William Taylor at the Ship in Pater-Noster-Row; Henry Clements at the Half-Moon in St Paul's Church-Yard; Richard Parker and Ralph Smith under the Piazza of the Royal Exchange; and John Morphew near Stationers-Hall
- 4to, 24 cm.; pp. [3] xvi, 890
- *Notes*

1. The text ends at the foot of p. 890, in mid-sentence.

- **1711**
- *Atlas geographus ... Vol. II*
- LONDON; in the Savoy ...
- 4to, 24 cm.; pp. [2], 891-1772
- *Notes*

1. The text on p. 891 begins with the continuation of the sentence left unfinished at the end of Vol. I. No description of the British Isles was included, but in an 'Advertisement' following the tp., the printing of a volume to be titled *Britannia & Hibernia Antiqua & Nova* was forecast to begin 'in less than six months.'

- **1712**
- *Atlas geographus* ... by the best artists. Asia is one volume, with thirty one maps, Sanson's tables, etc. as may be seen in the catalogue thereof annex'd to the preface. (Vol. III)
- LONDON; in the Savoy ...
- 4to, 23 cm.; pp. viii, 8, 851
- *Notes*

1. The printing of *Britannia and Hibernia* is now forecast to begin 'in about three months.'

- **1714**
- *Atlas geographus* ... to this time. Illustrated with about 17 new maps, cuts, Sanson's tables, etc. as may be seen in the catalogue thereof annex'd to the index: the maps done by Herman Moll, geographer; in which are all the latest observations of the Atlas for the whole world. Europe is two volumes, Asia a third, and this is the fourth. Vol. IV
- LONDON; in the Savoy ...
- 4to, 23 cm.; pp. [2], ii, 808

▸ *Notes*

1. Publication of the Introduction of *Great Britain and Ireland* is now promised for 'the first Monday in January next' (dated 30 Nov. 1713).

▸ **1717**

▸ *Atlas geographus* ... to this time; with about 30 new maps, cuts ... latest observations. Europe in two volumes, Asia the third, Africa the fourth, and this the fifth. Vol. V. To which is added a catalogue of the maps, cuts, and Sanson's tables in all the five volumes, and a description of Posnia, by omission left out in Europe

▸ LONDON; in the Savoy: printed by Eliz. Nutt, for John Nicholson at the King's Arms in Little Britain; and sold by John Morphew near Stationers-Hall. Where may be had Numb. 1, to 17 (price 1 s. each)

▸ 4to, 24 cm.; pp. iv, 807, [5]

▸ ***Libraries*** **L, O** : AzU, CSmH, CtW, CtY, CU, ICN, IEN, InU, LNT, MBAt, MdPB, MiD, MiU, MtHi, MWiW, NcU, NIC, NN, NPV, OClWHi, PBL, PBm, PHi, PPL, RPJCB : CaOTU

▸ *Notes*

1. In a note at the end, the printing of *Britannia* is said to have reached Gloucestershire (proceeding alphabetically through the counties).

   According to Cox (1938:350) this work appeared in at least five eds. This seems unlikely. His opinion seems to have been based on the fact that he regarded the 2nd ed. of COMPLETE SYSTEM (1744) as a subsequent ed. of the *Atlas geographus*. He was wrong about that, but having made the mistake, he could then have been influenced by the opinion of the compilers of the COMPLETE SYSTEM (1744), who saw their work as a continuation of a four-editon series that they identified as the *Complete Geographer*. See notes to THESAURUS GEOGRAPHICUS (1695).

   Whoever were the compilers of this publication, it was a major work, and was clearly intended to be the standard reference for all those seeking information about the location and current condition of all the countries of the world. References to sources are made in the text. In the Preface it is stated that work on the *Atlas* had begun seven years previously, i.e., in 1704.
2. MOLL (1739) may be another ed. of the *Atlas geographus*.

ATLAS MINIMUS **Y mSG?**

▸ **{1758}* (01)**

▸ *Atlas minimus*, or a new set of pocket maps of the several empires, kingdoms and states of the known world, with historical extracts relative to each. Drawn and engrav'd by J. Gibson from the best authorities, revis'd, corrected and improved, by Eman: Bowen, geographer to His Majesty

▸ LONDON; publish'd accorg. to Act of Parlt. Jany. 2d, 1758 & sold by J. Newbery

▸ 11 cm; Ff. 52

▸ ***Libraries*** : : CaOTP

▸ ***Source*** St. John (1975, 1:178), which implies that it is suitable for youth

▸ *Notes*

1. The relationship of this work to SELLER (1679) is not known.

AVITY, Pierre d' **A tr**

- **1615 (01)**
- *The estates, empires, and principallities of the world.* Represented by ye description of countries, maners of inhabitants, riches of provinces, forces, government, religion; and the princes that haue governed in euery estate. With the begining of all militarie and religious orders. Translated out of the French by Edw. Grimstone, Sargeant at Armes
- LONDON; printed by Adam Islip; for Mathewe Lownes, and John Bill
- Folio, 34 cm.; pp. [18], 1234
- ***Libraries*** C, DC, DU, EU, ExC, GP, GU, **L**, LG, MC, NcP, O, WnC, [IDT] : CoDu, CtY, CU, DFo, DLC, DW, ICN, MiU, MnU, NcD, NcU, NN, OClW, OrU, PSt, TxU
- ***Notes***

1. The first large-scale description of the countries of the world in English. Compared with ORTELIUS (1606), it has far more text, but lacks maps. The Preface contains reasons for the inclusion of the material found in the book. America is dealt with only briefly, as part of the Spanish Empire. This is characteristic, as countries are looked on as the estates of their rulers. According to the Preface of BOTERO (1630), much of Avity's material was taken from Botero, but the disparity in length of the two books, at least in their English versions, raises doubts as to the accuracy of this statement.
2. British Museum (1959-66) catalogues under Grimstone, Edward.
3. Discussed in Gilbert (1919). For the French original, see Broc (1980:95-6).

B

BAKER, W.R. **Y SG?**

- **{1882} (01)**
- *Geographical reader*
- LONDON
- ***Libraries*** n.r.
- ***Source*** Allford (1964:249)

BALDWYN, George Augustus **A**

- **1794? (01)**
- *A new, royal, authentic, complete, and universal system of geography*; or, a modern history and description of the whole world. Containing new, full, accurate, authentic, and interesting accounts and descriptions of Europe, Asia, Africa, and America, as consisting of continents, islands, oceans, seas, rivers, lakes, promontories, capes, bays, peninsulas, isthmusses, gulphs, etc. and as divided into empires, kingdoms, states, and republics. Together with a description of their limits, boundaries, climate, soil, natural and artificial curiosities and productions, religion, laws, government, revenues, forces, antiquities, etc. Also the provinces, cities, towns, villages, forts, castles, harbours, sea-ports, aqueducts, mountains, mines, minerals,

fossils, roads, public and private edifices, universities, etc. contained in each: and all that is interesting relative to the customs, manners, genius, tempers, habits, amusements, ceremonies, commerce, arts, sciences, manufactures, and language of the inhabitants. Together with an accurate and lively description of all the various kinds of birds, beasts, reptiles, fishes, amphibious creatures, insects, etc. including the substance and essence of the most remarkable voyages and travels, which have been performed by the navigators and travellers of different countries, particularly the latest discoveries in the South Seas, and towards the North Pole; with every curiosity that has hitherto appeared respecting the different parts of the universe. Comprising every interesting discovery and circumstance in the narratives of Captain Cook's voyages round the world. Together with all recent discoveries made in the Pelew Islands, New Holland, New South Wales, Botany Bay, Fort Jackson, Norfolk Island, north and west coasts of America, the interior parts of America, Africa, China, Caffraria, India East and West, Arabia, Madagascar, Russia, etc. etc. Carefully written and compiled from the late journals of the voyages and travels of Captains Phillips, King, Ball, Hunter, White, Dixon, Portlock, Mears, Patterson, Bruce, Anbury, Rochon, Morse, Blyth, Ross, Imlay, Keate, Brissot, Hodges, etc. etc. Also compendious histories of every empire, kingdom, state, etc. with their various revolutions. To which will be added a new, complete, and easy guide to geography, astronomy, the use of globes, etc. With an account of the rise and progress of navigation, its improvements and utility to mankind. The whole embellished and enriched with upwards of an hundred most elegant and superb copper-plates, engraved in such a manner as to do infinite honour to the respective artists by whom they are executed. These embellishments consist of views, maps, land and water perspectives; birds, beasts, fishes, etc. as also the various dresses of the inhabitants of different countries, with their strange ceremonies, customs, amusements, etc. etc. By ... Assisted by many gentlemen eminent for their knowledge in the science of geography; particularly Charles Andrew Roberstson, esq. - Clement Walley Oulton, esq. - and Henry Hogg, M.A.

- LONDON; sold by Alex Hogg, at (No. 16) Pater-Noster-Row, and by all the booksellers of Bristol ... York: and by all other booksellers in England, Wales, Scotland and Ireland
- Folio, 39 cm.; pp. [4], iv-cxii, 812, v-viii, [8] [the CaBVaU library catalogue notes some misnumbering of the pages]
- *Libraries* NcU : CtY, DLC : **CaBVaU**
- *Notes*

1. The tp. does not carry the date of publication, but this is presumed to be the date found at the foot of the frontispiece. In addition, many of the maps and illustrations carry the date 1794. The Preface takes the form of a prospectus intended to lead people to subscribe to the work, which was to be issued in eighty numbers. At the foot of p. 626, there is a note addressed to the subscribers assuring them that the claims made for the work would be fulfilled. There is a list of subscribers at the end of the book. It includes the names of booksellers, some of whom took up to 100 copies. In the note on p. 626, there are denigrating references to competitors.

   The description of the new discoveries in the Pacific, which includes some lengthy direct quotations, comes first, occupying pp. v-cxii. Relatively

little history is included, and current events tend to be dealt with in an off-hand way; for example, it takes a careful reading of the section on North America to reveal the fact that a new country had recently come into existence there. In a similar way, no reference is made to the revolution in France, whose ruler is still identified as Louis XVI; the description of that country is conducted in terms of the old provinces, though the map of France shows the departments 'as decreed by the National Assembly.'
2. For contemporary works of comparable scope and scale, see ADAMS, M. (1794?), BANKES (1788?), COOKE, G.A. (1790?), and PAYNE, J. (1791).

BANKES, Thomas, BLAKE, Edward Warren, and COOK, Alexander A
▸ **{[1787 (or 1788)]} (01)**
▸ *By the King's royal licence and authority. A new royal authentic and complete system of universal geography antient and modern*: including all the late important discoveries made by the English, and other celebrated navigators of various nations, in the different hemispheres; and containing a complete genuine history and description of the whole world, as consisting of empires, kingdoms, states, republics, provinces, continents, islands, oceans, etc. with the various countries, cities, towns, promontories, capes, bays, peninsulas, isthmusses, gulphs, rivers, harbours, lakes, aqueducts, mountains, volcanos, caverns, deserts, etc. etc. throughout Europe, Asia, Africa and America: together with their respective situations, extent, latitude, longitude, boundaries, climates, soil, natural and artificial curiosities, mines, metals, minerals, trees, shrubs, the various kinds of fruits, flowers, herbs and vegetable productions. With an account of the religion, laws, customs, manners, genius, tempers, habits, amusements, and singular ceremonies of the respective inhabitants: their arts, sciences, manufactures, learning, trade, commerce, military and civil governments, etc. Also exact descriptions of the various kinds of beasts, birds, fishes, amphibious creatures, reptiles, insects, etc. peculiar to each country; including every thing curious, as related by the most eminent travellers and navigators, from the earliest accounts to the present time. Likewise the essence of the voyages of the most enterprising navigators of different nations and countries, from the celebrated Columbus, the first discover of America, to the death of our no less celebrated countryman Captain Cook, etc. etc. Together with a concise history of every empire, kingdom, and state. Including an account of the most remarkable discoveries, settlements, battles, sieges, sea-fights, and various revolutions that have taken place in different parts of the world. The whole forming an authentic and entertaining account of every thing worthy of notice throughout the whole face of nature, both by land and water. With a great variety of curious articles, communicated by gentlemen who have travelled in various parts, and by captains of ships, etc. none of which ever appeared in print before. To which is added a complete guide to geography, astronomy, the use of the globes, maps, etc. with an account of the rise, progress, and present state of navigation throughout the known world. By the Reverend ... Vicar of Dixton in Monmouthshire, and author of the Christian's Family Bible. Edward ... Cook, A.M., teacher of geography, astronomy, and navigation. Enriched with upwards of one hundred beautiful

engravings, consisting of views, whole-sheet maps, plans, charts, antiquities, quadrupeds, birds, fishes, reptiles, vegetables, men, manners, customs, ceremonies, etc. the whole executed in a superior stile, by the first artists in the kingdom

- LONDON; printed for J. Cooke, No. 17, Pater-noster-row
- Folio
- ***Libraries*** : MeB
- ***Notes***

1. The number of issues and eds. of this work is not easy to establish. The basic problem is that the tps. carry no dates. If variations in the wording of the tps. are taken as a guide, there were at least twelve issues. However, comparisons of the texts of the copies held by BL and CaBViPA, each of which hold two, suggest that, judging by changes made to the body of the text, there may have been no more than two editions. The differences between these particular two are not great, being limited to some revisions to the first twelve pages. However, there is, in addition, some internal evidence that makes it possible to provide dates for a total of five editions or issues. The evidence is provided in notes to the appropriate subentries.
   On this occasion the entries in NUC are of little help in unravelling the problems on hand. It lists six separate entries, with copies located in sixteen libraries. However, when the tps. of the volumes held by the libraries are compared with the entries in NUC, discrepancies are found. The notes and American library holdings that follow are based on the examination of facsimiles of the tps. of their holdings, as well as on the five copies I was able to examine personally. In the UK, references to locations can be found in the catalogues of regional libraries located at BmP, BrP, E, LcP, MP.
2. The issue held by MeB is listed first because, according to the tp., it contains the fewest illustrations, and I presume that the number tended to increase rather than grow smaller. The MeB catalogue shows the date as 1787, but they state that it is taken from the royal licence to publish 'which is dated the fourth day of January, 1787.' Subsequently, they report that their copy has the illustrations dated 1788 that are discussed in note 4 of (02) below.
3. For contemporary works of a comparable scope and scale, see: ADAMS, M. (1794?), BALDWYN (1794?), COOKE, G.A. (1790?), PAYNE, J. (1791).

- **1788? (02)**
- *By the king's ... authority. A new ... system of universal geography antient and modern:* ... commerce, governments, etc. Also exact ... navigation throughout the known world. Published by the royal licence and authority of His Britannic Majesty King George III. And containing every important, interesting, valuable and entertaining discovery throughout the whole of Captain Cook's Voyages round the World. Together with those of all other modern as well as antient circumnavigators round the globe, particularly those of Byron, Mulgrave, King, Clerke, Gore, Carteret, Wallis, Bougainville, etc. (Performed by order of his Britannic Majesty.) As well as all other modern navigators and travellers who have published their discoveries in the various languages throughout the world. By ... one hundred and fifty beautiful ...

- ▸ LONDON; printed for J. Cooke ...
- ▸ Folio, 39 cm., 2 vols.; Vol. I, pp. [3], iv, 5-460; Vol. II, pp. [3], 461-990, [6]
- ▸ *Libraries* : : **CaBViPA**
- ▸ *Notes*

1. The tp. of Vol. II differs from that of Vol. I as follows: 'By ... A new, royal, authentic, and complete ... geography, antient ... Africa, and America ... languages throughout the world. Vol. II. By ... Printed for C. Cooke ...'
2. The new discoveries in the Pacific are presented first (to p. 106). Descriptions of Asia, Africa, America, and Europe are then presented in that order. The last-named has 414 pp. devoted to it, compared with 462 for the others combined. There is relatively little history, even in the section on the British Isles, and that is not disproportionately long.
3. The CaBviPA catalogue contains the following information: 'Royal licence for publication issued 1787. Vol. I published by John Cooke; v. 2, published by his son Charles Cooke, who succeeded his father as publisher in 1810. *cf.* Stephen, Dict. nat. biog.' On the basis of the information contained in the entries that follow, it seems unlikely that Vol. II was not published until after 1810.
4. As far as the date of this CaBViPA copy is concerned, the following evidence is available: (1) The engraving depicting the death of Captain Cook is dated 31 Oct. 1788. (2) The date 1788 also appears on 'A new and accurate chart of the Western or Atlantic Ocean.'

- ▸ **1788? (03)**
- ▸ *A new royal authentic and complete ... system of universal geography ...* engravings; consisting of ...
- ▸ LONDON; printed for J. Cooke, No. 17, Pater-noster Row; and sold by the booksellers of Bath, Bristol, Birmingham, Canterbury, Cambridge, Coventry, Chester, Derby, Exeter, Gloucester, Hereford, Hull, Ipswich, Leeds, Liverpool, Leicester, Manchester, Newcastle, Norwich, Nottingham, Northampton, Oxford, Reading, Salisbury, Sherborn, Sheffield, Shrewsbury, Worcester, Winchester, York; and by all other booksellers in England, Scotland and Ireland
- ▸ Folio, 38 cm.; pp. [1], iv, 5-990, [2]
- ▸ *Libraries* **L**
- ▸ *Notes*

1. Placed third because the wording of the tp. follows that of (02) rather than that of (01), and then carries in addition, at the foot of the page, the list of 30 towns and cities where booksellers are located.

- ▸ **{[1795]} (04)**
- ▸ *A new, royal, and authentic system of universal geography, antient ...* hemispheres, from the celebrated Columbus ... Cook, etc. and containing a genuine history ... lakes, mountains, volcanoes, deserts, etc. throughout ... Africa, and America ... shrubs, fruits ... genius, habits ... and ceremonies ... manufactures, trade ... insects, etc. Together with a complete history of ... state. Also an account of the most remarkable battles ... water. In which is introduced to illustrate the work, a considerable number of the most

accurate whole sheet maps, forming a complete atlas. To which is added, a ... known world. Likewise containing every important, interesting, and valuable discovery throughout the whole of Captain Cook's Voyages round the World. Together with those of other modern circumnavigators, particularly Byron, Carteret, Wallis, Clerke, Gore, King, Forrest, and Wilson. Also containing a particular account of the Pelew Islands. And the latest accounts of the English colony of Botany Bay: with a particular description of Port Jackson, Norfolk Island, etc. where the convicts are now settled. The whole forming a complete collection of voyages and travels. By the Rev. Thomas ... Cook, esq. and Thomas Lloyd. Embellished with near two hundred beautiful engravings, consisting of views, antiquities, customs, ceremonies, besides whole sheet maps, plans, charts, etc. executed in a much superior stile than any thing that has ever appeared in this kingdom

- ▸ LONDON; printed for C. Cooke ... [as in (03); i.e., 30 towns]
- ▸ Folio
- ▸ ***Libraries*** : OCl, WaS
- ▸ ***Notes***

1. According to WaS, this copy was published eight years after the first ed.

- ▸ **{1795}? (05)**
- ▸ *A new, royal, and authentic system* ... [as on the tp. of (04)]
- ▸ LONDON; printed for C. Cooke ... [as in (03); i.e., 30 towns]
- ▸ Folio, 2 vols.
- ▸ ***Libraries*** : NN
- ▸ ***Notes***

1. Because the tps. are identical, this is taken to be a variant of (04), bound in two vols.
2. The tp. of Vol. II follows: 'A new, royal, and complete system ... modern; including ... [as on tp. of (04)] ... water. To which is added ... [as on tp. of (04)] ... Willson. Also containing a particular account of the Pelew Islands, etc. Vol. II. By ... [as on tp. of (04)] ... antiquities, birds, fishes, reptiles, customs, ceremonies, besides whole sheet maps, plans, charts, etc. executed in a much superior stile than any thing that has ever appeared in this kingdom.'

   The place of publication and publisher are: London; printed for C. Cooke, No. 17, Pater-noster-row; and sold by the book-sellers of Bath ... Hereford, Ipswich, Liverpool ... Nottingham, Oxford, Salisbury, Worcester, York ... [i.e., 22 towns].

- ▸ **{1795}? (06)**
- ▸ *A new, royal and authentic system ... modern*; including ... Willson ... Port Jackson, where the ...
- ▸ LONDON; printed for C. Cooke ... [as in (03); i.e., 30 towns]
- ▸ Folio
- ▸ ***Libraries*** : InU
- ▸ ***Notes***

1. Because the tp. is very similar to that of (04) and (05), it is taken to be a variant, bound in one volume.

- **1797? (07)**
- *A new and authentic system* ... [as on tp. of (04)] ... Europe Asia ... fruits, herbs ... [as on tp. of (04)] ... ceremonies, etc. together with whole sheet maps, plans, charts, etc. executed in a much superior stile than any work of the like kind that has ever appeared in this kingdom
- LONDON; printed for C. Cooke ... [as in (03); i.e., 30 towns]
- 39 cm., 2 vols.; Vol. I, pp. [5], 506; Vol. II, pp. [2], 507-990, [2]
- ***Libraries*** L
- ***Notes***

1. Compared with (02), some changes have been made to the text of pp. 5-11. First, there is an Address to the Reader (a preface). It begins: 'Since the publication of a New System of Geography upwards of ten years have elapsed ...' Then, on p. 10, in a section headed 'Improvements in the Settlement and Country of Port Jackson,' there occur the words 'From the latest accounts, dated December 21, 1795, and received at the beginning of January 1797.'
2. The tp. of Vol. II begins as the tp. of Vol. II of (05): 'A new, royal, and complete system.'

- **1797? (08)**
- *By the king's royal licence and authority. A new, royal, authentic, and complete system* ... [as on tp. of Vol. II of (02), which is very similar to the tp. of Vol. I of (02); see note 1 to that entry]
- LONDON; printed for C. Cooke, No. 17, Pater-noster-row [no booksellers]
- Folio, 40 cm.; pp. [1], iv, ii, 5-990, [4]
- ***Libraries*** : : **CaBViPA**
- ***Notes***

1. The preface is the same as that of (07) above.
2. There is a second tp., bound after p. 460 and intended for Vol. II. It is identical to that of Vol. II of (05) above.

- **1797? (09)**
- *A modern, authentic and complete of universal geography*. Including all the late ... [as on the tp. of (04)] ... hemispheres; and containing a genuine history and ... [as on the tp. of (01)] ... lakes ... [as on the tp. of (04)] ... whole of Cooke's Voyages. Together with all the discoveries made by other mariners since the time of that celebrated circumnavigator. Also, a particular description of the improved state of the new colony formed at Port Jackson and Norfolk island, where the convicts are now settled. Including a particular account of the excursions and discoveries made in the interior parts of New Holland. The whole forming a complete collection of travels. By the Rev. Thomas Bankes, Vicar ... Bible. Embellished with near two hundred beautiful views, antiquities, customs, ceremonies, besides whole sheet maps, plans, charts, etc. executed in a much superior stile than any thing that has ever appeared in this kingdom
- LONDON; printed for C. Cooke ... [as in (03); i.e., 30 towns]
- Folio, 38 cm.; pp. [1], iv, 5-990, [2]
- ***Libraries*** OSg : MiU, NNC, TxU : CaBViP

▸ *Notes*
1. Maggs (1964:155) notes a work, published in 1796, with what seems to be this title, but only 'pp. 4, 106,' which he describes as being 'Cook supplement intended for adding to the Bankes Collection of Voyages published ten years previously.'

▸ **{1797}? (10)**
▸ *A modern* ... [as on the tp. of (09)]
▸ LONDON; printed for C. Cooke ... [as in (03); i.e., 30 towns]
▸ Folio, 2 vols.
▸ ***Libraries*** : CtY, NRU
▸ *Notes*
1. The date is assumed to be the same as that of (09), with this being a variant bound in two volumes.
2. The tp. of Vol. II differs from its counterpart in (05) and (08) as follows: 'A new and authentic system ... geography, antient and modern: including ... water. In which is introduced ... [as on the tp. of Vol. I of (04)] ... atlas. To which is added ... [as on the tp. of Vol. II of (05) and (08)] ... Wilson. Also containing ... [as on the tp. of Vol. I of (04)] ... Lloyd. Vol. II. Embellished with ... [as on the tp. of Vol. I of (07)].
3. Cox records this edition as a main entry. He gives its date as 1789. He then states: 'The date is approximate. An edition, London, dated 1790 and one dated 1791 (see below). The maps were done by Thomas Bowen, the well-known cartographer who died 1790' (Cox 1938:355). Later he continues: '1791 Bankes, T. (Rev). Geography, Antient and Modern, including all the latest discoveries from Columbus to the death of Captain Cook, a Genuine Guide to Geography, Astronomy, Navigation, the discoveries of Captain Cook, Byron, Carteret, Wallis, Forrest ... and an account of the Pelew Islands and Botany Bay. Maps by Brown and nearly 200 engravings by Grignon. 2 vols, fol. London. Date is approximate.' Cox gives no source.

▸ **{1797}? (11)**
▸ *A modern* ... [as on the tp. of (09)]
▸ LONDON; printed for C. Cooke ... [as in (03); i.e., 30 towns]
▸ Folio, 3 vols.
▸ ***Libraries*** : ICN
▸ *Notes*
1. The date is assumed to be the same as (09), with this being a variant bound in three volumes
2. The full text of the tp. of Vol. III reads as follows: 'A new and authentic system of universal geography, ancient and modern; containing a genuine history and description of the whole world. By The Rev. Thomas Banks [sic], Vicar of Dixton, in Monmouthshire. Edward Warren Blake, esq. Alexander Cook, esq. And Thomas Lloyd.'

▸ **{[1810?]} (12)**
▸ *A modern* ... [as on the tp. of (09)]
▸ LONDON; printed for C. Cooke ... [as in (03); i.e., 30 towns]
▸ Folio, 2 vols.

- ▸ *Libraries* : MiD
- ▸ *Notes*

1. MiD report that they believe this ed. 'was published around 1810.'
2. The tp. of Vol. II resembles the tp. of Vol. I of (07) very closely. It differs as follows: 'A new and authentic system ... Europe, Asia ... Lloyd. Vol. II. Embellished with ... to any work ...'

   The place of publication and the publisher are: London; printed by Jaques and Co. Lower Sloane-street, for C. Cooke, Paternoster-row, and sold by ... [as in (03); i.e., 30 towns].
3. A microform version of (09) is available from CIHM.

BARCLAY, John **mSG**

- ▸ **1631 (01)**
- ▸ *The mirrour of mindes*, or; Barclays icon animorum, Englished by T[homas] M[ay]
- ▸ LONDON; printed by Iohn Norton, for Thomas Walkley, and are to be sold at his shop, at the signe of the Eagle and Child in Britaines-Burse
- ▸ 12mo, 12 cm.; pp. [10], 322, 224
- ▸ *Libraries* C, ECP, **L**, O : CLU, CoDU, CtY, DFo, ICN, ICU, IEN, MH, MHi, MiU, MnU, NcD, NNUT, OrU, PPULC
- ▸ *Notes*

1. On the margins of special geography. 'The most impressive work on national character of [its] time is the *Icon Animorum* ... first published in 1614; a translation from the Latin to the English was made by Thomas May ...' (Glacken 1967:452). Also contains one of the first references to the beauty of an English landscape (p. 42, cited by Glacken). Puts forward an explanation of the varieties of human character or personality. Among the sources of variety are the *Spirit of the Age* and the *Spirit of the Region*. The general manner in which the latter operates is set out in ch. 2, together with some brief remarks on the nature of non-European peoples. Chs. 3-9 (pp. 65-322) describe the character of the peoples of the leading European countries, including the Turks and the Jews.
2. Glacken notes: 'For an exhaustive study of the work see Collignon, "Le Portrait des Esprits (*Icon animorum*) de Jean Barclay," *Mémoires de l'Académie de Stanislas, 1905-1906* Ser. 6, Vol. 3 (1906), pp. 67-140' (Glacken 1967:452); he also notes: 'For a less enthusiastic appraisal, see E.G.R. Taylor (1934, 134-136).' Taylor, incidentally, in the course of her discussion wrongly refers to the author as William Barclay (Taylor 1934:134).

BARNES, W. ?

- ▸ **{1847} (01)**
- ▸ *Geography and ethnography*
- ▸ [LONDON]; Longman
- ▸ 18mo
- ▸ *Libraries* n.r.
- ▸ *Source* BCB, Index (1837-57:109)

BARNES, William **Y SG? Am**
- **{1839} (01)**
- *Modern geography*; being entirely a new plan, on the classification system ... Adapted to the use of classes and private learners
- TROY, NY; J.M. Stevenson
- ***Libraries*** : MH
- ***Notes***
1. NUC records a further ed.: 1839, 2nd (DLC).

[BARROW, John] **Dy**
- **{1759-60} (01)**
- *A new geographical dictionary*. Containing a full and accurate account of the several parts of the known world, as it is divided into continents, islands, oceans, seas, rivers, lakes, etc. The situation, extent, and boundaries, of all the empires, kingdoms, states, provinces, etc. in Europe, Asia, Africa, and America. Their constitutions, revenues, forces, climate, soil, produce, manufactures, trade, commerce, cities, chief towns, universities, curious structures, ruins, antiquities, mountains, mines, animals, vegetables, minerals, together with the religion, learning, policy, manners, and customs, of the inhabitants. To which is prefixed, an introductory dissertation, explaining the figure and motion of the earth, the use of the globes, and doctrine of the sphere ...
- LONDON; J. Coote
- 2 vols.
- ***Libraries*** OSg (Vol. II only) : AU, CLU, InU, KyBgW
- ***Source*** NUC (NB 0150035)

- **1762, 2nd ... (see below) (02)**
- *A new geographical ... dictionary* ... sphere, in order to render the science of geography easy and intelligible to the meanest capacity. The second edition, revised and corrected. Illustrated with a new and accurate set of maps of all the parts of the known world, making a compleat atlas; the dresses of the inhabitants; and a great variety of plans and perspective views of the principal cities, towns, harbours, structures, ruins, and other pieces of antiquity
- LONDON; printed for J. Coote, at the King's Arm in Pater-Noster Row
- 39 cm., 2 vols.
- ***Libraries*** OSg (Vol. I only) : OrPR
- ***Notes***
1. Barrow is identified as the author in this ed.

- **{1762(?)} (03)**
- *A new geographical dictionary* ...
- LONDON; printed for J. Coote
- 35.5 cm., 2 vols.
- ***Libraries*** : MiD, ViU
- ***Source*** NUC (NB 0150039), which notes that the date is written MDCCLXIL

▸ *Notes*
1. Cox (1938) records an ed.: 1763, 2 vols.

BARTHOLOMEW, John **Dy (& At)**
▸ **{1871} (01)**
▸ *Zell's descriptive hand atlas of the world*
▸ PHILADELPHIA; T.E. Zell
▸ 36 cm.
▸ *Libraries* : DLC
▸ *Source* NUC (NB 0158698)
▸ *Notes*
1. NUC records further eds.: 1873 (CoU, DLC, MdBJ, MH, MnU, N, NB, NN, OCl, OrP, PHi, PPULC, PU, WaS); 1875 (Nh, NN); and [1881] (DLC, IU, MiU, NNC).

BAZELEY, Charles W. **Y Am**
▸ **{1830} (01)**
▸ *The juvenile scholar's geography* ... interspersed with more than 1500 questions on the principal maps: and accompanied with a miniature, but complete atlas ...
▸ PHILADELPHIA; the author
▸ 15 cm.; pp. viii, [9]-324
▸ *Libraries* : NNC, PPULC
▸ *Source* NUC (NB 0214199)
▸ *Notes*
1. NUC records a further ed.: 1850 (PPL, PPULC). Discussed in Nietz (1961: 223) and Sahli (1941, *passim*); cited by Elson (1964:399).

BAZELEY, Charles W. **SG? Am**
▸ **{1815(?)} (01)**
▸ *A system of ancient and modern geography*; or, an account of the past and present state of the several countries of the world ...
▸ PHILADELPHIA
▸ *Libraries* : PPULC

▸ **{1819} (02)**
▸ *A system of ancient and modern geography* ...
▸ PHILADELPHIA
▸ *Libraries* : PHi, PPL, PPULC, PU
▸ *Source* NUC (NB 0214207)

[BEAUMONT, John] **Y mSG**
▸ **1694 (01)**
▸ *The present state of the universe*, or an account of I. The rise, births, names, matches, children, and near allies of all the present chief princes of the world. II. Their coats of arms, motto's, devises, liveries, religions, and

languages. III. The names of their chief towns, with some computation of the houses and inhabitants. Their chief seats of pleasure, and other remarkable things in their dominions. IV. Their revenues. To which are added some other curious remarks; as also an account of common-wealths, relating to the foregoing heads

- ▸ LONDON; printed and are to be sold by Randall Taylor, near Stationers-Hall
- ▸ 4to, 19.5 cm.; pp. [4], 100
- ▸ *Libraries* L : **CtY**, IEN
- ▸ *Notes*

1. The dedication is signed by Beaumont, and makes it plain that the book is intended as a source of information for those who are not yet of 'mature years.'

   The text begins with Austria, because it has 'the largest extent of dominions among the European princes' (p. 1); see PINKERTON (1802). It contains information about countries in every continent, and so can be classified as a marginal special geography, intended for children.

- ▸ **{1696} (02)**
- ▸ *The present state of the universe*; or, an account ...
- ▸ LONDON; W. Whitwood
- ▸ 21 cm.; pp. 100
- ▸ *Libraries* : DLC

- ▸ **1697, 2nd ... (see below) (03)**
- ▸ *The present state of the universe*, or an account ... devices ... revenues, powers and strength. Also an account ... same heads. The second edition much emended and enlarged, with the addition of the styles or titles of the several potentates and republicks
- ▸ LONDON; printed by B. M. and are to be sold by W. Whitwood at the Crown in Little Brittain
- ▸ 12mo, 14.5 cm.; pp. [4], 151
- ▸ *Libraries* L : **CtY**, MiD
- ▸ *Notes*

1. In this ed., the dedication is replaced by 'Advertisement to the Reader,' which is unsigned. It contains references to similar and seemingly earlier works, denying plagiarism. The key sources are identified.
2. Smaller print is used in this ed. than in (01). The body of the text has shrunk in width from 10.5 to 6.7 cm. Some of the genealogies have additional information at the end. This is usually brief, so that, apart from titles and styles of the rulers, the text is not much enlarged, despite the claim on the tp.

- ▸ **{1701}, 3rd ... (see below) (04)**
- ▸ *The present state of the universe* ... strength. V. Their respective styles, and titles, or appellations. Also an account of commonwealths ... heads. The 3d ed. continu'd and enlarg'd
- ▸ LONDON; printed by B. Motte
- ▸ 12mo, 15.5 cm.; 6 p. l., 159, [1] p., 1 l., [5] p., 1 l.

- *Libraries* L : CLU, DFo, DLC, OCl
- *Source* BL and NUC (NB 0228031)

- **{1702} (05)**
- *The present state of the universe*; or, and account of the ... present chief princes of the world, their coats of arms ... chief towns, revenues, powers and strength, ... 12 copperplate portraits and 70 engravings of the ensigns, colors or flags of ships at sea, belonging to the several princes and states of the world
- LONDON
- 12mo
- *Libraries* n.r.
- *Source* Cox (1935)

- **{1704}, 4th ... (see below) (06)**
- *The present state of the universe* ... [as in 04] religions and ... appellations. VI. And an account ... heads. To this 4th ed. continu'd and enlarg'd, with several effigies wanting in the former impression; as also the various bearings of their several ships at sea; are added their several territories, which are distant from them in other parts of the world
- LONDON; printed by B. Motte
- 12mo, 15.5 cm.; 5 p. l., 164 p., 1 l., [4]
- *Libraries* L : DLC, MH
- *Source* BL and NUC (NB 0228033)

BEAUTIES ... **A**

- **{1763-4} (01)**
- *The beauties of nature and art displayed, in a tour through the world*; containing I. A general account of all the countries in the world ... II. A particular account of the most curious natural productions of each country ... III. An historical account of the most remarkable earth-quakes ... and other public calamities ... IV. Extraordinary instances of longevity ... V. Particular descriptions of the most remarkable public buildings ... VI. Curious remains of antiquity ... laws, customs, and traditions of the inhabitants; together with a summary view of the ... revolutions among them ...
- LONDON; printed for J. Payne
- 18mo, 14 cm., 14 vols.
- *Libraries* : ICU, IU, NN, PPL, PPULC
- *Source* NUC (NB 0229409)

- **1774-5, 2nd, greatly improved (02)**
- *The beauties of nature ... tour in the world*, remarkable for either natural or artificial curiosities; their situation, extent, and divisions; their rivers, air, soil, chief cities, etc. II. ... country, in the animal, vegetable, and fossil kingdoms; of remarkable mountains, caverns, and volcanoes; of medicinal and other singular springs; of cataracts, whirlpools, etc. III. ... earthquakes, inundations, fires, epidemic diseases, and other ..., which have at different times visited the inhabitants. IV. ... longevity, fertility, etc. among the inhabitants; together with an account of their most celebrated inventions,

discoveries, etc. V. ... buildings, and other singular productions of art. VI. ... antiquity; remarkable laws ... view of the most extraordinary revolutions among them. Illustrated and embellished with copper plates

- LONDON; printed for G. Robinson, in Pater-Noster-Row
- 12mo, 14 cm., 13 vols. bound in 6; Vol. I (1774), V. 1; pp. xii, 216; V. 2, pp. [4], 216; Vol. II (1774), V. 3, pp. [2], 216; V. 4, pp. [2], 216; Vol. III (1774), V. 5, pp. 216; V. 6, pp. 216; Vol. IV (1774), V. 7, pp. 216; V. 8, pp. 219; Vol. V (1774), V. 9, pp. 216; V. 10, pp. 216; Vol. VI (1775), V. 11, pp. 216; V. 12, pp. 216; V. 13, pp. 216
- ***Libraries*** **L** : Cst, CtY, PPULC, PU, RPB
- ***Notes***

1. V. 1 deals with Great Britain and Ireland; V. 2, Great Britain and Ireland (continued) including a history of England to 1485; V. 3, history of England to 1763; V. 4, pp. [2], 216, France; V. 5, history of France (continued), Spain, Portugal, and Italy; V. 6, Italy (continued), Germany, Bohemia, Hungary, and Switzerland, Netherlands, Denmark, Norway, and Sweden; V. 7, Denmark, Norway, and Sweden (continued), Poland, Russia in Europe, Turkey in Europe, Asia in general, Turkey in Asia; V. 8, Turkey in Asia (continued), Persia, India; V. 9, India (continued), China and Japan, Indian Islands, Africa, Barbary; V. 10, Barbary (continued), Negroland or Guinea, Abyssinia (includes the rest of the mainland), African Islands, America in general, South America, North America; V. 11, 12, 13, devoted between them to the voyages of Byron, Wallis, Carteret, and Cook in the southern hemisphere (being extracts from the original accounts).

   Rather more information is provided about natural history than in most special geographies.

BELL, James **A**

- **{1829-1832} (01)**
- *A system of geography, popular and scientific*, or, a physical, political, and statistical account of the world and its various divisions. [Illustrated by a complete series of maps, and other engravings ...]
- GLASGOW; Blackie, Fullarton, & Co.
- 6 vols. in 12 pts.
- ***Libraries*** : ICRL
- ***Source*** NUC (NB 0279372)
- ***Notes***

1. Allford (1964:243) records a 'System of Geography' by J. Bell in 1812, published in Glasgow. See SYSTEM OF GEOGRAPHY (1805) and GLASGOW GEOGRAPHY.

- **{1831} (02)**
- *A system of geography* ... or a physical ...
- GLASGOW; Blackie ...
- 22 cm., 6 vols.
- ***Libraries*** : N
- ***Source*** NUC (NB 0279373), which notes a change of date in the library catalogue entry from 1834 to 1831

- **1832 (03)**
- *A system of geography* ... engravings
- GLASGOW; Archibald Fullarton and Co.; W. Tait, Edinburgh; W. Curry, Jun., & Co., Dublin; Simpkin & Marshall, and W.S. Orr, London
- 8vo, 22 cm., 8 vols.; Vol. I, pp. [1], vii, [2], vi-cxxxii, 508; Vol. II (bound with I), pp. vi, 600; Vol. III, pp. vi, 558; Vol. IV (bound with III), pp. vi, 647; Vol. V, pp. vi, 627; Vol. VI (bound with V), pp. vi, 640
- ***Libraries* L** : CtY, MA, MiU, ODW, OCl, PPAN, PPULC
- ***Notes***

1. Vol. I describes Russia, Poland, Scandinavia, Germany, and Austria; Vol. II, Switzerland, the Netherlands, France, and Mediterranean Europe; Vol. III, Great Britain, and Africa; Vol. IV, Asia except China; Vol. V, China and north and central America; Vol. VI, South America, the West Indies, and Australia, and then ends with a history of geography.

   In a prefatory address, dated 1828, B. refers to A.F. Büsching as the father and founder of scientific geography; he also names other, more recent geographers on whose work he has drawn freely. General geography (mathematical, physical, political [principles of] occupy 125 pp.; apart from the concluding section of Vol. VI, the remainder of the text is devoted to special geography.

   This is the last British example of a massive description of all the countries in the world written by a single author. Its relationship to the GLASGOW GEOGRAPHY (1822) would seem to be worth investigating. It is comparable to CONDER (1824-30) and MALTE-BRUN (1822-33) in scale. Compared with the former, at least, Bell is both more systematic and more scholarly. He cites the authorities for his accounts, though in a rather condensed form, so that some research would now be needed before they can be identified with confidence.

   The section on the history of geography at the end of Vol. VI may be the first in English that is not a history of either cartography or exploration. Instead it is primarily a history of the attempts to provide an account (intermittently an explanatory account) of the countries of the world, separately or in total. Bell's knowledge of both Continental and Classical literature is impressive, although the latter can almost be taken for granted in a British scholar of the early nineteenth century. It is doubtful whether a better guide to the European literature of the 16th to 18th centuries that was intended to provide descriptions of the countries of the world, past and contemporary, is available in English.
2. NUC records further eds. (all 6 vols.): 1834, Glasgow (DLC); 1838, Glasgow (NB); 1842-4, Edinburgh (OCU, PP, PPULC); 1846-7, London (DLC, ICJ : CaBViP). BCB, Index (1837-17:110) records an ed. in 1849, [Edinburgh?]. NUC records further eds.: 1850-1 (MH); 1853, London (CtY, MBAt); and 1855, London (PU, PPULC).

[BELLEGARDE, Jean Baptiste Morvan de] **mSG tr**

- **{1708} (01)**
- *A general history of all voyages and travels throughout the old and new world*, from the first ages to this present time. Illustrating both the ancient and modern geography. Containing an accurate description of each country,

its natural history and product; the religion, customs, manners, trade, etc., of the inhabitants, and whatsoever is curious and remarkable in any kind. An account of all discoveries, hitherto made in the most remote parts, and the great usefulness of such attempts, for improving both natural and experimental philosophy; with a catalogue of all authors that have ever describ'd any part of the world, an impartial judgement and criticism on their works for discerning between reputable and fabulous relater; and an extract of the lives of the most considerable travellers. By Monsr. Du Perier of the Royal academy. Made English from the Paris edition[.] Adorn'd with cuts

- LONDON; printed for Edmund Curll and E. Sanger
- 8vo, 20 cm.; 4 p. l., 364, [8] p.
- ***Libraries*** L : CtY, CU, DFo, DLC, ICN, MBAt, MiU, NBu, NCU, MnU, MtU, NIC, NSyU, OFH, PPULC, RPJCB, TxU, ViU
- ***Notes***

1. NUC (NB 0284674). 'Aside from the introduction, the work relates only to America. French original published at Paris in 1707 under the title "Histoire universelle des voyages," with dedication signed "Du Perier." The Amsterdam edition of 1706 bears on the tp. the name of the Abbé de Bellegarde as the author.'
2. Beeleigh Abbey Books, *A catalogue of voyages and travels*, BA/23, W. & G. Foyle, Ltd, London, n.d., item 30, notes: 'A rare work which deals mainly with the early voyages of the Spaniards to America, and contains much information relating to the Indian tribes. The book was republished with a new title in 1711.'
3. NUC (NB 0284667) records the following title by Bellegarde, published in London in 1711 and printed for E. Curll: 'A complete collection of voyages made into North and South America, in due order as they happen'd ... to this present time. Accurately describing each country ... inhabitants, with whatsoever else is curious ... kind. To which is pre-fix'd an introductory discourse, shewing the great advantage of trading into those parts; with the improvements that have been made by our late navigators in order to the effecting speedy voyages, exactly describing the coasts, and illustrating both the ancient and modern geography. The whole extracted from the works of the most considerable travellers. By Monsr. l'abbe Bellegarde of the Royal academy. Translated from the French original ...'

   NUC then provides a cross reference to Bartolomé de las Casas, 'La découverte des Indes occidentales,' Paris, 1701, which is identified as being translated into French by Bellegarde. A search through the entries in NUC attributed to Casas reveals an English translation, published in 1699 in London, and printed for Daniel Brown ... and Andrew Bell; the English title begins: 'A relation of the first voyages and discoveries made by the Spanish in America ... To which is added the art of travelling, shewing how a man may dispose his travels to the best advantage' (NC 0180251). That entry in NUC contains the following note: 'First edition, first issue? of this translation. Also published in the same year [1699] under the title: An account of the first voyages and discoveries made by the Spanish in America ... A translation from the French, apparently from the *Relation des voyages et des découvertes que les Espagnols ont fait dans les Indes Occidentales* ... Amsterdam, 1698.'

   It seems that what appears at first sight to be a special geography published in London in 1708 is largely a second-hand translation of a book of voyages to America written in Spanish by a man who died in 1566.

BENTLEY, Joseph **A**

- **1839? (01)**
- *Modern geography*, for the student, the man of business, and all classes who wish to know something of the world we live in; containing all the topographical, physical, historical, commercial, and political facts worthy of notice, relating to every empire, kingdom, republic, state, and country in the world; especially of the British Empire. Also, a treatise on the newly-invented plane globe, containing a copious definition of scientific terms, and fifty-three problems with rules for working them on the plane globe and planetarian, so as to illustrate the pleasing science of geography, and some of the elements of astronomy, in the way usually done by the common globe; the whole accompanying the plane terrestrial globe
- MANCHESTER; printed and sold by Bancks [sic] and Co. Published and sold by the inventor [i.e., the author], and his agents; also by R. Leader, Sheffield
- Obl. 4to (24 × 24 cm.); pp. 1 [1], ii, ii, 140 (although square, as bound, all pp. except tp. are printed the normal shape, i.e., 24 × 12 cm., so on being opened the book presents 4 pp. to view. Done, presumably, so that the plane globe could be bound in the end papers)
- ***Libraries*** L (lacks pp. 1-4)
- ***Notes***

1. The approach taken is chorographical, even topographical, with information approaching that of a list set out in the form of prose.
2. BL supplies the date.

- **184–?, anr issue? (02)**
- *Modern geography* ... know something of the habitable globe; containing ...
- MANCHESTER; printed ... also by Wrightson and Webb, Birmingham; and H. Johnson, London
- 24 cm.; pp. [2], ii, ii, 140 (bound in the normal fashion so as to present two pages when open)
- ***Libraries*** OSg

BERNARD, John Augustine **Dy**

- **1693 (01)**
- *A geographical dictionary* representing the present and antient names and states of all the countries, kingdoms, provinces, remarkable cities, universities, ports, towns, mountains, seas, streights, fountains, and rivers of the whole world; their distances, longitudes, and latitudes, with a short historical account of the same, and a general index of the antient and Latin names. Together with all the market-towns, corporations, and rivers, in England, wanting in both the former editions. Very necessary for the right understanding of all antient and modern histories, and especially of the divers accounts of the present transactions of Europe. Begun by Edmund Bohun, esquire. Continued, corrected, and enlarged with great additions throughout,

and particularly with whatever in the geographical part of the voluminous Morery and Le Clerk occurs observable

- LONDON; printed for Charles Brome, at the Gun at the west end of S. Pauls
- Folio, 31 cm.; pp. [8], 437, [20]
- *Libraries* C, CrlC, EU, ExC, **L**, LvU, O, OMa, ShP, SoM, [IDT] : CLU, ICN, InU, KMK, MBBL, MH, MnU, NN, PBa, PPULC, RPJCB
- *Notes*

1. For the previous eds. referred to in the tp. see BOHUN. The additions to Bohun's Dictionary made by Bernard are sufficiently numerous to justify giving him a separate entry in this guide.

   Bernard provides a preface in which he relates his indebtedness, and the modifications he made, to '*Le grand dictionaire historique, etc.* ... of Lewis Morery, D.D. Printed at Utrecht, 1692, with the supplement of J. Le Clerc, D.D., in four tomes in folio, French.' Bernard has three major criticisms of Morery: (1) he includes much fabulous material, (2) he is excessively chauvinistic and detailed in his treatment of all things French, (3) he is too critical and too little-detailed, not to say erroneous, in things English.

- **1695, anr ed. (02)**
- *A geographical dictionary* ... observable. To which are added the general praecognita of geography and the doctrine of the sphere. As also a continuation of the most remarkable transactions that have any where lately happened, and which illustrate the geography and the history of places to the end of the campaign, 1694. With an an [sic] alphabetical table of the most noted rivers and mountains in England and Wales; shewing the rise, course and falls of the first, and the situation and extent of the latter. Never before published
- LONDON; printed ...
- Folio, 32 cm.; pp. [8], 23, 437, 4, 16, [17]
- *Libraries* ChC, **O**, PetC : CtY, DLC, MdBJ, MiD, PPULC
- *Notes*

1. The tp. is reset but, apart from the addition noted above, it is not significantly changed from (01). The body of the text seems unchanged; the additions were incorporated in three series of newly numbered pages added before and after the dictionary proper.

- **1710, anr ed. (03)**
- *A geographical dictionary* ...
- LONDON; printed for R. Bonwicke, W. Freeman, T. Goodwin, J. Walthoe, M. Wotton, S. Manship, J. Nicholson, R. Parker, B. Tooke, and R. Smith
- Folio, 30 cm.; pp. (8), 437, (21), 23, 16, 4
- *Libraries* **O** : ICRL
- *Notes*

1. The tp. is reset but, apart from the publishers, is not significantly changed from 1695. The text appears to be completely unaltered, although the additional materials have been bound in a different order.
2. Discussed in Bowen (1981:98-100).

BEVAN, William Latham **Y**

- **1869 (01)**
- *The student's manual of modern geography.* Mathematical, physical, and descriptive
- LONDON; John Murray
- 12mo, 19 cm.; pp. xiv, 674
- ***Libraries*** L
- ***Source*** BL notes: '[1868]' after giving the date as 1869, which is that given on the tp.
- ***Notes***

1. BL and NUC record further eds. (all 674 pp., except as noted): 1869 (: CoFS, NIC); 1871, 2nd (L); 1872, 3rd (: ViU); 1874, 4th (: MH); 1876, 5th (L); 1879, 6th, 675 pp. (L : MH); and 1884, 7th, 675 pp. (L : PPULC, PU).

BEZANT, J. **Y mSG?**

- **{[1857]}, New (01)**
- *Geographical questions*, classed under heads, and interspersed with historical and general information ... A new edition carefully revised and corrected by W. Cooke Stafford
- LONDON; Joseph Masters
- 12mo; pp. 101
- ***Libraries*** L

BICKHAM, George **mSG**

- **1743-54 (01)**
- *The British monarchy*; or, a new chorographical description of all the dominions subject to the king of Great Britain. Comprehending the British Isles, the American colonies, the Electoral states, the African & Indian Settlem'ts. And enlarging more particularly on the respective counties of England and Wales. To which are added alphabets in all the hands made use of in this book. The whole illustrated with suitable maps and tables; likewise, adorned with head-pieces, and other embellishments; and engraved by George Bickham
- LONDON; Publish'd according to act of parliament, october 1st, 1743, and sold by G. Bickham
- Folio, 32 cm.; pp. 188
- ***Libraries*** L, LG : CtY, DLC, MB, MH, NcU, OrP, PPL, PPULC, RPJCB
- ***Notes***

1. A marginal ornament to the stream of special geography. It is principally a case of penmanship by example. The book was the joint effort of three men of the same family. Both the illustrations and the letterpress were engraved on copperplate. The text is carried on recto pages only, and so is shorter than the physical size of the book would suggest. It was prepared with a second tp. for binding in two vols., but the pagination is continuous. Only the tp. was published in 1743. The rest of the work was published in parts, and was not finished until 1754. There were later eds. in 1748 and 1749. These show variations in the wording of the tp. A facsimile of the ed. of

1748 was published in 1967 by Frank Graham, Newcastle-upon-Tyne, UK. It is the authority for these notes. NUC records copies of the ed. of 1748 at CoMC, CtY, DLC, LU, MdBP, MH, MWA, MWiW, NN, OFH, PHi, PPULC, RPJCB, ViU; and of the edition of 1749 at MH, PPF, PPULC. Facsimile of 1967 at **CaAEU**.

BIDLAKE, J. **Y SG?**
- **{1808} (01)**
- *Geography, introduction*
- Place/publisher n.r.
- ***Libraries*** n.r.
- ***Source*** Peddie and Waddington (1914:226)

BIDLAKE, John Purdue **Y mSG?**
- **{1867} (01)**
- *Irving's catechism of general geography* ... rewritten and arranged by J.P. Bidlake, etc.
- Place/publisher n.r.
- 12mo
- ***Libraries*** L

BIGLAND, John **A**
- **{1805} (01)**
- *Geographical and historical view of the world*
- Place/publisher n.r.
- ***Libraries*** n.r.
- ***Source*** Peddie and Waddington (1914:226)

- **1810 (02)**
- *A geographical and historical view of the world*: exhibiting a complete delineation of the natural and artificial features of each country; and a succinct narration of the origin of the different nations, their political revolutions, and progress in arts, sciences, literature, commerce, etc. The whole comprising all that is important in the geography of the globe and the history of mankind
- LONDON; printed for Longman, Hurst, Rees, and Orme, PaternosterRow; Vernor, Hood, and Sharpe, Poultry; and J. Cundee, Ivy-Lane
- 8vo, 20 cm., 5 vols.; Vol. I, pp. xxxvi, xxxviii, 551; Vol. II, pp. [2], 664; Vol. III, pp. [2], 744; Vol. IV, pp. [2], 556; Vol. V, pp. [2], 675, [1]
- ***Libraries*** **L** : ICRL, Mi, NB, PPL, PPDrop, PPULC
- ***Notes***

1. Vol. I deals with England, predominantly historical; Vol. II covers France, Belgium, the Netherlands, and Spain; Vol. III the balance of Europe other than Turkey; Vol. IV, Turkey and all Asia except India; and Vol. V, India, Africa, and America.

   Intended as a scholarly work; a four-page list of authorities is provided after the Preface. It deserves comparison with MORSE (1789), especially the later eds., and with PINKERTON (1802).

- **1811, anr ed. (03)**
- *A geographical and historical view of the world ...* mankind. With notes, correcting and improving the part which relates to the American continent and islands. By Jedidiah Morse, D.D.[,] A.A.S.[,] S.H.S. Author of the American Universal Geography, etc.
- BOSTON; printed by Thomas B. Wait and co; sold by them, and by Mathew Carey, Philadelphia, and Samuel Pleasants, Richmond
- 23 cm., 5 vols.; Vol. I, pp. xiv, xxxiv, 405; Vol. II, pp. 511; Vol. III, pp. 574, [2]; Vol. IV, pp. 430, [2]; Vol. V, pp. xli, 535, [2]
- ***Libraries*** **L** : DLC, KyLx, MH, MHi, MiC, MiU, MWA, NBu, NcU, NjR, NjNbS, OCl, PPULC, PU, PSc, TU
- ***Notes***

1. BL and NUC record further eds.: 1812, London, 5 vols. (L); 1812?, 2nd American, Boston, 5 vols. (: KyBgW) [NB 0481591 notes: 'c1811']; 1812, 2nd American, Philadelphia & Boston (: CtY, DGU, ICU, MoU, Nh, NWM, PMA, ViLxW, ViU).

BIGLAND, John **Y SG?**

- **{1816} (01)**
- *System of geography for the use of schools*
- 12mo
- ***Libraries*** n.r.
- ***Source*** British Museum (1881-1900), not in BL (1959-66)
- ***Notes***

1. BL records further eds.: 1834, 11th; 1855, 14th, 144 pp.

BISSET, J. **Y SG?**

- **{1805} (01)**
- *The geographical guide*
- LONDON
- ***Libraries*** n.r.
- ***Source*** Allford (1964:242)

[BISSET, James] **Y OG**

- **{1818} (01)**
- *Juvenile geography; or, poetical gazetteer*: comprising a new and interesting description of the principal cities and towns in the United Kingdom of Great Britain and Ireland
- LONDON; printed for J. Souter, by Jas. Adlard and sons
- 14 cm.; pp. v, 114
- ***Libraries*** : : CaOTP
- ***Source*** St. John (1975, 2:799)
- ***Notes***

1. The full title shows that this book deals only with the British Isles.

BLACKIE, Walter Graham, comp. **A**

- **{1882} (01)**
- *The comprehensive atlas and geography of the world*: comprising an extensive series of maps; a description, physical and political, of all the countries of the earth; a pronouncing vocabulary of geographical names, and a copious index of geographical positions. Also numerous illus. ... and a series of coloured engravings representing the principal races of mankind. Comp. and engr. from the most authentic sources under the supervision of ...
- LONDON, Glasgow, Edinburgh, Dublin; Blackie & son
- Pp. ix, [3], 304, 96
- ***Libraries*** : OHi
- ***Notes***

1. OHi reports: 'The volume does contain approximately 300 pages of text giving rather extensive coverage of the countries that appear on its maps. The text is organized by continents, then by individual countries within each continent. It includes such subheadings as geography, flora and fauna, social customs, political system, etc. Illustrations of various sizes are intermixed with the text. The 96 pages of maps and statistical tables are bound after the body of the text.'
2. For a possible author of the text, see the note to CHISHOLM (1884).
3. 'This work was issued in parts from 1880' (Beeleigh Abbey Books, *A catalogue of voyages and travels*, Catalogue BA/23, W. & G. Foyle Ltd, London, n.d., item 8).

- ***Notes***

1. BL and NUC record further eds. (both London, 304, 99 pp.): 1883 (L) and 1884 (: IU).

BLACKIE, Walter Graham, ed. **Dy**

- **{1852} (01)**
- *The imperial gazetteer*; a general dictionary of geography, physical, political, statistical, and descriptive. [Compiled from the latest and best authorities]. Edited by ...
- GLASGOW; Blackie & son
- 4to, V. 1
- ***Libraries*** : DN, MHi, NN
- ***Source*** NUC (NB 0523304)
- ***Notes***

1. NUC records a further ed.: 1853, Glasgow (PPL, PPULC).

- **1855 (02)**
- *The imperial gazetteer* ... Edited ... With seven hundred illustrations, views, costumes, maps, plans, etc. ...
- GLASGOW; Blackie and son, Queen Street; South College Street, Edinburgh; Warwick Square, London
- 27.5 cm., 2 vols.; Vol. I, pp. [5], vi-xii, 1308; Vol. II, pp. [4], 1288, 59
- ***Libraries*** L, **[IDT]** : CoD, CtY, DLC, MiD, MiU, NcD, NN, NWM, OClND, OClW, PP, PPB, PPULC
- ***Notes***

1. Claims to 'notice all known towns, having not less than 1000 inhabitants,'

and to 'construct all the articles upon a uniform plan.' The appendix to Vol. II contains abstracts from the British census of 1851, giving the populations of all the parishes in the UK, as well as the populations of all villages, towns, and cities whose populations exceed 500. Also lists the areas of the parishes of England and Ireland.
2. BL and NUC record further eds.: 1856 (L : CU); 1868, 2 vols. (: ICU); 1873, 2 vols. in 4 (: PP, PPULC); 1874, 2 vols. (L : ICU, IU, ViU, WU); 1876, 2 vols. (: MH); 1878, 2 vols. (: OClStM).

BLAEU, Willem Janszoom **OG**
- **{1654} (01)**
- *A tutor to astronomy and geography*; or, an easie and speedy way to understand the use of both the globes, celestial and terrestrial. Laid down in so plain a manner that a mean capacity may at the first reading understand it and with a little practice, grown [sic] expert in those divine sciences. Translated from the first part of Gulielmus Blaeu, instituto astronomica. Published by J.M. Whereunto is annexed the ancient poetical stories of the several constellations in heaven
- LONDON; printed for Joseph Moxon, and are to be sold at his shop, at the signe of Atlas, in Cornhill; where you may also have globes of all sizes
- 4 p. l., 356 p.
- ***Libraries*** : CSmH
- ***Source*** NUC (NB 0526759), which identifies the publisher as Joseph Moxon
- ***Notes***

1. See MOXON (1659) and MOXON (1665).
2. Though Blaeu is a relatively minor figure in the history of cartography, the standard works on the subject usually discuss him briefly (e.g., Bagrow 1964, Crone 1978).

BLAGDON, Francis William **A n.c.**
- **[1808] (01)**
- *The modern geographer*: being a general and complete description of Europe, Asia, Africa, and America, with the oceans, seas, and islands, in every part of the world. Including an exposition of the civil and military governments, in the different empires, kingdoms, states, and colonies; comparative statements of revenues, commerce, arts, sciences, and manufactures; a digest of the laws by which each country is governed, and a copious explanation of the manners, costume, amusements, and religions, of the respective inhabitants. Prepared and digested, upon a new plan, from the latest and most accurate authorities, with notes, historical, critical, and explanatory. By ... author of the folio history of ancient and modern India, published under the patronage of His Majesty and the Court of Directors; of the quarto life of Lord Viscount Nelson, etc. etc. Volume I. Containing the geography of North and South America. Illustrated with maps, charts, and engravings
- LONDON; printed by Brimmer and Co. 21, Water-Lane, Fleet-Street, for

A. Whelier, No. 3, Paternoster-Row; and sold by all the booksellers in the United Kingdom

- 21 cm.; pp. [iii], iv-vi, [1], 8-621, [3]
- ***Libraries*** : CU, DLC, NBu, PPULC : **CaBVaU**
- ***Notes***

1. Only the first of the five volumes planned was ever published. The date of publication is established on internal evidence: B. refers (pp. 489 and 574) to events that took place in January and February 1807; then, in the course of a detailed review of the national accounts of the United States (p. 173), he refers to 1809 in the future tense.

   B. provides occasional extracts from works that are identified by reference to their authors; some of these run to two or more pages. Their purpose is to illustrate some general principle or pattern of behaviour, or some element of local or national character. He has a low opinion of many things American, especially those associated with republicanism. In the realm of natural history he gives the impression of being up to date. He provides much information about the natives of North America – possibly more than any other special geographer.

   The overall impression is that of a work by an enthusiastic and diligent reader with a wide range of interests.

BLAISDALE, Silas **Y mSG Am**

- **{1829}, 3rd (01)**
- *Primary lessons in geography*, consisting of questions adopted to Worcester's atlas
- BOSTON; Marsh & Capen
- ***Libraries*** : MH
- ***Source*** NUC (NB 0530659)
- ***Notes***

1. NUC records a further ed.: 1832, 4th, Boston, 36 pp. (MH, NNC).
2. Elson (1964:400) cites the ed. of 1832.

BLAKE, John Lauris **Y SG? Am**

- **{1833} (01)**
- *American universal geography, for schools and academies ...*
- BOSTON; Russell, Odiorne & co.
- 25.5 cm.; 2 p. l., 144 p.
- ***Libraries*** : DLC, MeB, Nh
- ***Source*** NUC (NB 0532086)
- ***Notes***

1. NUC records further eds.: 1833, 2nd (MH); and 1834, 2nd (MH).

BLAKE, John Lauris **Y Am**

- **{1826} (01)**
- *A geographical, chronological and historical atlas*, on a new and improved plan; or, a view of the present state of all the empires, kingdoms, states, and colonies in the known world

- NEW YORK; Cooke and Co
- 8vo; ix, 10-196 p.
- ***Libraries*** : CtY, DLC, ICJ, IdU, MH, MiU, MnU, MWA, NB, NBuC, NjR, NN, NNC, OClW, OClWHi, PHi, PP, PPAN, PPULC, PSC, PU
- ***Source*** NUC (NB 0532164). Cited by Elson (1964:398).

BLAKE, J[ohn] L[auris] **Y Am**
- **{1830} (01)**
- *A geography for children*. With eight copperplate maps and thirty wood cuts. By Rev ... A.M. [Author of First Book in Astronomy; Improvements in Conversation on Natural Philosophy, and other works of education]
- BOSTON; Richardson, Lord & Holbrook. No. 133, Washington Street
- 22.5 cm.; pp. 64
- ***Libraries*** : DLC, MB, MH, NNC
- ***Source*** NUC (NB 0532165)
- ***Notes***

1. BL and NUC record an ed. in 1831, Boston, 68 pp. (**L** : NFQC, PP); it is the source of the words in [ ].
2. The text takes the form of question and answer. There are sections devoted separately and explicitly to history. Emphasis is given to America, which is dealt with first among the continents.
3. Elson (1964:399) cites the ed. of 1831.

BLAKE, John Lauris **Y SG? Am**
- **{1814} (01)**
- *A text-book in geography and chronology*, with historical sketches. For schools and academies
- PROVIDENCE, RI; Robinson & Howland, booksellers; H. Mann & co., printers
- 18.5 cm.; pp. 219 (1)
- ***Libraries*** : Nh, NSyU, PMA, PPULC, PU, RP
- ***Source*** NUC (NB 0532220)

BLAKE, John Lauris **Y SG? Am**
- **{1821} (01)**
- *A text-book; or first lessons in modern geography*: on a new plan
- BOSTON; Richardson and Lord
- 31 cm.; pp. iv, (5)-28, (2)
- ***Libraries*** : DLC

BLAKE, John Lauris **Y OG Am**
- **{1845} (01)**
- *The wonders of the earth*; containing an account of its remarkable features and phenomena: to wit, minerals, mountains, caverns, earthquakes, volcanoes, storms, whirlwinds, and other miscellaneous matter. Designed for the introduction of young persons

- CAZENOVIA, NY; Henry & Sweetlands
- 16 cm.; 252 p.
- *Libraries* : DLC, UPB
- *Source* NUC (NB 0532227)

BLEASE, H.J. **Y SG?**

- **{1820} (01)**
- *A system of British geography*, for the use of schools ... To which are added, three hundred questions, etc.
- LONDON; Darton, Harvey & Darton
- 24mo; pp. iv, 227
- *Libraries* L
- *Notes*

1. May not be worldwide.

BLOME, Richard **A**

- **{1680-82} (01)**
- *Cosmography and geography in two parts*: the first, containing the general and absolute part of cosmography and geography, being a translation from ... Varenius. To which is added the much wanted schemes omitted by the author. The second part, being a geographical description of all the world, taken from the notes and works of ... monsieur Sanson ... to which are added about an hundred cosmographical, geographical and hydrographical tables ...
- LONDON; printed by S. Roycroft for R. Blome
- 35 cm., 2 vols. in 1
- *Libraries* : CSmH, DLC, NN, RPJCB : CaBVaU
- *Source* NUC (NV 0046816), which catalogues it under Varen

- **1683 (02)**
- *Cosmography and geography* ... translation from that eminent and much esteemed geographer Varenius. Wherein are at large handled all such arts as are necessary to be understood for the true knowledge thereof. To which is added ... and works of the famous monsieur Sanson late geographer to the French king. To which are added ... tables of several kingdoms and isles in the world, with their chief cities, sea-ports, bays, etc. drawn from the maps of the said Sanson. Illustrated with maps
- LONDON; printed by S. Roycroft, and are to be sold by William Abington at the Three Silk-worms in Ludgate-street
- Folio, 35 cm.; 2 pts; pp. [6], 364; [2], [t.p. (see below) and blank reverse], 493
- *Libraries* **L** : DLC, NPV, OClWHi
- *Notes*

1. The tp. to the second part reads: 'A geographical description of the world, taken from the works of the famous monsieur Sanson, late geographer to the present French king. To which are added, about an hundred geographical and hydrographical tables, of the kingdoms, countreys, and isles in the world, with their chief cities and sea-ports; drawn from maps of the said

monsieur Sanson and according to the method of the said description. Illustrated with maps. The second part. Printed in the year 1680.'

Although the order in which the continents are described differs from that in BLOME (1670), and although the material on some, though not all, countries has been altered or reorganized, there is sufficient similarity between this work and the former to justify identifying this as a second ed. of the earlier work. However, I have chosen, for the purposes of this guide, to treat this work as a different book.

2. The first part, as bound (for it was printed two, or even three, years later than the second, depending upon the impression; see above), contains the first translation of the *Geographia generalis* of Varenius, published in 1650, into English (see VAREN). Baker (1963a:108) gives the date of publication as 1693, which was that of the third impression; see below.

- **{1693} (03)**
- *Cosmography and geography* ... tables ... The 3rd impression ... To which is added the county-maps of England, drawn from those of Speed
- LONDON; printed by S. Roycroft for R. Blome
- 35 cm., 2 vols. in 1
- ***Libraries*** : CtY, DLC, MBAt, MiU, N, NcU, ViU
- ***Source*** NUC (NV 0046821)
- ***Notes***

1. Peddie (1933:255) records an ed. in 1652. As this would have been published 53 years before Blome's death, it seems unlikely, though it is not impossible.
2. Discussed in Bowen (1981:96-8) and E.G.R. Taylor (1937) [Note: Taylor was wrong in stating (p. 530) that Blome never published 'the translation of Varenius' Principles.']

BLOME, Richard **A**

- **1670 (01)**
- *A geographical description of the four parts of the world* taken from the notes & workes of the famous Monsieur Sanson, geographer to the French king, and other eminent travellers and authors. To which are added the commodities, coyns, weights, and measures of the chief places of traffick in the world; compared with those of England, (or London) as to the trade thereof. Also, a treatise of travel, and another of traffick, wherein the matter of trade is briefly handled: the whole illustrated with variety [sic] of useful and delightful mapps and figures. A work beneficial and acceptable to all men, especially to those that intend to spend some part of their time in other countreys, or desire to be informed of them here at home. Also very necessary for merchants, factors, and mariners: and which hitherto hath been undertaken by none
- LONDON; printed by T. N. for R. Blome, dwelling in the Savoy near the Kings Wardrobe, and for convenience are also sold by Nath. Brooks at the Angel in Cornhil, Edw. Brewster at the Crane in St Pauls Church-yard, and Tho. Basset at the George in Fleetstreet, near Cliffords-Inn
- Folio, 37 cm.; (in 5 parts) pp. [11], 113, [5], 82, [4], 138, [6], 56, [2], 55, [1]

- ▸ *Libraries* C, E, HeP, **L**, MRu, O : CtY, DAU, DLC, MH, MHi, MnHi, MnU, NcU, NhD, NjP, NN, NNC : CaBViP
- ▸ *Notes*

1. Contains, in addition to the descriptions of the countries of the world, chapters on travel and on commerce. The last contains a lengthy guide on how to keep financial records, and a review of the principal trading companies of the day. In the Preface it is stated that this volume is the first of four; the second is to contain a translation of the *General Geography* of Varenius; the third is to be devoted to hydrography, and the fourth to the geography, hydrography, and chorography of the British Isles.

   Blome did publish a translation of Varenius, the first in English (BLOME 1680-2). He also published *Britannia, or a geographical description of the kingdoms of England, Scotland and Ireland*, London, 1673, folio. There is nothing in BL that seems to correspond to the volume on hydrography in general, unless it is *A description of the island of Jamaica; with the other isles and territories in America, to which the English are related*, London, 1672, 8vo. According to E.G.R. Taylor, the publication of John Seller's *English Pilot* (copyright 1671) 'precluded any similar publication by Blome' (Taylor 1937:530).

- ▸ **1680, anr ed.**

See BLOME (1680-2).

BLOMFIELD, Ezekiel **A**

- ▸ **{1804} (01)**
- ▸ *General view of the world*
- ▸ BUNGAY
- ▸ Book 3, pp. 574-662
- ▸ *Libraries* : MH

- ▸ **{1805} (02)**
- ▸ *A general view of the world*, geographical, historical and philosophical; on a plan entirely new
- ▸ BUNGAY; Brightly and T. Kinnersley
- ▸ 28 cm., 2 vols.
- ▸ *Libraries* : ViU

- ▸ **1807 (03)**
- ▸ *A general view of the world ...*
- ▸ BUNGAY; printed and published by C. Brightly and T. Kinnersley
- ▸ 4to, 25 cm., 2 vols.; Vol. I, pp. [1], v, [1], 3-254, 1-827; Vol. II, pp. [2],695, [29]
- ▸ *Libraries* **L** : AU, CtY, DLC, MiU, OrP, Viu
- ▸ *Notes*

1. The principal sources are identified in the Preface. Vol. I, pp. 3-254, provides more in the way of natural history, as well as information on astronomy and the problem of location, than do most special geographies. Vol. I, pp. 1-92 (second series), provides summary descriptions of the four continents. The British Isles then occupy pp. 93-571; a very large part of this

space is devoted to history. The remainder of Vol. I is devoted to northern Europe. The rest of Europe, including a treatment of the Roman Empire, occupies Vol. II, pp. 1-308. The whole of Africa is dealt with in only 67 pp.

BLUNDEVILLE, Thomas **OG**

- **{1589} (01)**
- *A briefe description of vniuersal mappes and cardes*, and of their vse: and also the vse of Ptholemy his tables, etc.
- LONDON; R. Ward for T. Cadman
- 4to
- ***Libraries*** L : DFo, ICN, IU, MH, NN
- ***Notes***

1. BL notes: 'Subsequent editions of this work form pt. 8 of the second and later editions of "M. Blundeuile his Exercises"'; for which see the next entry.
2. Facsimile ed., 1972 (Johnson/TOT).

BLUNDEVILLE, Thomas **OG**

- **{1594} (01)**
- *M. Blundeuile his exercises, containing six treatises* ... which treatises are verie necessarie to be read and learned of all young gentlemen that haue not bene exercised in such disciplines, and yet are desirous to haue knowledge as well in cosmographie, astronomie, and geographie, as also in the arte of nauigation, in which art it is impossible to profite without the helpe of these, or such like instructions. To the furtherance of which arte of navigation, the said M. Blundevile speciallie wrote the said treatises ...
- LONDON; Iohn Windet
- 4to; ff. 350
- ***Libraries*** L : CCC, DLC, DN-Ob, MB, MdAN, MH, NjP, NN
- ***Source*** BL and NUC (NB 0566428)
- ***Notes***

1. Concerned with cartographical problems, in particular those that arise from the sphericity of the earth; not a special geography. There were later eds. in 1597 (L : CtY, ICN, MWiW, NIC, RPJCB, ViNeM); 1606 (L : DFo, MnU, NN); 1613 (L : DFo, DLC, MB, NN); 1622 (L : CtY, DFo, DLC, MH, MiU, MWiW, NNE, OCl, RPJCB, WU : CaAEU); 1636 (L : DFo, DN-Ob, DLC, DN, ICN, IaU, MiU, RPJCB, ViU); and 1638 (MeB, NN).
2. Facsimile ed., 1971 (Johnson/TOT).

BOARDMAN, A. **Y mSG?**

- **{1853}, 2nd (01)**
- *The pupil teacher's historical geography* ... Second edition, revised and enlarged
- LIVERPOOL; G. Philip & Son
- 8vo; pp. viii, 118
- ***Libraries*** L

▸ *Notes*
1. ECB, Index (1856-74:130) records a further ed.: 1869.

BOEMUS, Joannes (Johann) **mSG tr**
▸ **1611 (01)**
▸ *The manners, lawes, and cvstomes of all nations*. Collected out of the best writers by Ioannes Boemvs Avbanvs, a Dutch-man. With many other things of the same argument, gathered out of the Historie of N. Damascen. The like also out of the History of America, or Brasill, written by Iohn Lerius. The faith, religion and manners of the Aethiopians, and the deploration of the people of Lappia, compiled by Damianus à Goes. With a short discourse of the Aethiopians, taken out of Ioseph Scaliger his seuenth booke de Emendatione temporum. Written in Latin, and now newly translated into English by E. Aston
▸ LONDON; printed by George Eld.
▸ 4to, 18 cm.; pp. [16], 589, [3]
▸ *Libraries* C, **L**, O : CSmH, DFo, DLC, ICN, MB, MnU, MWiW, NIC, NN, NNC, RPJCB, WU
▸ *Notes*
1. 'Boemus' work is a compilation in which he describes the mores and habits of diverse peoples of the world. The first edition is from *Augustae Vindelicorum, in officina Sigismundi Grimm...*, 1520. It had such success that it was re-edited countless times in Latin, English and Spanish. Each publisher added something to his edition, reorganizing it, and bringing it up to date' (Borba de Moraes 1958, 1:96). The first English edition was published in 1555. It was entitled *The Fardle of Facions*, and contained descriptions of only Africa and Asia. Even this 1611, ed., which deals with at least some of the peoples of every continent does not aspire to contain a description of every country in the world. The editor does, however, begin the section on each country dealt with by giving its location, so that the work may be seen as belonging to the tradition of special geography.
2. Discussed in Hodgen (1953).

BOHN, Henry G. **A**
▸ **1861 (01)**
▸ *A pictorial hand-book of modern geography*, on a popular plan, compiled from the best authorities, English and foreign, and completed to the present time; with numerous tables and a general index. Illustrated by 150 engravings on wood, and 51 accurate maps engraved on steel
▸ LONDON; Henry G. Bohn, York Street, Covent Garden
▸ 8vo, 18 cm.; pp. [2], x, [2], 529
▸ *Libraries* **L**, O : DLC, MB, NB, OCU, OO
▸ *Notes*
1. Based on 'Malte-Brun and Balbi's System of Universal Geography, with additions of Mr James Laurie,' and on the Gazetteer of Keith Johnston (p. iv). Bohn sticks rigidly to description, eschewing any explicit theory.
2. See LAURIE, James (1842).

- **{1862}, 2nd, rev. (02)**
- *A pictorial hand-book ...*
- LONDON; H.G. Bohn
- 18 cm.; 1 p. l., × p., 1 l., 529 p.
- ***Libraries*** : DLC, M, MB, MdBP, MiD, WaS, WaTC
- ***Source*** NUC (NB 0599609)

- **{1865}, 3rd, rev. (03)**
- *A pictorial hand-book ...*
- LONDON; Bell & Daldy
- 18 cm.; 1 p. l., × p., 1 l., 529 p.
- ***Libraries*** : ICN, MH, ViU
- ***Source*** NUC (NB 0599610)

BOHUN, Edmund **Dy**
- **1688 (01)**
- *A geographical dictionary*, representing the present and ancient names of all the countries, provinces, remarkable cities, universities, ports, towns, mountains, seas, streights, fountains, and rivers of the whole world: their distances, longitudes and latitudes. With a short historical account of the same; and their present state: to which is added an index of the ancient and Latin names. Very necessary for the right understanding of all modern histories, and especially the divers accounts of the present transactions of Europe
- LONDON; printed for Charles Brome, at the Gun, at the west-end of St. Pauls
- 8vo, 18 cm.; pp. [806] [NUC]
- ***Libraries*** Brf, C, **L**, LDW, LIT, LvU, NP : CtY, CU-S, DLC, MiU, NNUT, OKentU, OU, PPULC, PU, ViU : CaBViPA, CaOKU
- ***Notes***

1. As the title suggests, this is essentially a dictionary. The preface presents a clear guide to the type of material to be found in the body of the work, the sources used, and a short history of previous works of this type. None of the latter, it is alleged, were in English, except for a 12mo translation of 'Monsieur du Val, geographer to the French king,' a work of poor quality in Bohun's opinion.
2. ESTC and NUC record further eds. (both London, Brome): 1688, 840 pp. (: MWiW, N, PP, PPULC); 1691, 2nd, [806] pp. [NUC] (C, **L**, O : CLU, CtY, IaU, ICN, NNR, RPJCB, WaU).
3. A microform version is available from CIHM.

- **1693, anr ed. (01)**

See BERNARD (1693).

- **1695, 4th (02)**

See BERNARD (1693).

- **1710, anr issue (03)**

See BERNARD (1693).

BOND, R. **Y SG?**
- **{1838} (01)**
- *Geography for children*
- ? Thomas
- 12mo
- ***Libraries*** n.r.
- ***Source*** BCB, Index (1837-57:110)

BOND, Rowland **Y SG?**
- **{1836} (01)**
- *Popular geography*: a companion to Thomas's library and imperial school atlases, etc.
- LONDON; Joseph Thomas; Simpkin & Marshall
- 16mo, 2 pts.
- ***Libraries*** L : CtY.
- ***Source*** BL and NUC (NF 0083486), which notes that this work is included as the text in Fenner, Rest [sic], *Pocket atlas of modern and ancient geography ...*
- ***Notes***

1. NUC records a further ed.: 1853 (CtY).

BONWICK, James **Y Aus**
- **1845 (01)**
- *Geography for the use of Australian youth*
- HOBART, Van Diemen's Land [Australia]; sold by S.A. Tegg, Hobart Town; James Darling, Launceston; and at Sydney, by W.A. Colman
- 12mo, 15 cm.; pp. vii, [1], 204
- ***Libraries*** **L** : CSt, CU
- ***Notes***

1. Six pp. on the earth in the solar system and matters relating to its sphericity at the beginning, and 38 on physical geography at the end; special geography occupies the remainder of the book. It begins with Australia, which receives 38 pp; there follow Asia, Europe, Africa, and America (with the islands in the Pacific included with the last). Description is provided at the national level.

BONWICK, James **Y Aus**
- **1859, 2nd (01)**
- *Geography for young Australians*. Used in the schools of the denominational and national Boards of Education in Victoria
- MELBOURNE; printed and published by Wm. Goohugh & Co., Flinders Lane East, and sold by all booksellers
- 16mo, 14 cm.; pp. [1], 66, [2]
- ***Libraries*** **L** : CSt
- ***Notes***

1. The information is highly condensed and presented in semi-tabular form.

- **1865, 5th (02)**
- *Geography for young Australians.* Sanctioned by the Commissioners of Education in Victoria
- MELBOURNE; Fergusson & Moore, 48 Flinders Lane East, and sold ...
- 12mo, 15 cm.; pp. [1], 72
- ***Libraries*** **L**
- ***Notes***

1. Most of the additional information has been added to the section on Australia.

BOOK OF THE WORLD ... **? tr?**

- **[1852]-3 (01)**
- *Book of the world.* A family miscellany for instruction and amusement. Edited [? translated?] by Dr Gaspey
- PHILADELPHIA; Weik and Wieck
- 4to, 2 vols.
- ***Libraries*** L (mislaid) : CSmH, NjR, NN
- ***Source*** NUC (NB 0640310), which notes: 'Probably a translation of Das Buch der Welt. Titles of landscape engravings in German.'
- ***Notes***

1. NUC's entry also provides Gaspey's full name, as well as identifying him as the editor of the work rather than as the translator. Other entries in NUC, however, show that he translated a number of other German books into English.

BORTHWICK, John Douglas **Dy NG Can**

- **1859 (01)**
- *Cyclopædia of history & geography*; being a dictionary of historical & geographical antonomasias, origins of sects, etc.; peculiar etymologies, and remarkable facts in history and geography, for the use of students and general readers. By ... High School Department of McGill College
- MONTREAL; published by R. & A. Miller, St. François Xavier St.
- 18 cm.; pp. [7], 8-251
- ***Libraries*** ICU, MH, NN : **CaOONL**
- ***Notes***

1. 'Antonomasias is a term applied to that form of expression, when the title, office, dignity, profession, science, or trade is used instead of the true name of the person or place; as, the Iron Duke [for the Duke of Wellington]' (7). Contains a very miscellaneous collection of information. There are, for example, entries for *three* (pp. 219-25!), *thumbscrew*, and *Tyre*, but not for Toronto, or even Ontario. Thus it falls well outside the genre of special geography.
2. A microform version is available from CIHM.

BOSCAWEN, Mary Frances Elizabeth **Y mSG?**

- **{1854} (01)**

- *Conversations on geography*: or, the child's first introduction to where he is, what he is, and what else there is besides
- LONDON; Longman & Co.
- 8vo; pp. xiv, 512
- *Libraries* L

[BOTERO, Giovanni] **tr**

- **1601 (01)**
- *The travellers breviat*, or an historicall description of the most famous kingdomes in the world: relating their situations, manners, customes, ciuill gouernment, and other memorable matters. Translated into English
- LONDON; imprinted at London by Edm. Bollifant, for Iohn Iaggard
- 4to, 18 cm.; pp. [4], 179, [1]
- *Libraries* **L**, LeU : MH, RPJCB
- *Notes*

1. In the Dedication the translator identifies himself as Robert Iohnson. The text opens with a general description of the world, but although the discovery of America is mentioned (p. 1), the only description of any part of the continent is contained within the description of Spain; see AVITY (1615). Compared with ABBOT (1599), there is more historical information, especially on the Roman Empire, and more information about the military strength of the countries described. The relations between countries are also dealt with in several cases. Indeed, compared with most special geographies the book is relatively political.

- **1601, anr ed. (02)**
- *The worlde*, or an historicall ... kingdomes and common-weales therein. Relating ... scituations ... memorable matters. Translated into English, and inlarged
- LONDON; imprinted at London by Edm. Bollifant, for Iohn Iaggard
- 4to, 19 cm.; pp. [4], 222
- *Libraries* CEm, **L**, O : CSmH, DFo
- *Notes*

1. The opening 10 pp. are new. They are devoted to the means by which rulers can add to the strength of their countries. There is also much new material on Italy and on Muscovy, and a new section on Switzerland, as well as other smaller additions to the descriptions of European countries. Besides the new material, paragraphs are reorganized in some places; this happens also in the case of some non-European countries, although outside Europe the facts presented seem little changed.

- **1603, anr ed. (03)**
- *An historicall description of the most famous kingdomes in the world* ... and enlarged with an addition of the relations of the states of Saxony, Geneva, Hungary and Spaine; in no language euer before imprinted
- LONDON; printed at London for Iohn Iaggard
- 4to, 18 cm.; pp. (4), 268
- *Libraries* BmU, CGc, **L**, LU, O : CSmH, NN, OCl, PPULC

▸ *Notes*

1. Not all the additions were noted in the title; the section on Moscovia, for example, is considerably lengthened. Besides new material, there is a change in the order in which some of the countries of Europe are dealt with. Spain is dealt with twice, the last thirty pages being devoted to 'another relation of the state of Spaine, later than the former, written in ... 1595, by Sin. Francisco Vendramino Embassadour from the State of Venice, to his Catholike maiestie' (p. 237).

▸ **1608, anr ed. (04)**

▸ *Relations, of the most famovs kingdoms and common-weales thorovgh [sic] the world.* Discoursing of their scituations ... customes, strengthes and pollicies. Translated ... of the estates of Saxony ... Hungary, and the East Indies, in any language never before imprinted

▸ LONDON; printed for Iohn Iaggard, dwelling in Fleetstreet, at the Hand and Starre, betweene the two Temple Gates

▸ 4to, 18 cm.; pp. [6], 330 (of which the first 112 are not numbered)

▸ ***Libraries*** C, CT, **L**, OT, [IDT] : DFo, ICN, MH

▸ *Notes*

1. Not all the increase in length is due to additional material; a reduction in the amount of letterpress per page is responsible for almost half the increase. Africa is increased by 8 to 9 pp., while America, which appears for the first time as a continent, is given 7 to 8 pp. (besides the information still given under Spain). Besides the new material, the contents have been considerably reorganized; the British Isles are now put first, followed by Scandinavia, and, among several other changes, Africa now follows Europe instead of being placed after Asia. Russia and European Turkey are included in Asia.

▸ **1611, anr ed. (05)**

▸ *Relations of the most famous kingdoms* ... through the world ...

▸ LONDON; printed for J. Jaggard ...

▸ 4to, 19 cm.; pp. [4], 437

▸ ***Libraries*** Crf, **O**, OW : CU-S, DFo, DLC, MiU, NN, OO, ViU

▸ *Notes*

1. The work is now divided into six books of uneven length: Book 1 contains observations on why some princes are powerful and others weak, plus instructions for travellers; Book 2 deals with Europe; Book 3 with Africa; Book 4 with part of Asia; Book 5 with India, China, Japan, and Siam; and Book 6 with America.

▸ **1616, anr ed. (06)**

▸ *Relations of the most famous kingdoms* ... strengths, greatnesse, and policies. Enlarged, according to moderne observation

▸ LONDON; printed for John Jaggard

▸ 4to, 20 cm.; pp. [4], 437, [4]

▸ ***Libraries*** C, NcP, **O** : CtY, DFo, ICN, MiD, MiU, NN, OU, RPJCB

▸ *Notes*

1. This ed. is not in Pollard and Redgrave (1946). The arrangement of the work is similar to that of 1611, but America has been allotted a second

book, for a total of seven. The two books on the New World comprise only 14 pp. between them. Although no mention of this is made on the tp., a 4-p. index of place names has been added. Although the sequences in which continents and countries are treated are basically the same as in 1611, considerable rewriting has taken place; e.g., the section on Great Britain has been expanded considerably while those on France and Spain have been shortened.

- **1630, anr ed. (07)**
- *Relations of the most famous kingdoms* ... policies. Translated out of the best Italian impression of Boterus. And since the last edition by R.I. now once againe inlarged according to modern observation; with addition of new estates and countries. Wherein many of the oversights both of the author and translator, are amended. And unto which, a mappe of the whole world, with a table of the countries, are now newly added
- LONDON; printed by John Haviland, and are to be sold by John Partridge at the signe of the Sunne in Pauls church-yard. (Entered in the Stationer's Register to J. Smithwick, 24 Feb. 1626 [STC])
- 4to, 19 cm.; pp. [8], 644, [3]
- ***Libraries*** BmU, C, **L**, **O**, SkPt : CtY, DFo, DLC, ICN, MWA, RPJCB, ViU, WU
- *Notes*

1. The Preface refers to this as the third edition, though it seems to be at least the sixth. The Dedication is replaced by a Preface, which identifies, for the first time, the author as Botero. The extensive new material has apparently been added by the anonymous editor, who is clearly not Johnson the translator. The new material is distributed fairly evenly throughout, though proportionately Book 5 has been expanded the least, and, in Europe, the description of Norway is unaltered. Despite the claim on the tp. to have added an index, the latter is not as good as that found in the 1616 edition.
2. Discussed in Gilbert (1919:330-1), and E.G.R. Taylor (1934:133).

BOURN, Thomas **Dy**

- **1807 (01)**
- *A concise gazetteer of the most remarkable places in the world*; with brief notes of the principal historical events, and most celebrated persons, connected with them: to which are annexed references to books of history, voyages, travel, etc. intended to promote the improvement of youth in geography, history, and biography
- LONDON; printed for the author; and sold by J. Mawman, in the Poultry; J. Harris, St. Paul's Church Yard; and Darton and Harvey, Gracechurch Street
- 8vo, 21 cm.; pp. [ca. 250]
- ***Libraries*** L
- *Notes*

1. NUC records further eds.: 1807 (FMU); 3rd, 1822, 984 pp. (MHi).
2. As the subtitle implies, it is a geographical dictionary rather than a gazetteer in the modern sense.

BOURNE, William **mSG**

- **1578 (01)**
- *A booke called the treasure for traueilers*, deuided into fiue bookes or partes, contaynyng very necessary matters, for all sortes of trauailers [sic], eyther by sea or by lande
- LONDON; imprinted for Thomas Woodcocke, dwelling in Paules Churchyarde, at the sygne of the Blacke Beare
- 4to, 18 cm., 5 pts.; Pt. I, 32 l., [2]; Pt. II, 25 l., [2]; Pt. III, 22 l., [4]; Pt. IV, 21 l., [2]; Pt. V, [2], l. 5-20 (i.e., a total of 132 l.)
- ***Libraries*** **L** (lacks leaves 1, 4, and 7 in Pt. I) : DFo, DLC, NN, PPULC, WU
- ***Notes***

1. Not a special geography. Pt. I is devoted to surveying; Pt. II is a non-alphabetically ordered gazetteer laid out in the form of prose. Most of Pt. V would now be classed as physical geography.
2. Cox has the following note: 'The contents of the five books is explained in the dedication as follows: "the fyrst is Geometrie perspectiue, the seconde Booke is appertainying [sic] unto Cosmographie, the thirde Booke is Geometrie general, the fourth Booke is Statick, and the fyfth and last Booke is appertayning vnto natural Philosophie ..." In the second book chapters eight and nine give the latitude and longitude and the length of the longest summer day of numerous places in America and the West Indies ... the concluding chapter of the fifth book contains the author's opinion as to how America and all the then newly discovered lands became peopled. He believed that America was part of the mythical island of Atlantis, and that the Atlantic Ocean was full of mud. - Quaritch' (Cox 1938:359-60).

BOWEN, Emmanuel
See COMPLETE SYSTEM ... (1747).

BOWRING, F. **Y SG?**

- **{1841} (01)**
- *Geography for children*
- ? J. Chapman
- 12mo
- ***Libraries*** n.r.
- ***Source*** BCB, Index (1837-57:110)
- ***Notes***

1. Given the next entry, it seems likely that the initial F. in the BCB Index is a misprint.

BOWRING, Thomas **Y SG?**

- **{1831} (01)**
- *The first book of geography for children*
- LONDON; J. Green; Bristol, J. Philip; Manchester, Forrest & Fogg
- 16 cm.; pp. iv, 127

- *Libraries* : : CaOTP
- *Source* St. John (1975, 2:800)

BRADLEY, John **Y SG?**
- **{1812} (01)**
- *Elements of geography for the use of schools*
- LIVERPOOL
- *Libraries* n.r.
- *Source* Allford (1964:243)
- *Notes*

1. British Museum (1881-1900) records a further ed.: 1819, 2nd.

BRADLEY, John **Y**
- **1818 (01)**
- *The preparatory geography*, in a series of lessons, with suitable interrogations, and six maps of reference, for the use of junior classes. In this is developed the author's method of reading and teaching his second edition of the Elements of Geography
- LONDON; printed for the author; and sold by Messrs. Lackington and Co., Finsbury Square; and Messrs. Whittaker, Ave-Maria-Lane
- 12mo, 17 cm.; pp. [2], 34
- *Libraries* **L**
- *Notes*

1. Begins with 27 numbered lessons devoted to parts of the world, and composed of numbered paragraphs. For every paragraph except the last there is a question; these ask, in effect, for the regurgitation of the information in the paragraphs. Then come lessons 28 to 70. These contain a list of approximately 200 places, divided into groups of six (more or less) per lesson; about each place 1 to 6 lines of information are given. The list is alphabetical, starting at Aalburg and ending at Ammer(!), 'A range of mountains in the south of Algiers.'

BREWER, Ebenezer Cobham **Y SG?**
- **{[1864]} (01)**
- *My first book of geography*
- LONDON; Cassell & Co.
- 12mo; pp. 72
- *Libraries* L

BRICE, Andrew **Dy**
- **1759 (01)**
- *The grand gazetteer*, or topographic dictionary, both general and special, and antient as well as modern, etc. Being a succinct but comprehensive geographical description of the various countries of the habitable known world, in Europe, Asia, Africa, and America; more especially of Great Britain and Ireland, and all the British settlements abroad, or where we

have trade, commerce, or correspondence. Shewing the situation, extent, and boundaries, of all the empires, kingdoms, republicks, provinces, cities, chief towns, etc. with their several climates, soils, produces, animals, plants, minerals, etc. the government, traffick, arts, manufactures, customs, manners, and religion, of the divers nations; and the vast many admirable (some of them stupendous) curiosities both natural and artificial; the most remarkable events, accidents, and revolutions, in all past ages; etc. etc. Aptly and requisitely interspers'd with many thousands of uncommon passages, strange occurrences, critical observations (as well sacred as profane), and proper relations; which most agreeably surprise, and delightfully inform. Diligently extracted, and as accurately as possible compiled, from the most esteemed voyagers, travellers, geographers, historians, criticks, etc. extant. A work in its form entirely new, very necessary for numbers, and serviceable to all degrees of readers,– (not excepting the most learned, and with libraries best furnish'd) – readers not only of news-papers, magazines, etc. etc. etc. but of histories of former ages or the present, the classicks, and even the sacred writ itself; the antique articles being collected either from original authors or the best translators, and divers learned commentators on the Bible, etc. etc

- ▸ EXETER; printed by and for the author, at his printing-house, in Northgate-street, Exon.
- ▸ Folio, 34 cm.; pp. [3], 1446
- ▸ ***Libraries*** BatP, BidP, C, EU, **L**, SAN : CtY, ICRL, MWA, NN, OkU
- ▸ ***Notes***

1. It is a dictionary, though the entries for countries are comparable in length with those of many large special geographies, e.g., FENNING, COLLIER, et al (1764-5).

   One of the last times *special geography* is used as a living term with the meaning given by Varen?

- ▸ **1759, anr ed. (02)**
- ▸ *A universal geographical dictionary; or, grand gazetteer*; of general, special, antient and modern geography: including a comprehensive view of the various countries of Europe ... especially of the British dominions and settlements throughout the world: describing their soil, extent and situation; their several productions, animal, vegetable and mineral; their government, arts, manufactures, traffic, genius, manners, and religion; as well as the many admirable and stupendous curiosities, (natural and artificial) to be found therein; with the most remarkable events, accidents, and revolutions of past ages. Interspersed with many thousand uncommon passages, strange occurrences and critical observations, both sacred and profane: diligently extracted, and compiled with the utmost accuracy, from the most approved travellers, geographers, historians, philologers etc. A work, not only agreeably amusing but also instructive, and of singular utility to persons of every rank and station. Illustrated by a general map of the world, particular ones of the different quarters, and of the seat of war in Germany
- ▸ LONDON; printed for, and sold by J. Robinson and W. Johnston, in Ludgate-Street; P. Davey and B. Law in Ave-Maria-Lane; and H. Woodgate and S. Brooks, in Pater-Noster Row
- ▸ Folio, 35 cm., 2 vols.; Vol. I, pp. [3], 648; Vol. II, pp. [2], 649-1446

- *Libraries* **EU, OSg** : CtY, DLC, PPL, PPULC
- *Notes*

1. Despite the revision to the full title, not many changes seem to have been made to the text, although a note on p. 1441 shows the Exeter ed. to be the earlier of the two. The book at EU is bound in one volume.

- **{1760} (03)**
- *The grand gazetteer* or topographic dictionary, etc.
- EXETER
- Folio; pp. 1446
- *Libraries* : MH
- *Source* NUC (NB 0789373), which notes: 'Titlepage wanting'
- *Notes*

1. Discussed in Bowen (1981:313).

BROOKES, Richard **Dy**

- **{1762} (01)**
- *The general gazetteer: or, compendious geographical dictionary ...*
- LONDON; J. Newbery
- 8vo, 21.5 cm.; pp. viii, xxxii, [756]
- *Libraries* L : DLC, IEN, ViU, ViW
- *Source* BL and NUC (NB 0839206)
- *Notes*

1. BL and NUC record further eds. (both London): 1766, 2nd, [648] pp. (: CtY, DLC, PHi, PPULC); 1773, 3rd (L : CtY, MiD).

- **{1776}, 4th, revised ... (see below) (02)**
- *The general gazetteer* ... dictionary. Originally compiled by R. Brookes, revised, corrected, and greatly improved by W. Guthrie and E. Jones. 4th ed.
- DUBLIN; printed for J. Williams
- Unpaged
- *Libraries* : ICU, OClW
- *Source* NUC (NB 0839209)

- **{1778}, 4th, improved ... (see below) (03)**
- *The general gazetteer* ... Embellished with nine maps. 4th ed., improved with additions and corrections, by Alex. Bisset
- LONDON; J.F. and C. Rivington, [etc.]
- 21.5 cm.; pp. xvi, [624]
- *Libraries* : DLC, ViU
- *Source* NUC (NB 0839210)

- **1782, 5th (04)**
- *The general gazetteer* ... dictionary. Containing a description of all the empires, kingdoms, states, republics, provinces, cities, chief towns, forts, fortresses, castles, citadels, seas, harbours, bays, rivers, lakes, mountains, capes, and promontories in the known world; together with the government, policy, customs, manners, and religion of the inhabitants; the extent,

bounds, and natural productions of each country; and the trade, manufactures, and curiosities, of the cities and towns; their longitude, latitude, bearing and distances in English miles from remarkable places; as also, the sieges they have undergone, and the battles fought near them. Including an authentic account of the counties, cities, and market-towns in England and Wales; as also the villages, with the days on which the fairs are kept. Embellished with nine maps. The fifth edition, improved, with additions and corrections

- LONDON; printed for J.F. and C. Rivington, T. Carnan, and J. Johnson, in St. Paul's Church Yard; S. Crowder, G. Robinson, and R. Baldwin, in Pater-Noster-Row; B. Law, in Ave-Maria-Lane; T. Lowndes, and J. Murray, in Fleet-Street; C. Dilly in the Poultry
- 8vo, 22 cm.; pp. [2], iii-xvi, [674]
- ***Libraries* L**
- *Notes*

1. The full title suggests a bias in favour of military information.
2. BL and NUC record a further ed.: 1786, 6th, London, [691] pp. (L : CtY, DLC, MB, MiU, OO, OU).

- **1791, 7th (05)**
- *The general gazetteer* ... cities, market-towns, and principal villages in Great-Britain and Ireland, with the several fair days in each town and village. This edition includes, not only an accurate description of the new discovered islands in the South Seas, from the most authentic accounts - but also a very considerable number of articles respecting America, and several hundreds peculiarly relative to Ireland, never in any former edition. Embellished with maps. Originally compilled ... Revised, corrected, and greatly improved by W. Guthrie, and E. Jones
- DUBLIN; printed by P. Wogan, No. 23, Old Bridge
- 22 cm.; unpaged
- ***Libraries*** : ICN, KyU, MB, MH, MHi, NjR, OClW, PHi, PPULC : **CaBViPa**
- *Notes*

1. BL and NUC record further eds. (London, except as noted): 1793 (: KyLx); 1794, 8th, [756] pp. (: CtY, NIC); 1795, 9th, [460] pp. (L : DLC, KyLx, M, MB, MeB, NcD); [CIHM records an ed. 1796, London, B. Law (: : CaOLU)]; 1797, 10th, [669] pp. (L : DLC, MB, PPULC, PPWa); 1800, 11th, [722] pp. (: MB, MeWC, ViU); 1801, 12th, Montrose [Scotland], 306 l. (: MA, MHi, NN : CaBVaU); 1802, 12th, [756] pp. (L : MnHi, NIC, NWM); 1807, 13th (: ICRL, PHi, PPULC); 1808, 8th, Dublin, (BfQ : NcD); and 1809, 14th [776] pp. (: DLC : CaQMM).

- **1815, 16th, with ... (see below) (06)**
- *The general gazetteer* ... states, provinces, cities, towns, forts, seas, harbours, rivers, lakes, mountains, capes, etc. in the known world; with the extent, boundaries and natural productions of each country; the government, customs, manners, and religion of the inhabitants; the trade, manufactures, and curiosities, of the cities and towns, with their longitude and latitude, bearing and distance in English miles from important places; and the remarkable events by which they have been distinguished. Illustrated ... The

sixteenth editions, with very considerable additions and improvements, from the best and most recent authorities

- LONDON; printed for F.C. and J. Rivington, G. Wilkie, W. Lowndes, Scatcherd and Letterman, J. Cuthell, J. Nunn, C. Law, Longman, Hurst, Orme, and Brown, Cadell and Davies, B. Crosby and co. Clarke and sons, J. and A. Arch, Newman and co. Black, Parry, and co. J. Richardson , J.M. Richardson, Lackington, Allen, and co. R.S. Kirby, R. Scholey, T.H. Hodgson, J. Booth, R. Baldwin, Sherwood, Neely, and Jones, J. Mawman, T. Hamilton, W. Baynes, J. Asperne, T. Tegg, Craddock and Jay, Gale, Curtis, and Fenner, G. and S. Robinson, J. Walker and co. J. Robinson, Wilson and son, York, and Constable and co. Edinburgh
- 22 cm.; pp. [4], iv-xvi, [799] [829 (NUC {NB 0839226})], [2]
- ***Libraries*** EU, **[IDT]** : DLC, IU
- ***Notes***

1. The Preface (dated Nov. 1814) identifies the 8th ed. of 1794 as having been a major revision, with further revisions being made in each succeeding ed.
2. NUC records further eds. (London, except as noted): 1816, 3rd, Berwick-on-Tweed [England] (IU); 1818, 4th (PP, PPULC); 1819, 15th, 776 pp. (MB, PHi, PPULC); 1820, 17th (CLU, ICU); 1824, 15th, 776 pp. (DLC); 1825, 15th (AzU).

   There was an 18th ed. in 1827, London [595] pp. (EU, **L** : MiD, MiU, NcD). The Advertisement (i.e., the Preface) states that many revisions have been made to keep the book up to date; the editor notes that, apart from the one map that forms the frontispiece, those included in previous eds. have been dispensed with; the omission is justified by the ready availability of good atlases.
3. In addition to the foregoing, there were numerous other versions of this work. In alphabetical order they are as follows.
   (a) *Brookes' general gazetteer abridged* ... London, 1796, B. Law (NB 0839196 and NSB 0116530)
   (b) *Brookes' general gazetteer improved*; or ... dictionary: containing ... Ireland: together with a succinct account of, at least, seven hundred cities, owns and villages in the United States ... The 1st American ed. Philadelphia, 1806, Jacob Johnson (NB 0839201); the 2d American ed., Philadelphia and Richmond, 1812, Johnson and Warner (NB 0839203)
   (c) *Brookes' general gazetteer improved*; or, compendious geographical dictionary; in miniature. Containing ... J.H. Bain ... The 1st American ed., rev. and cor. in every part. Baltimore, 1815, Joseph Cushing (NB 0839204). This was not the first 'miniature' ed.; see (h) below.
   (d) *Brookes' universal gazetteer*, re-modelled and brought down to the present time. By John Marshall, with numerous additions by the American editor, and the population of every town, county, territory and state, according to the census of 1840. Philadelphia, E.H. Butler, 1843 (KMK : CaBVaU) [NSB 011637]. See (k) for what may be the same series.
   (e) *Darby's edition of Brookes' universal gazetteer*; or, a new geographical dictionary ... To which are added the constitution of the United States, and the constitutions of the respective states. Illustrated by a ... map of the United States. The 3d American ed., with ample additions and

improvements by William Darby ... Philadelphia, 1823, Bennett & Walton (NB 0839184). For the first two American eds. see (b) above. There were further eds.: 1827, 2nd [sic] (NB 0839185); 1843, 3rd [sic] (NB 0839189); 1845, 4th (NB 0839186); and 1850 (NB 0839188).

(f) *A general gazetteer*; or, compendious geographical dictionary ... The whole revised, and corrected to the present period, by A.G. Findlay ... London, 1842, W. Tegg (NB 0839235). There were further eds.: 1849, 6th (NB 0839236); 1851, new (NB 0839237); and 1863, New (NB 0839238) [Blackwell's Rare Books, *Travel, Architecture, the Fine Arts*, Catalogue A71, 1984, item 63, asserts the last ed. to have been 1876].

(g) *The general gazetteer and geographical dictionary* ... New and enl. ed. edited by James A. Smith, L[ondon], 1816 (NB 0839197). There was a later ed. in 1876, Tegg (NB 0839198). Also in 1876, an ed. with a similar title was published in London by Ward, Lock (NB 0839199).

(h) *The general gazetteer ... dictionary, in miniature* ... Originally written by R. Brookes, M.D. The first edition. Montrose, printed by D. Buchanan, and sold by booksellers in London, Edinburgh, etc, 1804, 14 cm.; 356 pp. (NB 0839222); also 1806, 3rd, London (ICN, from NSB 0116532).

Subsequent versions of the 'miniature' ed. appeared in: 1816, now much improved, and brought down to the present time under the direction ... of the Rev. Jedidiah Morse, Boston, Melville Lord (NB 0839228); 1823, abridged ... by A. Picquot, London (NB 0839232); and 1854, [with] Supplement by ... A.G. Findlay, London (NB 0839205).

In addition, BL records: 'A general gazetteer in miniature ... The whole revised ... by A.G. Findlay, London, 1843.'

(i) *The London general gazetteer*; or, compendious geographical dictionary ..., London, Tegg, 1833 (NB 0839254).

There was another ed. in 1834 (: : **CaViPA**); the full title is similar to (06) but differs as follows: '... a description of the nations, empires ... states, provinces, cities, towns, forts, seas, harbours, rivers, lakes, canals, mountains, capes, etc. of the known world: with the extent, boundaries, and natural ... country; the government, customs, manners ... inhabitants; the trade ... towns, with their longitude and latitude, bearing and distance in English miles, from remarkable places; and the various historical events by which they have been distinguished. Illustrated with maps. Originally compiled ... The whole re-modelled, and the historical and statistical department brought down to the present period. By John Marshall).'

There were further eds: 1836 (NB 0839255) and 1839 (NB 0839256).

(j) *The modern gazetteer* ... 3rd ed. Dublin, 1771, W. and W. Smith (NB 0839258)

(k) *A new universal gazetteer*, containing ... The whole re-modelled and the historical and statistical department brought down to the present period, by John Marshall, esq. Illustrated with two hundred engravings. With numerous additions by the American editor, including the population of the United States for 1830; a description of the various Indian tribes in North America; and a view of the missionary stations in all parts of the world. And containing a brief dictionary of commerce ... New York, W.W. Reed & co., 1832 (NB 0839270).

There were further eds.: (Boston, except as noted): 1835, New York (NB 0839272); 1839, Philadelphia (NB 0839273); 1840 [1839] (NB 0839274); 1844, Philadelphia (NB 0839275); 1845 (NB 0839276); 1847 (NB 0839278); 1848 (NB 0839279); 1850 [c1839] (NB 0839280); 1853 [c1839] (NB 0839281); 1855 [c1839] (NB 0839283).

(l) *A new universal gazetteer or geographical dictionary.* No place, 1821? [Thought to be a copy of Darby's first revision] (NB 0839284)

(m) *The union gazetteer*; or, Brookes & Walker improved ... London, 1806, Oddy and co. (NB 0839285)

(n) *Universal gazetteer*, or, a new geographical dictionary: containing a description of the empires, kingdoms, states, provinces, cities, town, forts, seas, harbours, rivers, lakes, mountains capes, etc. in the known world; with the government, customs, and manners, of the inhabitants ... 2nd ed., with ample additions and improvements. Philadelphia, Bennett & Walton, 1827 (FTaSU, CarbS, PP [NSB 0116536])

4. Microform versions of the eds. of 1796, 1801, and 1809 of the original series are available from CIHM.

BROOKS, John H. **Y Am**

- **{1838} (01)**
- *A geography on a new plan with appropriate questions and answers* ... exhibiting the prevailing religions, forms of government, degrees of civilization, and the comparative size of mountains ...
- JOHNSTOWN, NY; printed by W. Clarke
- 13 cm; 86 p.
- ***Libraries*** : DLC

BROTHERS OF THE CHRISTIAN SCHOOLS **Y Can**

- **{1876}, 3rd (01)**
- *The intermediate illustrated geography*
- MONTREAL; P.L. Lesage
- ***Libraries*** : : CaBVaU

BROTHERS OF THE CHRISTIAN SCHOOLS **Y Can**

- **[1876] (01)**
- *The new intermediate illustrated geography* for the use of the Christian Schools in the Dominion of Canada
- MONTREAL; J. Chapleau & Son
- Pp. [5], 2-54.
- ***Libraries*** : : CaBVaU, CaOONL
- ***Notes***

1. Special geography occupies pp. 8-37. The information about places is provided in numbered paragraphs; it is followed by maps and related questions intended to test the pupil's knowledge.
2. A microform version is available from CIHM.

BROTHERS OF THE CHRISTIAN SCHOOLS **Y Can**

- **{1878}, 2nd (01)**
- *The new primary illustrated geography* for the use of the Christian Schools
- MONTREAL
- ***Libraries*** : : CaBVaU
- ***Notes***

1. Presumed to be special geography on the basis of the similarity of this title to the preceding one.

BRUCE, E[dward] and BRUCE, J[ohn] **mSG**

- **{1801}? (01)**
- *Introduction to geography and astronomy*
- ***Notes***

1. Known from the Preface to the second ed.; see (02) below.

- **{1805} (02)**
- *Introduction to geography and astronomy*
- ***Source*** The Preface to the second ed., reprinted in the third ed.; see (03) below

- **1810, 3rd, with ... (see below) (03)**
- *An introduction to geography and astronomy*, by the use of globes and maps. To which are added, the construction of maps, and a table of latitudes and longitudes. The third edition, with considerable additions and improvements. By ... teachers of geography and the mathematics, Newcastle upon Tyne
- LONDON; printed for C. Cradock and W. Joy, 32, Paternoster Row, by C. Stower, Paternoster Row
- Pp. xx, 448
- ***Libraries*** : CU
- ***Notes***

1. A letter from CU included a facsimile of the tp. and of both the Preface to the second ed. and the Advertisement to the third. The authors enlarged the second ed. by 'prefixing a compendium of geography' (p. x); it was devoted chiefly to 'natural geography.' 'The political situation of Europe is [so] frequently varying [that] little ... has been said on this head: only it was necessary to enumerate the principal states. The boundaries are not described, that being given as an exercise to the pupil.' The book grew out of the experience of teaching geography, 'and as the greatest part of it was drawn up at a time when there was no intention of publishing it, care has not always been taken to make proper acknowledgements. This defect can no otherwise be remedied than by stating that to Varenius's Geography [VAREN], Martin and Brandsby on the Globes, Dr Hutton's Mathematical Dictionary and Translations of Montucla's Mathematical Recreations, the Companion to Arrowsmith's Map of the World [ARROWSMITH, Aaron, Sr, 1794], Pinkerton's Geography [PINKERTON 1802], and the Scientific Dialogues, this work is much indebted' (pp. xivxv). The additions to the third ed., which 'occupy nearly a hundred pages' provide information about the British Isles and, to a lesser extent, about Europe; typically, it consists

of 'a catalogue of the chief towns ... the rivers on which they stand, their population, and remarks on the trade of each' (p. xviii). Trade, and in particular the goods traded, command attention.
2. BCB, Index (1837-57), BL, and NUC record further eds.: 1812, 4th (L); 1813, 4th (: DLC); 1845 (n.r.); 1846, 10th (L : OkU); 1850, 11th (L); 1859, 12th (L).

BRYCE, James **Dy**

- **1856 (01)**
- *A cyclopaedia of geography*, descriptive and physical forming a new general gazetteer of the world and dictionary of pronunciation, by ... master of the Mathematical Department in the High School of Glasgow. With numerous illustrations
- LONDON and Glasgow; Richard Griffin and Company, publishers to the University of Glasgow
- 8vo, 20 cm.; pp. vi, [2], 823
- ***Libraries*** **L** : CtY, CU, M

- **1859, anr ed. (02)**
- *The library gazetteer*; or, dictionary of descriptive and physical geography, compiled from the most recent authorities. With an introductory treatise on physical geography. By ... Glasgow. Illustrated by maps and numerous engravings
- LONDON and Glasgow; Richard Griffin ...
- 8vo, 23.5 cm.; pp. [v], vi-xlviii, 823
- ***Libraries*** **L** : MB, NN
- ***Notes***

1. Apart from the addition of the treatise on physical geography, spot-checking revealed no changes in the text. The change in title is thus misleading.

- **1862, 3rd, thoroughly revised (03)**
- *The family gazetteer*; and atlas of the world ... The atlas by W. and A.K. Johnston
- LONDON; Charles Griffin and co. Sold for them by William Wesley Queen's Head Passage, Paternoster Row
- 8vo, 21 cm.; pp. [5], vi-xii, 819, [3]
- *Libraries* L, **[IDT]**
- ***Notes***

1. The date is taken from the Preface. The atlas consists of maps bound in the body of the text. According to BL, this work was published in parts, continuing into 1863.

- **1880, anr ed., thoroughly rev. and greatly enl. (04)**
- *The library cyclopaedia of geography*, descriptive, physical, political and historical, forming a new gazetteer of the world. By ... and Keith Johnston
- LONDON and Glasgow; William Collins, sons & company, limited
- 8vo, 25cm; pp. [5], 6, [3], 18-929, [4]
- *Libraries* EU, L : MdBJ

- *Notes*

1. The Preface notes that Bryce died during the course of revising this work, having reached the letter M. Revision was finished by Johnston, who states that the number of places described has more than doubled. He also identifies some collaborators, and implies that there were others.
2. NUC records a similar title, London, n.d., Collins (OC), and also an ed. in 1897(?) (DLC).

BRYCE, R.J. **Y OG**

- **1826 (01)**
- *First principles of geography; simplified*, and arranged in a natural order, for the use of beginners
- BELFAST; printed by Simms and M'Intyre, Donegall Street
- 16.5 cm.; pp. [3], iv, [1], 6-16
- ***Libraries*** **[IDT]**
- ***Notes***

1. A pamphlet devoted largely to the definition of terms used in geography. Pp. 12-14 provide verbal descriptions of the location and form of the coastlines of the 'eastern' and 'western' continents.

BULLEN, H. St J. **Y SG?**

- **{1797} (01)**
- *Geography*
- Place/publisher n.r.
- ***Libraries*** n.r.
- ***Source*** Allibone (1854)

- **{1799} (02)**
- *Elements of geography*, expressly designed for the use of schools
- LONDON; printed for T. Hurst, etc.
- ***Libraries*** : MH
- ***Notes***

1. That this is (02) of the previous is a presumption.

BULLOCK, Robert **Y mSG**

- **1810 (01)**
- *Geography epitomized, or a companion to the atlas*; comprising a series of lessons, proper for the first course of geographical instruction in schools. With copious examinations corresponding to the lessons, so arranged as to form at the same time a series of amusing geographical games. Also an appendix, containing some easy instructions and problems relative to the practical use of maps
- WIGAN [England]; printed by D. Lyon, for the author; sold by Mawman, Poultry, and Crosby and Co., in Paternoster Row, London; Wilson and Son, York; and Mozley, Gainsborough
- 4to, 24 cm.; pp. vi, 82
- ***Libraries*** L

- *Notes*

1. The tp. provides an accurate guide to the book. The brief text is little more than a series of lists in the form of prose. Discussed in Vaughan (1972).

BURBURY, Frances E. **Y SG?**

- **{1867} (01)**
- *Mary's geography*: companion to Mary's grammar illustrated with stories, etc.
- LONDON; Longmans & Co.
- 16mo; pp. viii, 388
- ***Libraries*** L
- *Notes*

1. BL records a further ed.: 1873.

BÜSCHING, Anton Friedrich **A tr n.c.**

- **1762 (01)**
- *A new system of geography*: in which is given, a general account of the situation and limits, the manners, history, and constitution, of the several kingdoms and states in the known world; and a very particular description of their subdivisions and dependencies; their cities and towns, forts, sea-ports, produce, manufactures and commerce. Carefully translated from the last edition of the German original. To the author's introductory discourse are added three essays relative to the subject. Illustrated with thirty-six maps, accurately projected on a new plan
- LONDON; printed for A. Millar in the Strand
- 4to, 27 cm., 6 vols.; Vol. I, pp. [2], iii-xlviii, 668; Vol. II, pp. [8], 3-609; Vol. III; pp. [11], 4-816; Vol. IV, pp. 11-592; Vol. V, pp. [9], 4-680; Vol. VI, pp. [5], 2-622, [112]
- ***Libraries*** GU, **L**, SanU : DLC, MBAt, PPL, PPULC, RPB, ViU
- *Notes*

1. The subtitles of the volumes are as follows: 'Volume the first. Containing, Denmark, Norway, Greenland, Sweden, Russia and Poland. Volume the second. Containing, Hungary, Transylvania, Sclavonia, Dalmatia, Turkey in Europe, Portugal, Spain and France. Volume the third. Containing, Italy, Sardinia, Naples, Sicily, England, Wales, Scotland, Ireland, United Netherlands, and Swizzerland. Volume the fourth. Containing, part of Germany, viz. Bohemia, Moravia, Lusatia, Austria, Burgundy, Westphalia, and the circle of the Rhine. Volume the fifth. Containing, part of Germany, viz., circles of the Upper Rhine, Swabia, Franconia, and Upper-Saxony. Volume the sixth. Containing, in Germany, part of Brandenburg, Pomerania, the circles of Upper-Saxony, Swabia, Franconia, and the county of Glatz.'
2. As the subtitles to the volumes show, this is a translation of only that part of Büsching's *Erdbeschreibung* that deals with Europe (which is all that he had written by that time). No wonder that Pinkerton calls him 'prolix' (PINKERTON 1802, 2:781); this on the basis of 'the last French edition' (1:344). He does, however, look on him as the best single source of information about Europe. It seems likely that few places with any pretensions to

being a town rather than a village are not mentioned. Perhaps because Büsching was a clergyman, frequent references to churches are found.

NUC (NB 0934282) notes that the translation was made by P. Murdoch.

3. 'Translated from the last edition of the German original. Bindley says, "Though the minuteness of Büsching is generally tiresome and superfluous, yet we can pardon it for the accuracy of its details"' (Beeleigh Abbey Books, *A catalogue of voyages and travels*, Catalogue BA/23, W. & G. Foyle, London, n.d., item 14.
4. Discussed in Büttner and Jakel (1982), and Wilcock (1986).

BÜSCHING, Anton Friedrich
An introduction to the study of geography ... See WYNNE (1778).

BUTLER, Charles **Y**

- **{1868}?, 20th, revised ... (see below) (01)**
- *Charles Butler's young people's guide to geography* for the use of schools and private instruction. 20th edition, revised, and arranged from Dr Farr's guide to geography, by Robert Henry Mair
- LONDON; Dean & Son, 65, Ludgate Hill, E.C.
- 12mo; pp. iv, 278, [18]
- ***Libraries*** L : : **CaQQS**
- ***Notes***

1. It is possible that this book, in terms of its content and organization, represents the final stage of a process of development that began with the next entry.
2. CaQQS records the date as 1800?, but ECB, Index (1856-76:131) gives it as 1868. 'Dr Farr' may have been Edward Farr (see FARR 1850, 1870). The last 18 pp. carry advertisements for books published in the series: 'Corner's Historical Library.'
3. The text consists of numbered paragraphs organized within chapters, one chapter per country. At the end of each chapter, there is a list of questions to be put to students. More information is provided about the countries of Africa than in many other school geographies of the period.
4. A microform version of the 20th ed. is available from CIHM.

BUTLER, Charles **Y SG?**

- **{1846}? (01)**
- *An easy guide to geography, and the use of the globes*
- LONDON; T. Dean & Co.
- 12mo; pp. 200
- ***Libraries*** L
- ***Notes***

1. The date is confirmed by BCB, Index (1837-57:110).

BUTLER, Frederick **Y mSG? Am**

- **{1825}, 2nd (01)**

- ▸ *Elements of geography and history combined*, in a catechetical form, for the use of families and schools. By ... A.M. Accompanied with an atlas
- ▸ WETHERSFIELD, [CT]; Deming & Francis
- ▸ 18 cm.; vi, [7]-408 p.
- ▸ ***Libraries*** : CSmH, CtY, MH, NNC, OClWHi, PPULC
- ▸ ***Source*** NUC (NB 1008780 and 1008781)
- ▸ ***Notes***

1. NUC records further eds.: 1827, 3rd, New York, Robbins, 407 pp. (KyU, OClWHi, OO); 1828, 4th, Wethersfield, 420 pp. (DLC, OU, TU).
2. Elson (1964:398) cites (01); Hauptman (1978:431) quotes directly from (01).

BUTLER, George **At**

- ▸ **{1872}, New (01)**
- ▸ *Geography, modern*
- ▸ [LONDON]; Longmans
- ▸ ***Libraries*** n.r.
- ▸ ***Source*** ECB, Index (1856-76:131)
- ▸ ***Notes***

1. NUC also notes it, but provides a cross reference to Appleton's *Hand Atlas of Modern Geography in 31 maps ... Edited, with an introduction on the study of geography by the Rev. ...* According to the information provided in NA 0361090, this work is an atlas not a special geography.

BUTLER, John Olding **Y**

- ▸ **{1826} (01)**
- ▸ *The geography of the globe*, containing a description of its several divisions of land and water, to which are added problems on the terrestrial and celestial globes, etc.
- ▸ LONDON; printed for the author & proprietors
- ▸ 12mo; pp. xii, 356
- ▸ ***Libraries*** L

- ▸ **{1841}, 5th, with ... (see below) (02)**
- ▸ [*The geography of the globe* ...] 5th ed., with alterations and additions, by J. Rowbotham
- ▸ LONDON; Simpkin, Marshall & Co.
- ▸ 12mo; pp. vii, 349
- ▸ ***Libraries*** L
- ▸ ***Notes***

1. BL records further eds.: 1850, 8th, 360 pp.; 1852, 9th, 362 pp.; 1855, 10th, 362 pp. (L, [DTI]); 1858, 11th, 367 pp.; 1864, 12th, 373 pp.; and 13th, 1870, 374 pp.

BUTLER, John Olding **Y SG?**

- ▸ **{1854}, 17th (01)**

- *A new introduction to geography* ... The seventeenth edition. With an appendix containing problems on the globes, etc.
- LONDON; William Walker
- 8vo; pp. 185
- ***Libraries*** L
- ***Notes***

1. See NEW INTRODUCTION (1803)*.

- **{1860}, 18th (02)**
- [*New introduction to geography* ...]
- LONDON; William Walker
- 8vo; pp. iv, 185
- ***Libraries*** L
- ***Source*** BL gives the date as '1860 (1859),' and records 19th ed., 1866, 192 pp.

BUTLER, Samuel **Y**

- **{1836} (01)**
- *Abridgement of Dr. Butler's modern and ancient geography* ... Arranged in the form of question and answer ... By Mary Cunningham
- LONDON; Longman & Co.
- 12mo; pp. 124
- ***Libraries*** L
- ***Notes***

1. According to the tp. of BUTLER, S. (1855), below, the author was a Bishop of Lichfield, and not the well-known author.

BUTLER, Samuel **Y mSG?**

- **{1809} (01)**
- *Butler's geographical and map exercises*, [designed for the use of young ladies and gentlemen]. Corrected and improved by Stephen Addington
- PHILADELPHIA; M. Carey
- 18 cm.; 1 p. l., vi, [7]-51 p.
- ***Libraries*** : CtY, DLC, ViW
- ***Source*** NUC (NB 1011397)
- ***Notes***

1. NUC records a further ed.: 1813, 2nd, Philadelphia, 55 pp. (CtY, DLC, ViW).
2. See note 3 to BUTLER, William (1798).

BUTLER, Samuel **Y**

- **{1812}* (01)**
- *A sketch of modern and ancient geography*
- LONDON
- ***Libraries*** : NWM
- ***Source*** NUC (NB 1011409), which gives the date as 18–?

- **{1812} (02)**
- *A sketch of modern and antient geography* for use of schools
- SHREWSBURY; W. Eddowes
- 23 cm.; pp. xxix, 246
- ***Libraries*** : IU
- ***Notes***

1. NUC records further eds.: 1813, 2nd (NN); 1818, 4th, 260 pp. (NcU, PPULC, PU); 1825, 7th (MH); 1828, 8th, 345 pp. (CU, PPF, PPULC); 1830, 9th, 345 pp. (ViU); 1833, 10th, 364 pp. (DLC); 1838 (NWM); 1849, new, rev. by his son, 406 pp. (MiD); 1852, (PPL, PPULC); and 1885 (PP)
2. The author died in 1840 (Allibone 1854).

BUTLER, Samuel **Y**

- **1855, New ... (see below) (01)**
- *A sketch of modern geography, for the use of schools*. By ... D.D. Late Lord Bishop of Lichfield, and formerly Head Master of Shrewsbury School. New edition, revised by Rev. Thomas Butler
- LONDON; Longman, Brown, Green, and Longman
- 12mo, 19cm.; pp., xviii, 264
- ***Libraries*** L
- ***Notes***

1. By the standards of works of this class, the text has relatively large amounts of genuine prose and few lists in prose form. The changes referred to on the tp. were made, presumably, only to the modern geography, hence the change in the title.

BUTLER, William **Y OG**

- **{1798} (01)**
- *Geographical and biographical exercises*
- LONDON
- ***Libraries*** n.r.
- ***Source*** Allford (1964:241). See note to (02) below.
- ***Notes***

1. NSTC records for W.B.: *Exercises on the globes; interspersed ... historical information ... use ... ladies*, 1808, 3rd (L, O); 1808, 4th (E, L).

- **1811, 7th (02)**
- *Geographical and biographical exercises*, designed for the use of young ladies. The seventh edition
- LONDON; printed for the author, and J. Harris, corner of St. Paul's Church-Yard; J. Mawman, Ludgate Street; and Darton and Harvey, Grace-church-Street
- 8vo, 17 cm.; pp. x, 34, [2]
- ***Libraries*** L
- ***Notes***

1. The preface provides insight into the pedagogical methods of the time. The book contains lists of places that pupils are expected to be able to locate on maps.

2. For this author NUC records: '(Butler's) geographical and map exercises for the use of young ladies and gentlemen. Corr. and improved by Stephen Addington ... 2nd ed. Philadelphia, M. Carey, 1813, viii, 9-55 p., 18 cm' (NB 1011940; MnU). MnU report, however, that there is nothing in the book to indicate that it is by *William* Butler. Given the existence of entry (02) of BUTLER, Samuel (1809), above, it seems likely that NB 1011940 is a ghost.
3. St. John (1975, 2:801) records an ed.: 1816, London, 48 pp. (: : CaOTP). She notes that: 'This booklet was intended to be accompanied by maps. Butler was a teacher of writing, arithmetic, and geography in "ladies" schools, in conjunction with his son-in-law, Thomas Bourn.'

BUTLER, ?, ed. **Y SG?**

- **{1825} (01)**
- *Introduction to geography*
- ? Sael
- 12mo
- ***Libraries*** n.r.
- ***Source*** British Museum (1881-1900)
- ***Notes***

1. Might be the work identified in Peddie and Waddington (1914:226) as 'Introduction to G[eography], *Sael*, [18]25.'

C

C., T. **mSG**

- **1698 (01)**
- *The new atlas: or travels and voyages in Europe, Asia, Africa, and America*, thro' the most renowned parts of the world, viz., from England to the Dardanelles, thence to Constantinople, Aegypt, Palestine, or the Holy Land, Syria, Mesopotamia, Chaldea, Persia, East-India, China, Tartary, Muscovy, and by Poland; the German empire, Flanders and Holland, to Spain and the West Indies; with a brief account of Aethiopia, and the pilgrimages to Mecca and Medina in Arabia, containing what is rare worthy of remarks in those vast countries; relating to building, antiquities, religion, manners, customs, princes, courts, affairs military and civil, or whatever else of any kind is worthy of note. Performed by an English gentleman in nine years travel and voyages, more exact than ever
- LONDON; printed for J. Cleave in Chancery-Lane near Serjeant's Inn, and A. Roper at the Black Boy in Fleet-street
- 8vo, 18 cm.; pp. [8], 236
- ***Libraries*** L, LA, OMa : CSmH, DFo, DLC, NN, OCl, OClWHi, RPJCB
- ***Notes***

1. 'A little volume that seems to be made out of some collections of books and travels rather than any real voyage. - Churchill, *Introduction*' (Cox 1935: 77). The source quoted by Cox is presumed to be: Churchill, Awnsham,

and John Churchill, *A Collection of Voyages and Travels*, London, 1732, 6 vols. According to this source, the book was published in 1699, and has a title that differs from the one given above in minor ways. In the text the emphasis is on the places visited, rather than on the adventures of the author/traveller. The locations of the places visited are given, and for some countries, such as Poland, the information is equivalent to that found in special geographies.

CALKIN, J.B. **Y SG? Can**
- **{1875} (01)**
- *Nova Scotia school series: the world: an introductory geography*
- HALIFAX; A. & W. MacKinlay
- ***Libraries*** : : **CaBVaU**
- ***Notes***

1. CaBVaU also holds an ed. of 1886.

CALKIN, J.B. **Y SG? Can**
- **{1869} (01)**
- *School geography of the world*
- HALIFAX; A. & W. Mackinlay
- ***Libraries*** : : CaNSWA
- ***Source*** CIHM, which has a microform version
- ***Notes***

1. **CaBVaU** holds a copy of an ed. of 1881, Toronto.
2. CIHM reports a new ed., 1893.

CALKIN, J.B. **Y SG? Can**
- **{1878} (01)**
- *The world: an introductory geography*
- TORONTO; James Campbell and Son
- ***Libraries*** : : **CaBVaU**

CAMP, David Nelson **Y SG? Am**
- **{1857} (01)**
- *Camp's geography*, embracing the key to Mitchell's series of outline maps
- HARTFORD; O.D. Case & co.
- 22.5 cm.; 2 p. l., [vii]-viii, [9]-200 p.
- ***Libraries*** : DLC
- ***Notes***

1. NUC records further eds.: 1859, 208 pp. (OO); 1860 (MH); and 1861, 200 pp. (NcU).

CAMP, David Nelson **Y Am**
- **{1862} (01)**

- *Camp's higher geography*, prepared to accompany Mitchell's series of outline maps ...
- HARTFORD; O.D. Case & co. Chicago; G. Sherwood
- 23.5 cm.; pp. vi, [7]-20
- ***Libraries*** : DLC, OO
- ***Source*** NUC (NC 0068370). NUC records a further ed.: 1864 (CtY).

CAMP, David Nelson **Y SG? Am**

- **{1863} (01)**
- *Camp's primary geography*, prepared to accompany Mitchell's series of outline maps, and designed for primary schools and classes ...
- HARTFORD; O.D. Case & co.
- 8vo; pp. 64
- ***Libraries*** : DLC
- ***Notes***

1. NUC records further eds.: 1864 (OClWHi); and 1865 (OClWHi).
2. Discussed in Culler (1945, *passim*).

CAMPANELLA, Tommaso **mSG tr**

- **1654 (01)**
- *A discourse touching the Spanish monarchy*. Laying down directions and practises whereby the king of Spain may attain to an universal monarchy. Wherein also we have a political glasse, representing each particular country, province, kingdom, and empire of the world, with wayes of government by which they may be kept in obedience. As also the causes of the rise and fall of each kingdom and empire. Newly translated into English, according to the third edition of this book in Latine
- LONDON; printed for Philemon Stephens, and are to be sold at his shop at the Gilded Lion in Paul's Church-Yard
- 4to, 19 cm.; pp. [8], 232
- ***Libraries*** CEm, GlP, **L**, MaPL, NP, O, PetC, [IDT] : CLU, CSmH, CtY, DFo, DLC, FU, InU, MB, MH, MiU, NjP, NN, OCU, PPL, PU, TxDaM
- ***Notes***

1. Divided into 32 chs., of which 1 to 19 are concerned with the principles and methods of government, presented as being both universal and yet particularly appropriate for the King of Spain. Ch. 32 is concerned with the raising and maintenance of a navy that would allow that monarch to control the oceans of the world. Chs. 20 to 31 deal with the countries of Europe, Africa, and America, plus Persia and Turkey, but do so exclusively in terms of the political relations existing between them and Spain, together with the policies that would either lead to their being subordinated to Spain, or to their being rendered ineffective as its enemies. Unlike the usual special geographies, no attention is paid to the location of countries or to their boundaries.

   According to the translator, who is not identified in this issue (see below), the original Latin edition was published in either 1599 or 1600. The overtly political nature of the book is made apparent in the title of (02) below.

- **1659, anr ed. (02)**
- Thomas Campanella an Italian friar and second machiavel. His advice to the King of Spain for attaining *the universal monarchy of the world*. Particularly concerning England, Scotland and Ireland, how to raise division between King and Parliament, to alter the government from a kingdome to a commonwealth. Thereby embroiling England in civil war to divert the English from disturbing the Spaniard in bringing the Indian treasure into Spain. Also for reducing Holland by procuring war betwixt England, Holland, and other sea-faring countries, affirming as most certain, that if the King of Spain become master of England and the Low Countries, he will quickly be sole monarch of all Europe, and the greatest part of the New World. Translated into English by Ed. Chilmead, and published for awakening the English to prevent the approaching ruine of their nation. With an admonitorie Preface by William Prynne of Lincolnes-Inn esquire
- LONDON; printed for Philemon Stephens at the Gilded Lyon in St. Paul's Church-Yard
- 4to, 20 cm.; pp. [14], 232
- *Libraries* **L** : CtY, DLC, ICN, MBAt, MH, NcU, RPJCB
- *Notes*

1. Despite the great change in the title, the body of the book remains unaltered from 1654.

[CAMPBELL, Douglas H.] **Y SG?**

- **{1853} (01)**
- *The essentials of geography*: or, a companion to all geographies ... on a new plan. [By the Master of St. Chloe school, Woodchester]
- LONDON; A. Hall, Virtue & Co.
- 16mo; pp. 39
- *Libraries* L

CAMPBELL, William C. **Y SG? Can**

- **{1885} (01)**
- *The new illustrated geography and atlas* ... With thirty-six full page colored maps prepared by J. Bartholomew
- TORONTO; C.B. Robinson
- *Libraries* : : CaOTU
- *Source* CIHM, which has a microform version

CARPENTER, Nathanael **OG**

- **1625 (01)**
- *Geography delineated forth in two bookes*. Containing the sphæricall and topicall parts thereof. By ... fellow Exeter Colledge in Oxford
- OXFORD; printed by Iohn Lichfield and William Tvrner, printers to the famous Vniversity, for Henry Cripps
- 4to, 22 cm., 2 vols. in 1; pp. [18], 274, [4]; [16], 286, [4]
- *Libraries* AbU, BrU, C, **L**, LU, LUU, MaPL, O, [IDM] : CSmH, CU, DFo, DLC, MH, MiU, NjP, RPJCB, WU

- *Notes*

1. Deals with the form of the earth and its relationship to the universe, including a consideration of the Copernican hypothesis (Vol. I), and with those branches of systematic physical and human geography that were of contemporary interest.
2. Facsimile ed., 1976 (Johnson/TOT).

- **{1635}, 2nd, corr. (02)**
- *Geographie delineated ...*
- OXFORD; printed by J. Lichfield, for H. Cripps
- 19 cm., 2 vols. in 1
- ***Libraries*** AbrwthWN, C, L, O : CtY, DFo, IU, MB, MBAt, MH, NN, NNC, RPJCB, TxU
- ***Source*** NUC (NC 0154092)
- ***Notes***

1. Discussed in Baker (1928), Livingstone (1988), and E.G.R. Taylor (1934).

CARR, William **mSG**

- **1695 (01)**
- *The travellours guide*, and historians faithful companion: giving an account of the most remarkable things and matters relating to the religion, government, custom, manners, laws, policies, companies, trade, etc. in all the principal kingdoms, states, and provinces, not only in Europe, but other parts of the world; more particularly England, Holland, Flanders, Denmark, Sweeden, Hamburg, Lubeck, and other of the principal cities and towns of the German Empire, Italy, and its provinces, Rome, France etc and what is worthy of note to be found and observed in them: as to rivers, cities, pallaces, fortifications, strong towns, castles, churches, antiquities, and divers remarks upon many of them. Instructions how we ought to behave our selves in travelling: the prises of land and water passages, provisions: and how thereby to avoid many ilconveniences: with a catalogue of the chief cities, etc. And the number of houses every one of them are said to contain. With many other things worthy of note. Being the 16 years travels of William Carr, gentleman, sometimes Consul for the English at Amsterdam, in Holland
- LONDON; E. Tracy [NUC (NC 0157402); in the BL copy, the bottom of the tp. is missing]
- 12mo; 13 cm.; pp. [12], 210, [6]
- ***Libraries*** L, O : ICN
- ***Notes***

1. Relative to the contents of the book, this title makes the most exaggerated claims of any work listed in this guide. As BL notes, this is the reissue of: *Remarks of the government of severall parts of Germanie, Denkmark, Sweedland, Hamburg, Lubeck, and Hansiactique townes, but more particularly of the United Provinces, with some few directions how to travell in the States Dominions. Together with a list of the most considerable cittyes in Europe, with the number of houses in each citty*. Written by Will: Carr ... Amsterdam. Printed in Amsterdam Anno Dom. 1688. This title does give a

reasonable idea of the book's contents. The preface of 1695, which is as effusive in its claims for the book as is the title, is signed by E.T.
2. In addition to the first ed. of 1688, there were others in 1691, 1725 (5th), and 1744 (7th). Their titles are all variations of that of 1688, avoiding the inflated claims made in 1695.

CARTÉE, Cornelius Sowle **Y Am**
- **{1855} (01)**
- *Elements of physical and political geography.* Designed as a text book for schools and academies, and intended to convey just ideas of the form and structure of the earth, and the principal phenomena affecting its outer crust, the distribution of animals, and man upon its surface; together with its present political divisions. By ... A.M., Principal of Harvard School, Charlestown, Mass. Illustrated by wood engravings
- BOSTON; Hickling, Swan, and Brown
- 8vo, 20 cm.; pp. [3], 4-342
- *Libraries* L : DLC, MBAt, MiU, OCl, OCU, OO, PPULC, PU : **CaAEU**
- *Notes*

1. Contains more information about elements of the physical environment than is generally the case in special geographies, but it does fall within that tradition. Opens with some advice to the teachers who are expected to use the book; though Pestalozzi is not mentioned, his strategy of having the students begin by studying their immediate locality (in this case learning how to make a map by surveying and plotting the dimensions of their classroom) is recommended.

   The format of the book is unusual, with information being provided about countries in two separate sections; first there is 'special' geography (pp. 50-129), by which term Cartée means elements of the physical environment; after that (pp. 217-330) there is a section devoted to 'political' geography, which corresponds roughly to special geography in the terminology of this guide, including the provision of further information about features of the physical environment as observed in local places. Between the two sections, pp. 142-93 are devoted to explicating the physical processes that are at work in the environment.
2. NUC records further eds.: 1856 (MB, MH); 1859 (MH); 1860 (MH); and 1861 (DLC).

CARVER, Jonathan **A**
- **1779 (01)**
- *The new universal traveller.* Containing a full and distinct account of all the empires, kingdoms, and states, in the known world. Delineating, not only their situation, climate, soil, and produce, whether animal, vegetable, or mineral, but comprising also an interesting detail of the manners, customs, constitutions, religions, learning, arts, manufactures, commerce, and military force, of all the countries that have been visited by travellers or navigators, from the beginning of the world to the present time. Accompanied with a description of all the celebrated antiquities, and an accurate history of every nation, from the earliest periods. The whole being intended to convey

a clear idea of the present state of Europe, Asia, Africa, and America, in every particular that can either add to useful knowledge, or prove interesting to curiosity

- ▸ LONDON; printed for G. Robinson, in Paternoster-Row
- ▸ Folio, 37 cm.; pp. [3], iv, 668, [6]
- ▸ *Libraries* BrP, EU, **L** : CtY, IU, M, MiU, MnS, NN, PPULC, PSC : CaAEU
- ▸ *Notes*

1. Europe is the subject of approximately 62 per cent of the text; of that share approximately 36 per cent is devoted to the British Isles. Attention is focused on chorography and even topography, rather than on history or economic activity.

CATECHISM ...

- ▸ **1847, 9th**
- ▸ *Catechism of geography ...*

See MANNING (1847) and PINNOCK (1821, 1831a, 1831b).

CATHOLIC TEACHER, A **Y Am**

- ▸ **{[1875]}a (01)**
- ▸ *Sadlier's excelsior geography. No. 1 ...*
- ▸ NEW YORK; W.H. Sadlier
- ▸ 17.5 cm.; pp. 69
- ▸ *Libraries* : DHEW, DLC
- ▸ *Source* NUC (NS 0017274)
- ▸ *Notes*

1. The verso of the cover page of No. 3 in the series states that No. 1 is 'for primary classes.' Granted that this is the most significant difference between this book and the others in the series, then this work is a special geography of the most elementary kind. See CATHOLIC TEACHER, 1875c, below.

CATHOLIC TEACHER, A **Y Am**

- ▸ **{[1875]}b (01)**
- ▸ *Sadlier's excelsior geography. No. 2 ...*
- ▸ NEW YORK; W.H. Sadlier
- ▸ 23.5 cm.; pp. 104
- ▸ *Libraries* : DHEW, DLC, WaSpG
- ▸ *Source* NUC (NS 0017276)
- ▸ *Notes*

1. The verso of the cover page of No. 3 in the series states that No. 2 is 'for intermediate classes.'
2. NUC records an ed. [1877], Rev., New York, W.H. Sadlier, 104 pp. (DHEW, DLC).

CATHOLIC TEACHER, A **Y Am**

- ▸ **[1875]c (01)**

- *Sadlier's excelsior geography. Number Three.* Including local, physical, descriptive, historical, mathematical, comparative, topical, and ancient geography, with synoptical reviews, map-drawing, and relief maps
- NEW YORK; W.H. Sadlier, publisher, No. 11 Barclay Street
- 31 cm.; pp. 118
- ***Libraries*** : DHEW, DLC
- ***Source*** NUC (NS 0017279)
- ***Notes***

1. On the verso of the cover page, No. 3 in the series is described as being intended for 'advanced classes.' In the Preface it is stated that the series was prepared for use in Catholic schools to supply 'a want long felt ... viz., ... a perfectly truthful geography; one that would deal fairly and justly with Catholic countries and Catholic peoples ... The plan of the work is in accordance with the latest and best methods of teaching ... The maps, with a number of illustrations and portions of the text in Monteith's popular Series of Geographies, have been taken ... and used by the author.'

   Judging by the full title and size of *Monteith's comprehensive geography*, as recorded in entry (03) to MONTEITH (1872), it seems likely that it was the source from which this book, i.e., No. 3 in Sadlier's series, was derived.

   NUC records further eds. (all New York, W.H. Sadlier): [1878], New, 120 pp. (DLC); 1881, Ed. for the Pacific coast, 120, [2], 18 pp. (DLC) [NS 0017281 notes: 'Geography of the Pacific coast region. Prepared by H.B. Norton: 18 p. at the end.' N. was also responsible for the 'Geography of the Pacific coast region' that accompanied *Monteith's comprehensive geography*; see entry (02) to MONTEITH (1872)]; 1882, Rev., 124 pp. (DHEW, DLC).
2. Elson (1964:404) cites the ed. of 1878.

[CATLOW, Maria E.] **OG**

- **{1855} (01)**
- *Popular geography of plants*; or, a botanical excursion round the world ... Edited by Charles Daubeny
- LONDON; Lovell Reeve
- 8vo, 17 cm.; pp., xl, 370
- ***Libraries*** L : DLC, FMU, ICJ, IU, MBaT, MBH, MH, NN, OO, OU, PPULC, PU, RPB, TU
- ***Source*** BL and NUC (NC 0231226)
- ***Notes***

1. Only the very short title used by Allford (1964:246) could suggest the possibility of this being special geography.

CELLARIUS **Y**

- **1789 (01)**
- *General view of geography, ancient and modern*: or, an attempt to impress on the mind of a school-boy, a general idea of Cellarius's Ancient, and Guthrie's Modern Maps: to be repeated, at least in substance, by boys, who

may first draw the bare coasts, or out-lines of the maps referred to, with the rivers here mentioned, and afterwards point out the provinces, cities, etc. in the blank space, according to directions afforded in this book. For the use of Rugby School

- COVENTRY; printed by N. Rollason, in Hight-Street, and sold by him, and S. Clay in Rugby
- 8vo, 16? cm.; pp. 72
- ***Libraries*** L (pages cropped)
- ***Notes***

1. A school textbook. General and astronomical geography occupies the first 29 pp., special geography the remainder. The information is highly condensed.

CHAMBAUD, Louis **Y**

- **{1751(?)} (01)**
- *Geography methodized*, for the use of young gentlemen and ladies. Containing a true account of the world ... Illustrated with a dictionary, explaining and describing the things signified by the names of the productions of nature and art, mentioned in the description of the world: and making a compendium of natural history ...
- LONDON; Printed for A. Linde
- 17 cm.; pp. xi, 357
- ***Libraries*** : NNC, PPULC, PU
- ***Source*** NUC (NC 0281527)
- ***Notes***

1. Despite NUC, MWA no longer has a copy of this book.

CHAMBERLAYNE, Peregrine Clifford **Y**

- **1685, 2nd (01)**
- *Compendium geographicum*: or, a more exact, plain, and easie introduction into all geography, than yet extant, after the latest discoveries, or alterations; very useful, especially for young noblemen and gentlemen. The like not printed in English. By ... of the Inner Temple, gent. The second edition
- LONDON; printed for William Crook at the Green Dragon without Temple Bar
- 12mo, 14 cm.; pp. [16], 186, [1]
- ***Libraries*** L : CLU, CtY, NIC
- ***Notes***

1. Based on the *Introduction to Geography* (in French) of M. de Launay. Essentially a reference work. There is a short section on maps and globes, and then another short one on geographical terms, including place names in continents other than Europe. Only for Europe are many places identified (pp. 18-60), and the information in that section is often little more than a list in the form of text. The remainder of the book is taken up with indices of Latin and modern place names.

   According to an entry in Cox (1938:345), the first ed. was probably published in 1682. See CLIFFORD (1682).

CHAMBERS, Richard **Y mSG**

- **{1835} (01)**
- *The geographical and biographical compendium*, containing concise memoirs of illustrious persons; a gazetteer of remarkable places, and forming ... a key to the author's geographical questions and exercises
- LONDON; Sherwood, Gilbert & Piper; Harris
- 12mo; pp. 107
- ***Libraries*** L
- ***Notes***

1. BL and NUC note a further ed.: 1838 (L : CtY). BCB, Index (1837-57: 109) notes an ed. in 1848.

CHAMBERS, Richard **Y mSG?**

- **{1818} (01)**
- *Geographical questions and exercises*, blended with historical and biographical information
- LONDON; Sherwood, Neely & Jones
- 12mo; pp. vi, 72
- ***Libraries*** L
- ***Notes***

1. BL records further eds.: 1825, 2nd; 1834, 4th; and 1837, 5th. BCB, Index (1837-57:109) notes an ed. in 1843.

CHARLTON, James **Y SG?**

- **{1829} (01)**
- *An introduction to geography and the use of the globes*, for the use of schools
- NEWCASTLE; T. & J. Hodgson
- 8vo; pp. 230
- ***Libraries*** L : IU
- ***Source*** BL and NUC (NC 0316106)

CHASE, G[eorge] A[mbrose] **Y Can**

- **1887 (01)**
- *High school geography with maps and illustrations, part I., physical; part II., descriptive.* Authorized for use in the high schools and collegiate institutes by the Department of Education
- TORONTO; Canada Publishing Company (Limited)
- 23.5 cm.; pp. [4], v-vii, 190, 5 p. l.
- ***Libraries*** : : **CaAEU**, CaBVaU, CaOTP
- ***Notes***

1. Biased towards physical geography, which is implied by the full title, and towards Canada, which is not, though the bias is justified in the Preface. The authorities cited in the Preface as the source for the physical geography are Lyell, Dawson, Geikie, Dana, and Huxley. Only one author is identified by name as a source for the descriptive portion of the special geography,

A.R. Wallace. Much of Pt. II is devoted to a description of the physical geography of the countries being reviewed.
2. BL records eds. in 1903 and 1906.
3. A microform version of (01) is available from CIHM.

CHEVREAU, Urbain **NG tr**

- **1676 (01)**
- *The mirror of fortune*: or, the true characters of fate and destiny. Wherein is treated of the growth and fall of empires, the destruction of famous cities, the misfortunes of kings, and other great men, and the ill fate upon virtuous and handsome ladies. Whereunto is added moral, politick, and natural reflections upon several subjects. Written in French by Monsieur Chevreau, and newly translated into English by D. Decoisnon
- LONDON; in the Savoy: printed by T. N. and are to be sold by Sam. Lowndes ...
- 8vo, 17 cm.; pp. [14], 326, [2]
- ***Libraries*** **L**, O : CLU, ICN
- ***Notes***

1. The second section, 'Mischances caused by the elements' (pp. 102-32), can be classed as an early survey of natural hazards.

CHISHOLM, George Goudie **A**

- **1882 (01)**
- *The two hemispheres*: a popular account of the countries and peoples of the world. By ... translator of 'Switzerland: its scenery and its people.' Illustrated by above 300 engravings printed in the text
- LONDON; Blackie & son, 49 & 50 Old Bailey, E.C.; Glasgow, Edinburgh, and Dublin
- 22 cm.; pp. [vii], viii-xvi, 992
- ***Libraries*** : DLC, **MB**, NSyU
- ***Notes***

1. Coming very nearly at the end of the period covered by this guide, the preface is worth quoting almost in its entirety: '[This] work ... contains a complete account of the surface of the earth, its products, and inhabitants in greater or less detail. In the Introduction the general relations of land and water and the form and depth of the ocean-bed are considered; and the body of the work then opens with an account of the Polar regions both ... north and south. This is followed by a description of ... Europe as a whole, its superficial configuration, rivers and lakes, geology and minerals, climate, vegetation and zoology, people and political divisions; after which the different countries and regions composing the continent are described separately, and under each the same topics are treated in greater detail, while in addition notices are supplied relating to trade and commerce, means of communication, government and defence, the chief towns, and the historical vicissitudes of the territories which now compose the various political divisions of Europe. In the remainder of the work the other great land-masses of the earth, with the countries which they comprise, are each in turn treated of in the same manner ... Pictorial illustrations ... accompany the

descriptions ... to convey definite ideas ... of interesting localities, notable physical features, and characteristic groups of people representing the habits and social and out-door life of natives of various countries. The author, while endeavouring to make an interesting and readable volume suitable for ... the family circle, has been careful to arrange the information in such a systematic manner, as also to adapt the work for use as a book of reference, and to fit it for a place in the study, the library, and the counting house. A copious Index has been added' (p. [v]).

Occasional references to sources are given.

2. Discussed in Maclean (1988).

CHISHOLM, George Goudie **A**

- **{1884} (01)**
- *The world as it is*; a popular account of the countries and peoples of the earth. By ... M.A., B.Sc., editor and translator of 'Switzerland: its scenery and its people.' Illustrated by numerous engravings on wood and coloured maps and plates
- LONDON; Blackie & Son, 49 & 50 Old Bailey, E.C.; Glasgow, Edinburgh, and Dublin
- 23 cm., 2 vols.; (cx, 1052 p)
- *Libraries* : CU, DLC, ICU, NB
- *Source* NUC (NC 0384840)
- *Notes*

1. DLC reports: 'It appears from a brief examination of [this work and *The Two Hemispheres* (see above)] that they are essentially the same. Although the second work is issued in two columns, the paging is the same as in the first, with the exception of a longer introduction and an expanded index.'

The last sentence of the Preface reads: 'The main portion of the text of this work is included in the letterpress of the *Comprehensive Atlas and Geography*, but the physiographical introduction has been greatly extended and practically re-written for "The World as It Is"' (p. vi). For a work with that title, published in 1882, see BLACKIE (1882).

In 1886, presumably on the basis of these two works, Chisholm published *Longman's School Geography* (NC 0384818), which reached a 5th ed. by 1897 (NC 0384821). Then in 1889 he published *Handbook of Commercial Geography*, London, Longmans, 8vo, 23 cm.; pp. x, [2], 515. This was not a special geography. In a double sense it marks the end of that tradition. First, because it was not intended to be 'a mere repertory of the where and whence of commodities of all kinds'; rather, it was intended 'to impart an "intellectual interest" to the study of geographical facts relating to commerce' (p. iii). It begins by identifying the agents of production; then deals with the things produced, and ends with the places of production. As the latter are dealt with on a country-by-country basis, the roots in special geography are clear, but by limiting his discussion to the products of economic activity, Chisholm is able to provide, though as much by implication as by formal exposition, a theoretical frame to his work. That the explanation and what it explains are still of interest today is shown by the fact that the *Handbook*, now in its 20th ed. and revised by the second in a line of successors to Chisholm, is still in print.

The second reason for seeing this book as the end of the tradition of special geographies is the fact that Chisholm himself had written one. Even with a change of title, it did not go beyond a second ed. The contrast in sales is marked.

CHURCHILL, A. and J. **Gt**
- **{1709}**
- *The compleat geographer ...*
- Place/publisher n.r.
- ***Libraries*** n.r.
- ***Source*** Cox (1938:349)
- ***Notes***

1. This is almost certainly a ghost. No other source records it. See (03) of THESAURUS GEOGRAPHICUS.

CLARK, Charles R. **Y SG? Am**
- **{1866} (01)**
- *School geography*
- SAN FRANCISCO; H.H. Bancroft & Co.
- ***Libraries*** n.r.
- ***Notes***

1. Discussed in Culler (1945, *passim*).

CLARK, S. **A**
- **1689 (01)**
- *A new description of the world.* Or a compendious treatise of the empires, kingdoms, states, provinces, countries, islands, cities and towns of Europe, Asia, Africa and America: in their scituation, product, manufactures, and commodities, geographical and historical. With an account of the natures of the people, in their habits, customs, wars, religions and policies, etc. As also of the rarities, wonders and curiosities, of fishes, beasts, birds, mountains, plants, etc. With several remarkable revolutions, and delightful histories. Faithfully collected from the best authors
- LONDON; printed for Hen. Rhodes next door to the Swan tavern, near Brides-Lane, in Fleet-Street
- 12mo, 14 cm.; pp. [7], 232
- ***Libraries*** L, LC : IU, MB, MdBJ, MiU, MnU, RPJCB : CaBViPA
- ***Notes***

1. Despite the difference in the spelling of their names, BL catalogues this work under Clarke, (Samuel) Minister of St. Benet Fink; NUC follows suit. A quick comparison of the texts does not support the supposition that only one author is involved.

   The use of a small typeface and narrow margins allows a longer text than the small size of the book suggests. See CLARKE, Samuel (1657).
2. IU catalogues under the title.

- **{1696}, 2nd (02)**
- *A new description of the world* ... The second edition
- LONDON; printed for Hen. Rhodes at the Star the corner of Brides-Lane, Fleet-Street
- 12mo, 15 cm.; 3 p. l., 218, (2) p.
- ***Libraries*** : CLU, RPJCB
- ***Source*** NUC (NC 0464333)

- **{1708} (03)**
- *A new description of the world.* Or, a ... manufactures and commodities ... nature of the people ... birds, rivers, mountains ... histories
- LONDON; printed for Henry Rhodes ...
- 15 cm.; 3 p. l., 218 p.
- ***Libraries*** : CLU
- ***Notes***

1. NUC records further eds.: 1712, London, 220 pp. (OSg : CU, DLC : CaBViPA); 1719, London (RPJCB).
2. A microform version of (01) is available form CIHM.

CLARK, T. **?**

- **{1823} (01)**
- *Modern geography and history*
- LONDON
- ***Libraries*** n.r.
- ***Source*** Allford (1964) and Peddie and Waddington (1914:227)
- ***Notes***

1. The author's name is spelt Clarke, according to Vaughan (1972).
2. Discussed in Hart (1957), according to Vaughan (1972:130).

CLARKE, Charles Baron **Y**

- **{1878} (01)**
- *A class-book of geography*
- LONDON; Macmillan & Co.
- 8vo, 17 cm.; pp. vii, 280
- ***Libraries*** L : ICU
- ***Source*** BL and NUC (NC 0460088)

- **1883 (02)**
- *A class-book of geography.* With eighteen coloured maps
- LONDON; Macmillan and Co
- 16.5 cm.; pp. [5], vi-vii, [1], 280
- ***Libraries*** **GP**
- ***Notes***

1. No preface or introduction. Opens with definitions of geographical terms and then the main divisions of the globe; pp. 16-280 are devoted to special geography. Predominantly factual in tone; cautious about such geographical 'theories' as those that explain the commercial enterprise and achievements of Europeans by the length of the coastline and the great interpenetration of

land and sea, noting the counterfactual example of southeast Asia. Avoids judgements on the qualities and worth of peoples.

2. NUC records a further ed.: 1886, London, 280 pp. (NN); BL records an ed. in 1889.

CLARKE, J.W. **Dy n.c.**

- **1814 (01)**
- *A new geographical dictionary*; containing a description of all the empires, kingdoms, states, and provinces, with their cities, towns, mountains, capes, seas, ports, harbours, rivers, lakes, etc. in the known world: with a correct account of the government, customs, manners, and religion, of the inhabitants - the extent, boundaries, and natural productions, of each country - their trade, manufactures, and commerce - the longitude and latitude of all cities and towns; with their bearings and distances from remarkable places - and the various events by which they have been distinguished. Together with the population of the counties, cities and towns in England, Scotland, and Wales, according to the latest census taken by order of Parliament in 1811. Illustrated with plates and maps. The whole compiled from the best authorities
- LONDON; printed for Nuttall, Fisher, and Co. Caxton Buildings, Liverpool
- 27 cm., 2 vols.; Vol. I, pp. [4], iv-xxviii, 975; Vol. II, pp. 1036
- *Libraries* L, **GP** : KBB, NN
- *Notes*

1. The editor compares his work with that of 'Dr Brookes, and the Rev. C. Crutwell,' and lists his major sources. His title is strongly reminiscent of BROOKES (1762). Vol. II ends at Sorento in mid-sentence. However, BL and both NUC (NC 0461915 and 0461916) agree that only two vols. were ever published. They also concur in the date of publication. See CRUTT-WELL (1798).

- **{1825}, 2nd (02)**
- [*A new geographical dictionary*]
- 2? vols.
- *Libraries* L
- *Notes*

1. BL gives the date of publication as, '[1814, 25],' and adds: 'Vol. 2 is of the second edition.'

CLARKE, Samuel **A**

- **1657 (01)**
- *A geographicall description of all the countries in the known world.* As also of the greatest and famousest cities and fabricks which have been, or are now remaining: together with the greatest rivers, the strangest fountains, the various minerals, stones, trees, herbs, plants, fruits, gums, etc. which are to bee found in every country. Unto which is added, a description of the rarest beasts, fowls, birds, fishes, and serpents which are least known amongst us. Collected out of the most approved authors, and from such as

were eye-witnesses of most of the things contained herein. By Sa: Clarke, Pastor of the Church of Christ in Bennet Finck, London

- LONDON; printed by R. I. for Thomas Newberry, at the Three Lions in Cornhill, over against the Conduit
- 4to, 29 cm.; pp. [4], 225, [8]
- ***Libraries*** LC, **O** : CLU, DLC, MH, MiU, NBuG, NN, RPJCB
- ***Notes***

1. Countries are described in the first 195 pages, with the cities of England receiving a disproportionately large share of attention. The last 30 pp. are devoted to natural curiosities, wonderful plants and animals, and diverse human works.
2. This work was also issued bound with Clark's *A Mirrour or Looking-Glasse both for Saints and Sinners*, 3rd ed., T.R. and E.M. for Thos. Newberry, London, 1657 (information from BL).

- **1671, anr ed. (02)**
- *A geographical description* ... are now remaining: whereunto are now added, an alphabetical-description of all the counties in England, and Wales; and of the four chiefest English plantations in America. Together with the rarest beasts ... contained herein
- LONDON; printed by Tho. Milbourn for Robert Clavel, Tho. Passinger, William Cadman, William Whitwood, Tho. Sawbridge, and William Birch
- Folio, 30 cm.; pp. [4], 293, 85, 35
- ***Libraries*** **L**, LRUS, O : CU, DLC, ICN, MBAt, MH, MnU, N, NcU, NNUT, OClW, Vi, ViU
- ***Notes***

1. This issue is included as the second, third, and fourth parts of Vol. I of the 4th ed. of Clark's *Mirrour*. There are 90 additional pp. on the counties and towns of England and Wales, and 18 additional pp. on America (excluding the additions on the English plantations, which are paginated separately); otherwise there are very few changes. The section on natural history, entitled 'Examples of the wonderful works of God in the creatures,' is also paginated separately in this issue, where it had been included in the regular pagination in 1657. See CLARK, S. (1689).
2. Discussed in Bowen (1981:95-6).

CLAVEY, William **Y**

- **1806 (01)**
- *An introduction to modern geography*: familiarized by a set of skeleton maps and references
- BATH; printed and sold for the author, by M. Gye, Market-Place: sold also by Champante and Witrow, Jewry-Street, Aldgate, and Johnson, St. Paul's Church-Yard, London; and W. Browne, Tolzey, Bristol
- 4to, 21 cm.; pp. [5], 6-63
- ***Libraries*** **L**
- ***Notes***

1. Provides outline maps of countries and continents, together with lists of named places located in the regions mapped. The places are chiefly features

of the physical environment. The book ends with lists of questions to be put to students concerning the location of places.

- **{1812} (02)**
- *Modern geography*
- Place/publisher n.r.
- ***Libraries*** n.r.
- ***Source*** Peddie and Waddington (1914:227)

- **1847, 3rd ed., revised to the present time (03)**
- *An introduction to modern geography*, familiarized ... references, amply extended by copious descriptions, and confirmed by numerous examination questions
- BATH; printed for the author, by John and James Keene
- 4to, 21 cm.; pp. [1], [5], 6-104
- ***Libraries*** L
- ***Notes***

1. Begins with a brief statement of Clavey's method of teaching geography, which is essentially that of getting students to locate places on outline maps. Some pages of text containing conventional special geography have been added to the maps and lists of place names.
2. See CLEGG (1795).

CLEGG, John **Y mSG**

- **1795 (01)**
- *Elements of geography*, or an easy introduction to the use of the globes and maps; consisting of a treatise on the astronomical part of geography, a blank atlas, and a large collection of geographical questions for the exercise of the pupil on each of the maps, with the answers printed separately
- LIVERPOOL; printed for J. M'Creery
- 4to, 28 cm.; pp. [1], [1]-21, [1], 23 l., [1]-8
- ***Libraries*** L
- ***Notes***

1. In its own way it is a special geography, because the exercise of filling in the blank maps would probably get the students to remember places as effectively as the rote memorization favoured by other authors.
2. See CLAVEY (1806).
3. NUC records a further ed.: 1796 (DLC).

CLEOBURY, Miss **Y SG?**

- **{1815} (01)**
- *Practical geography*; in a series of exercises ... With numerous outline maps and a copious appendix of the chief places, etc.
- NOTTINGHAM; J. Dunn
- Folio; pp. 51
- ***Libraries*** L

CLIFFORD, Peregrine **Gt**

- **{1682}**
- *Compendium geographicum*: or, a more exact, plain, and easie introduction into all geography than yet extant, after the latest discoveries or alterations. Very useful, especially for young noblemen and gentlemen, the like not printed in England
- ***Source*** Cox (1938:345). Not in BL, NUC, or Wing (1972). Almost certainly a ghost. See CHAMBERLAYNE (1685).

- **{1684}, 2nd, with additions (02)**
- ***Source*** Cox (1938:345)

CLUTE, John J. **Y SG? Am**

- **{1833} (01)**
- *The school geography*
- NEW YORK; Samuel Wood & sons
- 18.5 cm.; 1 p. l, v-viii, [5]-363 p.
- ***Libraries*** : AU, DLC, MB
- ***Source*** NUC (NC 0497148) and Sahli (1941:381)
- ***Notes***

1. Discussed in Sahli (1941, *passim*); cited in Elson (1964:400).

CLUVER (Cluverius), Philip **A tr**

- **1657 (01)**
- *An introduction into geography, both ancient and moderne*, comprised in six bookes by ... together with severall incidentall remarques, newly added
- OXFORD; printed by Leonard Lichfield, Printer to the University for Rob. Blagrave
- 15 cm.; pp. [4], 341 [many irregularities; actually 354]
- ***Libraries*** O : DFo, **MH**, RPJCB
- ***Notes***

1. The dedication is by the translator, and is signed 'H.S.'

   Comparable in size and scope to HEYLYN (1621) and the later eds. of ABBOT (1599), though the 'first booke' contains more information on cartography and the means of establishing location, on the distribution of the seas, and on the division of the land into continents than they do.
2. Discussed in Bowen (1981:71-2).

CLYDE, James **Y SG?**

- **{1860} (01)**
- *Elementary geography*
- EDINBURGH; Constable's educational series
- 8vo; pp. vii, 156
- ***Libraries*** L
- ***Notes***

1. BL questions the date; it is confirmed by ECB, Index (1856-76:130).
2. BL records further eds.: 1861, 3rd; 1864, 9th; 1866, 10th; 1868, 11th;

1870, 12th; 1871, 13th; 1872, 14th; 1874, 15th; 1875, 16th; 1876, 17th; 1877, 18th; 1878, 19th; 1881, 20th; 1882, 21st; 1883, 22nd; 1885, 23rd; and also 1889 and 1892.
3. [IDT] has a copy of one of the eds. of this work.

CLYDE, James **Y SG?**
- **{1859} (01)**
- *School geography*
- EDINBURGH; T. Constable & Co.
- 8vo; pp. xii, 457
- ***Libraries*** L
- ***Notes***
1. BL questions the date; it is confirmed by ECB, Index (1856-76:131).
2. BL records further eds. as follows: 1864, 8th; 1866, 9th; 1866, 10th; 1867, 11th; 1870, 12th; 1871, 13th; 1872, 14th; 1874, 16th; 1876, 17th; 1877, 18th; 1878, 19th; 1880, 20th; 1881, 21st; 1883, 22nd; 1886, 23rd; and also 1890, 1895, and 1898.

COBBIN, Ingram **Y SG?**
- **{1828} (01)**
- *Elements of geography*
- Place/publisher n.r.
- ***Libraries*** n.r.
- ***Source*** Peddie and Waddington (1914:226)

- **{1836}, 5th (02)**
- *Elements of geography, on a new plan* ... adapted to the capacities of young children ... Illustrated with maps and cuts ...
- LONDON; Frederick Westley & A.H. Davis
- 12mo; pp. viii, 118
- ***Libraries*** L
- ***Notes***
1. BCB, Index (1837-57:110) records ed. of 1852.

COLBY, Charles Galusha **A Am n.c.**
- **{1857a [c1856]} (01)**
- *The diamond atlas*. With descriptions of all countries: exhibiting their actual and comparative extent, and their present political divisions, founded on the most recent discoveries and rectifications
- NEW YORK; S.N. Gaston
- 20 cm., 2 vols.
- ***Libraries*** : FMU
- ***Notes***
1. NUC (NM 0798812), which is the source, notes: 'Contents. - [V. 1] *The Western Hemisphere, by Charles Colby*. - [V. 2] The Eastern Hemisphere, by Charles W. Morse. Each vol. has added t.p.: Morse and Gaston's diamond atlas' (emphasis added).

2. NUC (NC 0525618) also records an ed. in 1857, but the Western hemisphere only; copies at (CtY, DLC, ICarbS, IU, MH, MiU, N, NN, OClWHi, OrU, TxDaM).
3. For the volume describing the Eastern hemisphere, see MORSE, Charles W. (1840).

- **{1858} (02)**
- *The diamond atlas*. With descriptions of all countries: ... The western hemisphere
- NEW YORK; S.N. Gaston
- 8vo; 239 pp.
- ***Libraries*** : DI-GS, MB, MH
- ***Source*** NUC (NC 0525619)

- **{1859} (03)**
- *Diamond atlas*, with ... countries, exhibiting ... divisions ...
- NEW YORK; S.N. Gaston
- 239 p.
- ***Libraries*** : ICRL, PP, PPULC
- ***Source*** NUC (NC 0525620)
- ***Notes***

1. For a smaller version, see note to (02) of COLBY (1857b), below.

COLBY, Charles Galusha **A Am**

- **{1856} (01)**
- [*Morse's*] *general atlas of the world*, containing seventy maps [drawn and engraved from the latest and best authorities]. With descriptions and statistics of all nations to the year 1856
- NEW YORK; D. Appleton
- [Folio] 44 cm; 72 p. [3 p. l, [33] l.]
- ***Libraries*** : DLC, MB, ViU
- ***Source*** NUC (NM 0798816, and 0798817)
- ***Notes***

1. NUC follows the library cataloguers who treat the man presumably responsible for the maps, rather than the author of the text, as the primary author in cataloguing this title under Morse, Charles W.
2. The words in the title set in [ ] are taken from NM 0798817. The two entries in NUC are treated here as referring to a single title on the grounds that the only differences between them seem to be those present in the form of the catalogue entries.

COLBY, Charles Galusha **A Am n.c.?**

- **{1857b} (01)**
- *The world in miniature*, with descriptions of every nation and country; together with a treatise on physical geography. The Western hemisphere
- NEW ORLEANS; A.B. Griswold
- 20 cm.; 239 p.
- ***Source*** NUC (NC 0525628)

- *Libraries* : AU, DLC, F, MnHi, NcA, NcD, NcRSh, NcU, NPV, ScNC, TNF

- **{1861} (02)**
- *The world in miniature* ... To which is prefixed an important treatise on physical geography ... Universal
- NEW ORLEANS; Bloomfield & Steel
- 12mo; 1 p. l., [142] pp.
- *Libraries* : DLC, MBAt
- *Source* NUC (NC 0525629), which notes: 'Cover-title: Morse and Gaston's diamond atlas.'

COLLEGE ... GEOGRAPHY ... **Y mSG**
- **1850? (01)**
- *The college elementary geography*. Compiled and the maps produced for the proprietors of the 'The College Atlas.'
- LONDON; published (for the proprietors), by H.G. Collins, 22 Paternoster Row
- 23.5 cm.; pp. [4], 5-32 [incl. maps]
- *Libraries* **[IDT]**
- *Notes*

1. The cover acts as the tp. Only 17 pp. of text: 2 for each continent; 2 for England; 1 for seasons, climate, and tides; 2 for preface and introduction; and 2 for the two hemispheres. A manuscript note on the cover gives 1850 as the date of publication.

COLLIER, J.
See FENNING, COLLIER, and others (1764-5), in particular the note to entry (04).

COLTON, George Woolworth, and COLTON, Joseph Hutchinson
- *Notes*

1. The relationship between the books that are attributed to these two authors is not clear. The information that follows under each of them is derived largely from NUC.

COLTON, George Woolworth
- *American school quarto geography* ... (see COLTON, Joseph H. 1865)

COLTON, George Woolworth
- *Atlas of the world* ... (see FISHER, Richard S. 1854)

COLTON, George Woolworth
- *Colton's atlas of the world* ... (see (03) of FISHER, Richard S. 1854)

COLTON, George Woolworth
▸ **{1877} (01)**
▸ *Colton's common school geography* ... (see (02) of COLTON, Joseph H. 1868)

COLTON, George Woolworth
▸ *Colton's general atlas* ... (see FISHER, Richard S. 1857)

COLTON, George Woolworth
▸ *Colton and Fitch's introductory school geography* ... (see FITCH, George W. 1856)

COLTON, George Woolworth
▸ *Colton and Fitch's modern school geography* ... (see FITCH, George W. 1855)

COLTON, George Woolworth **Y Am**
▸ **{1860} (01)**
▸ [*Colton & Fitch's primer of geography*]
▸ [NEW YORK? Ivison?]
▸ ***Libraries*** n.r.
▸ ***Notes***
1. Known from the copyright statement of (03) below.

▸ **{1864} (02)**
▸ *Colton & Fitch's primer of geography*. Illustrated by 16 maps, and numerous engravings
▸ NEW YORK; Ivison, Phinney, Blakeman. Chicago; S.C. Griggs & co
▸ 19.5 cm.; pp. 80
▸ ***Libraries*** : DLC

▸ **1866 (03)**
▸ *Colton & Fitch's primer of geography* ...
▸ NEW YORK; Ivison, Phinney, Blakeman
▸ 18.5 cm.; pp. [5], 6-80
▸ ***Libraries*** : **NN**
▸ ***Notes***
1. Takes the form of question and answer; respondents are expected to be able to identify rivers, seas, gulfs, mountain ranges, etc., and also capital cities; much concerned with relative location, especially outside the United States. Resembles CORNELL (1858a).
2. Culler (1945) may deal with (02).

COLTON, George Woolworth **Y Am**
▸ **{1871} (01)**

- *New introductory geography*
- St. PAUL, MN; Sheldon
- 24 cm.; pp. 85
- ***Libraries*** : N
- ***Notes***

1. NUC (NC 0569647) notes that the copy is imperfect, with the tp. missing and the date taken from the Preface.
2. On the basis of a mutilated copy at **CaAEU**, the book is written at an elementary level. The text is descriptive, rather than in the form of question and answer, though questions are provided for use in conjunction with maps. There is a strong bias in favour of information about the U.S., and some signs of racism.
2. NUC records further eds.: 1873, 85 pp. (OO); 1875, 85 pp. (DLC, ICU); 1879, 85 pp. (MiU); 1880, (MH, OClWHi); and 1881 (OO).
3. Culler (1945) may deal with this book.
4. See COLTON, Joseph H. (1872).

COLTON, George Woolworth and R.S. Fisher

- *Atlas of the world* ... (see FISHER, Richard S. 1854)

COLTON, George Woolworth and R.S. Fisher

- *Colton's general atlas* ... (see FISHER, Richard S. 1857)

COLTON, George Woolworth, and R.S. FISHER

- *Colton's illustrated cabinet atlas and descriptive geography* ... (see FISHER, Richard S. 1859)

COLTON, H. **mSG**

- **1698 (01)**
- *The artist's vade mecum*: or the most useful arts and sciences improv'd and made easie. Containing 1. The curious art of dialing, in drawing and placing all sorts of sun-dials by a truer or more exact rule than hitherto found out. 2. Geometry applied to the most profitable arts of surveying, measuring timber, or any solid bodies; gauging casks, brewers tuns, wine-vessels, etc. 3. Finding the length and circumference answering to any arch, in degrees and decimal parts. 4. The area or segments of a circle, whose whole area is unity, to the ten thousandth part of the diameter; with many other useful tables, ready stated. 5. A compleat body of astronomy, or a view of the cælestial globe; places of the sun, moon, and fixed stars, the names of the most noted stars, in what signs they are posited; their longitude and latitude, etc. The doctrine of the *primum mobile*, and the account of time rectified and freed from error; compared with the Julian and Gregorian calenders. To which is added, a compleat body of geography; describing all the empires, kingdoms and states in the known parts of Europe, Asia, Africa and America. The like never before made publick; illustrated with 14 copper-plates

- LONDON; printed for Eben Tracy, at the Three Bibles on London-bridge
- 8vo, 18 cm.; pp. [16], 510 [i.e., 512 according to NUC], [14]
- ***Libraries*** : **MH**
- ***Notes***

1. Opens with problems related to cartography flowing from the sphericity of the earth, and with technical terms used in geography. Pp. 426-53 are devoted to a condensed special geography, with attention focused on the subdivisions of the land masses and their boundaries. These pages contain chs. 3, 5, and 4 (in that order) of the section devoted to geography. There follows ch. 15[!] (pp. 453-8), which is devoted to maps, more particularly to problems of projections and of scale. In short, it lies on the margin of special geography.

COLTON, Joseph Hutchins **Y Am**

- **{1865} (01)**
- *American school quarto geography*: comprising separate treatises on astronomical, physical, and civil geography. With an atlas of more than 100 steel plate maps, profiles and plans, on 42 large sheets, drawn on a new and uniform system of scales, by G. Woolworth Colton
- NEW YORK; Ivison, Phinney, Blakeman
- 36 cm.; pp. 118
- ***Libraries*** : DLC, MB
- ***Source*** NUC (NC 0569867)
- ***Notes***

1. NUC Vol. 116, p. 588, provides an entry that has a very similar title, varying only in a central phrase, as follows: 'American ... geography, comprising the several departments of mathematical, physical ... scales.' It is listed among the entries for COLTON, George Woolworth, but is cross referenced to COLTON, Joseph Hutchins, and presumed to be the work to which this is a note.
2. This may be a condensed version of MORGON (1863).

COLTON, Joseph Hutchins **Y mSG? Am**

- **{1877-9} (01)**
- *Colton's common school geography*. Supplements
- NEW YORK
- ***Libraries*** : DLC

COLTON, Joseph Hutchins **Y Am**

- **{1872} (01)**
- *Colton's new introductory geography ...*
- NEW YORK; Sheldon & Co.
- Pp. 85
- ***Libraries*** : DHEW, DLC, MH
- ***Source*** NUC (NC 0570003), which notes: 'Colton's new series.'
- ***Notes***

1. Comparing this entry with COLTON, G.W (1871), i.e., *New Introductory*

Geography, above, suggests that this pair of entries may refer to the same book. However, the fact that Culler (1945) distinguishes between *Colton's New Introductory Geography*, New York, Sheldon & Co., 3rd ed., 1881, and Colton, J., *New Introductory Geography*, New York, Sheldon & Co, 1881, supports the view that these two works are distinct from each other.

In his general discussion of Colton's schoolbooks, Carpenter (1963:266), does not distinguish between the two *Introductory Geographies*; this seems a fair indication of the care he gives to detail.

COLTON, Joseph Hutchins **Y Am**

- **{1868} (01)**
- *J.H. Colton's common school geography.* Illustrated by numerous engravings and twenty-two maps, by G. Woolworth Colton
- NEW YORK; Ivison, Phinney, Blakeman. Chicago; S.C. Griggs & Co.
- 31 cm.; pp. 104
- ***Libraries*** : DLC
- ***Notes***

1. NUC record further eds.: 1872, 130 pp. (DLC); 1874 (NN); 1876, 134, 7 pp. (FTaSU, MiEM) [the presence of two series of numbered pages suggests that this ed. includes the description of a particular state].

- **{1877} c1872 (02)**
- *Colton's common school geography.* Illustrated ... twenty-two study maps, drawn expressly for this work, and specially adapted to the wants of the class-room. To which are added two full-paged railroad maps, showing the chief routes of travel, and a complete series of twelve commercial and reference maps of the United States
- NEW YORK; Sheldon & Co., 8 Murray Street
- 31.5 cm.; pp. 134
- ***Libraries*** : DLC
- ***Source*** NUC (NC 0569587) and letter from DLC. OClJC report (letter) that they no longer hold a copy.
- ***Notes***

1. NUC records further eds.: 1878 (MnU, OClWHi, OO) [Culler discusses (1945, *passim*), identifying it as the Michigan ed. (1945:307)]; 1880, (MH, ViU); 1880, Ohio ed., 134 pp. (OO); 1880, Pennsylvania ed. [library n.r.; in Culler (1945:307), not NUC]; [NSC 0092646 lists what appears to be an additional copy of this ed. (IU, OkU), but attributes it to G.W. Colton]; 1881, 140 pp. (DHEW, DLC, OClWHi); and 1883, 124 pp. (MoU) [NC 0569882 seems to look on this as an issue of the 1880 ed.; it adds a note: 'For 1877 edition see Colton, G.W.'].
2. Elson (1964:404) cites the ed. of 1878.

COLTON, Joseph Hutchins **Y mSG Am**

- **{1860} (01)**
- *J.H. Colton's historical atlas.* A practical class-book of the history of the world. Comprising, in a series of inductive lessons, the origin and progress

of nations, their history, chronology, and ethnology, combined with their ancient and modern geography, with ... maps ... and plates, etc.
- NEW YORK; Ivison, Phinney, Blakeman
- Folio; pp. 52
- ***Libraries*** L : DLC, ICRL, ODW
- ***Source*** NUC (NC 0569901)

COMPENDIOUS GEOGRAPHICAL ... GRAMMAR ... **A**
- **1795 (01)**
- *A compendious geographical and historical grammar*: exhibiting a brief survey of the terraqueous globe; and shewing, the situation, extent, boundaries, and divisions of the various countries; their chief towns, mountains, rivers, climates, and productions; their governments, revenues, commerce, and their sea and land forces; likewise, the religion, language, literature, customs, and manners of the respective inhabitants of the different nations: and also, a concise view of the political history of the several empires, kingdoms, and states. Embellished with maps
- LONDON; printed for W. Peacock, No. 18, Salisbury Square
- 12mo, 13 cm.; pp. xxiii, 404, [7]
- ***Libraries*** **OSg** : CoU, DLC, InU, NcD, NjP, OC : CaBViPA
- ***Notes***

1. A small-scale counterpart to the contemporary and very popular *Grammar* of GUTHRIE (1770). More than half the text is devoted to Europe. Two years earlier, the same publisher produced *A compendious geographical dictionary* (see below).

- **{1802}, 2nd, corrected and considerably improved (02)**
- *A compendious geographical grammar* ... with maps
- LONDON
- 12mo, 13 cm.; pp. xxviii, 403, [7]
- ***Libraries*** L, **OSg** : CtY, DLC, OClWHi, PU
- ***Notes***

1. Catalogued by BL under *Grammar*.

COMPENDIOUS GEOGRAPHICAL DICTIONARY ... **Dy**
- **{1793} (01)**
- *A compendious geographical dictionary*, containing a concise description of the most remarkable places, ancient and modern, in Europe, Asia, Africa, and America, interspersed with historical anecdotes. To which is added a table of the coins of the various nations, and their values in English money. To the whole is prefixed, an introd., exhibiting a view of the Newtonian system of the planets ...
- LONDON; printed for W. Peacock
- ***Libraries*** : CtY, DLC, NIC
- ***Source*** NUC (NC 0599785)

- **{1795}, 2nd (02)**
- *A compendious geographical dictionary* ... To which are added, a chrono-

logical table from the creation to the present time; a monthly list of all the fixed fairs in England and Wales; and a table of coins ... Embellished with maps

- LONDON; W. Peacock
- 14 cm.; pp. iv, 33 [367]
- ***Libraries*** : CtY, ICRL, IU, LNT, MChB, NcD, NjR, NN, OC, OCH, PPL, ViU, WaU
- ***Source*** NUC (NC 0599786 and 0599788)

- **{1804}, 3rd (03)**
- *A compendious geographical dictionary* ... added, tables showing the state of the representation of Ireland, in the united Parliament; the population of England and Wales, and the coins ... value ... money; to ... introduction ... planets, etc. ... maps, including those of Australasia and Polynesia
- LONDON; printed for W. Peacock and sons, by C. Rickaby
- 14 cm.
- ***Libraries*** : TxU
- ***Source*** NUC (NC 0599789)

COMPENDIUM OF GEOGRAPHY ... Y

- **{1854}, new, revised (01)**
- *Compendium of geography*; being an abridgment of the larger work; entitled an epitome of geographical knowledge, ancient and modern. Compiled for the use of teachers and advanced classes of the national schools of Ireland. New edition, revised
- DUBLIN; Her Majesty's Stationary Office
- ***Libraries*** n.r.
- ***Notes***

1. Bowen (1981:318) identifies the authors as U.K. Commissioners of National Education in Ireland.
2. See EPITOME ... (1845).

COMPLEAT GEOGRAPHER, The
See (03) of THESAURUS GEOGRAPHICUS (1695).

COMPLETE SYSTEM ... A

- **{1744} (01)**
- *A complete system of geography*. Being a description of all the countries, islands, cities, chief towns, harbours, lakes, and rivers, mountains, mines, etc. of the known world. Shewing the situation, extent, and boundaries of the several empires, kingdoms, republics, principalities, provinces, etc., their climate, soil, and produce; their principal buildings, manufactures, and trade; their government, policy, religion, manners, and customs; and the distance and bearing of all the principal towns from one another. Comprehending the history of the universe, both ancient and modern; and the most material revolutions and changes that have happen'd in it, either by conquest or treaties; with whatever is curious and remarkable in the works of

nature or art. To which is prefixed, an introduction to geography, as a science: an explanation of maps: the doctrine of the sphere: the system of the world: and a philosophical treatise of the earth, sea, air, and meteors. The whole illustrated with seventy maps, all new-drawn and ingraved by Emanuel Bowen, according to the latest discoveries and surveys; and printed on distinct half-sheets, the full size of the book; making, of themselves, a complete atlas, for the use of all gentlemen, merchants, mariners, and others, who delight in history and geography. This work, extracted from several hundred books of travels and history, is brought down to the present time; preserving all that is useful in the fourth and last edition of the Complete Geographer, publish'd under the name of Herman Moll, etc. 2 vols.

- [LONDON]; printed for William Innys, Richard Ware, Aaron Ward ...
- Folio; Vol. I, pp. [iv+] xxviii+ 1,013 [+iii]; Vol. II, pp. xii+ 804 [+xxiv]
- *Libraries* ShP
- *Source* ESTC. Title, publisher, and pagination as recorded in B.H. Blackwell's Catalogue No. 949 (p. 247) [c1970].

- **1747 (02)**
- *A complete system of geography* ... and boundaries, of the several empires ... provinces, etc. their climate ... one another. Including the most material revolutions and changes that have happened in every state, either by conquest or treaties; and comprehending whatever is curious ... works of art or nature. To which is prefixed ... and meteors. In two volumes. The whole illustrated with seventy maps, by Emanuel Bowen, Geographer to his Majesty, being all new-drawn and ingraved according to the latest discoveries and surveys; making, of themselves, a complete atlas, for the use of all gentlemen, merchants, mariners, and others, who delight in history and geography. This work ... Complete Geographer, published under the name of Herman Moll, etc.
- LONDON; printed for William Innys, Richard Ware, Aaron Ward, J. and P. Knapton, John Clarke, T. Longman and T. Shewell, Thomas Osborne, Henry Whitridge, Richard Hett, Charles Hitch, Stephen Austen, Edmund Comyns, Andrew Millar, James Hodges, Charles Corbett, and Jo. and Ja. Rivington
- Folio, 41 cm., 2 vols.; Vol. I, pp. xii, xxviii, 1013, [3]; Vol. II, pp. [4], 804, [24]
- *Libraries* **L** : DLC, MeB, NcD, NcU, NN, OClWHi, PP : CaBVaU
- *Notes*

1. Catalogued by BL under Bowen and under System.
2. Physically massive, it contains a preface justifying the length and summarizing the contents. Vol. I is devoted to Europe, including Turkey. The subdivisions of all countries in Europe except Norway are dealt with. Those used vary considerably; in England, Wales, and Scotland they are the counties, in Switzerland the cantons, in France the provinces, in Germany and Italy the many independent kingdoms and principalities.
3. See THESAURUS GEOGRAPHICUS (1695), in particular (03) and (04).

COMPREHENSIVE GEOGRAPHY ... **Y Am**
- **{[1876]}a (01)**
- *The comprehensive geography. Number one*; or, first lessons in geography
- NEW YORK; P. O'Shea
- 72 p.
- ***Libraries*** : PPiU

COMPREHENSIVE GEOGRAPHY ... **Y Am**
- **{1876}b (01)**
- *The comprehensive geography. Number 2 ...*
- NEW YORK; P. O'Shea
- 24.5 cm.; 112 p.
- ***Libraries*** : DLC

COMPREHENSIVE GEOGRAPHY ... **Y Am**
- **{1876}c (01)**
- *The comprehensive geography. Number 3*. Including mathematical, physical, topical, political, historical, sacred, and classical geography ...
- NEW YORK; P. O'Shea
- 32 × 25 cm.; 136 p.
- ***Libraries*** : DLC, OClJC
- ***Source*** NUC (NC 0602174). Cited in Elson (1964:404), who notes: '(For Roman Catholic schools).'

CONCISE SYSTEM ... **Y SG?**
- **{1774} (01)**
- *A concise system of geography*, wherein the first principles of the science are laid down in a plain and easy manner suited to the capacities of youth
- DEVIZES [England]; printed by T. Burrough
- ***Libraries*** : MH

CONCISE SYNOPSIS ... **Y**
- **{1813} (01)**
- *Geography, concise synopsis*
- Place/publisher n.r.
- ***Libraries*** n.r.
- ***Source*** Peddie and Waddington (1914:226)

- **1829, 3rd, corrected and considerably enlarged (02)**
- *A concise synopsis of geography* for the use of the junior department of the Royal Military College of Sandhurst
- LONDON; printed for T. Egerton, Bookseller to the Ordinance Military Library, Whitehall
- 8vo; pp. [3], 2-102
- ***Libraries*** L, **[IDT]**

▸ *Notes*

1. No preface or introduction. Pp. 5-74 contain information about places (population of countries, height of mountains, length of rivers) essentially in tabular form; pp. 75-98 are concerned with the description and use of the terrestrial globe, and with problems associated with its use; pp. 99-102 are a glossary of technical terms with their etymology, including Greek script where necessary.

CONDER, Josiah **Dy**

▸ **{1834} (01)**
▸ *A dictionary of geography, ancient and modern ...*
▸ LONDON; Thomas Tegg and son; [etc.]
▸ 12mo, 19 cm.; 2 p. l., vii-viii, 724 p.
▸ ***Libraries*** L : DLC, OO
▸ ***Source*** BL and NUC (NC 0608908)

CONDER, Josiah **A**

▸ **{1824-30} (01)**
▸ *The modern traveller*; a description of the various countries of the globe
▸ LONDON; printed for T. Tegg and J. Duncan
▸ 16 cm., 30 vols.
▸ ***Libraries*** : DLC, IEN, MiU
▸ ***Source*** NUC (NC 0608929)

▸ **1830 (02)**
▸ *The modern traveller*. A description, geographical, historical, and topographical, of the various countries of the globe. In thirty volumes
▸ LONDON; James Duncan, 37, Paternoster-Row
▸ 12mo, 15 cm., 30 vols.
▸ ***Libraries*** **L** : MH, N, NN, RPB, ViU [assuming that this issue corresponds to NUC (NC 0608928)]
▸ *Notes*

1. The set at L consists of volumes of the original and new eds. of various dates, each preceded by an additional title page bearing the date 1830 (from BL).
2. The individual volumes contain as many as four tps. The first is always the one given above. The subsequent ones identify the continent or countries described. They read as follows:

   Vol. 1, 2nd tp.: *The modern ... globe. Thirty-three volumes. Volume the First. Palestine.* Printed for Thomas Tegg and son, 73, Cheapside, and James Duncan, Pater-noster-Row, [ND]. Pp. [2], 372

   Vol. 2, 2nd tp.: *The modern traveller. Volume the second. Syria and Asia Minor. Vol. I*

   3rd tp.: *The modern traveller. A description ... globe. Syria and Asia Minor. Vol. I.* New edition. 1831. James ...

   4th tp.: The modern traveller. Syria and Asia Minor. Vol. I. New edition. 1831. Pp. [6], iv, 354

   Vol. 3, 2nd tp.: *The ... third. Syria ... Vol. II*

3rd tp.: *The modern traveller. A description ... globe. Syria ... II.* New edition. 1831. James ... Pp. [8], [i-iii], iv, 356

Vol. 4, 2nd tp.: *The ... fourth. Arabia*

3rd tp.: *The modern traveller. A popular description, geographical, historical, and topographical, of the various countries of the globe. Arabia.* LONDON; printed for James Duncan; Oliver and Boyd, Edinburgh; M. Ogle, Glasgow; and R.M. Tims, Dublin. 1825. Pp. [4], [i-iii], iv, 362

Vol. 5, 2nd tp.: *The ... fifth. Egypt, Nubia, and Abyssinia. Vol. I*

3rd tp.: *The modern traveller ... globe. Egypt ... Abyssinia. Vol. I.* 1827. LONDON ... Dublin. Pp. [4], [i-iii], iv, 366

Vol. 6, 2nd tp.: *The ... sixth. Egypt ... Vol. II*

3rd tp.: *The modern traveller ... Abyssinia. Vol. II.* 1827. LONDON ... Dublin. Pp. [4], [i-iii], iv, 364

Vol. 7, 2nd tp.: *The ... seventh. India. Vol. I*

3rd tp.: *A popular description ... India. Vol. I.* Printed for James ... Dublin. Pp. [4], [i-iii], iv, 359

Vol. 8, 2nd tp.: *The ... eighth. India. Vol. II*

3rd tp.: *The modern traveller ... India. Vol. II.* 1828. Printed ... Dublin. Pp. [4], [i-iii], iv, 372

Vol. 9, 2nd tp.: *The ... ninth. India. Vol. III*

3rd tp.: *The modern traveller ... India. Vol. III.* 1828. Printed ... Dublin. Pp. [4], [i-iii], iv, 390

Vol. 10, 2nd tp.: *The ... tenth. India. Vol. IV.*

3rd tp.: *The modern traveller ... India. Vol. IV.* 1828. Printed ... Dublin. Pp. [4], [i-iii], iv, 376

Vol. 11, 2nd tp.: *The ... eleventh. Burmah, Siam, etc.*

3rd tp.: *The modern traveller ... Birmah [sic], Siam, and Annam.* 1826. LONDON. Printed ... Dublin. Pp. [4], [i-iii], iv, 367

Vol. 12, 2nd tp.: *The ... twelfth. Persia and China. Vol. I*

3rd tp.: *The modern traveller ... Persia and China. Vol. I.* 1827. LONDON ... Dublin. Pp. [4], [i-iii], iv, 372

Vol. 13, 2nd tp.: *The ... thirteenth. Persia and China. Vol. II*

3rd tp.: *The modern traveller ... Persia and China. Vol. II.* 1827. LONDON ... Dublin. Pp. [4], [i-iii], iv, 340

Vol. 14, 2nd tp.: *The ... fourteenth. Turkey*

3rd tp.: *The modern traveller ... Turkey.* 1827. LONDON ... Dublin. Pp. [4], [i-iii], iv, 356

Vol. 15, 2nd tp.: *The ... fifteenth. Greece. Vol. I*

3rd tp.: *The modern traveller ... Greece. Vol. I.* 1826. LONDON ... Dublin. Pp. [4], [i-iii], iv, 375

Vol. 16, 2nd tp.: *The ... sixteenth. Greece. Vol. II*

3rd tp.: *The modern traveller ... Greece. Vol. II.* 1826. LONDON ... Dublin. Pp. [4], [i-iii], iv, 336

Vol. 17, 2nd tp.: *The ... seventeenth. Russia*

3rd tp.: *The modern traveller ... Russia.* 1825. LONDON ... Dublin. Pp. [4], [i-iii], iv, 338

Vol. 18, 2nd tp.: *The ... eighteenth. Spain and Portugal. Vol. I*

3rd tp.: *The modern traveller ... Spain and Portugal. Vol. I.* 1826. LONDON ... Dublin. Pp. [4], [i-iii], iv, 369

Vol. 19, 2nd tp.: *The ... nineteenth. Spain ... Vol. II*

3rd tp.: *The modern traveller ... Spain and Portugal. Vol. II.* 1826. LONDON ... Dublin. Pp. [4], [i-iii], iv, 342

Vol. 20, 2nd tp.: *The ... twentieth. Africa. Vol. I*

3rd tp.: *The modern traveller ... Africa. Vol. I.* 1829. LONDON ... Dublin. Pp. [4], [i-iii], iv, 374

Vol. 21, 2nd tp.: *The ... twenty-first. Africa. Vol. II*

3rd tp.: *The modern traveller ... Africa. Vol. II.* 1829. LONDON ... Dublin. Pp. [4], [i-iii], iv, 348

Vol. 22, 2nd tp.: *The ... twenty-second. Africa. Vol. III*

3rd tp.: *The modern traveller ... Africa. Vol. III.* 1829. LONDON ... Dublin. Pp. [4], [i-iii], iv, 323

Vol. 23, 2nd tp.: *The ... twenty-third. United States of America and Canada. Vol. I*

3rd tp.: *The modern traveller ... North America. Vol. I.* 1829. LONDON ... Dublin. Pp. [6], [i], ii-v, [3], 372

Vol. 24, 2nd tp.: *The ... twenty-fourth. United ... Vol. II*

3rd tp.: *The modern traveller ... North America. Vol. II.* 1830. LONDON ... Dublin. Pp. [4], [i-iii], iv, 335

Vol. 25, 2nd tp.: *The ... twenty-fifth. Mexico and Guatemala. Vol. I*

3rd tp.: *The modern traveller ... Mexico and Guatemala.* 1825. LONDON ... Dublin. Pp. [4], [i-iii], iv, 371

Vol. 26, 2nd tp.: *The ... twenty-sixth. Mexico ... Vol. II*

3rd tp.: *The modern traveller ... Mexico and Guatemala. Vol. II.* 1825. LONDON ... Dublin. Pp. [4], [i-iii], iv, 320

Vol. 27, 2nd tp.: *The ... twenty-seventh. Columbia*

3rd tp.: *The modern traveller ... Columbia.* 1825. LONDON ... Dublin. Pp. [4], [i-iii], iv, 356

Vol. 28, 2nd tp.: *The ... twenty-eighth. Peru and Chile*

3rd tp.: *The modern traveller ... Peru. - Chile.* 1829. LONDON ... Dublin. Pp. [4], [i-iii], vi-vi, 360

Vol. 29, 2nd tp.: *The ... twenty-ninth. Brazil and Buenos Ayres.* Vol. I

3rd tp.: *The modern traveller ... Brazil and Buenos Ayres.* Vol. I. 1825. LONDON ... Dublin. Pp. [4], [iii-v], vi-vii, [3], 350

Vol. 30, 2nd tp.: *The ... the thirtieth. Brazil ... Vol. II*

3rd tp.: *The modern traveller ... Brazil and Buenos Ayres. Vol. II.* 1825. LONDON ... Dublin. Pp. [8], 340

3. The Preface contrasts the essence of special geography (as understood in this guide) with reports of travel and exploration in these words: 'In that work [i.e., the best extant collection of voyages and travels] the reader would in vain seek for an accurate account of the present state of a single country on the face of the globe. Amusing and often valuable as are the recitals of the older travellers, they of necessity abound with obsolete errors; they are generally barren of scientific information; and much of the information they contain, is out of date ... To give the results of modern discovery, combined with our previous stock of information, in a succinct and popular form, so as to exhibit, at one view, the present state of our knowledge with regard to each particular country traversed by European travellers, was the object proposed in undertaking the present work. The only publication in which this had been attempted with any degree of competent ability, is the Geography of M. Malte Brun. In his steps, it were no

disgrace to follow. His plan, however, is somewhat different; the topographical description is for the most part extremely brief and hurried, while the dissertations are extended beyond what might seem proper in a popular work' (pp. 7-8).

The Preface also contains the following claims: 'It has been a marked feature of the present work, that authorities are specifically and minutely cited for every statement ... Another feature of the work upon which much care has been bestowed, is the statistical tables, exhibiting the corresponding ancient and modern territorial divisions, the extent of surface, population, etc. Of the tables which add so materially to the value of M. Malte Brun's work, the Editor has of course availed himself; but, in most cases, he has found it necessary to compile them from more recent information, and to follow his own judgement in the geographical arrangements. It is to be lamented, that no work on geography in the English language, (the translation of M. Malte Brun excepted) has any claim to scientific accuracy' (p. 15).

To my knowledge no comparison of CONDER (1824-30) and MALTE-BRUN (1822-33) has ever been made. It would seem to be a worthwhile undertaking, though it should be noted that Conder did not include descriptions of either the British Isles or the countries of the Grand Tour (on the grounds that more than sufficient information was already available on them; however, see (04) below for three volumes on Italy). It may also be asked whether BELL (1829-32) does not also deserve to be compared with Conder and Malte-Brun.

- **{1824-31} (03)**
- *The modern traveller ...*
- LONDON; printed for Thomas Tegg and Son
- 15 cm., 33 vols.
- ***Libraries*** : GHi, NBuG, OO
- ***Source*** NUC (NC 0608930)

- **1824-34 (04)**
- *The modern traveller*
- LONDON; printed for Thomas Tegg, 73, Cheapside; and James Duncan, Paternoster Row
- 14 cm., 33 vols.
- ***Libraries*** L (incomplete, lacks Vols. 3, 5-7, 14, 16, 18, 24, 30, 31) : KyU : CaAEU (incomplete, lacks Vols. 31-33)
- ***Notes***

1. Vols. 32 and 33 of the BL set are Vols. II and III of Italy; presumably the missing Vol. 31 is Vol. I of Italy.
2. NUC records incomplete sets at (DS, MB, NN) [NC 0608933], and at (DLC, MH, NjR, NN, RP) [NC 0608934].
3. The vols. at CaAEU are catalogued separately, not as a set.

COOKE, George Alexander A

- **{[1790?]} (01)**
- *Modern and authentic system of universal geography*, containing an accurate

and entertaining description of Europe, Asia, Africa, and America ... Being a complete and universal history and description of the whole world ...
- LONDON; printed for C. Cooke ... by J. Adlard
- 27 cm., 2 vols.
- ***Libraries*** : CU
- ***Source*** NUC (NC 0669145). See note to (04) below.

- **{[18-?]} (02)**
- *Modern ... system of universal geography ...*
- LONDON; D. Cock
- 28 cm., 2 vols.
- ***Libraries*** : MnU
- ***Source*** NUC (NC 0669146)

- **{1806?} (03)**
- *Modern ... system of universal geography* ... America; continents, islands, provinces ... Being a complete and universal description of the whole world ... To illustrate the work are introduced a considerable number of new and accurate maps, charts, etc. comprising a complete atlas; and to render it more interesting are given upwards of one hundred engravings ...
- LONDON; C. Cooke, etc.
- 27.5 cm., 3 vols.
- ***Libraries*** : NNC
- ***Source*** NUC (NC 0669149)

- **1807 (04)**
- *Modern ... system of universal geography* ... America: as consisting of continents, islands, provinces, countries, cities, towns, promontories, capes, bays, peninsulas, isthmuses, gulphs, lakes, rivers, harbours, forts, mountains, volcanos, etc. Being a complete ... world, as divided into empires, kingdoms, states, republics, etc. Together with their respective situations, latitudes, longitudes, extent, boundaries, climates, soils, minerals, vegetable productions, beasts, birds, fishes, amphibious animals, reptiles, insects, etc. with many natural and artificial curiosities: likewise a particular description of the inhabitants, with an account of their military and civil government, revenues, laws, trade, commerce, arts, sciences, manufactures, religion, customs, manners, genius, habits, ceremonies, amusements, etc. With details of the most remarkable battles, sieges, sea-fights, and revolutions, that have taken place in different parts of the world: including interesting narratives from all the navigators that have made new discoveries, as well as those of a more remote period, particularly Columbus, Magellan, Bhering, Mandana, Fernandez, Forbisher [sic], Drake, Cavendish, Hudson, Quiros, Le Maire, Schouten, Tasman, Cowley, Dampier, Rogers, Clipperton, Roggewein, Anson, Byron, Wallis, Carteret, Cook, Furneaux, Forest, Clerke, King, Maurelle, Wilson, Perouse, Portlocke, Bligh, Edwards, Vancouver, Missionaries, D'Entrecasteaux. Likewise copious accounts from the most celebrated travellers, as Thunberg, Sparman, Addison, Shaw, Le Comte, Pocock, Hanway, Wood, Bruce, Wraxhall, Moore, Coxe, Render, Forster, Brissot, Macartney, Browne, Pallas, Mungo Parke, Hearne, Mackenzie, Weld, Barrow, Sonnini, Collins, Denon, Rochefoucault. Form-

ing a complete collection of voyages and travels, and comprising an authentic and entertaining account of every thing worthy of notice throughout the whole face of nature. To illustrate the work are introduced a considerable number of new and accurate maps, charts, etc. comprising a complete atlas; and to render it more interesting, are given numerous engravings, executed in a superior style of elegance, forming a complete series of superb embellishments, consisting of views, antiquities, edifices, portraits, habitations, canoes, implements, customs, ceremonies, amusements, etc.

▸ LONDON; printed for C. Cooke, 17, Paternoster Row, by Macdonald and son, 46, Cloth Fair, West Smithfield; and sold by all the booksellers in the United Kingdom

▸ 25.5 cm., 2 vols.; Vol. I, pp. [7], ii-xx, [2], 7-816 (321-4 inserted upside down); Vol. II, pp. [2], 3-768, [4]

▸ ***Libraries*** : NjR : CaAEU

▸ ***Notes***

1. The date is taken from the frontispiece; none is given on the tp. Seventeen major sources are discussed briefly in the Preface. The text starts with an account of recently discovered islands in the Pacific (pp. [5]-364). Europe is dealt with at the end of Vol. II. Placing that continent last among the four makes this book almost unique in that respect. The placement is justified by the need 'to afford time and opportunity for an ample and accurate display of the important events and memorable revolutions that have taken place in that quarter during the series of the last ten years' (note at foot of Vol. I, p. [5]).

   Occasional long, direct quotations are included in the text. Overall the work is a large, unsystematic collection of facts, presenting, apparently, whatever caught the author's attention in the course of his reading. For example, in the section on England there is an unusually lengthy discussion of the form of government; this topic is construed so broadly as to include not only some remarks on the gradations of rank among the aristocracy, and the nature of the legal system, but also the national debt.

▸ **1810? 1811? (05)**

▸ *Modern ... system of universal geography* ... America; as consisting of ... together with their respective ... amusements, etc. with details of ... Rochefoucault: forming ...

▸ LONDON; printed for C. Cooke, 17, Paternoster Row, by Macdonald and son, 46 Cloth Fair, West Smithfield; and sold by all the booksellers in the United Kingdom

▸ 4to, 26 cm., 2 vols.; Vol. I, pp. [5], xx, 816; Vol. II, pp. 768, [4]

▸ ***Libraries*** L : NBuG, NN, WaS, WaU

▸ ***Notes***

1. BL gives the date as c1810, but there is a reference to an event in January 1811 (Vol. II, p. 706).
2. NUC records further eds. (both London): [1813?], 2 vols. (CtY); 1817, 2 vols. (DLC, MHi, OrU, PU.
3. For contemporary works of a comparable scope and scale, see ADAMS, M. (1794?), BALDWYN (1794?), BANKES (1788?), and PAYNE, J. (1791).
4. Discussed in Downes (1971:386).

COOKE, John **?**
- **{1801} (01)**
- *The circular atlas and compendious system of geography*
- Place/publisher n.r.
- 4to
- ***Libraries*** n.r.
- ***Source*** Watt (1824)

COOKE, John **Y OG**
- **1812 (01)**
- *A general synopsis of geography*, with the projection of maps and charts; to which is prefixed, an historical introduction to the sciences of geometry, astronomy, and geography: the whole illustrated by twenty copper-plates, explanatory of the geometrical figures and problems, definitions, etc. with an easy and regular method of drawing maps. By ... geographer, and late engineer to the Admiralty
- LONDON; printed for James Cooke, 57 High-Street, Bloomesbury[,] and sold by Longman and Co. Paternoster Row, S. Bagster, Strand; Rees and Curtis, and Nettleton and Son, Plymouth; Woolmer and Uph[,] Exeter; J.C. Motley, Portsmouth; S. Hazard, Bath; W. Brown and W. Sheppard, Bristol; Milner, Liverpool; Brodie, Dowdon and Luxford, Salisbury; and Darcy and Hildyard, Hull
- 26 cm.; pp. [1], [ix], x-lvii, [1], 53
- ***Libraries*** **L**
- ***Notes***

1. Not special geography. For the author, geography was the making of maps, and it is to this subject that this book is devoted, with a prefatory address in which it is made clear that it is intended for students.

COOKE, William **?**
- **{1812} (01)**
- *Geography*
- LONDON
- 4to
- ***Libraries*** n.r.
- ***Source*** Allibone (1854)

CORNELL, Sarah S. **Y Am**
- **{1858}a (01)**
- *Cornell's first steps in geography*
- NEW YORK; D. Appleton & company
- 17 cm.; pp. 66, [2]
- ***Libraries*** : DLC, IC, NcD, NcU, NjR, OClWHi, WaS
- ***Source*** NUC (NC 0706243)
- ***Notes***

1. (: NN : CaAEU) hold copies of this title with c1858 that have pp. [3], 4-72.

2. The text consists of questions and answers. It starts with elements of physical geography; then proceeds to special geography. Devoted to relative location and the identification of features of physical geography visible on small-scale maps, plus capital cities. There are occasional items of climate and primary production. The last 4 pp. carry advertisements. Very similar to COLTON, G.W. (1860).
3. NUC records further eds. (all New York, Appleton): 1859 (MH); 1863? c1858 (MH); 1868? (MH); 1874? (MH); 1876, 66 pp. (CtY, MoU : CaBVaU); 1882? (MH); and 1890?
4. Culler discusses (01) (1945, *passim*), as does Carpenter (1963:266).

CORNELL, Sarah S. **Y Am**

- **{1858}b (01)**
- *Cornell's grammar-school geography*: forming part of a systematic series of school geographies ...
- NEW YORK; D. Appleton and company
- 29 cm.; pp. 108
- ***Libraries*** : DLC, MH, OO
- ***Source*** NUC (NC 0706254)
- ***Notes***

1. On the basis of an examination of an ed. held by NN that lacks the tp., this work is found to contain greater detail and to present more precise statistics than the *Primary* (CORNELL 1854) and the *Intermediate* (CORNELL 1853) *Geographies*.

- **{1859} (02)**
- *Cornell's grammar-school geography* ... geography, forming ... geographies, embracing an extended course and adapted to pupils of the higher classes in public and private schools
- NEW YORK; D. Appleton & company
- ***Libraries*** : MH, ODW
- ***Source*** NUC (NC 0706255)
- ***Notes***

1. NUC records further eds. (all New York, Appleton): 1860, 108 pp. (MH, ViU); 1862, pp. n.r. (MH); 1863, pp. n.r. (MH); 1865, pp. n.r. (MH); 1866, pp. n.r. (MH); 1868, rev., 122 pp. (DLC, NN); 1869, 122 pp. (DLC, OCU); 1870, 122 pp. (CtY); and 1878 (NN) [NC 0706261 notes: 'reprint of 1868 ed.'].
2. Dryer (1924:120-1) provides a general discussion; Culler discusses the ed. of 1870 (1945, *passim*); Hauptman (1978:434) quotes from the ed. of 1878, and (1978, n. 56) treats briefly of the author; Elson (1064:403) cites the ed. of 1870.

CORNELL, Sarah S. **Y Am**

- **{1856} (01)**
- *Cornell's high school geography*: forming part third of a systematic series of school geographies, comprising a description of the world; arr. with

special reference to the wants and capacities of pupils in the senior classes of public and private schools ...
- NEW YORK; D. Appleton & company
- *Libraries* : OO

- **1857, c1856 (02)**
- *Cornell's high school geography* ... arranged ... schools. Embellished by numerous engravings, and accompanied by a large and complete atlas, drawn and engraved expressly for this work. By ... corresponding member of the American Geographical and Statistical Society
- NEW YORK; D. Appleton and Company, 346 & 348 Broadway
- 19 cm.; pp. [3], 4-405
- *Libraries* : CU, DLC, MH, OClWHi, OO, PU : **CaAEU**
- *Notes*

1. The Preface contains an analysis of the shortcomings of contemporary school geography textbooks. They are said to suffer from an 'unphilosophical arrangement ... Hard labor may enable a pupil to learn the government of a country, the population of a city, the length of a river, and other facts equally dry and repulsive. But geography is something more than a mere collection of detailed facts: it is a science founded on fixed principles ... which must be thoroughly understood before [the details] can be profitably learned' (p. [3]). It is then claimed with respect to this book: '1. That it is arranged on the true inductive system, commencing with elementary principles, and proceeding by way of gradual advances from deduction to deduction and from step to step until the whole is ground is covered. 2. The arrangement is clear and practical ... 3. It is interesting. 4. It facilitates the teacher's task, by neither requiring reference to Tables nor asking questions which the learner cannot answer without aid. 5. It contains as many facts as can be advantageously remembered, while it eschews an embarrassing multiplicity that would be forgotten as soon as learned' (p. 4).

   The apparent form of the relationship implied between deductions and the 'true inductive system' turns out to be a hint that this Preface is a fog of words designed rather to sell the book than to provide an accurate idea of what is to come. The body of the book is divided into three parts; the first, which occupies pp. 5-322, 'embraces descriptive geography and exercises on the maps in the accompanying atlas' (p. 4). These pages are divided into 160 numbered lessons; these contain a mixture of textual information and questions, some with and some without the answers. In the latter case the answer is presumably to be found on the atlas. By implication, the others are facts that 'can be advantageously remembered.' Thus question 10 of lesson 8, 'Lessons on the map of North America,' asks, 'What waters surround Cornwallis Island? [Ans.] It has not been ascertained, except that Wellington Channel is on the east, and Barrow Strait on the south' (p. 21).

   'Part II treats of mathematical geography [i.e., the problems presented by the sphericity of the earth with respect to the setting up of a graticule]. Part III is an outline of physical geography' (p. 4). It seems that whatever 'fixed principles' may give order to geography, they are not ones that entail understanding the nature of the physical environment as a prerequisite to studying the facts of special geography.
2. NUC records further eds.: (New York, Appleton, 405 pp., except as

noted): 1859 (DLC, MH, N) [Culler (1945:307) notes: 2nd ed.]; 1860 (CtY, MH, OFH); 1861, pp. n.r. (IU, MH, MiU, PU); 1862, pp. n.r. (MH); 1864 (MB, MH, PPiU); 1866 (CtY, DLC, NN, PU); 1867, pp. n.r. (MH); 1870 (DLC); 1872 (ViU); 1877, pp. n.r. (MH); and 1880 (MH).
3. Nietz discusses (01) (1961:230-1), and Dryer (1924:121) provides a general discussion.

CORNELL, Sarah S. **Y Am**

- **{1854} (01)**
- [*Cornell's primary geography ...*]
- ***Libraries*** n.r.
- ***Notes***

1. This ed. is known from the copyright statement (1854, 1857) in entry (03) below.

- **{1855} (02)**
- *Cornell's primary geography*, forming first of a systematic series of school geographies
- NEW YORK; D. Appleton & company
- 22.5 cm.; pp. 96
- ***Libraries*** : DLC

- **1857, revised (03)**
- *Cornell's primary geography*, forming the first part of a systematic series of school geographies
- NEW YORK; D. Appleton & company, 443 & 445 Broadway
- 17.5 cm.; pp. [3], 6-98
- ***Libraries*** : MH, **NN**
- ***Notes***

1. The text is arranged in the form of question and answer. Simple terms (e.g., island, sea) have illustrations as well as definitions. Cartographic terms are avoided, east and west being replaced by references to sunrise and sunset. Concerned with information visible on small-scale maps (see CORNELL 1858a). Starts with simple facts and proceeds step by step to more complex topics, ending with the United States 'by virtue of its complicacy' (p. 6). More detailed than CORNELL (1858a).
2. According to NUC (NC 0706306) the book is 22.5 cm. high. NUC records further eds. (all New York, Appleton): 1858, 98 pp. (DLC); 1859 (MB, MH); [1865? c1857] (MH); [1867] (DLC, MB, MH); [1871] (MB, MH, NN, OO); [1875], 99 pp. (DLC, MoU); [1877?, c1875] (MH); and 1888.
3. Dryer (1924:120) provides a general discussion; Nietz (1961:198, 201, 230-1) discusses (01); Culler (1945, *passim*) discusses (03).

CORNELL, Sarah S. **Y Am**

- **{1853} (01)**
- *Intermediate geography*
- NEW YORK; D. Appleton & Co.

- *Libraries* n.r.
- *Source* Culler (1945:307)

- **1855 (02)**
- *Cornell's intermediate geography*: forming part second of a systematic series of school geographies. Designed for pupils who have completed a primary or elementary course of instruction in geography
- NEW YORK; D. Appleton and company, 346 & 348 Broadway. Cincinnati; Rickey, Mallory & company
- 28.5 cm.; pp. [3], 4-84
- *Libraries* : DLC, MB, MH, MoU, **NN**, OO, Wa
- *Notes*

1. Introduces mathematical geography, i.e., the division of the earth's surface by lines identified by means of astronomical observation. Divides 'mankind' into five races and four grades, the former, by implication, on the basis of the colour of the skin, the latter, explicitly, on the basis of attainments in arts, science, and economic activity. In special geography information is provided in two forms: first, text, divided into short paragraphs under subheadings, with information on geographical position, soil, climate, productions, inhabitants, etc.; second, maps, with which are associated questions whose answers can be gleaned from the maps. No judgements are made in the text as to the grade of the people who inhabit a country, but the information provided allows a reader to make such judgements.
2. NUC records further eds. (all New York, Appleton, except as noted): 1856, 88 pp. (DHEW, DLC); 1857 (MH); 1859? c1855 (MB, MH); 1860?, c1855 (MB, MH); 1860, Memphis, 88 pp. (MB) [NC 0706285 notes: 'Southern Education series']; 1867, 96 pp. (DLC, OClWHi, OO); 1875, 100 pp. (OClWHi); 1876 (MH, OO); 1877, no pub., 100 pp. (MiU); 1878 (MH); 1881 [c1867], 102 pp. (MiD [NC 0706291 notes: 'includes pp. [i]-viii: Geography of Michigan']); and 1885, 102 pp. (NN); and 1888.
3. Culler (1945, *passim*) and Nietz (1961:230-1) discuss (01); Hauptman (1978:431, 434-5) thrice refers to the ed. of 1885 in general discussion.

CORNWELL, James **Y SG?**

- **{1858} (01)**
- *Geography for beginners*
- LONDON; Simpkin, Marshall & Co; Hamilton, Adams & Co.
- 12mo; pp. 94
- *Libraries* L
- *Notes*

1. According to BL the date is uncertain. It is confirmed by ECB, Index (1856-76:130). BL notes: 'Later editions form part of "Dr. Cornwell's Educational Works".' Under that entry, BL records eds. in 1869, 27th; and 1904, 70th.

CORNWELL, James **Y SG?**

- **{[1847]} (01)**
- *A school geography*

- LONDON; Simpkin, Marshall & Co.; Hamilton, Adams, & Co.
- 12mo; pp. 317
- ***Libraries*** L
- ***Notes***

1. BCB, Index (1837-57:110) records ed. of 1848. (: : CaBVaU) has ed. of 1871, 360 pp. BL and NUC record further eds.: [c1859] [sic], 5th (: MH); [185–], 15th, 317 pp. (: DLC); 1855? 19th, 336 pp. (L); 1862, 31st, 338 pp. (: ICU); 1868, 44th, 360 pp. (: MH); 1878, 63rd, 336 pp. (L); 1880, 67th, 360 pp. (: MB); 1881, 68th, 336 pp. (L); 1881, 69th (: MH); 1882, 72nd (: NN); and 1904.

[COTTINEAU de KLOGUEN, Denis Louis] **Y Am**

- **{1806} (01)**
- *Geographical compilation for the use of schools*; being an accurate description of all the empires, kingdoms, republics and states, in the known world: with an account of their population, government, religion, manners, literature, universities, history, civil divisions, ecclesiastical hierarchy, principal cities (with an account of their importance, remarkable monuments, illustrious citizens, commerce and population) etc. etc. etc. The whole arranged in catechetical form. Compiled from the best American, English, and French authors, by D.L.C. teacher of geography ...
- BALTIMORE; printed for the compiler, by John West Butler
- 16.5 cm., 2 vols.
- ***Libraries*** : CSmH, DLC, InU, IObT, MdBE, MdBG, MdBS, MdHi, MWA, NbHi, NN, NNC, PP, PPAmP, PPL, RNR, ScU, ViU
- ***Source*** NUC (NC 0736066)
- ***Notes***

1. NUC notes that 'copyright notice indicates Louis Denis Cottineau as author.' Cited by Elson (1964:394-5).

CRAMPTON, Thomas, and TURNER, Thomas **Y SG?**

- **{1857-60} (01)**
- *The geographical reading book*; being a series of inductive lessons in geography
- LONDON; Groombridge & Sons
- 3 pts.
- ***Libraries*** L
- ***Notes***

1. ECB, Index (1856-76:130) records ed. of 1861.

CRAWLEY, William John Chetwoode **Y mSG?**

- **{1871} (01)**
- *A manual of historical geography* for the use of civil service students, etc.
- LONDON; George Philip & Son
- 8vo; pp. ix, 151
- ***Libraries*** L

CRISP **Y SG?**
- **{17-? 180-?} (01)**
- *Crisp's lessons in geography*
- LONDON
- ***Libraries*** n.r.
- ***Source*** London Catalogue (1811)

[CROKER, John Wilson] **Y SG?**
- **{1829} (01)**
- *Elements of geography*; for the use of young children
- 12mo; pp. 94
- ***Libraries*** L
- ***Notes***

1. BL identifies the author and records further eds.: 1835, 3rd; and 1847, 4th.

CROKER, J.W. **?**
- **{1858} (01)**
- *Useful geography*
- LONDON
- ***Libraries*** n.r.
- ***Source*** Allford (1964:246)

CRUIKSHANK, James **Y Am**
- **{1867} (01)**
- ... *A primary geography*
- NEW YORK; William Wood and company
- 25 cm.; pp. vi, 88
- ***Libraries*** : DLC, NN, OO
- ***Source*** NUC (NC 0814162) and Culler (1945:307)
- ***Notes***

1. Discussed in Culler (1945, *passim*), and by Dryer (1924:121-2).

[CRULL, Jodocus?] **NG tr**
- **{1705} (01)**
- *An introduction to the history of the kingdoms and states* of Asia, Africa, and America both ancient and modern, according to the method of Samuel Puffendorf, Counsellor of State to the late King of Sweden
- LONDON; printed by R.J. for T. Newborough, at the Golden Ball, J. Knapton, at the Crown, in St. Paul's Church-Yard; and R. Burrough, at the Sun and Moon in Cornhill
- 19.5 cm.; pp. [24], 621, [9]
- ***Libraries*** LvU, **[IDT]** : InU, IU, MH, MiD, NN, NNC, NNUT, RPJCB
- ***Notes***

1. Despite the title, not a special geography. NUC (NC 0814433) attributes the work to J. Crull on the basis of the BL entry for the 1733 ed. of Samuel von Pufendorf, *Einleitung zu der Historie der vornehmsten Reiche und*

Staaten. The entry in BL reads: 'PUFENDORF (S. von) ... Einleitung ... Staaten, so jetziger Zeit in Europa sich befinden, etc. (Vierter Theil zu Herrn Samuel, Frey-Herrns von Puffendorf, Einleitung zu der Historie der vornehmsten Reiche und Staaten von Asia, Africa und America, welche ein gelehrter Englander [i.e., J. Crull?] kurtzlich abgehandelt und beschreiben. Anjetzo ... ubersetzt von C.J.W.). 1733, etc. 8vo.' BL's attribution is based, presumably, on that provided by the translator (i.e., C.J.W.) in his 'introduction to the most gentle reader.' The attribution made by C.J.W., however, is also a surmise, not a statement of fact.

2. The book was reviewed in the *Journal des Scavans, avec les supplemens pour les mois d'Octobre, Novembre, Decembre*, 1708, 42 (1709) pp. 240-2. The reviewer noted that, despite the title, there was a great difference between the method used by Puffendorf in his *Introduction à l'histoire des principaux royaumes et états de l'Europe* and that of his self-professed follower. Overall, the work is dismissed as merely another abridgement of universal history.
3. NUC (NP 0641183) records another ed. in 1736 (PPT, Vi : CaOTP).
4. Both BL and NUC spell Pufendorf's name with a single f, but give cross references to it from Puffendorf, that spelling appearing on the tps. of some published versions of his books. See PUFFENDORF (1695).

CRUTTWELL, Clement **Dy**

- **1798 (01)**
- *The new universal gazetteer*; or, geographical dictionary; containing a description of all the empires, kingdoms, states, provinces, cities, towns, forts, seas, harbours, rivers, lakes, mountains, and capes in the known world ...
- LONDON; printed for G.G. and J. Robinson ... and G. Kearsley
- 8vo, 23 cm., 3 vols.
- ***Libraries*** L, SanU : CU, DLC, MB, MeB, MWA, OCl, OClWHi, PPL : CaBViPA
- ***Notes***

1. 'Our plan is to include every part of the known world, that is capable of designation or description; pointing out its situation, particular character, its form of government, or a reference to the government to which it is subject; its commerce, and productions; and the manners, dress, and peculiarities of the people, where those are distinguishable as a class' (Preface). The principal sources are listed. A geographical dictionary rather than a gazetteer in the modern sense.

- **{1800} (02)**
- *The new universal gazetteer* ... Together with an atlas containing 26 whole-sheet maps
- DUBLIN; J. Stockdale
- 28 cm.; pp. vi, 1032
- ***Libraries*** : NcD, OClWHi
- ***Source*** NUC (NC 0815821)

- **1808, 2nd (03)**
- *The new universal gazetteer* ... dictionary. Containing ... world; with the

government, customs, manners, and religion of the inhabitants. The extent, boundaries, and natural productions, of each country; the trade, manufactures, and curiosities, of the cities and towns, collected from the best authors; their longitude, latitude, bearings, and distances, from the best and most authentic charts. With twenty-eight whole sheet maps ... Second edition. In four volumes

- LONDON; printed for Longman, Hurst, Rees, and Orme, Pater-Noster-Row; and Cadell and Davies, Strand
- 8vo, 22 cm., 4 vols.; Vol. I, pp. [5], viii-xv, [668]; Vol. II, pp. ca. 670; Vol. III, pp. ca. 670; Vol. IV, pp. ca. 690
- ***Libraries*** L, **[IDT]** : DLC, MBAT, MH
- ***Notes***

1. Contains the preface of the first ed. See EDINBURGH GAZETTEER (1822) and LONDON GAZETTEER (1825).

CUMMINGS, J[acob] A[bbot] **Y Am**

- **{1818} (01)**
- *First lessons in geography and astronomy ...*
- BOSTON; Cummings and Hilliard
- 14.5 cm.; 1 p. l., [vii]-viii, [9]-82 p.
- ***Libraries*** : DLC, MHi, OClh, PU
- ***Source*** NUC (NC 0832404)

- **{1819}, 2nd (02)**
- *First lessons in geography and astronomy*, with seven plain maps and a view of the solar system, for the use of young children, as preparatory to ancient and modern geography
- BOSTON; Cummings and Hilliard
- 14.5 cm.; pp. viii, [9]-82
- ***Libraries*** : MH, NNC
- ***Source*** NUC (NC 0832405)
- ***Notes***

1. NUC and BL record further eds.: 1822, 3rd (: CtY, MH); and 1823, 4th (L : CtY, MH, NNC, ViU).

- **1825 (03)**
- *First lessons in geography* ... system. For the use ...
- NEW YORK; Collins & Hannay, No. 230 Pearl Street. And Cummings, Hilliard, & co. Boston. Stereotyped by T.H. Carter & co. Boston
- 14.5 cm.; pp. [6], viii, 9-83
- ***Libraries*** : **NN**, NNC
- ***Notes***

1. A special geography for school children; gives relatively little historical information about the country in which it was intended for use (i.e., the U.S.). Besides giving an erratic supply of contemporary facts about other countries, it goes so far as to praise some of them. The section on astronomical/mathematical geography is poorly organized, and so may have been one of the works that inspired CORNELL (1854) to adopt her step-by-step presentation.

2. NUC records further eds.: 1825, Boston (MH, PU); 1826, 5th, Bellows Falls, 82 pp. (MH, NcRS, NN, Vt); 1826, Concord, NH (MH); 1826, Hallowell [ME], 82 pp. (MH, NNC); 1826, New York and Boston (MBAt); 1827, Boston (MH); 1828, 11th, New Haven, 82 pp. (NN); 1830, 14th, New Haven (MH); and 1840, New Haven (CtY).
3. Brigham and Dodge (1933:13) discuss (01), as does Carpenter (1963:253); Elson (1964:397-8) cites (02) and eds. of 1823, 1825, and 1826; Nietz (1961:208, 218, 222) are probably references to this work.

CUMMINGS, Jacob Abbot **Y Am**

- **{1813} (01)**
- *An introduction to ancient and modern geography*, on the plan of Goldsmith and Guy; comprising rules for projecting maps. With an atlas
- BOSTON; Cummings & Hilliard
- 12mo; pp. xx, 26
- *Libraries* : MB, MH
- *Source* NUC (NC 0832418)
- *Notes*

1. BL and NUC record further eds. (all Boston, Cumings & Hilliard): 1814, 2nd, 256 pp. (: CtHT, DLC, MHi, MnHi, NSyU); 1815, 3rd, 322 pp. (: CtY, DLC, KyU, MB, MHi, MnU, NSyU, PPAN, PU) [Sahli (1941, *passim*)]; 1817, 4th, 316 pp. (: DLC, NSyU, ViU); and 1818, 5th, 316 pp. (: CtY, DLC, KyBgW, MB, MHi, NSyU).

- **1818, 6th (02)**
- *An introduction to ancient and modern geography ...*
- BOSTON; published and sold by Cummings and Hilliard, No. 1. Cornhill. Cambridge, Hilliard & Metcalfe
- 17 cm.; pp. [iii], iv-xx, 316
- *Libraries* : DLC, MB, MHi, MiU, NSyU, ViU : **CaBVaU**
- *Notes*

1. In addition to Goldsmith (PHILLIPS, R. 1810?) and Guy (GUY, Joseph 1810?), Cummings gives credit to Pinkerton (PINKERTON 1805?), Walker (WALKER, J. 1788?), Adams (ADAMS, Daniel 1814?), Reeves, Evans (EVANS, John 1809?). Special geography of the conventional type occupies pp. 1-132; it is followed by a chronology (pp. 132-6), ancient geography (pp. 137-90), a survey of selected topics (forms of government, religions, natural curiosities, and a brief view of the universe, pp. 191-227), a section on maps and projections (pp. 228-55); ends with questions for examination and an index of toponyms.
2. BL and NUC records further eds. (Boston, Cummings & Hilliard, except as noted): 1820, 7th, 319 pp. (: AU, CtY, DLC, FMU, InU, KyU, MB, MHi, MiU, MWH, N, NjNbS, NSyU, OOxM, PV, ViU); 1821, 8th, 328 pp. (: CtY, ICJ, MB, MHi, MiD, NN, NNC); 1822, New York, n.p. (: PPeSchw); 1823, 9th, 340 pp. (L : DLC, MH, PU); 1823, 9th rev., 340 pp. (: CtY, DLC, MiD, MsU, NcU, ViU); 1825, 10th, 204 pp. (L : DLC, ICU, MH, OClW, OClWHi, ViU) [Sahli (1941, *passim*)]); 1826, 10th, Philadelphia, Tower & Hogan (: MH); and 10th, 1827, Boston, Hilliard, etc., 202 pp. (: OWoC, PPAmP, PPL).

3. Nietz (1961:205) provides a facsimile of pp. 40-1, but wrongly identifies the author as Cummins.
4. Sahli discusses (02) (1941, *passim*); provided that the omission of the first three words of the title is not significant, Carpenter (1963:253-4) provides a general discussion; Elson (1964:397) cites 1820, 7th.

CUNINGHAM, William **mSG**

- **1559 (01)**
- *The cosmographical glasse*, conteinying the pleasant principles of cosmographie, geographie, hydrographie, or navigation
- LONDON; imprinted at London by John Day, dwelling ouer Aldergate, beneath Saint Martins
- Folio, 30 cm.; pp. [12], 202 (with some misnumbering)
- ***Libraries*** C, **L**, O : DFo, DLC, ICN, MB, NN
- ***Notes***

1. In the Preface the study of cosmography and geography is justified, and they are distinguished from each other and from chorography. The first book is concerned with the lines identified in the sky by astronomers by means of relating the movements of the sun to the background formed by the pattern of the stars. PROCLUS is the chief authority referred to. The second book deals with the corresponding lines ascribed to the surface of the earth by projecting down on to it the lines identified in the sky. The third book is largely devoted to the principles of cartography, especially those concerned with transferring the graticule from a spherical to a plane surface. The fourth deals with problems met by sailors in navigation, including tides as well as finding location and setting a course. The fifth provides a brief description of the four continents, as well as of some countries, and gives lists of the latitudinal and longitudinal coordinates of numerous places. At the end there is a six-page index - a rare feature in so early a geography.
2. 'Although similar to the standard works of the period and particularly to that of Apianus, it cannot in fairness be ignored [in the history of English geography]' (Baker 1928 and 1963a:1).

   Cox contains the following note: 'This interesting book, which is very rarely found complete, is one of the most artistic productions of Day's press - both in typography and illustrations: indeed, it is superior to any book which had appeared in English up to this time. It is dedicated to Robert Dudley ... afterwards the Earl of Leicester ... The last three pages of the text are devoted to "A particular description of suche parts of America, as are by trauaille founde out", in which Columbus is not mentioned, the discovery being attributed to Vespucci - Quaritch' (Cox 1938:330-1).

   Parks says of this work: '[It] was an elementary treatise, mainly on mathematical geography. It was by no means up to date, and it was speedily superseded. None the less, it remains the first substantial English work on the subject, the first book to prove that the English had at last taken up geography' (Parks 1961:18).

   E.G.R. Taylor called it 'The first original English work on [cosmography] ... Naturally he follows closely the continental models provided by the Cosmographies of Apian and Orontius Finaeus ... and the *Cosmo-*

graphical Glasse contains little of purely geographical interest, but it is notable for its exposition ... of the method of survey by triangulation' (Taylor 1930:26). She also notes that, 'like so many of his contemporaries, Cuningham realized that his theoretical knowledge of astronomical methods was superior to the instrumental accuracy he could obtain' (Taylor 1930: 27).

Bricker notes that C. defined geography as 'the imitation, and description of the face, and picture of th'earth' (Bricker 1976:9), thus embracing both verbal description and cartographic depiction.

3. This work is analyzed at length by Allford (1964). See also Livingstone (1988).

CUNNINGHAM, Mary
See BUTLER, Samuel (1836).

CURTIS, John Charles **Y SG?**
- **{1875} (01)**
- *A first book of geography*
- LONDON; Simpkin, Marshall & Co.
- 8vo; pp. 24
- ***Libraries*** L

CURTIS, John Charles **Y SG?**
- **{1867} (01)**
- *Outlines of geography*. For school and home use
- LONDON; Simpkin, Marshall & Co.
- 8vo; pp. 48
- ***Libraries*** L

D

DAVIDSON, R. **?**
- **{1738} (01)**
- *A new guide to geography*
- LONDON
- ***Libraries*** n.r.
- ***Source*** Allford (1964:240)

DAVIDSON, Robert **Y mSG**
- **1787 (01)**
- *The elements of geography, short and plain*. Designed as an easy introduction to the system of geography in verse ... Designed for the use of schools. With or without seven copper plates

- LONDON; printed and sold by T. Wilkins, No. 23, Aldermanbury
- 16.5 cm.; pp. [3], 4-24
- *Libraries* L : MB : CaOTP
- *Notes*

1. The attribution of this title to Davidson is misleading; he is the author of *Geography Epitomized* (see DAVIDSON 1784, below). This work, which is anonymous, is, as the tp. is intended to say, the introduction or preface to Davidson's book, which is a special geography written in verse.

The anonymous author, or editor, as he would be better described, begins by addressing 'British youth, and the young people at Quebec, New Brunswick, and Nova Scotia' (p. [3]). The short work that follows thus has some claim to being the first Canadian special geography. Its general theme is that geography, when well taught, can be an agent making for peace between the people of different nations. More specifically, it introduces Robert Davidson, who is 'a tutor in one of the American seminaries of learning: he may truly be stiled an excellent teacher of geography. We the people of England, are greatly obliged to him for his care and labour to improve us: and I trust we have humility and good sense enough to learn from an American. True genius is confined to no one climate or quarter of the globe: it knows no party in politics, it scorns all the quarrels of kingdoms and states, as utterly unworthy of a moment's regard. It aims to enlighten and polish the minds of youth, and spread just ideas of religion and science, and by mingling pleasure with instruction, diffuse a clear knowledge of the stupendous works of the deity amongst mankind. Geography comprehends four extensive channels for knowledge to the mind of the rising age' (pp. [3]-4).

The four 'channels' are: (1) the obliquity of the earth's axis, which gives rise to the seasons of the year; (2) the division of the earth's surface into land and sea, which makes communication over great distances possible, and also provides for the observation of tides; (3) the variety, from place to place, of climate, soil, and minerals, which makes for 'mutual dependence' (p. 4). 'The fourth grand use of geography is this: it is of vast importance to religion and history' (p. 4).

The praise of geography continues for a further 4 pp., ending with what most academic geographers of our day would regard as a backhanded compliment: 'And who dares to stile himself scholar that is unacquainted with the *easiest* science in the world' (p. 8, emphasis added). There follows a condensed special geography, which amounts to little more than a list of place names and some astronomy deemed to be relevant.

The *Elements* is followed by a work whose tp. is that of the next entry, i.e., DAVIDSON (1784), with the difference that it is dated 1786; for more, see the next entry.

[DAVIDSON, Robert] **Y Am**

- **{1784} (01)**
- *Geography epitomized*; or, a tour round the world; being a short but comprehensive description of the terraqueous globe: attempted in verse (for the sake of memory;) and principally designed for the use of schools. By an American

- ▸ PHILADELPHIA; Rrinted [sic] and sold by Joseph Crukshank
- ▸ 12mo, 16.5 cm.; pp. vi, [1], 8-60
- ▸ ***Libraries*** : NN, PPL, PPULC, RPB
- ▸ ***Source*** NUC (ND 0067899)
- ▸ ***Notes***

1. NUC also records ed. of 1786 (RPB); this is almost certainly the work listed above; see DAVIDSON 1787.

- ▸ **{1787} (02)**
- ▸ *The elements of geography, short and plain.* Designed as an easy introduction to the system of geography. In verse, by ... Designed for the use of schools. With or without copper plates
- ▸ LONDON; T. Wilkins
- ▸ 16.5 cm.; pp. 24, 4 [v]-vi, [7]-60
- ▸ ***Libraries*** : MB
- ▸ ***Source*** NUC (ND 0067897)
- ▸ ***Notes***

1. The author is identified on the tp. of this ed. NUC also notes: 'added tp.: "Geography epitomized: or, a tour around the world: being a short but ... schools. By an American. Philadelphia, J. Crookshank; London, T. Wilkins, 1786".'
2. NUC records further eds.: 179–?, 60 pp. (CtY, NNC, PP); 1791, Burlington, 64 pp. (DLC, ViU. [This is the first American ed. in which the author is identified on the tp.]); 18–?, [Leominster, MA], (GU) [NSD 0010145 notes that this ed. is anonymous]; [Elson (1964:394) records 1800, Chapman Whitcomb]; 1803, Morris-Town, 60 pp. (NNC, RPB); and 1805, Stanford, 72 pp. (MH, MWA).
3. Hauptman (1978, n. 23) records an ed. in 1974; from his comments about it, however, it seems likely that he means 1784.

DAVIES, Benjamin **Y? Am**

- ▸ **{1805}, 2nd (01)**
- ▸ *A new system of modern geography*; or, a general description of the most remarkable countries throughout the known world[: their respective situations ... together] with their principal historical events [and political importance in the great commonwealth of nations, complied from the most modern system of geography and the latest voyages and travels, and containing many important additions to the geography of the United States that have never appeared in any other work of the kind;] illustrated with eight maps [comprising the latest discoveries, and engraved by the finest American artists] ...
- ▸ PHILADELPHIA; J[acob] Johnson & co.
- ▸ 17.5 cm.; pp. xxv, [26]-628
- ▸ ***Libraries*** : DLC, KyU, MB, MH, PSt, KyU : CaQQS
- ▸ ***Source*** NUC (ND 0068847)
- ▸ ***Notes***

1. The words in [ ] are derived from CIHM.
2. Nietz (1961:228) refers to a 'geography textbook' published by this author in 1804. Whether a book of this size and published at a relatively early date

was originally intended to be a schoolbook is a question that deserves further investigation.
3. NUC records a further ed.: 1805, 2nd, 2 vols. in 1 (DLC).

- **{1813}, 3rd, carefully corr. and rev. (02)**
- *A new system of modern geography* ... world ...
- PHILADELPHIA; Johnson and Warner, and for sale at their bookstores in Philadelphia and Richmond (Vir)
- 19 cm.; pp. 447
- ***Libraries*** : CtY, MMeT, NN, OClWHi, PPeSchw, PPL, Vi, ViU
- ***Source*** NUC (ND 0068850)
- ***Notes***

1. NUC records a further ed.: 1815, 4th, 447 pp. (KyU, MB, PHi, ViLxW).
2. Nietz (1961:212, 214-6, 218) and Sahli (1941, *passim*) discuss (01).
3. A microform version of (01) is available from CIHM; it was published in 2 vols.

DAVIS, Henry H. **mSG**

- **1830 (01)**
- *Brief outlines of descriptive geography*: to which is subjoined, a table of latitudes and longitudes
- [LONDON]; printed for John Taylor, bookseller and publisher to the University of London, 30 Upper Gower-Street
- 8vo, 16.5 cm.; pp. [3], iv-viii, 89, [3]
- ***Libraries*** : **MH**
- ***Notes***

1. The Preface contains some strictures on current methods of teaching geography. It also identifies published works then available on which teachers could draw. The information in the body of the text is provided in tabular and near-tabular form. This is a consequence of the fact that D. intended his book to be a source of information to be used by teachers, not a text to be given to students.

DAVITY, Pierre
See AVITY, Pierre d'.

DELL, Barton **?**

- **{1844}, 4th ed. ... (see below) (01)**
- *A geographical summary* ... The fourth edition, much improved, and corrected to the present time
- LONDON; Hamilton, Adams & Co. Bristol; Lavers & Ackland
- 16mo; pp. 32
- ***Libraries*** L

DEMARVILLE **Y**

- **{1757} (01)**

- *The young ladies geography*: or, compendium of modern geography; a work equally useful to youth, and masters and mistresses of schools; illustrated with all necessary maps and cuts ... by Mr. ...
- LONDON; printed by J. Haberkorn and sold by H. Chapelle
- 21 cm., 2 vols.
- ***Libraries*** : ICU, NjP
- ***Notes***

1. NUC (ND 0157162) notes: 'Added t.p. in French, text in English and French on opposite pages.'

- **1765 (02) (anonymous in this ed.)**
- *The young lady's geography*; containing an accurate description of the several parts of the known world; their situation, boundaries, chief towns, air, soil, manners, customs, and curiosities. Compiled from the writings of the most eminent authors, with particular attention to the modern state of every nation. To which is prefixed, an introduction to geography; wherein the terms made use of in that science, and the method of speedily acquiring a thorough knowledge of maps, are explained in so concise a manner, as to render the whole perfectly easy to be attained, without the assistance of a teacher. Also, an astronomical account of the motion and figure of the earth, the vicissitudes of night and day, and the four seasons of the year. Dedicated to her Majesty Queen Charlotte
- LONDON; printed for R. Baldwin, in Pater-Noster-Row, and T. Lownds, in Fleet-Street
- 12mo, 17 cm.; pp. [12], xxi, 253
- ***Libraries* L** : CtY, ICU, NN, NNC
- ***Notes***

1. This ed. is anonymous, but the author is identified by both BL and NUC. A school text; unusual in having more space devoted to America than to Europe.
2. BL records another issue of this ed. in 1766.

DESCRIPTION ... **Gt?**

- **{1620} (01)**
- *Description of the world*
- LONDON
- ***Libraries*** n.r.
- ***Source*** Cox (1935:72). Might be (07) of ABBOT (1599) with the information derived from a secondary source, and hence a ghost.

DESCRIPTION ... **?**

- **{1695}* (01)**
- *A description of the four parts of the world*, viz. Europe, Asia, Africa, and America. Giving an account of their dominions, religions, forms of government, and metropolitan cities. Also, how America was first discovered by the Europeans, and what purchases they have made therein. Collected from the writings of the best historians. Reprinted in the year MDCXCV
- EDINBURGH

- 8vo; pp. 23
- ***Libraries*** n.r.
- ***Source*** Sabin (1873, 5:367). Also Wing (1972), where this title is listed as item 1158, and the location given as 'F.S. Ferguson, London'; an additional note states: 'mostly in National Library of Scotland (Advocates'), Edinburgh.'

- **{1770} (02)**
- *Description of the four parts of the world* ... America
- Place/publisher n.r.
- 12mo
- ***Libraries*** n.r.
- ***Source*** Sabin (1873, 5:367)
- ***Notes***

1. See also NEW DESCRIPTION (1689).

DESCRIPTION ... **mSG**

- **[1800?] (01)**
- *A description of the four parts of the world*, viz. Europe, Asia, Africa, America, with the several kingdoms, etc. contained therein. Together with the religion, nature of the air, soil, and different traffick of each province or kingdom. To which is added for the more easy understanding the form of the world
- LONDON; printed and sold [in London]
- 12mo, 15 cm.; pp. 24
- ***Libraries*** L
- ***Notes***

1. A pamphlet, bound with chap books in BL. It is a popular special geography. There is no description of the British Isles. The last two pages carry 'An hymn on the creation of the world.'

DEVEREUX, Marion **Y SG?**

- **{1866} (01)**
- *Geography in rhyme*, adapted for young pupils, and the use of schools. With questions for examination
- LONDON; T.J. Allman
- 8vo; pp. 211
- ***Libraries*** L

DEVEREUX, Robert, and others **NG**

- **1633 (01)**
- *Profitable instructions*: describing what speciall obseruations are to be taken by trauellers in all nations, states and countries; pleasant and profitable. By the three much admired, Robert, late Earl of Essex. Sir Philip Sidney. And Secretary Davison
- LONDON; printed for Benjamin Fisher, at the signe of the Talbot without Aldergate

- 12mo, 12 cm.; pp. [9], 103
- ***Libraries*** L (cropped) : CSmH, CtY, DFo, DLC, ICN
- ***Notes***

1. Very short; fewer than 40 words per page. The instructions by Davison come first (pp. 1-24). He provides a list of classes of information to be gathered by travellers. If gathered, such information would make possible the compilation of a special geography. The rest contains advice to travellers, some practical, some moralistic.

DICK, Archibald Hastie **Y SG?**

- **{1869} (01)**
- *Geography, elementary*
- ? Murby
- ***Libraries*** n.r.
- ***Source*** ECB, Index (1856-76:130)

- **{[1870]} (02)**
- *An elementary geography for schools*
- LONDON; Thomas Murby
- 8vo, 2 pts.; pp. 196
- ***Libraries*** L
- ***Notes***

1. BL notes: '[Excelsior School Series].'

DICTIONARY ... **Gt**

- **{1592}**
- ***Notes***

1. E.G.R. Taylor lists *A Dictionary, Historical, Geographical, Astronomical and Poetical*, and adds: 'Licensed, 26 June, 1592. Probably that of Charles Estienne, not published until 1670' (Taylor 1934:203). Wing (1948, Vol. 2) lists 'Estienne, Charles. Dictionarium historicum, 1670,' and cites Madan 2865. Madan's entry No. 2865 is: 'Stephanus, Carolus. Dictionarium Historicum, Geographicum, Poeticum, Authore Carolo Stephano ... Editio novissima; ... Recensuit, supplevit, locisque pene infinitis emaculavit Nicolaus Lloydins ... A general dictionary of Classical and Biblical proper names, with the geographical part extended over medieval times' (Madan 1912:247-8). Though the edition of 1670 was the first printed at Oxford, others were published at Paris at least as early as 1553 (NUC).

   Whatever the source of Taylor's entry, it seems likely that no dictionary with the English title that she lists was ever published; but see GEOGRAPHICAL DICTIONARY (1602) and THORIE.

DIONYSIUS (Periegetes) **Y tr**

- **{1572} (01)**
- *The Surueye of the world*, or situation of the earth, so muche as is inhabited. Comprysing briefly the generall partes thereof, with the names both new and olde, of the principal countries, kingdoms, peoples, cities, towns,

postes, promontories, hils, woods, mountains, valleyes, rivers and fountains therein conteyned. Also of seas, with their clyffes, reaches, turnings, elbows, quicksands, rocks, flattes, shelues and shoares. A work very necessary and delectable for students of geographie, saylers, and others. First written in Greeke by Dionise Alexandrine, and now Englished by Thomas Twine, Gentl.

- LONDON; imprinted at London, by Henrie Bynneman
- 8vo, 15 cm.; pp. [92] of which 6 are preface and 84 text
- ***Libraries*** L, O : DFo, MiU, PPRF, PU
- ***Notes***

1. 'The original of this work, known as Periegesis, was a description of the habitable world written in Greek hexameters by Dionysius Periegetes, also known as Lybicus or Africanus. Nothing is known of the date of writing or of the nationality of the author, but he is supposed to have been an Alexandrian of the time of Hadrian [ca. 100 A.D.]. The work was popular in ancient times as a school book. - from Encycl. Brit., 13th edit.' Cox (1938: 334). FREE (1789) also provided a translation, in the form of blank verse.

DONNE B. **?**

- **{1806} (01)**
- *Treatise of modern geography*
- Place/publisher n.r.
- 8vo
- ***Libraries*** n.r.
- ***Source*** Peddie and Waddington (1914:227)

DOUGLAS, J[ames] **Y SG?**

- **{1852a} (01)**
- *An introductory geography*[, for the use of junior pupils;] intended as an introduction to the text-book of geography
- LONDON
- ***Libraries*** n.r.
- ***Source*** Allford (1964:246)
- ***Notes***

1. That this is James D. is a surmise; Allford records only the initial J.
   The words in the title set in [ ] are taken from the BL entry for the ed. of 1869.
2. BL and NUC record further eds. (all Edinburgh): 1867 (L); 1869, 88 pp. (: DLC); 1870 (L); 1871 (L); 1872 (L); 1876 (L); 1878, 88 pp. (L); 1881, 88 pp. (L); and also 1888, 1891, 1892, 1895, and 1899.

DOUGLAS, James **Y SG?**

- **{1871} (01)**
- *Progressive geography for the use of schools*, etc.
- EDINBURGH
- 8vo
- ***Libraries*** L

▸ ***Notes***
1. BL records further eds. (all Edinburgh; all after 1880 have 160 pp.): 1872, 1875, 1876, 1878, 1879, 1880, 1882, 1883, 1884, and 1885. There were further eds. in 1889, 1892, 1898, and 1901.

DOUGLAS, J[ames D.] **Y SG?**
▸ **{1852b} (01)**
▸ *A textbook of geography*
▸ LONDON
▸ ***Libraries*** n.r.
▸ ***Source*** Allford (1964:246). That this is James is a surmise; Allford records only the initial J.

▸ **{[1869]} (02)**
▸ *A text-book of geography for the use of schools*
▸ EDINBURGH
▸ 8vo
▸ ***Libraries*** L
▸ ***Notes***
1. Assumed to be the successor to (01).
2. BL records further eds. (all Edinburgh; all after 1876 have 348 pp.): 1870; 1871; 1873, 3rd; [1875] 1876, 4th; 1880, 6th; 1883, 7th; 1885, 8th; and 1887, 9th.

DOWER ?
▸ **{1838} (01)**
▸ *Geography, political, Dower's*
▸ ? Orr
▸ Imp. 4to
▸ ***Libraries*** n.r.
▸ ***Source*** BCB, Index (1837-57:110)

DOWLING, J. **Y SG?**
▸ **{1832} (01)**
▸ *An introduction to Goldsmith's grammar of geography*
▸ LONDON
▸ 12mo
▸ ***Libraries*** L
▸ ***Notes***
1. BL records a further ed.: 1851, new.

DOWLING, J. **Y mSG**
▸ **{1831}a (01)**
▸ *Key to the five hundred questions on the maps* of Europe, Asia, Africa, etc
▸ LONDON

- 12mo
- *Libraries* L

DOWLING, J. **Y mSG**
- **{1831}b (01)**
- *Questions on Goldsmith's geography.* Five hundred questions on the maps of Europe, Asia, Africa, etc. Principally compiled from the maps in the last edition of Goldsmith's grammar of geography
- LONDON
- 12mo
- *Libraries* L
- *Notes*

1. See PHILLIPS, R. (1803).

DOWNE, B. **?**
- **{1804} (01)**
- *Modern geography*
- Place/publisher n.r.
- *Libraries* n.r.
- *Source* Allibone (1854)

[DRUMMOND, M. Gawin] **Y**
- **1708 (01)**
- *A short treatise of geography general and special.* To which are added tables of the principal coins in Europe and Asia; with those mentioned in the Holy Scriptures, and ancient Roman authors. Collected from the best authors upon that subject, for the use of schools
- EDINBURGH; printed by Mr. Andrew Symson
- 8vo, 15.5 cm.; pp. [4], 90
- *Libraries* **L** : DFo
- *Notes*

1. Intended for the 'young scholar.' Presents an elaborate classification of the peoples of the world depending on their location relative to one another: in the same or opposite hemispheres (in terms of both latitude and longitude), and relative to the location in the sky of the sun at noon in different seasons of the year (equatorward or poleward of the people in question). This is part of general geography, as defined by Varen. D. ends this discussion of the classification of people with the observation that they are also 'distinguished into a great many nations and people, as French, German, Italian, etc.' (p. 14). He thus demonstrates the desire to base his classifications on a universal principle, and, simultaneously, to take empirical observations into account.
2. ESTC records a similar work of this date at EU, but lists the author as Drummond, M. George.

- **{1714}, 2nd (02)**
- *A short treatise of geography general and special.* To which is added, a brief introduction to chronology with tables ...
- EDINBURGH
- 16 cm.; 4 p. l., 136 p.
- ***Libraries*** : NNC

- **{1740}, 3rd (03)**
- *A short treatise of geography general and special.* The third edition. To which is added, brief introduction to chronology, with tables of the principal coins in Europe and Asia, with those ... [as in 01] authors upon these subjects, for the use of schools
- EDINBURGH; printed for T. and W. Ruddimans and sold by Gawin Drummond, at his shop in the Parliament-house and other booksellers in town and country
- 15 cm.; pp. [4], 131
- ***Libraries* L**
- ***Notes***

1. The author's full name is given in this ed. Claims to be revised since the last ed.
2. Discussed in Bowen (1981:314).

DRURY, Luke **Y mSG? Am**

- **{1822} (01)**
- *A geography for schools*, upon a plan entirely new, consisting of an analytical arrangement of all the great features, particularly adapted to an atlas of forty luminous and concise maps, wherein every important article of natural geography, and the chief towns, are plainly and distinctly denoted by references exactly corresponding to the analysis; which also gives tabular and comprehensive views of statistical geography, dimensions, altitudes, population, etc., etc., volcanoes, cataracts: and also of ancient geography, upon the same plan
- PROVIDENCE; printed by Miller & Hutchins
- 26 cm.; 42 p.
- ***Libraries*** : CtY, DLC, LU, NBuC, NNC, PU, RPB
- ***Source*** NUC (ND 0385485). Cited by Elson (1964:397).

DUFAUZET, A.G. **mSG**

- **1733 (01)**
- *Geography epitomiz'd*
- LONDON; D. Cole, sculp.; sold by Thomas Bonles [sic] in St. Paul's Church-Yard, and John Bonles at the Black Horse in Cornhill
- A single sheet, approximately 51 cm square (creased)
- ***Libraries* [IDT]**
- ***Notes***

1. In effect a poster that, in a series of cartouches and tables, presents a sum-

mary of the information contained in the general and special geographies of the time. At the left foot of the sheet, Dufauzet (sp.?) is identified as the 'inv[entor?].'

DUNN, Samuel **At**

- **1774 (01)**
- *A new atlas of the mundane system*; or of geography and cosmography: describing the heavens and the earth, the distances, motions and magnitudes, of the celestial bodies; the various empires, kingdoms, states, republics, and islands, throughout the known world. The whole elegantly engraved on sixty-two copper plates. With a general introduction to geography and cosmography; in which the elements of these sciences are compendiously deduced from original principles, and traced from their invention to the latest improvements
- LONDON; printed for Robert Sayer, map and print-seller, (No. 53) in Fleet-Street
- Folio, 47 cm.; pp. [4], 20, 4
- *Libraries* **L** : DLC, IU, OClWHi, PPAmP
- *Notes*

1. Essentially an atlas. The second series of four unnumbered pp. contains the geographical part of the text. On pp. 3-4, a summary description, characteristically 50 to 80 words, is provided for the countries displayed in the 42 maps that follow.
2. There were later eds. (L) and (SanU) have 1778, as do (: CtY, ICN, ICU, MiD, MiU, NjP, PPL). There were also: 1788, 2nd (SanU : MiU, NN); 1786-9, 3rd (: DLC, MB); 1796, 4th (: DLC, PPAmP); 1800, 5th (: DLC); 1810, 6th (BfQ : LNT, MdBP, MH); 1816, 6th (: PV).

DU PERIER (Monsieur)
See BELLEGARDE (1708).

DUVAL, Pierre **Y tr**

- **1685 (01)**
- *Geographia universalis*: the present state of the whole world: giving an account of the several religions, customs, and riches of each people; the strength and government of each polity and state; the curious and most remarkable things in every region; with other particulars necessary to the understanding history [sic], and the interest of princes. Written originally at the command of the French king, for the use of the dauphin, by the Sieur Duval, geographer in ordinary to his majesty; and made English by Ferrand Spence
- LONDON; printed by H. Clark, for F. Pearce, and are to be sold by Benjamin Cox, at the Princes Arms in Ludgate-Street
- 8vo, 17 cm.; pp. [20], 373 [428, which is misprinted as 482, according to the pagination, but 298 is followed by 353], [2]
- *Libraries* C, **L**, OMa : DLC, MH, N, ViU

▸ *Notes*
1. Begins with northern lands, and then presents the continents in the unusual order of America, Africa, Asia, Europe; includes with the last the island colonies of European powers in the East and West Indies. (This may have been the French tradition; See AVITY 1615.) About half the book is devoted to Europe.

▸ **{1690}, 2nd, corrected ... (see below) (02)**
▸ *Geographia universalis* ... to his majesty. The second edition, corrected and enlarged, by R.M. M.D
▸ LONDON; printed by J. Rawlins and H. Clark, and are to be sold by most booksellers
▸ 8vo, 18 cm.; pp. [8], 376
▸ ***Libraries*** CE, O : NN
▸ ***Source*** NUC (ND 0465495), which notes: 'The incorrect page references in the index reflect the paging errors in the first edition ... This 1690 edition is correctly paged. The catchword "an" on p. 376 is apparently a printer's error. It does not appear in the British Museum copy of a similar 'second edition, corrected and enlarged by R.M. M.D." published in 1691.' See (03) below.

▸ **1691 (03)**
▸ *Geographia universalis* ... enlarged, by R.M. M.D
▸ LONDON; printed for Tho. Newborough at the Golden Ball in S. Paul's Church-Yard
▸ 4to, 19 cm.; pp. [8], 376
▸ ***Libraries*** L, O : CLU, NSyU, ViU
▸ ***Notes***
1. The BL copy is apparently a proof copy for it has two blank leaves between every pair of pages. Spot-checking suggests that the text is virtually the same as that of 1685. The only change noticed was the omission of two lines on p. 249.

DUVAL, Pierre
See GEOGRAPHICAL DICTIONARY (04).

DWIGHT, Nathaniel **Y Am**
▸ **{1795} (01)**
▸ *A short but comprehensive system of the geography of the world*: by way of question and answer. Principally designed for children, and common schools
▸ HARTFORD, [CT]; printed by Elisha Babcock
▸ 17 cm.; pp. 240
▸ ***Libraries*** : CSmH, InU, MB, MH, N, NN
▸ ***Source*** NUC (ND 0470313)
▸ ***Notes***
1. See notes following (03) below.

- **{[1795]}, 2nd Connecticut, enl. and improved (02)**
- *Short but comprehensive system of the geography of the world ...*
- HARTFORD; printed by Hudson & Goodwin. Sold by them, at their book store, opposite the North meeting-house. By I. Beers, New Haven. By B. Talmadge, & co., Litchfield. By T.C. Green, New-London; and by Andrew Huntingdon, Norwich
- 16.5 cm; 221 p., 1 l.
- ***Libraries*** : DLC, MH, NjP, RPJCB
- ***Source*** NUC (ND 0470314)
- ***Notes***

1. NUC records further eds.: [1795], Philadelphia, 214 pp. (MB); and [1796?], 2nd Connecticut, Hartford, 225 pp. (NN) {ND 0470317 notes: 'Evans 286078; misdated [1795]; contains reprint of a letter from the Boston, [1796] edition'}; Elson (1964:394) records 1796, 2nd, Boston, David West.

- **{[1796]}, 2nd, enlarged ... (see below) (03)**
- *Short but comprehensive system of the geography of the world ...* The first Albany edition, from the second Hartford edition, enlarged and improved
- ALBANY; printed by Charles R. and George Webster (according to Act of Congress). Sold at their book-store, in the white house, corner of State and Pearl-streets ...
- 10.5 cm.; pp. 192
- ***Libraries*** : DLC
- ***Notes***

1. This book had a complicated publishing history. BL and NUC record further eds.: 179–, 1st Albany ed., from 2nd Hartford, Albany, NY, 189 pp. (: DLC); 1796, 2nd, Boston, 215 pp. (: CSt, CtY, MH, MiU, MWA, NNC, OO, PU, RPJCB); 1796, Hartford, 215 pp. (: DLC, InU, MHi, NSyU, OClWHi); 179–(?), 3rd Connecticut, Hartford, 215 pp. (: CtY, DLC, MH, NjP, OClWHi); 1797, 4th, Boston, 215 pp. (: MB, MH, MWA, NSyU, RPJCB); 1798, 1st Albany from 2nd Hartford, 187 pp. (: DLC); 1799, 2nd Albany from 2nd Hartford, 187 pp. (: CoU, MWA); 1800, 4th Connecticut, Hartford, 214 pp. (: DLC, MH, MiU, OO, PP, RPJCB : CaBVaU); 1801, 6th, Boston, 215 pp. (: DLC, MB, MH, MHi, PU); 1801, 1st New Jersey, Elizabethtown, [228] pp. (: CTY, NN, PU); n.d., 5th Connecticut, Hartford, 214 pp. (: DLC, InU, MB); 1802, Philadelphia, 214 pp. (: CtHT, ViW); 1803, 2nd New Jersey, Elizabeth-town, [228] pp. (: CtY, DLC, IaU, NjP); 1804, 3rd Albany from 2nd Hartford, 216 pp. (: ICN, MB, OHi); 1805, 6th Connecticut, Hartford, 216 pp. (: CtH, CtW, GMM, InSNHi, MWA, NN, NSyU, OClWHi, WGr); 1805, 3rd New Jersey, New York (: AU, CtY, MH); 1805, 1st Northampton, Northampton, MA, 209 pp. (: CtY, MH, NSyU); 1805, 2nd Northampton, 212 pp. (: MH, NNC, NSyU, ViU); 1806, 3rd Albany, 215 pp. (: N, NSyU); 1806, 3rd Northampton, 216 pp. (L : CtY, DLC, MH, MHi, MWH, NBU, NSyU, PU, ViU); 1807, 7th Connecticut, Hartford, 216 pp. (: CtHT, DLC, FE, MWA, NBAb, NN, NNC, PU); 1807, 4th Northampton, 216 pp. (: CSt, InU, MH, NcA, OClW, OClWHi); n.d., 4th New Jersey, New York, 216 pp. (: DLC, NN); 1809, 4th Albany from 2nd Hartford, 206 pp. (: MiD, MiKC, MiU, NPV, NSyU, PHi); 1810, 5th New

Jersey, New York, 216 pp. (: NjP); 1811, 5th Northampton, 216 pp. (L : MH, MHi, NcD, OClWHI, PP); n.d., 5th Albany, 214 pp. (: CtHT, DGU, DNLM, MiPon, MWA, N, NRU, NSyU, PLatS); 1812, 6th Northampton, 216 pp. (: CtY, MH, NNC, NSyU, OC); 1813, 5th Albany, 218 pp. (: MPB, MWA, NRU, NSyU); 1813, 6th, New York, 216 pp. (: MH, NjR, NNC); n.d., 6th Albany from 6th Northampton, 215 pp. (: N); and 7th Northampton, 216 pp. (: CSmH, CSt, DLC, MH, NcD, NSyU)
BL records 3rd Northampton, 1806, and 5th Northampton, 1811.
2. The following notes are based on 4th Connecticut, 1800 (**CaBVaU**). Uses 'the form of a catechism' because 'it admits of it being much more comprehensive, and more easily understood by children, than any of the small geographies which have been heretofore designed for them' (p. [3]). In terms of the amount of information provided there is a bias in favour of Europe, but it is the United States that is praised most highly. Relative to the genre as a whole, there is a preference for cultural rather than utilitarian information. There is also an idiosyncratic flavour, as shown by the following extract from a 'concise description of London,' where there are 'one hundred and thirty-one charity schools, two hundred and seven inns ... five thousand and seventy-five ale-houses ... four hundred [sedan] chairs ... and one hundred and fifty thousand dwelling-houses. There are no elegant royal palaces in England. Windsor Castle is the best' (p. 33). Though ethnocentric the tone is dispassionate.
3. Carpenter provides a general discussion (1963:249-51); Nietz (1961:206, 227) discusses (01), one of the eds. of 1796 (1961:214), and the 3rd, Northampton, 1806 (1961:216, 218, 220); his information is probably derived from Sahli (1941, *passim*). Brigham and Dodge (1933:7) discuss Northampton, 1807; Elson (1964:394-5) cites five eds., and Hauptman (1978:430) draws on the 6th of Boston, 1801.

DWIGHT, Nathaniel **Y Am**

- **{1815} (01)**
- *A system of universal geography, for common schools*: in which Europe is divided according to the late act of the Congress of Vienna ... By the Rev. ...
- NEW YORK; E. Duyckinck
- 12mo; 212 p.
- ***Libraries*** : NN, NSyU, OO
- ***Source*** NUC (ND 0470353)
- ***Notes***

1. NUC records further eds.: 1816, Northampton, MA, 213 pp. (DLC, NN, PHi); 1817, Albany, 210, [2] pp. (IRoC, N, NBuG); 1817, Northampton, 216 pp. (CtY, CU, MH, NNC, NSyU). Cited by Elson (1964:396); Hauptman (1978:431) draws on the ed. of 1816.

E

EACHARD, Laurence **Dy**

- **{1692} (01)**
- *The gazetteer's, or newsman's interpreter*: being a geographical index of all the considerable cities, towns, ports ... etc. in Europe
- LONDON; printed for Tho. Salusbury at the Kings Arms next St. Dunstan's church in Fleet-Street
- 15 cm.; pp. [225]
- ***Libraries*** : CtY, IU, NcD, PU
- ***Source*** NUC (NE 0019337)
- ***Notes***

1. On the basis of later eds., this is a dictionary; it is not a gazetteer as that word is used today.
2. The OED recognizes the title of this work as the first published use of the word *gazetteer* with the meaning of geographical index or dictionary. It quotes the Preface (p. 1) of the 2nd ed. (see below), where E. writes: 'The title was given to me by a very eminent person, whom I forbear to name.'
3. NUC and OED spell the author Echard.

- **{1693}, 2nd, corrected, much enlarg'd and improved (02)**
- *The gazetteer's: or ...* cities, patriarchships, bishopricks, universities, dukedomes, earldoms, and such like ... in Europe
- LONDON
- 15 cm.; n.p.
- ***Libraries*** O : ICN.

- **1695, 3rd (03)**
- *The gazetteer's* ... like; imperial and Hance, towns, ports, forts, castles, etc. in Europe. Shewing in what kingdoms, provinces, and counties they are; to what prince they are now subject; upon, or nigh what rivers, bays, seas, mountains, etc. they stand; their distances (in English miles) from several other places of note; with their longitude and latitude, according to the best and approved maps. Of special use for the true understanding of all modern histories of Europe, as well as the present affairs, and for the conveniency of cheapness and pocket-carriage, explained by abbreviations and figures. The third edition, corrected, much enlarged and improv'd beyond the two first editions
- LONDON; printed for John Salisbury, at the Rising Sun in Cornhill
- 12mo, 15 cm.; [237 pp.] [from NUC 0019340]
- ***Libraries*** C, **L** : CLU, CtY, IU, NN
- ***Notes***

1. The publishing history of this book is complex. From the 3rd ed. on it was sometimes prepared in two parts; these were sometimes revised separately, but whether they were issued separately or together is not clear. As will be

seen in what follows, several libraries have copies containing issues of different dates bound together.
2. ESTC and NUC record further eds. as follows (all London): 1695, 1704, 2 vols., Pt. 1 is 3rd, Pt. 2 is 1704 (: NcD); 1700, 4th [256] pp. (: CtY, NUC); 1702, 5th, [257] pp. (C : NcU); and 1703, 6th, [246] pp. (: FU, IU, MdBP, PHi, PPL).

- **1704, 7th ... (see below) (04)**
- *The gazetteer's* ... in Europe. The seventh edition, corrected and very much enlarged with the addition of a table of the births, marriages, etc. of all the kings, princes and potentates of Europe
- LONDON; printed for John Nicholson ... and Samuel Ballard ...
- 15.5 cm., 2 vols. in 1
- ***Libraries*** LinL : CLU, MH
- ***Source*** NUC (NE 0019347), which notes: 'Part 2 has title: "..." London: Printed for Thomas Newborough ... and George Sawbridge'
- ***Notes***

1. DLC has 2 vols. in 1, Pt. 1 is 8th, 1706; Pt. 2 is 1704. ESTC and NUC record further eds. (all London): 1706, 8th (C); 1706, 1707, 2 vols. in 1, Pt. 1 is 8th, Pt. 2 is 2nd (: MB, PCarlD); 1707, 9th, [390] pp. (: MB, MWA, PV, WaPS); 1707, 1709, Pt. 1 is 10th, Pt. 2 is 2nd, 1707 (: InU, KU); 1709, 1710, 2 vols. in 1, Pt. 1 is 10th, Pt. 2 is 3rd (EU, LRIS : CtY, DLC, IU, MH); 1710, 1716, 2 pts. in 1 vol., Pt. 1 is 11th, Pt. 2 is 3rd (: InU, IU); 1716, 11th (: CU, ICN, NjP); 1716, 1718, Pt. 1 is 11th, Pt. 2 is 4th (: AU, CtY); 1724, 2 pts. in 1 vol., Pt. 1 is 12th, Pt. 2 is 5th, London, F. Knapton (HIU : CoU, DLC, MB, RPJCB).

- **1724, 12th ... (see below) (05)**
- *The gazetteer's;* ... in Europe. The twelfth edition, corrected, and very much enlarged with the addition of all the towns in Great-Britain, which send members to Parliament; and of the towns and other places that give titles to the nobility; with the counties they lie in, and their distances from London
- LONDON; printed for J. Knapton ... R. Robinson ... and S. Ballard
- 17 cm., 2 vols. in 1
- ***Libraries*** : CLU, InU
- ***Source*** NUC (NE 0019360), which notes: 'Part 2 has title: "..." The fifth edition'
- ***Notes***

1. ESTC and NUC record further eds. (all London except as noted): 1724, 1732, 2 vols. in 1, Pt. 1 is 12th, Pt. 2 is 6th (: MoU); 1731, 1732, 2 vols. in 1, Pt. 1 is 13th, Pt. 2 is 6th (: DLC, NjR : CaBViP); 1738, 2 vols. in 1, Pt. 1 is 14th, Pt. 2 is 7th (: CLU, CtY, ICN, OGK); 1740, 2 vols. in 1, Pt. 1 is 15th, Pt. 2 is 8th, Belfast, with an appendix containing a description of Ireland (: NN); 1740, 1741, 2 vols. in 1, Pt. 1 is 15th, Pt. 2 is 8th, Dublin (: IU, ViU); 1741, 2 vols. in 1, Pt. 1 is 15th, Pt. 2 is 8th (: CtY, DLC, MB, MBAt, MiU, NcD, RPJCB); 1744, 2 vols. in 1, Pt. 1 is 16th, Pt. 2 is 9th (: CtY, RPJCB, TNJ); 1746 (: RPJCB); and 1751, 2 pts in 1, Pt. 1 is 17th, Pt. 2 is 10th (: DFo, MBAt).

EACHARD, Laurence **mSG**

▸ **1691 (01)**
▸ *A most compleat compendium of geography, general and special*; describing all the empires, kingdoms, and dominions, in the whole world. Shewing their bounds, situation, dimensions, ancient and modern names, history, government, religions, languages, commodities, divisions, subdivisions, cities, rivers, mountains, lakes, with their archbishopricks, and universities. In a more plain and easie method, more compendious, and (perhaps) more useful than any of this bigness. To which are added, general rules for making a large geography very necessary for the right understanding of the transactions of these times. Collected according to the most late discoveries, and agreeing with the choicest and newest maps. By ... of Christ's Colledge in Cambridge
▸ LONDON; printed for Thomas Salusbury at the sign of the Temple near Temple-Bar in Fleet-street
▸ 12mo, 13 cm.; pp. [36], 168
▸ *Libraries* **L**
▸ *Notes*

1. A pocket reference book intended for a university audience. Facts are presented in a relatively systematic way; they are generally limited to the names of the districts composing large political units, their relationship to places in the same general locality but known in earlier times by other names, their political affiliation (if this is not obvious, e.g., Dukedom of Milan ... now under King of Spain), and their chief town(s).

▸ **1691, 2nd, corrected ... (as below) (02)**
▸ *A most compleat compendium of geography;* ... archbishopricks, bishopricks, and universities ... compendious and useful than any of the lesser sort. Together with an appendix of general ... geography, with the great uses of that science. Very necessary ... the latest discoveries ... maps. 2nd edition, corrected, much improv'd, and enlarged above one quarter
▸ LONDON; printed for Tho. Salusbury ...
▸ 12mo, 14 cm.; pp. [16], 207, [25]
▸ *Libraries* **L**, O : AU, CLU, CtY, InU
▸ *Notes*

1. Europe receives 10 more pp., Asia 3, Africa 8, and America 7. A gazetteer of the principal cities of Europe has also been added.
2. BL and NUC records further eds. (all London, [15], 236, [12] pp.): 1693, 3rd (: CtY); 1697, 4th (L, O : CLU, ICN); [1700], 5th (: NN, RPJCB); 1704, 6th (L, OSg :); 1705, 7th, (L : IU); 1713, and 8th (L : CLU).

EARLE, John **NG**

▸ **1628 (01)**
▸ *Microcosmography*; or a peece of the world discovered; in essays and characters
▸ LONDON
▸ *Notes*

1. Not geography. The essays provide sketches of a variety of people.

EASTON, W. **Y SG?**
- **{1870} (01)**
- *Geography and grammar, introduction to*
- [LONDON]?; Groombr[idge]
- ***Libraries*** n.r.
- ***Source*** ECB, Index (1856-76:130)

EASTON, William **Y**
- **[1868] (01)**
- *A short introduction to geography.* By ... schoolmaster, Hereford
- LONDON; Groombridge and sons, Paternoster Row
- 16mo, 14 cm.; pp. 24, pt. 1
- ***Libraries*** C, **L**, O
- ***Notes***

1. Little more than a pamphlet. The cover acts as the tp. At three halfpence, the cheapest of any special geography whose price was noted. Deals only with Europe, and chiefly with the UK (pp. 3-18), but see note below. Many sections are lists of place names in the form of prose.

- **1870, anr ed. (02)**
- *A short introduction to geography*
- LONDON; Groombridge ...
- 16mo; pp. 48
- ***Libraries*** **L**
- ***Notes***

1. Pp. 1-24 are duplicates of the preceding (i.e., EASTON 1868). Pp. 25-33 provide information on the countries of Europe, with Asia, Africa, America, and Oceania occupying the rest.

EASY AND CONCISE **Y SG n.c. Can**
- **{1841} (01)**
- *Easy and concise introduction to modern geography*: containing an enlarged account of the British North American colonies, particularly Lower and Upper Canada, for the use of Canadian schools
- QUEBEC; printed and sold by W. Cowan & Son
- 18 cm.; pp. [1-3], 4-36
- ***Libraries*** : : CaOONL, CaOTP
- ***Source*** BoC 7631 and CIHM, which has a microform version
- ***Notes***

1. Takes the form of a catechism. Information is provided only about the Americas, with pp. 9-25 being devoted to British North America.
2. CaBVaU holds a copy of an ed. of 1842.

EASY GEOGRAPHY **Y SG?**
- **{1863} (01)**
- *Easy geography*
- LONDON

- *Libraries* n.r.
- *Notes*

1. Allford (1964:247) notes that the author is identified as 'an inspector of schools.'

EDINBURGH GAZETTEER ... **Dy**

- **1822 (01)**
- *The Edinburgh gazetteer*, or geographical dictionary: containing a description of the various countries, kingdoms, states, cities, towns, mountains, etc. of the world; an account of the government, customs, and religion, of the inhabitants; the boundaries and natural production of each country, etc. etc. Forming a complete body of geographical, political, statistical, and commercial. In six volumes
- EDINBURGH; printed for Archibald Constable and co. London; Longman, Hurst, Rees, Orme and Brown, and Hurst, Robinson & Co
- 8vo, 22 cm., 6 vols.
- *Libraries* L : CtY, DLC, NIC, NNUT, OClJC, PPL
- *Notes*

1. Supersedes *Cruttwell's Gazetteer*, the publishers having bought the copyright of that work (Advertisement, p. viii). See LONDON GAZETTEER (1825).

- **{1827} (02)**
- *The Edinburgh gazetteer* ... mountains, etc. of the world. With addenda, etc.
- LONDON; Longman, Rees, Orme, Brown, and Green
- 8vo, 22.5 cm., 6 vols.
- *Libraries* L : DNW
- *Source* NUC (NE 0038178) and BL, which calls it a reissue

EDINBURGH GAZETTEER ... **Dy**

- **{1824}, abridged (01)**
- *The Edinburgh gazetteer*, or compendious geographical dictionary, containing ... mountains, seas, rivers, harbours, etc. ... customs and religion ... inhabitants: the boundaries ... productions ... country, etc., forming ... commercial, abridged from the larger work, accompanied by maps constructed by A. Arrowsmith
- EDINBURGH; printed for A. Constable and co.; [etc.]
- 23.5 cm.; pp. [i], xx, 816
- *Libraries* : ICU, PP
- *Source* NUC (NE 0038176) supplemented by A63 *Antiquarian Books on Travel and Topography from Blackwell's*, 1984

- **{1829}, 2nd, brought down to the present time (02)**
- *The Edinburgh gazetteer* ... abridged ...
- EDINBURGH; Longman, Rees, Orme, Brown, and Green
- 8vo; pp. xx, 818
- *Libraries* L

ELEMENTARY GEOGRAPHY **Y SG?**
- **{1874} (01)**
- *Elementary geography*
- LONDON
- ***Libraries*** n.r.
- ***Notes***

1. Allford (1964:248) notes: '(Holborn Series).' BL records, under Holborn Series, 'The Holborn series [of schoolbooks].'

ELEMENTS OF GEOGRAPHY **Y Am**
- **{1789} (01)**
- *Elements of geography and astronomy*. Designed for young students in these sciences. Describing the system of the universe; the figure, motions and magnitude of the earth; its real and imaginary divisions; the situation and extent of the several empires, kingdoms, and states; their religion, and government, etc. With an explanation of maps, and the description and use of the globes ...
- PHILADELPHIA; printed and sold by John M'Culloch
- 16.5cm.; 40 p.
- ***Libraries*** : NNC
- ***Source*** NUC (NE 0080782). Cited in Elson (1964:393).

EMERSON, Frederick and Silas BLAISDALE **Y Am**
- **{1840} (01)**
- *The first view of the world*, combining geography and history
- BOSTON
- 17.5 cm.; pp. 144
- ***Libraries*** : DLC
- ***Source*** NUC (NE 0109314)
- ***Notes***

1. NUC (NE 0109446) notes that this work was republished in 1841 under the title *Outlines of geography and history* (DLC, MH). NUC records further eds. (both Philadelphia, Hogan & Thompson): 1843, 144 pp. (OClW); 1846 (MH).
2. Discussed in Culler (1945, *passim*).

EMERSON, Frederick and Silas BLAISDALE
- **{1841} (02)**
- *Outlines of geography and history*
- ***Notes***

1. See note to the previous entry.

[ENGLAND, John] **Y n.c.**
- **1820 (01)**
- *An historical geography*, containing a description of the various countries of the known world

- DUBLIN
- 12mo; pp. iv, 5-380
- ***Libraries*** **L**
- ***Notes***

1. Originally envisaged as a work of several volumes. It was intended for Roman Catholic students, the only books then available being held to be prejudiced against their religion (from preface: 'To the Catholics of the United Kingdom of Great Britain and Ireland'). The body of the text opens with a section of special geography devoted to the world as a whole, Europe, and England, predominantly the last (pp. 5-19). The balance of the book is devoted to English history, 'principally from Hume, Rapin, and the Rev. Alban Butler' (p. 19).

ENNIS, Francis **A**

- **1816 (01)**
- *A complete system of modern geography*; or, the natural and political history of the present state of the world. Illustrated with maps and engravings. Embracing the improvements of Goldsmith, Guthrie, Smith, and Brooke
- DUBLIN; printed by James Charles. No. 57, Mary Street
- 4to, 26.5 cm.; pp. xl, 1084, [16]
- ***Libraries*** **[IDT]**
- ***Notes***

1. Occasional use is made of long quotations, some with their sources identified, others not. Shows more interest in details of government and law than do most special geographies. Ireland occupies pp. 662-752. It is given less space than England, but more than any other single country; Russia and the U.S. also have more space devoted to them, in relative terms, than is usual in English works.

- **1816 (02)**
- *A complete system of modern geography ...*
- DUBLIN; printed ...
- 4to, 26.5 cm.; pp. xlviii, 1084, [8]
- ***Libraries*** **[IDT]**
- ***Notes***

1. Differs from (01) in carrying a list of some 600 subscribers. Of these 16 were booksellers who took more than one copy each; total sales amounted to approximately 1070 copies. The place of residence of the subscribers is given.

ENTERTAINING ACCOUNT ... **NG**

- **1752, 3rd (01)**
- *An entertaining account of all the countries of the known world* describing the different religions, habits, tempers, customs, traffick, and manufactures, of their inhabitants; the remarkable curiosities, amongst them, either of art or nature; the great variety of strange land and sea-animals, extraordinary plants, minerals, etc. Adorn'd, occasionally, with cuts of the habits of the

several people, uncommon animals, ceremonies, buildings, etc. all curiously engraved. The third edition
- SHERBORNE; printed by R. Goadby, in Sherborne; and sold by W. Owen bookseller, at Temple-Bar, London
- 8vo, 20 cm.; pp. 266
- ***Libraries*** **L** : DLC, Nc, NjR, NN
- ***Notes***

1. Not a special geography despite the title. Contains accounts of five voyages of travel and exploration.
2. Three other issues or eds. in 1752 according to NUC, and another in 1755.

EPITOME ... **Y**
- **{1845} (01)**
- *Epitome of geographical knowledge; ancient and modern.* Compiled for the use of the teachers and advanced classes of the National Schools of Ireland
- DUBLIN; Commissioners of National Education in Ireland
- ***Libraries*** : MH

- **1850, revised (02)**
- *Epitome of geographical knowledge, ancient ...*
- DUBLIN; published by the direction of the Commissioners ... Ireland, at their office in Marlborough-street. Sold by W. Curry, Jun., and Co., Dublin; R. Groombridge and Sons, London; George Philip, Liverpool; Fraser and Co., Edinburgh; Armour and Ramsey, and Donoghue and Mantz, Montreal, Canada; and Chubb and Co., Halifax, Nova Scotia
- 12mo, 17 cm.; pp. xii, 633
- ***Libraries*** **L**
- ***Notes***

1. Contains systematic physical and political as well as special geography. Approximately 150 pp. are devoted to historical information about countries, and roughly twice that to their contemporary condition. It would be interesting to know the author of what appears to be a fairly serious work. See note to (03) below.
2. Catalogued by BL under 'Ireland, Commissioners of National Education.'

- **{1854}, new (03)**
- *Epitome of geographical knowledge ...*
- DUBLIN; Thorn
- Sm 8vo; pp. xvi, 592
- ***Libraries*** : MB, MH
- ***Source*** NUC (NE 0142270)
- ***Notes***

1. See COMPENDIUM (1854).

- **{1857}, new, rev. (04)**
- *Epitome of geographical knowledge ...*
- DUBLIN; Alex. Thorn & sons
- 18 cm.; pp. xvi, 602

- ***Libraries*** : MH, NIC
- ***Source*** NUC (NSE 0019980)

EVANS, J.C. ?
- **{1838} (01)**
- *Geography, concise*
- ? Harvey
- 12mo
- ***Libraries*** n.r.
- ***Source*** BCB, Index (1837-57:110)

EVANS, John ?
- **{1801} (01)**
- *An epitome of geography*
- Place/publisher n.r.
- ***Libraries*** n.r.
- ***Source*** Watt (1824)

- **{1802}, 2nd (02)**
- *An epitome of geography*
- LONDON
- 16mo; pp. 208
- ***Libraries*** : MWA, PPL
- ***Source*** NUC (NE 0204344)

EVANS, John, ed. **A**
- **{1809} (01)**
- *New geographical grammar*. New system geography and univ. history of the known world
- LONDON
- 2 vols.
- ***Libraries*** : PPWa

- **{1810} (02)**
- *The new geographical grammar*; or, companion and guide through the various parts of the known world ... to which is prefixed a concise system of astronomical geography by James Ferguson
- LONDON; Albion Press, printed for James Cundee
- 8vo, 21 cm., 2 vols.
- ***Libraries*** O
- ***Source*** 19th C STC, and Howes Bookshop Ltd., 3 Trinity Street, Hastings, Sussex, UK; Catalogue 178, c January 1972

- **1811 (03)**
- *The new geographical grammar* ... world. Comprehending a copious and accurate account of the ancient and modern state of Europe, Asia, Africa, and America. With the situation, extent, and population of their islands,

peninsulas, isthmusses, capes, mountains, oceans, gulfs, bays, lakes, and rivers, also, a delineation of their climate, air, soil, vegetable productions, metals, minerals, natural and artificial curiosities; a correct historical account of the former and present state of the polity, revenues, arts, sciences, manufactures, and commerce, in the several empires, kingdoms, states, and colonies; to which ... Ferguson, esq. F.R.S. including the late discoveries of Dr. Herschel, and other eminent astronomers, in illustration of the work. With statistical tables, and a valuable ornamental atlas, accurately engraved. Concluding with an alphabetical index to the geography; and a chronology of remarkable events from the earliest periods of history to the present time. The whole superintended and revised by the Rev. ... A.M. Master of a seminary for a limited number of people, Pullin's Row, Islington; and author of 'A sketch of all religions,' 'Juvenile Tourist,' etc. In two volumes

- LONDON; printed for James Cundee, Ivy-Lane; Paternoster Row
- 8vo, 22 cm., 2 vols.; Vol. I, pp. [1], xl, 559, [1]; Vol. II, pp. [3], 560-1264
- ***Libraries*** **L** : DLC
- ***Notes***

1. Intended for the general public. Evans refers to himself as the editor of the work. Much space is devoted to history. Explicitly Eurocentric, and narrower still in outlook, as revealed by the assertion that 'the immortal Galileo lay rotting in the dungeons of the Inquisition for having dared to assert *the rotundity of the world*' (p. x, emphasis added). The section on British history ends with a tendentious account of the debates in the Irish Parliament prior to its dissolution in 1801. Overall, the work conveys a strong note of partiality and prejudice.

   The print is small, so the amount of material included is probably comparable with PINKERTON (1802). Vol. I is devoted to continental Europe, Vol. II to the British Isles (ca. 190 pp.) and the rest of the world. Asia occupies pp. 749-1089, Africa pp. 1090-163, and America pp. 1163-240. There follows a gazetteer and a chronology. Extensive amounts of historical material are provided.
2. According to NUC (NE 0204370), 'Vol. 1 has added t.-p., engr., dated 1809.'

EVANS, John and FORBES, Archibald **Y**

- **{1810} (01)**
- *A new system of geography and universal history of the known world*; comprehending a copious and accurate description of the ancient and modern state of Europe, Asia, Africa, and America; with the situation, extent, and population of their islands, peninsulas, isthmusses, capes, mountains, oceans, gulfs, lakes and rivers, also, a delineation of their clime, air, soil, vegetable productions, metals, minerals, natural and artificial curiosities. A correct historical account of the former and present state of the polity, revenue, arts, sciences, manufactures and commerce, in the several empires, kingdoms, states and colonies; the genius, customs, manners, religion, general character and pursuits of the respective inhabitants, to which is prefixed a concise system of astronomical geography, by James Ferguson,

esq. F.R.S. including the late discoveries of Dr. Herschel, and other eminent astronomers, in illustration of the work, with statistical tables; concluding with an alphabetical index to the geography; and a chronology of remarkable events from the earliest periods of history to the present time. The whole superintended and revised by ... author of 'A sketch of all religions,' 'Juvenile tourist,' etc. and ... author of 'The general gazetteer'
- [LONDON; T. Tegg]?
- [2 vols.]?
- ***Libraries*** n.r.
- ***Notes***

1. No copy of this ed. is known. The title given above is taken from (02), which follows. The date is found in the dedication.

- **1814(?), new (02)**
- *A new system of geography ...*
- LONDON; printed for Thomas Tegg, Cheapside; Khull and Co. Glasgow; and all other booksellers
- 8mo, 21.5 cm., 2 vols.; Vol. I, pp. [v] vi-xl, 559; Vol. II, pp. [iii], 560-1264
- ***Libraries*** : **MH**
- ***Notes***

1. The date is taken from the map facing the tp. At the end of the Contents the following statement appears: 'The former, half-sheet, containing the Title and the Advertisement, is to be cancelled, and that now given placed in its stead' (p. xii).
2. There are extensive direct quotations, with the author, but not the title or any other information, provided. It is very detailed for a book that is explicitly intended for students (in the dedication), with information about buildings and even furnishings; also on sports, pastimes, and education, and information of types now provided in guidebooks intended for tourists.

EVANS, Jonas **OG**
- **1821 (01)**
- *Geographical and astronomical definitions and explanations.* In two parts
- HAVERHILL; printed by Burill & Hersey
- 14.5 cm.; pp. [3], 4-36
- ***Libraries*** : **NN**
- ***Notes***

1. Provides information of the type commonly found in the opening section of special geographies. There are no illustrations, and only a reader who already knew the rudiments of the subject (i.e., not a young child) would be able to follow the text.

EVANS, Lewis **OG Am**
- **{1754} (01)**
- *Geographical, historical, political, philosophical and mechanical essays.* The first containing analysis of a general map of the middle colonies in America ...

- PHILADELPHIA; Frankli [sic]
- ***Libraries*** : PU
- ***Source*** NUC (NE 0204908)
- ***Notes***

1. Probably not special geography; there were further eds. in 1755 and 1756.

EWALD, Alexander Charles **Y**

- **1870 (01)**
- *A reference-book of modern geography.* For the use of public schools and civil service candidates. By ... Foreign correspondent of the Society of Antiquaries of Normandy; author of 'The Civil Service Guide,' 'Last Century of Universal History,' 'Our Constitution,' etc.
- LONDON; Longmans, Green and Co.
- 8vo, 17.5 cm.; pp. [5], vi-xi, [1], 330, [1]
- ***Libraries*** **L**
- ***Notes***

1. Less exhaustive than many contemporary geographical dictionaries, 'it aims at being fuller and more complete than the ordinary School Geographies' (p. v). It belongs to the tradition of special geography, though much of the information is conveyed in the form of lists.

EWING, Thomas **Y Can**

- **{1843} (01)**
- *The Canadian school geography*, by ... author of Principles of Elocution, Rhetorical Exercises, the English Learner, A System of Geography and Astronomy, and a New General Atlas
- MONTREAL; Armour & Ramsay; Kingston, Ramsey, Armour & Co.; Hamilton, A.H. Armour & Co.
- 17 cm.; 2 p. l., [11]-65, [4] p. (BoC 5256); pp. [6], 12, [3], 14-65, [5] (CIHM)
- ***Libraries*** : : CaBVaU (incomplete), CaOTP, CaQMNB
- ***Notes***

1. 'In the course of different visits which the author ... has paid to Canada, he has been struck with the great diversity, and for the most part, the antiquated and inferior character of the books used in the schools throughout the Province. In no branch of study was this more observable than in that of geography:- the greater portion of the text-books appear to be imported from the United States. The author has endeavoured to supply the deficiency by producing ... the present work which is upon the same principles as his "System of Geography"' (from the Preface).
2. A microform version is available from CIHM.

EWING, T[homas?] **?**

- **{1833} (01)**
- *Geographical vocabulary*
- Place/publisher n.r.

▸ *Libraries* n.r.
▸ *Source* Peddie and Waddington (1914:226)

EWING, Thomas **Y**
▸ **1816 (01)**
▸ *A system of geography*, for the use of schools and private students, on a new and easy plan; in which the European boundaries are stated as settled by the Treaty of Paris and Congress of Vienna: with an account of the solar system, and a variety of problems to be solved by the terrestrial and celestial globes. By ... teacher of English, geography, and history, in Edinburgh; author of 'Principles of Elocution,' the 'English Learner,' and a 'New General Atlas.'
▸ EDINBURGH; printed and published by Oliver and Boyd; and sold by Law and Whittaker, London; John Cumming, Dublin; and William Turnbull, Glasgow
▸ 16mo, 18.5cm; pp. [5], 300
▸ *Libraries* L
▸ *Notes*
1. Systematic in its organization, providing for every country a description of its 'historical,' 'political,' 'civil,' and 'natural' geography. These are defined much as in PINKERTON (1802). Introduces each country by identifying its location and boundaries, and listing the place names to be learned by the students. There follows a selection of 'interesting' facts associated with individual places within the country.

▸ **{1817} (02)**
▸ *Geography, system of*
▸ Place/publisher n.r.
▸ *Libraries* n.r.
▸ *Source* Peddie and Waddington (1914:227)
▸ *Notes*
1. NUC records one further ed.: 1823, 4th, Edinburgh, 288 pp. (CtY). BL records further eds. (all Edinburgh): 1826, 7th; 1839, 15th; 1844, 16th; 1848, 17th; 1854, 18th, rev.; 1859, 19th, rev.; 1868, 20th, rev.; 1870, 21st; 1871, 22nd; 1873, 24th; 1874, 24th; and 1878, 25th.

EWING, Thomas **Y Am**
▸ **1820 (01)**
▸ *Universal geography*, on a new and easy plan; in which the European boundaries are stated, as settled by the Treaty of Paris and Congress of Vienna: with an account of the solar system, and a variety of problems to be solved by the terrestrial and celestial globes. The whole revised, and the American part written entirely anew, being a sequel to Picketts' Geographical Grammar, by William Darby ...
▸ NEW YORK; printed and sold by Charles N. Baldwin, bookseller, No. 1 Chamber, corner of Chatham-St. - Sold also at the American Book Warehouse, No. 192 Greenwich-Street
▸ 16.5 cm.; pp. [9], 14-328

- *Libraries* : CtY, DLC, **NN**
- *Notes*

1. A typical special geography of the period. It has a motto on the tp. that reads: 'To enjoy the world as a rational being, is to know it, to be sensible of its greatness and beauty, to be sensible of its harmony, and, by these reflections, to obtain just sentiments of the Almighty Mind that framed it.' Explicitly shuns question-and-answer because Ewing wants students to think rather than learn by heart because 'the exploded method of learning words has too long dominated the thinking powers, and rendered the study of geography dry and uninteresting.'
2. NUC (NE 0216435) notes: 'Pickets' American school class book, no. 7.' See PICKET (1816).

F

FADEN, William **Y**

- **1777 (01)**
- *Geographical exercises*; calculated to facilitate the study of geography, and, by an expeditious method, to imprint a knowledge of the science on the minds of youth. With a concise introduction, explaining the principles of geography
- LONDON; printed for the proprietor (successor to the late Mr Jefferys, Geographer to the King) at the corner of St. Martin's Lane, Charing-Cross
- Oblong folio, 42 cm. [by 53]; pp. [6] [then 9 maps on verso pages, with corresponding graticule on facing recto page]
- *Libraries* L
- *Notes*

1. Gives credit in the Preface (p. [5]) to Henry Peacham, A.M., for first suggesting that the best way to learn geography was by drawing maps (in *The Compleat Gentleman*, 1622). As noted above, the book provides nine maps and nine blank graticules on which to copy the maps. The intention is that the student, by drawing the lines and symbols that form the maps of regions on the graticules provided in the book, and naming them, will learn of the existence of the regions and their location more effectively than they would by rote learning.

   The maps portray the four continents, the four hemispheres (North, South, East, West), and the whole world. The opening numbered pages contain an introduction to cartography and the art of drawing.

FAGE, Robert **A**

- **1658 (01)**
- *A description of the world*, with some general rules touching the use of globe, wherein is contained the situation of several countries. Their particular and distinct governments, religions, arms, and degrees of honour used among them. Very delightful to be read in so small a volume

- LONDON; printed by J. Owsley, and sold by Peter Stent, at the White-horse in Guilt-spur-street, between Newgate and Pye-corner
- 8vo, 16 cm.; pp. [1], 70, 2 of advertisements
- ***Libraries*** **L,** O : DLC, MB, NN, RPJCB
- ***Notes***

1. A short work in which Europe figures largely. Devotes much attention to knightly orders and their coats of arms.

- **{1663} (02)**
- [*Cosmography or, a description of the whole world ...*]
- Place/publisher n.r.
- ***Libraries*** n.r.
- ***Source*** Sabin (1873, 4:337)

- **1666, anr ed., enlarged with very many rare additions (03)**
- *Cosmography or, a description of the world*, represented (by a more exact and certain discovery) in the excellencies of its scituation, commodities, inhabitants, and history; of their particular ... used amongst them. Enlarged with very many and rare additions. Very delightful ...
- LONDON; printed by S. Griffin for John Overton at the White-Horse in Little Brittain, next door to Little St. Bartholomews-Gate
- 8vo, 16 cm.; pp. [6], 166, 2 of advertisements
- ***Libraries*** O
- ***Notes***

1. More than double the length of the first ed. Two-thirds of the new material is devoted to the world outside Europe, chiefly America. The bulk of the new matter consists of chronologies of recent events, but some is concerned with the nature of countries and their people.

   At first glance NUC seems to treat this as a different work from (01), but a close reading of NUC's entry shows that the two titles belong to different eds. of the same book.

- **{1667} (04)**
- *Cosmography* ... situation ... history: of their ...
- LONDON; printed by S. Griffin for John Overton ...
- 16.5 cm.; 4 p. l., 3-166 p.
- ***Libraries*** : CLU, MWiW
- ***Source*** NUC (NF 0011197) and Sabin (1873, 6:337). It is of this ed. that Sabin notes: 'Mostly related to America, especially from page 112 to 166.'

- **{1671} (05)**
- *Cosmographie or, a description ...*
- LONDON; printed by H.B. for John Overton at the White-Horse without Newgate at the corner of the Little Old Bailey near Fountain Tavern
- 17 cm.; 4 p. l., 3-166 p.
- ***Libraries*** : CSmH, N
- ***Source*** NUC (NF 0011195)

FAIRMAN, William **A**

- **1788 (01)**
- *A treatise on geography, the use of the globes, and astronomy*; in the order which the mutual connection and dependence of the several parts require, towards a perfect understanding of the whole
- LONDON; printed for J. Johnson, No. 72, St. Paul's Church-Yard
- 8vo, 21 cm.; pp. viii, [1], 368
- ***Libraries*** EU, **L** : CtY
- ***Notes***

1. Special geography occupies pp. 51-202, with a section on the principal cities of Europe at the end.

FALCONER, William **OG**

- **1781 (01)**
- *Remarks on the influence of climate*, situation, nature of country, population, nature of food, and way of life, on the disposition and temper, manners and behaviour, intellect, laws and customs, form of government, and religion, of mankind
- LONDON; printed for C. Dilly, in the Poultry
- 4to, 27 cm.; pp. xvi, 552, [28]
- ***Libraries*** EU, **L** : CtY, CU, CU-A, DLC, DNLM, ICJ, ICU, MH, NIC, NNNAM, PBL, PPL, PPPH, ViU, WU
- ***Notes***

1. Written by a medical doctor rather than a geographer. Though it is not special geography, it is included in this bibliography because most geographers are likely to see it as belonging to their tradition.

FALMOUTH, Viscountess
See BOSCAWEN (1854).

FARR, Edward **Y**

- **{1850} (01)**
- *Manual of geography for schools*
- LOND[ON]; A. Hall, Virtue & Co.
- 12mo/16mo
- ***Libraries*** n.r.
- ***Source*** Allibone (1854); BCB, Index (1837-57:110); and letter from MH reporting that they no longer have a copy
- ***Notes***

1. ECB, Index (1856-76:131) records 'Geography, manual of [pub.] Hall, 1858.'

- **{1861}, new ... (see below) (02)**
- *The manual of geography*, physical and political ... New edition. Edited by R. Saunders
- LONDON, Norwich [printed]
- 16mo
- ***Libraries*** L

▸ *Notes*
1. That (02) is a later ed. of (01) is a surmise.

FARR, Edward ?
▸ **{[1870]} (01)**
▸ *Our world*: its cities, peoples, mountains, seas and rivers ... With ... illustrations
▸ LONDON, Norwich [printed]
▸ 8vo
▸ ***Libraries*** L

FENNING, Daniel **Y**
▸ **{1754} (01)**
▸ *A new and easy guide to the use of the globes, and the rudiments of geography.* Wherein the knowledge of the heaven and earth is made easy to the meanest capacity: first, by giving a short and concise account of the four quarters of the world ... and by the solution of seventy useful problems, in geography, astronomy, navigation, etc. ... To which is annex'd three useful tables ...
▸ LONDON; J. Hodges
▸ 18 cm.; xxii, p., 1 l., 244 p.
▸ ***Libraries*** CoP : CtY, DLC
▸ ***Source*** NUC (NF 0083606)

▸ **1760, 2nd, with improvements by the author (02)**
▸ *A new and easy guide to the use of the globes; and* ... heavens ... world, with the distance and situation of the most principal islands and inland places, and by the solution ... etc. Written in familiar dialogues, in order to render it more easy, pleasant, and diverting to the learner; with some observations on Mr. Neale's patent globes. To which is annex'd, three useful tables. I. Shews the latitude and longitude of the principal places from the meridian of London. II. Shewing the sun's place, declination, time of rising and setting; length of days and nights, and beginning and ending of twilight every week, according to the new style. III. Shews the latitude, longitude, right ascension, and declination of the most eminent fixed stars. Recommended by several mathematicians
▸ LONDON; printed for S. Crowder, at the Looking-Glass, facing St. Magnus Church, London Bridge
▸ 12mo, 17 cm.; pp. xix, 240
▸ ***Libraries*** : N, PPL
▸ ***Notes***
1. Special geography occupies pp. 4-66. Information is given primarily with respect to location and length of day; occasionally miscellaneous items are included. The most striking of these is a list of twenty-four articles of faith attributed to Roman Catholics that is included within the description of Italy.

▸ **1769, 3rd, with ... (see below) (03)**
▸ *A new and easy guide to the use of the globes; and ...* places; and secondly by the solution ... etc. To which are annexed, three ... II. Shews ... Mathematicians. The third edition, with large corrections and improvements, particularly six large and accurate maps of the world and its different parts
▸ DUBLIN; printed for James Williams, at No. 5, in Skinner-Row
▸ 12mo, 17 cm.; pp. x, [2], 240
▸ ***Libraries*** **L**, LSM : MB
▸ *Notes*
1. Judging by the pagination no change has been made to the body of the book since 1760.

▸ **1770, 3rd, with large corrections ... (04)**
▸ *A new and easy and easy guide to the use of globes ...*
▸ LONDON; printed for S. Crowder, in Paternoster-Row
▸ 12mo, 17 cm.; pp. [1], x, 2, 180
▸ ***Libraries*** **L** : PU, WaU
▸ *Notes*
1. The use of smaller type and the elimination of much of the miscellaneous material accounts for the reduction in the number of pages.

▸ **1779, 4th (05)**
▸ *A new and easy and easy guide to the use of globes ...* by giving a concise account ... solution of upwards of seventy ... navigation, and dialling. In which are inserted three ... stars. To the whole is subjoined an appendix; containing a short account of the solar system, and of the comets and fixed stars. Recommended ...
▸ LONDON; Printed ...
▸ 12mo, 18 cm.; pp. [1], x, [2], 192
▸ ***Libraries*** **L** : CtY, MiU, NPV, RPJB
▸ *Notes*
1. The increase in length since 1770 is accounted for by the addition of the appendix. Spot-checking suggests that the rest of the text is unchanged.

▸ **1785, 5th, with large corrections and improvements (06)**
▸ *A new and easy and easy guide to the use of globes ...* latitude of the principal places, and their longitude from the meridian ... whole are subjoined, I. An appendix ... stars: II. A supplement; exhibiting a brief view of the figure and magnitude of the earth, of the nature of the atmosphere, of the theory of the tides, and a concise system of chronology. Recommended by the Rev. J. Warneford, M.A. and several eminent mathematicians. In this edition are now first inserted a view of the Copernican or true system of the universe, and the coelestial globe, neatly engraved
▸ LONDON; printed ...
▸ 12mo, 17 cm.; pp. [1], ix, [3], 213, [1]
▸ ***Libraries*** **L** : ViLxW
▸ *Notes*
1. There is more text on each page than in 1777; the special geography (apparently unchanged) now ends on p. 46. The appendix starts on p. 163, the supplement on p. 174.

▸ **{1787}, 3rd, with improvements by the author (07)**
▸ *A new and easy and easy guide to the use of globes ... geography ...* with some observations on Mr. Neale's patent globes. To which is annex'd ...
▸ DUBLIN; printed for H. Chamberlaine
▸ 18 cm.; × p., 1 l., 244 pp.
▸ ***Libraries*** : DLC, MA, MWA, RPJCB, ViU
▸ ***Source*** NUC (NF 0083612)

▸ **1792, 6th, corrected and improved, by ... (see below) (08)**
▸ *A new and easy and easy guide to the use of globes ...* inserted four useful ... London. II. Shews the number of miles contained in a degree of longitude, at every degree of latitude. III. Shews where every climate ends, with the length of the longest day at the end of each. IV. Shews the right ascension, declination, latitude, and longitude of the most eminent fixed stars. To the whole are subjoined ... stars: II. A supplement ... chronology. The sixth edition. Corrected ... by Joseph Moon, mathematician, Salisbury. Recommended ... edition is inserted ...
▸ LONDON; printed for S. Crowder ...
▸ 12mo, 17 cm.; pp. [1], ix, 213, [3]
▸ ***Libraries*** **L** : DLC, MBAt, MnU
▸ ***Notes***
1. An advertisement by Moon identifies the changes he has made; they are minor.

▸ **{1796}, 6th, corr. and impr., by J. Moon (09)**
▸ *A new and easy guide to the use of globes ...*
▸ DUBLIN; J. Rice
▸ ***Libraries*** : InU

▸ **{1797}, 7th, corr. and improved, by ... Moon (10)**
▸ *A new and easy guide to the use of globes ...*
▸ DUBLIN; printed by P. Wogan
▸ 18 cm.; pp. ix, [23], 213
▸ ***Libraries*** : DLC

▸ **1798, 7th, corrected ... (as 1792) (11)**
▸ *A new and easy guide to the use of globes ...* chronology. The seventh edition ... Salisbury. In this edition ...
▸ LONDON; printed for J. Johnson
▸ 12mo, 18 cm.; pp. [1], ix, [3], 213, [3]
▸ ***Libraries*** **L** : MH
▸ ***Notes***
1. Still carries the advertisement to the sixth ed.; that and the pagination suggest that no changes have been made since 1792. Presumably the London publishers ignored the Dublin eds. in identifying this as the seventh.

▸ **1804, 8th, corrected and greatly improved by the editor (12)**
▸ *A new and easy guide to the use of globes, and ... geography.* In which the knowledge ... is rendered simple and easy; first, by a concise account ... world, and all the principal islands; and secondly, by the solution of a great

variety of useful problems in geography ... navigation, dialling, etc. Illustrated with ten plates and maps. To the whole are subjoined two appendices: the I. contains ... system, of the comets, and fixed stars. The II. containing dissertations on the figure ... earth; - on the atmosphere; - on the tides; - and a short system of chronology

- LONDON; printed ... Johnson, G. and J. Robinson, F. and C. Rivington, G. Wilkie, Scatcherd and Letterman, Longman and Rees, and C. Law
- 12mo, 18 cm.; pp. [1], viii (vii misnumbered vi), 233, [3]
- ***Libraries*** **L**
- ***Notes***

1. According to the advertisement to this ed., the material on Europe has been reorganized to bring it into line with the new political boundaries obtaining on that continent. The special geography occupies pp. 15-50.

- **1809, 9th ed., corrected ... (13)**
- *A new and easy guide to the use of globes ...*
- LONDON; printed ... Johnson; F. and C. Rivington; Wilkie and Robinson; G. Robinson; Scatcherd and Letterman; C. Law; Longman, Hurst, Rees, and Orme; B. Crosby and Co.; J. Harris; and Sherwood and Co.
- 12mo, 18 cm.; pp. [1], viii (vii misnumbered as vi), 234, [1]
- ***Libraries*** **L** : CtY
- ***Notes***

1. Special geography occupies pp. 15-53; some changes have been made necessary by the outcome of wars in Europe.

- **{1827}, 1st American (14)**
- *A new and easy guide to the use of the globes*: to which is added, a plain method of constructing maps. First American ed., revised and corrected by Jos. Walker. Sr. and Jos. Walker, Jr. ...
- BALTIMORE; Fielding Lucas, Jr. No. 138, Market-Street
- 18 cm.; × p., 1 l., 244 p.
- ***Libraries*** : DGU, ViW
- ***Source*** NUC (NF 0083619)
- ***Notes***

1. According to NUC (NF 0083620), there was an ed. in 1875, but MWA, who are alleged to hold it, report that they have no record of such a book. See (06) above for a possible source.

FENNING, Daniel **Y mSG**

- **1764 (01)**
- *The young man's book of knowledge*: being a proper supplement to the young man's companion. In five parts, viz. Part I. Of knowledge in general ... Part II. Geography in a manner entirely new: containing, (by question and answer) 1. A general description of the four quarters of the world. 2. The situation, extent, and chief cities of the several kingdoms and countries of each quarter. 3. The nature and description of the globes, and explanation of the terms used in geography. 4. Tables of the latitude and longitude of several principal places; with many useful and necessary problems on the

terrestrial and celestial globes. Part III. Geometry ... Part IV. Natural philosophy in general. Part V. Theology ...

- LONDON; printed for S. Crowder, in Pater-Noster-Row; and B. Collins, in Salisbury
- 8vo, 17 cm.; pp. [3], viii, [4], 380 (93-110 twice), [4]
- ***Libraries*** **L** : ICU, NjP
- ***Notes***

1. The special geography occupies pp. 93-118 (second numbering). Gives the relative location and political subdivisions of the countries, plus a little miscellaneous information. The use of the globes and related topics occupy pp. 119-56.

- **{1774}, 3rd, with additions (02)**
- *The young man's book of knowledge ...*
- LONDON; printed for S. Crowder and B. Collins
- 8vo, 17 cm.; pp. xii, 376
- ***Libraries*** : CU

- **{1786}, 4th, rev., corrected and greatly improved (03)**
- *The young man's book of knowledge ...*
- LONDON; S. Crowder. Salisbury; B.C. Collins
- 12mo; pp. xiv, 381
- ***Libraries*** : ICU

- **{1794}, 5th, with the addition of the chronological part (04)**
- *The young man's book of knowledge ...*
- LONDON; S. Crowder, and B.C. Collins of Salisbury
- 18 cm.; pp. xv [1], 432
- ***Libraries*** : InU, NNC
- ***Source*** NUC (NF 0083706)

FENNING, Daniel, COLLIER (COLLYER in later eds.), J., and others **A**

- **1764-5 (01)**
- *A new system of geography: or, a general description of the world.* Containing a particular and circumstantial account of all the countries, kingdoms, and states of Europe, Asia, Africa, and America. Their situation, climate, mountains, seas, rivers, lakes, etc. The religion, manners, customs, manufactures, trade, and buildings of the inhabitants. With the birds, beasts, reptiles, insects, the various vegetables, and minerals, found in different regions. Embellished with a new and accurate set of maps, by the best geographers; and great variety of copper-plates, containing perspective views of the principal cities, structures, ruins, etc.
- LONDON; printed for S. Crowder at the Looking-Glass, in Pater-noster-Row; and sold by Mr. Jackson, at Oxford; Mr. Merril, at Cambridge; Mess. Smith, in Dublin; and all other booksellers in Great Britain and Ireland
- Folio, 35 cm., 2 vols.; Vol. I (1764), pp. [5], xxxviii (misnumbered xxxivii], [6], 5-519 (519 misnumbered 119), [8]; Vol. II (1765), pp. 784, [16]

▸ *Libraries* AbU, CoP, MrU, NwP, **OSg** : DLC, FU, ICRL, MsU, NNC, OCl, PPT, ViU
▸ *Notes*
1. The approach is essentially chorographical, even topographical in the case of the British Isles. There are no sections devoted wholly to history or chronology. Vol. I contains Asia (ca. 60 per cent) and Africa; Vol. II contains Europe (almost 80 per cent) and America. Being closely comparable in size and date with BRICE (1759), the relative merits of organization offered by geographical dictionaries and special geographies could be assessed by a comparison of the two.

PAYNE, in his Preface to (1791), acknowledged that he laid out his book on roughly the same plan as that used by Collyer in his book of 1765. I have been unable to find a special geography published in 1765 of which Collyer (or Collier) was identified as the principal author on the tp. However, given the statement made by Collyer in the Preface to the ed. of 1771 (see note to (04) below), it seems very probable that the work in question was the one to which this is a note.

▸ **1766-5, anr ed. (02)**
▸ *A new system of geography ...*
▸ LONDON; printed ... Pater-Noster-Row ...
▸ Folio, 36 cm., 2 vols.; Vol. I, pp. [5], xxxviii (misnumbered xxxivii), [6], 5-519, [9]; Vol. II, pp. 784, [20]
▸ *Libraries* **L**
▸ *Notes*
1. Vol. I, published 1766; Vol. II, published 1765.

▸ **1771, 3rd, revised, corrected, and improved (03)**
▸ *A new system ... geography*; or a ... minerals found ...
▸ LONDON; printed for J. Payne, and sold by J. Johnson, No. 72, St. Paul's Church-Yard
▸ Folio, 35 cm., 2 vols.; Vol. I, pp. [4], xxxviii, 519, [9]; Vol. II, pp. 788, [20]
▸ *Libraries* E, **L** : NcD, PU, Vi, ViW
▸ *Notes*
1. The tp. no longer acknowledges the contribution of 'others'; Fenning and Collyer now have the credit to themselves.

▸ **1772, 4th (04)**
▸ *A new system of geography ...*
▸ LONDON; printed for J. Payne, and sold by J. Johnson, No. 72, in St. Paul's Church-Yard
▸ 36 cm., 2 vols.; Vol. I, pp. [5], ii-xxxviii, [1], 6-519, [9]; Vol. II, pp. [3], 4-795, [18]
▸ *Libraries* **[IDT]** : CLSU, DLC, MiU, RPJCB
▸ *Notes*
1. According to the tp. of Vol. II, it was of the 3rd (1771) ed., but the pagination differs from that recorded in (03). Vol. I contains Asia and Africa, while Vol. II contains Europe and America, and, in an appendix, New

Zealand. At the end of the text, there is an index entitled 'Index to the second volume, containing Asia and Africa'; it is, in fact, the index to Europe and America, and so is in the appropriate volume.

The Preface was written by Collyer; it implies that Fenning, who was no longer alive, wrote only the Introduction to the 'former edition.'

- **{1778}, 5th (05)**
- *A new system* ... *world* ... by the late D. Fenning ...
- LONDON; printed for J. Johnson
- Pp. [1-4], i-xxxviii, 5-519, [520-8]
- ***Libraries*** : KU [Vol. I only]
- ***Source*** NUC (NSF 0010092)

- **{1785}, new, revised (06)**
- *A new system of geography* ...
- LONDON
- Folio
- ***Libraries*** L
- ***Notes***

1. From BL, which notes: 'A new edition revised by T. Hervey. The account of North America, corrected and improved by Captain Carver (Appendix to volume the first containing the voyages of Byron, Wallis, Carteret and Cook), Vol. I.' BL notes further: 'Imperfect; containing only pp. 1-202 of Vol. I and pp. 521-620 of the appendix.' The attribution of the revision to T. Hervey is incorrect; it was made by Frederic Hervey; see next entry.

- **{1787}, new, rev., enl., and improved (07)**
- *A new system* ... Containing a circumstantial ... America ... and an historical account of all the voyages of discovery performed by British navigators, during the last twenty years. Embellished ... improved by Frederick Hervey, esq.
- LONDON; printed for J. Johnson, and G. and T. Wilkie
- 37 cm.
- ***Libraries*** : NNEC
- ***Notes***

1. Facsimile ed. of Book Four (Vol. II, pp. 625-782), 1976, London, Unwin; with an introduction by G.R. Crone. He identifies this ed. as the third. He also states: 'Volume I is dated 1785 and Volume II, 1786, with passages in the text relating to the year 1787; some maps in this volume are also dated 1787. It seems therefore that the work was issued in parts, a practice which was then quite common.' The information provided on the tp., which, like the rest of the text, is reproduced in facsimile, makes it clear, however, that the extract published by Unwin comes from this ed., not from that of 1771, i.e., (03) above.

   Crone's last words were: 'Taken as a whole the account provided the reading public with a fair and balanced view of North America, as then known, after what was recognized to be a critical revolution in political affairs.'

FER, Nicholas de **A tr**

- **1715? (01)**
- *A short and easy method to understand geography.* Wherein are describ'd, the form of government of each country, its qualities, the manners of its inhabitants, and whatsoever is most remarkable in it: to which are added observations upon those things of importance that have happen'd in each state. With an abridgment of the sphere and the use of geographical maps. Made English by a gentleman of Cambridge. From the French of Mr. A.D. Fer, geographer to the French king
- LONDON; printed for H. Banks, at the Golden Key, over against St. Dunstan's Church, and F. Woodward, near the inner Temple Gate, both in Fleet-street
- 8vo, 20 cm.; pp. [7], v, [13], 312
- ***Libraries*** L, O : DFo
- ***Notes***

1. Still presents Ptolemy's cosmology 175 years after Copernicus and 25 to 30 after Newton's *Principia*.

   British Museum (1971-2) catalogues under Fer, and gives the date of publication as ca. 1715. Prior to that time BL had catalogued under G., B., the initials of the translator, as recorded at the end of the dedication. Presumably a translation of *Introduction à la géographie avec une description historique sur toutes les parties de la terre, par N. de Fer*, Paris, G. Paulus-du-Mesnil, 1708, 12mo, 297 pp. (from CGBN). Cox (1938:348) gives the date as ca. 1700, but admits that it is approximate.

FERNANDEZ, Joseph **Y SG?**

- **{1868} (01)**
- *Henry's junior geography*
- LONDON, Guildford [printed]; Bean
- 16mo
- ***Libraries*** L
- ***Source*** BL and ECB, Index (1856-76:130)

FERNANDEZ, Joseph **Y SG?**

- **{1867} (01)**
- *Henry's school geography*
- LONDON, Guildford [printed]; Bean
- 12mo
- ***Libraries*** L
- ***Source*** BL and ECB, Index (1856-76:131)
- ***Notes***

1. ECB, Index (1856-76:130) and (1874-80:67) record '10th ed. [pub.] Bean, 1874.'

FISHER, Gilman Clark **OG Am**

- **{1885} (01)**

- *The essentials of geography.* [1st] 2nd annual publication (Western hemisphere)
- BOSTON; New England Publishing Co.
- 8vo, 2 vols.
- ***Libraries*** : MB, MeWC, MH
- ***Source*** NUC (NF 0163084), which notes: 'The first volume is of the second edition'
- ***Notes***

1. Judging by later entries in NUC, this work was an annual publication providing 'the geographical news of the year.' NUC records further eds.: 1887, 3 vols. (PU); 1887 (OClWHi, Vi); and also 1888, 1890, 1892, 1893, 1893, 1894, and 1899.

FISHER, Richard S. (joint author with G.W. Colton) **A Am**

- **{1854} (01)**
- *Atlas of the world: illustrating physical and political geography.* Accompanied by descriptions geographical, statistical, and historical
- NEW YORK; J.H. Colton
- 49 cm.
- ***Libraries*** : DLC
- ***Source*** NUC (NC 0569581)
- ***Notes***

1. On the basis of other joint works by this pair of authors, the text is assumed to be the work of Fisher; see (02) below.

- **1855-6 (02)**
- *Atlas of the world, illustrating ... historical.* By G ... W ... C ... Accompanied by descriptions, geographical, statistical, and historical. By R ... S ... F ...
- NEW YORK; J.H. Colton and company, No. 172 William, Corner Beekan Street. London: Trubner and company, No. 12 Paternoster Row
- 46.5 cm., 2 vols.; Vol. I, pp. [6] (then maps and text interleaved); Vol. II, pp. [9] (then maps and text interleaved)
- ***Libraries*** : **CtY**, DLC, ICN, MiD, MiU, NB, NN, OClWHi, OO, PP, PPL, PPULC, RPB, T, TxU
- ***Notes***

1. In general each map is faced by an accompanying page of text. Most of the maps portray either continents or political jurisdictions, including some cities. In addition, however, there are maps showing the worldwide distribution of climate, plants, and animals. Except for the United States, where there are 7 pp. of tables of statistics derived from the Census of 1850, the text is generally descriptive in nature. The information is generally of the type now found in national handbooks and *The Statesman's Yearbook.* In the case of Africa, South America, and the United Kingdom, there are more pages of text than of maps.

- **{1858} (03)**
- *Colton's Atlas of the world*; illustrating physical and political geography
- NEW YORK; J.H. Colton and company. Baltimore; James Waters

- 48 cm., 2 vols.
- ***Libraries*** : CSmH, DLC, ICJ, KyLoF, MB, MBAt, NWH, NN, OCl, ODW, OHi, OrP, PHi, PPAN, PPULC, ViU, WaPS : CaBViP
- ***Source*** NUC (NC 0569583)

- **{1859} (04)**
- *Colton's atlas of the world ...*
- NEW YORK; J.H. Colton and company
- 48 cm., 2 vols. in 1
- ***Libraries*** : CU

- **{1878} (05)**
- *Colton's atlas of the world ...*
- NEW YORK; G.W. and G.C.B. Colton & Co.
- Folio; 1 p., 1 ., 142 f., 1 pl.
- ***Libraries*** : NN

FISHER, Richard Swainson **A Am**

- **{1849} (01)**
- *The book of the world*; being an account of all republics, empires, kingdoms, and nations, in reference to their geography, statistics, commerce, etc. ... With an index to all the countries, cities ... lakes, rivers, etc. mentioned on Colton's illustrated map of the world. Illustrated with maps and charts
- NEW YORK; J.H. Colton
- 8vo, 25 cm., 2 vols.
- ***Libraries*** : DLC, MB, MHi
- ***Source*** NUC (NF 0164872), which notes: 'Title varies slightly'

- **{1849} (02)**
- *The book of the world ...*
- NEW YORK; J.H. Colton
- 25 cm., 2 vols.
- ***Libraries*** : DLC
- ***Source*** NUC (NF 0164873), which notes: 'Another issue published the same year contains "Reference index to Colton's illustrated steel plate map of the world": [2], lxxxvi p. at the end of Vol. I'

- **1850-1, 2nd (03)**
- *The book of the world*: being ... commerce, etc. Together with a brief historical outline of their rise, progress, and present condition, etc., etc., etc. ... Illustrated with maps and charts
- NEW YORK; J.H. Colton, No. 86 Cedar-Street
- 25 cm., 2 vols.; Vol. I, pp. [5], vi-viii, 624; Vol. II, pp. [3], iv-vi, 705
- ***Libraries*** : CU, DLC, **NN**
- ***Notes***

1. Vol. I is devoted to the Americas, working from north to south. Each state of the U.S. is dealt with individually. Ignores physical geography by and large, especially in the U.S. The approach is discursive rather than system-

atic, though statistics on population and commercial activity are supplied consistently for the states of the U.S. Vol. II surveys the rest of the world.
An index to the work itself is provided for the first time.

- **1852, [3rd (NUC)] (04)**
- *The book of the world ...*
- NEW YORK; J.H. Colton ...
- 24 cm., 2 vols.; Vol. I, pp. [5], vi-vii, 624; Vol. II, pp. [3], iv-vi, 721
- ***Libraries*** : DLC, ICRL, **MH**, PV, WaWW
- ***Notes***

1. NUC lists a separate version of this title for the year 1852 as being held by CU; as the only apparent differentiating factor is the omission of an explicit statement that the latter is the 3rd ed., and as the tp. of the MH copy, which NUC claims to be 3rd ed., does not carry such a statement, it seems likely that NUC (NF 0164875) is a ghost. It is treated as such here.
   A page of 'critical notices' takes the place of a tp. Most of them imply that the reviewers perceived the book to be a work of reference.

FISHER, Richard S. (joint author with G.W. Colton) **A Am**

- **1857a (01)**
- *Colton's general atlas*, containing one hundred and seventy steel plate maps and plans, on one hundred imperial folio sheets by G.W. ... C. ... Accompanied by descriptions, geographical, statistical, and historical, by R. ... S. F. ...
- NEW YORK; J.H. Colton and company. London; Trubner and company
- 44 cm.; 8 p. l., [282] p.
- ***Libraries*** : DLC, ICN, IHi, MB, **MH**, MHi, MiU, MsSU, NB, OO, PP, PPL, PPULC, ViU
- ***Notes***

1. The arrangement and contents are very similar to those of *The Atlas of the World* (FISHER, R.S. 1854, above). There are, however, 4 pp. of additional text devoted to 'General view of the world.' They deal primarily with elements of the physical environment.
2. There is a map of Palestine, but in the text devoted to the Turkish Empire there is only a section on Palestine. Jerusalem is mentioned, but in the historical section there is no reference to Christianity.

- **{1858} (02)**
- *Colton's general atlas ...*
- NEW YORK; J.H. Colton & Co.
- Folio; 8 p. l., [256] p.
- ***Libraries*** : DLC, OClWHi, OHi, OO
- ***Source*** NUC (NC 0569597)

- **{1859} (03)**
- *Colton's general atlas ...*
- NEW YORK; J.H. Colton and company
- 44 cm.; 8 p. l., [282] p.
- ***Libraries*** : CtY, NIC, NjR, NN, ODW, PPCC, ViU

▸ ***Source*** NUC (NC 0569598)
▸ ***Notes***
1. NC 0569730 records: 'Colton (J.H.) and Company, New York. Colton's general atlas, by J.H. Colton and company. Accompanied by descriptions, geographical, statistical, and historical, by R.S. Fisher. New York; Johnson and Browning. 1859, 1 v., illus., maps.' Copies at (MH, MiU, NcD).
2. NUC records further eds. (all New York): 1860, [310] pp. (CtY, DLC, MiD, MnU, OC); 1862, 35 l. [sic] (CtY, MB, NN, WaS); 1863, 2 vols. (NB); 1864, 153 l. (MB, MH, NN); 1866, [305] pp. (DLC, DSI, NN, PHi, PPULC); 1868, [313] pp. (CtY, DLC, MdBP, NcD, NjR, WaU); 1870, [313] pp. (DLC, OO, PPC, PPULC); 1872, [313] pp. (DLC, IdU, PP, PPL, PPULC); 1873, [302] pp. (DLC, PP, PPF, PPULC); 1874, 165 l. (DLC, DN, MB, NN, OCU, PCC, PPC, PPULC); 1876, [282] pp. (DLC, MdBP, OCl); 1877, [288] pp. (DLC, OClWHi); 1878, 282 pp. (DLC, MH, NN, PP, PPULC, RPB); 1880, [276] pp. (DLC); 1881, [286] pp. (DLC, PP, PPULC); 1883, [282] pp. (DLC, DN, NN); 1884, [286] pp. (DLC, DN-Ob, Nb, NN); and 1887, 144 pp. pl. (NN, OC).
3. Given information provided by OC, the ed. with 'one hundred and eighty ... maps' recorded in NUC (NC 0569600) is a ghost.

FISHER, Richard S. (joint author with George W. Colton) **A Am**
▸ **{1859} (01)**
▸ *Colton's illustrated cabinet atlas and descriptive geography*. Maps by G. Woolworth Colton. Text by Robert Swainson Fisher
▸ NEW YORK; J.H. Colton
▸ 37 cm.; 6 p. l., 9-400 pp., 1 l.
▸ ***Libraries*** : CtY, MeB, MH, MHi, NNQ, OClWHi, PP
▸ ***Source*** NUC (NF 0164881)

▸ **1862 (02)**
▸ *Colton's illustrated cabinet atlas ...*
▸ NEW YORK; J.H. Colton, No. 172 William Street
▸ 4to[?], 37 cm.; 1 l., pp. 6, 2 l., pp. [9], 10-400, 1 l.
▸ ***Libraries*** L
▸ ***Notes***
1. Intended to be superior to the atlases and geographies then being used in schools without being excessively large and expensive; to blend 'the dignity of science and the easy colloquy of ordinary life' (p. 5). Claims to provide more physical geography than had been customary: 'This portion of the work has been based on the excellent treatise on the same subject found in the Atlas of Milner and Petermann, recently published in London. North America, in particular the United States, receives most attention.'

FISHER, Richard Swainson **A Am**
▸ **{1857}b (01)**
▸ *A general geography and history of the world*: containing comprehensive accounts of all countries and nations; their resources, industries, religions,

customs, etc.; the whole illustrated by a complete series of statistics, and by new and accurate steel plate maps
- NEW YORK; J.H. Colton & Co.
- 23 cm., 2 vols.; Vol. I, pp. 842; Vol. II, pp. 895
- ***Libraries*** : DLC
- ***Source*** NUC (NF 0164886) except for the pagination, which is taken from an advertisement in (02) of FISHER, R.S. (1859), above

FISHER, Richard Swainson **A Am**
- **1864 (01)**
- *Johnson's new illustrated* (steel plate) *family atlas*, with physical geography, and with descriptions geographical, statistical, and historical, including the latest Federal census, a geographical index, and a chronological history of the Civil War in American. By ... M.D., author of 'Colton's General Atlas of the World,' the 'Gazetteer of the United States,' and other statistical works; and late editor of the Journal of the American Geographical and Statistical Society. Maps compiled, drawn, and engraved under the supervision of J.H. Colton and A.J. Johnson. The new plates, copyright by A.J. Johnson, are made exclusively for Johnson's New Illustrated Family Atlas. Others are the same as used in 'Colton's General Atlas.'
- NEW YORK; Johnson and Ward, successors to Johnson and Browning (successors to J.H. Colton and company), No. 113 Fulton Street
- 41 cm.; pp. [6], 5-123 (full-page illustrations, text, and maps, interleaved)
- ***Libraries*** : CtY
- ***Notes***

1. Devoted primarily to the U.S. (22 pp. of text, primarily tables of statistics, and a 33-p. index of locations).

FITCH, George William **Y Am**
- **{1856} (01)**
- *Colton and Fitch's introductory school geography*. [Illustrated by twenty maps, and numerous engravings. Maps on a new and uniform system of scales, constructed expressly for this work by G. Woolworth Colton]
- NEW YORK; J.H. Colton and company [etc.]
- 21.5 cm.; 1 p. l., [5]-98 p.
- ***Libraries*** : DLC, NN, OO
- ***Source*** NUC (NF 0168835)
- ***Notes***

1. The words in [ ] are taken from the ed. of 1858, c1856, held at (**CaAEU**). On the basis of that copy, the text, which is at the elementary level, is composed of questions and answers. Features of world-wide occurrence are dealt with first, the treatment of race being dispassionate. The countries of the world are described on pp. 23-98, with the United States being treated first and in greatest detail.

- **{1857} (02)**
- ... *Introductory school geography* ...
- NEW YORK; J.H. Colton and co., Sheldon, Blakeman and co.

- 1 p. l., [5]-98 p.
- ***Libraries*** : MB, MH, Nh
- ***Source*** NUC (NF 0168836)
- ***Notes***

1. NUC records further eds. (all 'revised,' New York): 1859 (MH, OO); 1862, 98 pp. (MH, Nh, NN, WaU); 1863, 98 pp. (MH, NN); 1864, 98 pp. (CtY, MH); 1866, 98 pp. (MH, N); and 1867 (MH).
2. Discussed in Culler (1945, *passim*).

FITCH, George William **Y Am**

- **{1855} (01)**
- *Colton and Fitch's modern school geography.* [Maps on a new and uniform system of scales, constructed expressly for this work by George Woolworth Colton]
- NEW YORK
- 25 cm.; pp. 134
- ***Libraries*** : IEdS
- ***Source*** NUC (NSF 0019892). See note to (04).
- ***Notes***

1. The words in [ ] are taken from (03).

- **{1856} (02)**
- *Colton and Fitch's modern school geography ...*
- NEW YORK; J.H. Colton and company
- 25.5 cm.; pp. 123
- ***Libraries*** : DLC, MH, OO
- ***Source*** NUC (NF 0168843)

- **{1857} (03)**
- ... *Modern school geography ...*
- NEW YORK; J.H. Colton and co., etc.
- ***Libraries*** : MH
- ***Source*** NUC (NF 0168844)
- ***Notes***

1. NUC records eds. (all New York): 1858 (MH, PPL, PPWa); 1859 (MH); 1860, 134 pp. (MH, OClWHi, PPeSchW); 1861, 134 pp. (DHEW, MH); 1862, 134 pp. (MB); 1863 (MB); and 1864, 123 pp. (MH, RPB).

- **1866, Rev. ... (see below) (04)**
- ... *Modern school geography.* Illustrated by forty maps, and numerous engravings. Maps {as in (01)} ... Colton. Revised and corrected, according to the latest information by Charles Carrol Morgan
- NEW YORK; Ivison, Phinney, Blakeman & co., 48 & 50 Walker St. Chicago: S.C. Griggs & co., 39 & 41 Lake Street
- 24 cm; pp. [3], 4-134
- ***Libraries*** : DLC, MB, MH : **CaAEU**
- ***Notes***

1. The format and overall organization of the text is similar to that of the *Introductory Geography* (see FITCH 1856, above), but more information is

provided. 'In preparing this work, it has been the aim of the author to furnish a text-book upon geography, in all respects adapted to the wants of our Common schools.'
2. NUC records further eds. (all New York): 1867 (MH); 1868 (MH); and 1875, 134 pp. (CtY). In addition, (CaAEU) has ed. of 1886, 134 pp.
3. Culler (1945, *passim*) discusses (03).

FITCH, George William **Y Am**
- **{1853} (01)**
- *Introductory lessons in geography*
- NEW YORK; G. Savage; Philadelphia, Lippincott, Grambo & co.
- 26.5 cm.; 1 p. l., 38 p.
- ***Libraries*** : DLC
- ***Notes***
1. Discussed in Culler (1945, *passim*).

FLINT, John **Y**
- **1855 (01)**
- *Geography of production and manufactures*; with appendices. By ... one of the organizing members of the National Society
- LONDON; Hope and Co., 16 Great Marlborough Street
- 12mo, 14 cm.; pp. [3], iv-vi, [7]-53, [1]
- ***Libraries*** **L**
- ***Notes***
1. Explicitly intended for schoolchildren.
2. The title, with its exclusive reference to economic activity, is a sign of changing attitudes. There is a corresponding change in content, with major sections devoted to commercially important minerals and crops. In addition, there is a move away from the simple accumulation of facts: 'It is presumed that most persons admit that *geographical instruction should deal very largely in ideas*, and be something more than a dry repetition of names.' Later: 'It too often happens that the more rational and useful part of Geography, that which treats of the *natural resources* and wealth of countries, and the seats of the great manufactures and commerce, is entirely overlooked' (pp. v-vi; emphasis added; an early use of the phrase *natural resources*). The clearest sign of a desire to go beyond the gathering of facts in a Baconian spirit is found in the attempt to relate agricultural production to climate and to the isolines of temperature. A belief in the conditioning influence of climate is also suggested by the inclusion, in one of the appendices, of information about the trade winds. In short, the end of special geography is in sight.

- **1856, anr ed. (02)**
- *Geography of ... manufactures*. By ... Assistant Inspector of Schools in the Diocese of Lichfield, late organising master of the National Schools
- LONDON; published by the National Society, and sold at the Depository, Sanctuary, Westminster
- 12mo, 14 cm.; pp. [3]-36

▸ *Libraries* L
▸ *Notes*
1. The Preface is omitted, and with it the expressed interest in methodology. Also gone is the section on the relation of crops to climate. The appendix that was devoted to the trade winds has been revised and the information incorporated in the text.

FOGGO, David **Y SG?**
▸ **{1822} (01)**
▸ *Elements of modern geography*, with a great variety of problems on the ... globes, etc.
▸ EDINBURGH
▸ 12mo
▸ *Libraries* L

FOGGO, David **Y**
▸ **{1831}, new, completely re-modelled (01)**
▸ *Geography and the use of globes* ... New edition, completely re-modelled, etc.
▸ EDINBURGH [printed] and London
▸ 12mo
▸ *Libraries* L
▸ *Notes*
1. The Preface of the ed. of 1841 is dated 14 August 1830.

▸ **1841, new, completely ... (see below) (02)**
▸ *Geography* ... *globes*, designed for seminaries and private students. New edition, completely re-modelled, and illustrated with maps. By ... teacher of grammar, history, geography, and astronomy, Edinburgh
▸ EDINBURGH; published for Stirling & Kenney, Edinburgh; and James Duncan, and George Cowie & Co. London
▸ 15 cm.; pp. [iii], iv-xii, [1], 2-144
▸ *Libraries* : : **CaBVaU**
▸ *Notes*
1. A highly condensed special geography; much of the text takes the form of lists of toponyms. The ambiguity associated with the term *geography* is implicit in the statement in the Preface that 'the geography of each country is followed by a brief account of its extent, population, religion, government, productions, commerce, etc.' (p. iv).

FOSTER, Alexander Frederic **Y**
▸ **1862* (01)**
▸ *Elements of geography; for schools and families*. By ... late Assistant Commissioner on Popular Education; author of A History of Spanish Literature; A Latin-English Dictionary; Treatise on Geography; System of Geographical Pronunciation, etc. With national costumes
▸ LONDON; Chapman and Hall, 193 Piccadilly

- 8vo, 18.5 cm.; pp. [5], vi-xv, [1], 282
- ***Libraries*** L : : CaBViP
- ***Notes***

1. This book was intended to be a 'popular readable account of the world' for 'middle-class and popular institutions.' It contains descriptions of 'principal physical features, and so much of the social and political condition of each country as every fairly educated person ought to know' (p. [5]). It takes the form of continuous prose, though with headings. It is almost free of tables and lists.

   F. also states: 'About ten years ago I composed a general treatise on Geography, which has found considerable favour, not only in high-class schools, as a manual of instruction, but among adults, as book for reading and reference' (p. [5]). The next entry records a *General Treatise on Geography* by F. It was published in 1836. That implies that the 1st ed. of the *Elements* should have been published ca. 1846.
2. ECB, Index (1856-76:130) records: 'Geography, elements [pub.] Chapman, 1866.'

FOSTER, Alexander Frederic **A**

- **{1836} (01)**
- *General treatise on geography*
- 8vo
- ***Libraries*** L
- ***Notes***

1. Classed as adult special geography on the basis of a statement made in FOSTER (1862); see note to (01) above.
2. Allibone (1854) and BCB, Index (1837-57:110) record an ed. 1852, [pub.] Orr.

FOWLE, William B. **Y Am**

- **{1843} (01)**
- *The common school geography ...*
- BOSTON; William B. Fowle & N. Capen
- 18.5 cm.; pp. vi, [7]-228
- ***Libraries*** : DLC
- ***Notes***

1. NUC records a further ed.: 1847, Boston, 228 pp. (MB, MH).
2. Culler discusses (01) (1945, *passim*).

FOWLE, William B[entley] **Y Am**

- **{[c1849]} (01)**
- *An elementary geography*; being also a key to the new series of outline maps ...
- BOSTON; W.R. Fowle
- 14.5 cm.; pp. 79
- ***Libraries*** : CtY, DLC, MH, NN
- ***Source*** NUC (NF 0269911)

- ▸ *Notes*

1. Discussed in Culler (1945, *passim*).

FOWLE, William Bentley **Gt**

- ▸ **{1830} 3rd**
- ▸ *Modern practical geography*, on the plan of Pestalozzi
- ▸ BOSTON; Lincoln & Edmands
- ▸ 15.5 cm.; pp. 162, 24
- ▸ ***Libraries*** : MB, MBAt, MH, NNC, RPB
- ▸ ***Source*** NUC (NF 0269948), which records the title as: *Modern practical geography* ..., and notes: 'Cover-title: Fowle's practical geography, as taught in the Monitorial school, Boston. Comprising modern topography.'
- ▸ *Notes*

1. It seems probable that the difference in title between this entry and the next one should be disregarded; in other words, this title should be seen as the 3rd ed. of FOWLE (1824), below, rather than as a separate work, which is the way it is treated in NUC.
2. Cited by Elson (1964:399).

FOWLE, William Bentley **Y mSG Am**

- ▸ **{1824} (01)**
- ▸ *Practical geography*, as taught in the monitorial school, Boston. Part first
- ▸ BOSTON; published by T.P. & J.S. Fowle, No. 45, Cornhill. Crocker & Brewster, Printers
- ▸ 12mo, 14.5 cm.; pp. [3], 4-108
- ▸ ***Libraries*** L : DLC : **CABVaU**
- ▸ *Notes*

1. On the margin of special geography. The text consists of questions to be put by 'the monitor or instructor to the pupils' (p. 3). To answer the questions in Section I (pp. [3]-13), students must acquire the vocabulary of geography, i.e., they must know the meanings of words such as strait, peninsula, and hill; they also need to know the form and size of the earth, and the means of identifying locations on it through the use of a graticule. All subsequent questions have to do with the location of places, and, as such, could be answered by the pupil pointing to the place on a map. 'The Second and Third Parts, comprising statistiks, comparative geography, astronomy, as connected with geography, and ancient geography, will be published as soon as possible' (p. [2]).
2. According to Sahli (1941:381) the publishers were Crocker & Brewster.

- ▸ **1827, 2nd (02)**
- ▸ *Practical geography* ... first. Comprising modern topography
- ▸ BOSTON; published by Wait, Greens, and co.
- ▸ 14.5 cm.; pp. [3], 4-162
- ▸ ***Libraries*** : MH : **CaBVaU**
- ▸ *Notes*

1. 'The constant use of the first edition ... has enabled the author to improve this second edition very essentially, both in regard to the matter and ar-

rangement' (p. [2]). 'The ancient geography ... is very nearly ready for the press, and will be published very speedily' (p. [2]).
2. For the 3rd ed. see the previous entry.
3. Sahli (1941, *passim*) discusses (01), and Elson (1964:398) cites it.

FOWLE, W[illia]m B[entley] and Asa FITZ **Y mSG Am**
- **1845 (01)**
- *An elementary geography for Massachusetts children*
- BOSTON; Fowle and Capen, No. 138½ Washington Street
- 16mo, 15 cm.; pp. [3], 4-224
- ***Libraries*** L : DLC, MH, MWA, RPB : **CaBVaU**
- ***Notes***

1. Prepared in the conviction that, while children should learn geography, it is not only information about 'their native town, county, and state' (p. [3]) that they can be expected to want to know, but also that it is such information that they ought to know. Information about Massachusetts is provided on pp. 19-27 and 71-222. The bulk of the information provided in the latter section concerns the origin and locations of the towns in the state, arranged in order of decreasing population, on a county-by-county basis.

FRANCIS, F. **?**
- **{1812} (01)**
- *An introduction to geography* ... upon a new and easy principle
- LONDON; E. Lloyd
- 12mo; pp. xii, 81
- ***Libraries*** L
- ***Notes***

1. BL records a further ed. in 1818.

FRANSHAM, John **A**
- **{1740} (01)**
- *The world in miniature*: or, the entertaining traveller. Giving an account of every thing necessary and curious; as to situation, customs, manners, genius, temper, diet, diversions, religious and other ceremonies; trade, manufactures, arts and sciences, government, policies, laws, religions, buildings, beasts, birds, fishes, plants, reptiles, drugs, cities, mountains, rivers, and other curiosities belonging to each country. With several curious and useful tables
- LONDON; printed and sold by John Torbuck [etc.]
- 12mo, 17 cm., 2 vols.
- ***Libraries*** GU : CtY, PHi, MiD
- ***Source*** ESTC, NUC (NF 0343440), and Beeleigh Abbey Books, *A catalogue of voyages and travels*, BA/23, W. & G. Foyle Ltd., London, n.d., item 35

- **1741, 2nd, ... (see below) (02)**
- *The world in miniature* ... manners, religious and other ceremonies ... arts,

and sciences; government ... laws, buildings; beasts ... tables. The second edition, enlarg'd: also the addition of a new sett of cutts

- LONDON; printed, and sold by John Torbuck, in Clare Court, near Drury Lane; Mess. Astley and Austen, in St. Paul's Church Yard; T. Osborne, in Gray's Inn; A. Millar, in the Strand; and J. Hodges and T. Harris, on London Bridge
- 12mo, 17 cm., 2 vols.; Vol. I, pp. [2], 336; Vol. II, pp. [2], 275, [21]
- ***Libraries*** **L** : DLC, NN, RPJCB
- ***Notes***

1. Vol. I contains Asia, Africa, and Europe except for the British Isles; Vol. II contains America and the British Isles.
2. Cox (1935:79) reports eds. in 1745 and 1752; neither is in BL or NUC.

- **{1751} (03)**
- *World in miniature ...*
- DUBLIN
- ***Libraries*** AbU
- ***Source*** ESTC

- **{1767} (04)**
- *The entertaining traveller; or, the whole world in miniature.* Giving a description of ... country ... To this new edition is added, an account of the gigantic Pategonians lately discovered ...
- LONDON; printed for Henry Holmes
- 17 cm., 2 vols.
- ***Libraries*** : MHi, PPG
- ***Source*** NUC (NF 0343439)

FREE, John **mSG**

- **1789 (01)**
- *Tyrocinium geographicum londinense, or, the London geography*, consisting of Dr. Free's, Short Lectures; compiled for the use of his younger pupils, published chiefly for the information of genteel young citizens not attending the Gresham Professors, and very proper for the upper forms of great schools, for gentlemen of the first year in the universities; and for ladies that read history. To which is added, by the editor, translated from the Greek into English blank verse the Periegesis of Dionysius, the geographer, from the edition of Dr. Wells, containing the antient and modern science
- LONDON; printed for the author, and sold by Mr. Brown, corner of Essex-Street, Strand, Messrs, Rivington, St. Paul's Church Yard; Mr. Richardson, Royal Exchange, Messrs. Egerton, No. 32, Charing-Cross, Mr. Woodhouse, Lower Brook-Street, Grosvenor-Square; Mr. Ash, Little Tower-Street, Mr. Ginger, College-Street, Westminster, and by booksellers of Oxford and Cambridge. Price three shillings
- 12mo (18 cm.), 2 pts.; Pt 1, pp. 94; Pt 2, [2], 63, [3]
- ***Libraries*** **L** : IU
- ***Notes***

1. The first part is concerned with the divisions of the earth's surface that geographers and astronomers have devised from observations made on the

earth's astronomical relations. There is no special geography. The second part contains the Periegesis (DIONYSIUS 1572).

FRESNOY, Du, Mr. L'Abbé
See LENGLET DU FRESNOY (1737, 1742).

FRIEND TO YOUTH **Y SG? Am**

- **{1803} (01)**
- *An easy introduction to geography*, with a brief view of the solar system, etc. by way of question and answer adapted to the use of schools and private tuition, by a friend to youth
- CHARLESTON; printed for the Editor by J.J. Evans
- 15 cm.; pp. 60
- *Libraries* : CtY, NcU
- *Source* NUC (NE 0010481)

FRIEND TO YOUTH **Y**

- **1815 (01)**
- *An epitome of modern geography*, as determined at the Congress of Vienna; with historical notices, and questions for examination.
- LONDON; printed at the Mentorian Press, by S. Maunder for W. Pinnock; and sold by him; also by Law and Whittaker, 13 Ave-Maria-Lane, London; and by all other booksellers in the kingdom
- 12mo, 14 cm.; pp. [1], 405
- *Libraries* OSg
- *Notes*

1. In the Preface the author identifies himself as the author of the 'Epitome of Ancient Geography, lately published.' Special geography occupies all but pp. 1-21. The information is presented in numbered paragraphs, with history receiving almost as much space a chorography. Forty-seven per cent of the space is devoted to Europe, 17 per cent to Asia, 13 per cent to Africa, 7 per cent to North America, 6 per cent to South America, and 3 per cent to islands in the Pacific (including Australia).

G

GADESBY, Richard **Y**

- **1776 (01)**
- *A new and easy introduction to geography*, by way of question and answer, divided into lessons. Principally designed for the use of schools. Containing a description of all the known countries in the world; of their respective situations, divisions, mountains, rivers, principal cities and towns, forms of

government, religion, etc. Likewise several useful problems on the terrestrial globe, with an explanation of the vicissitudes of the seasons
- LONDON; printed for the author, and sold by S. Bladon, No. 16, Paternoster Row; and J. Walter, Charing-Cross
- 12mo, 18 cm.; pp. xii, 182
- ***Libraries*** **L**
- ***Notes***

1. Intended for use in schools; acknowledges borrowing from Guthrie's *Geographical Grammar* (GUTHRIE 1770). Though not taking the form of a normal special geography, it contains equivalent information. Europe is treated at relatively great length.

- **1783, 2nd, improved and enlarged (02)**
- *A new and easy introduction to geography* ... seasons. To which is now added, a new geographical table
- LONDON; printed ... No. 13 Paternoster Row
- 12mo, 18 cm.; pp. xii, 191, [1]
- ***Libraries*** **L**
- ***Notes***

1. Very few changes made since 1776. The independence of the United States is acknowledged, but not dwelt on.
2. Later eds. are known (all London): 1787, 3rd [cited by Bowen (1981)]; 1792, 191 pp. (: MB, NSyU); 1802, 265 pp. (: CLU).

GALVÃO, Antonio [or GALVANO] **OG tr**

- **1601 (01)**
- *The discoveries of the world from their first originall* vnto the yeere of our Lord 1555. Briefly written in the Portugall tongue by Antonie Galvano, Gouernour of Ternate, the chief island of the Malucos: corrected, quoted, and now published in English by Richard Hakluyt, sometimes student of Christchurch in Oxford
- LONDON; impensis G. Bishop
- 4to, 19 cm.; pp. [12], 97
- ***Libraries*** L, O : DLC, ICN, ICRL, MB, MHi, MiU, MnU, MWiW, NIC, NjP, RPJCB, ViU
- ***Notes***

1. 'Galvano's *Tratado* gives an account of all the discoveries, ancient and modern, which had been made up to the year 1550. In this respect it is the very first book of its kind. The matter is arranged in chronological order, the pre-Columbian discoveries occupying leaves 1-22, the greater part of the work (leaves 23-80) being devoted to voyages of discoveries made during the years 1493-1550. In this latter part the writer includes his own experiences' (Maggs 1962:33).

   The author starts by defining his precise objective, namely, to learn 'who were the first discouerers since the time of the flood' (p. 2). He acknowledges that there is much evidence in support of the view that 'it seemeth, that they which first came to be sailers were those which dwell in the east in the prouince of China' (p. 3). He notes that others claim the title, and

concludes: 'But omitting all iars [sic] and differences thereabouts, I will apply my selfe to my purposed discourse, and speake of that which histories have left in record' (p. 4). His starting point is the voyage of Tubal (of Ethiopia) to Spain, 143 years after the Flood. His authority on this point is 'Berosus.'
2. Facsimile ed., 1969 (Johnson/TOT).
3. BL records further eds.: 1745, 1803, 1809, and 1812. Reprinted by the Hakluyt Society, together with the original Portuguese, 1862; facsimile ed., 1969, New York; B. Franklin.
4. Discussed in Parks (1928 and 1961:215-16).

GARDINER, Alfonzo **Y**

- **{[1873-4]} (01)**
- *J. Heywood's standard lesson series in geography.* For standard IV (-VI)
- MANCHESTER; John Heywood
- 16mo., 3 pts.
- ***Libraries*** L
- ***Notes***

1. According to Lochhead (1980:108) only the material for standard VI would be special geography.

GARNIER, James **Y**

- **1748 (01)**
- *Geography made easy*: or, that valuable science comprehended in a compendious treatise, wherein are contained a particular etymological explanation, and clear definition of all the names, terms, etc. of that science, interspersed with the most remarkable events of history, and their chronology; with a particular account of the ancient and present state of all the chief cities of the whole world. To which is added an appendix, containing, a short view of astronomy. Including whatever is most necessary in that science. The whole by way of question and answer, in French and English. For the use of schools
- LONDON; printed for the author, and to be had at the French booksellers
- 8vo, 21 cm.; pp. [4], iv-viii, [1], iv-viii, [7], 2-189 (twice, in parallel)
- ***Libraries*** L : CLU
- ***Notes***

1. Bilingualism in the 18th century. The unusual pagination is explained by the fact that the entire text appears in both English and French. Apart from the preface, which is given completely first in French and then in English, the two versions are given on facing pages (on even-numbered pages, the French is on the left, the English on the right; on odd-numbered pages, the positions are reversed). The opening pages are devoted to the earth as a spherical body. Pp. 30-176 are special geography, of which over 100 are devoted to Europe. England and France receive nearly equivalent amounts of space, though Great Britain as a whole is allotted 20 pp. to France's 8. No explanation of the bilingual form is given, but, unusually in a book of this modest size, subscribers were obtained to support the publication; their names appear at the front. See LORIOT (1797).

GASKIN, J.J. **?**
- **{1857} (01)**
- *Geography made interesting*
- Place/publisher n.r.
- ***Libraries*** n.r.
- ***Source*** BCB, Index (1837-57, Index to 1857)
- ***Notes***

1. ECB, Index (1856-76:130) records an ed. in 1862.

GASPEY, Thomas W. **? tr Am**
- **[1852]-3 (01)**
- *Book of the world.* A family miscellany for instruction and amusement
- PHILADELPHIA; Weik and Wieck.
- 2 vols.
- ***Libraries*** L (mislaid) : CSmH, NjR, N
- ***Source*** NUC (NB 0640310), which notes: 'Probably a translation of Das Buch der Welt. Titles of landscape engravings in German.'

GAULTIER, Aloisius Edouard Camille **Y tr?**
- **{1792} (01)**
- *A complete course of geography, by means of instructive games*, invented by the Abbé Gaultier
- LONDON. Printed for the author
- 37 cm.; pp. viii, 12
- ***Libraries*** : DLC
- ***Source*** NUC (NG 0081917), which notes: 'Contents Pt. I. Geography of the British Islands. Pt. II. Geography of Europe. Pt. III. Geography of Asia, Africa, America and new-discovered islands.' See note to (02) below.
- ***Notes***

1. NUC records further eds. (both London): 1795, 2 vols. in 1 (DLC); and 1800, 4th, 50 pp. (DLC). Maggs (1964:323) records an ed.: 1806, London, 50 pp. NUC records an ed.: 1815, London, 50 [i.e., 52] pp. (CtY).

- **{1817}, anr, corrected ... (02)**
- *A complete course of geography ... instructive ...* Gaultier. A new edition ... parts. The first part, containing the game of simple geography, for teaching the names and situations of the different countries, and places of the earth. The second part, containing a geographical game, illustrative of ancient and modern history. To which is prefixed a treatise or short account of the artificial sphere
- LONDON; printed for J. Harris
- 39 cm.; pp. 52
- ***Libraries*** : MnU
- ***Notes***

1. 'Although the object of the game was the teaching and memorizing of the traditional information, the method and spirit [not being authoritarian] were

far removed from the traditional ones' (Vaughan 1972:144). Discussed in Vaughan (1972:143-5).

- **{1821} (03)**
- *A complete course of geography* ... Gaultier. Collated with the author's last Paris edition, and digested for Europe conformably to the territorial arrangements of the pacification of 1815, by Jehoshaphat Aspin
- LONDON; J. Harris
- 39 cm.; pp. 59
- ***Libraries*** : MoSW
- ***Source*** NUC (NG 0081924), which notes: 'Contents: – pt. 1. The game of simple geography – pt. 2. A concise treatise on the artificial sphere. – pt. 3. A geographical game, illustrative of ancient and modern history

- **{1832}, new (04)**
- *A course of geography by means of instructive games* invented by the Abbé Gaultier; collated with the author's last Paris edition, and digested for Europe conformably to the latest territorial arrangements. A new edition with numerous improvements, by John Olding Butler
- LONDON; J. Harris
- 42 cm.; pp. 59
- ***Libraries*** : IU
- ***Notes***

1. See PICTET and ST. QUENTIN (1791).

GAULTIER, Aloisius Edouard Camille **Y tr?**

- **{1824?} (01)**
- *An epitome of geography, designed as a companion to a geographical game,* invented by the Abbé Gaultier
- EYE, Suffolk; printed and sold for the author by B.C. White
- 15.5 cm.; pp. 93
- ***Libraries*** : MdBJ
- ***Source*** NUC (NG 0081930), which adds: 'An abridgment of the preparatory lessons of Gaultier's work (for the use of public seminaries).'

GAULTIER, Aloisius Edouard Camille **Y tr?**

- **1826 (01)**
- *Familiar geography*: by the Abbé Gaultier. Introductory to a complete course of geography, by means of instructive games
- LONDON; John Harris, St. Paul's Church-Yard
- 16mo, 13 cm.; pp. [2], vi, 185 (79 misnumbered 7), [1]
- ***Libraries*** **L** : MH
- ***Notes***

1. A special geography intended for young children. It provides information by means of question and answer.
2. St. John (1975, 2:804) records an ed.: 1828, 2nd, London, 185 pp. (: : CaOTP).

- **1832, 4th (02)**
- *Familiar geography ...*
- LONDON; John Harris ...
- 16mo, 13 cm.; pp. [2], vi, 186
- ***Libraries*** **L**
- ***Notes***

1. The only changes since (01) revealed by spot-checking are a reduction in the number of South American countries dealt with from eight to seven (associated with a slight lengthening of the relevant text), and slight changes in the information about Tasmania.

- **1844, 10th, revised and corrected (03)**
- *Familiar geography* ... Gaultier. With a concise treatise on the artificial globe
- LONDON; Grant and Griffith, successors to John Harris, corner of St. Paul's Church-Yard
- 16mo, 14 cm.; pp. [4], 254, [16]
- ***Libraries*** **L**
- ***Notes***

1. The text has been reorganized considerably.
2. BCB, Index (1837-57:110) records an ed. in 1852; ECB, Index (1874-80: 67) records 17th ed. in 1874.

GAULTIER, Aloisius Edouard Camille **Y mSG? tr?**

- **{1838} (01)**
- *Geographical and historical questions*, referring by characteristic and distinguishing marks, to the most remarkable places in the world. Forming a sequel to familiar geography, by the same author
- LONDON; J. Harris
- 14.5 cm.; pp. (4) [ix]-xi (1) 112
- ***Libraries*** : MH

GAZETTEER ... **Dy**

- **1850-7 (01)**
- *A gazetteer of the world*, or, dictionary of geographical knowledge, compiled from the most recent authorities, and forming a complete body of modern geography, physical, political, statistical, historical, and ethnographical. Edited by a member of the Royal Geographical Society. Illustrated with numerous woodcuts and one hundred and twenty engravings on steel
- EDINBURGH; A. Fullarton & co., Stead's Place, Edinburgh; 106 Newgate Street, London; and 22 Eustace Street, Dublin
- 26 cm., 7 vols.; Vol. I, pp. [12], 856; Vol. II, pp. [2], 896; Vol. III, pp. [6], 880; Vol. IV, pp. [2], 864; Vol. V, pp. [2], 874; Vol. VI, pp. [2], 848; Vol. VII, pp. [2], 860, [1], iv-vi, [2]
- ***Libraries*** L, **[IDT]** : DLC, DNLM, IU, MdBP, NN, PPDrop
- ***Notes***

1. Vol. I opens with a prospectus addressed to potential subscribers. The first

editor was James BELL (1829-32), but he died in the course of preparing the work. His successor is not named. Despite wide margins, the small size of the print makes for a massive work. Vol. VII concludes with an appendix that includes a guide to the pronunciation of geographical names, a guide to the pronunciation of the 'more important European and Oriental languages,' and an outline of general ethnology. It concludes with a preface that suggests that the work is being made available for general sale. See also note to (02) below.

2. BL notes: 'Published in thirty-two parts.'

- **1859 (02)**
- *A gazetteer* ... or dictionary ... authorities. And forming ...
- EDINBURGH; A. Fullarton & co., Stead's Place, Leith Walk; and 73 Newgate Street, London
- 8vo, 26 cm., 6 vols.; Vol. I, pp. [3], iv-vi, 896; Vol. II, pp. [2], 896; Vol. III, pp. [2], 880; Vol. IV, pp. [2], 864; Vol. V, pp. [2], 873; Vol. VI, pp. [2], 860
- ***Libraries*** : : **CaAEU**
- ***Notes***

1. The work ends with five appendices. Two relate contemporary and Classical place names; one deals with the pronunciation of current place names, and one with the pronunciation of major contemporary languages; the fifth contains an 'Outline of general ethnology,' and is largely devoted to a presentation of 'Pickering's classification,' which divides the human species into eleven races.
2. BL records an ed. of 7 vols. in 1860?; NUC records an undated ed. in 14 vols. at (ICRL) (NG 0092253).

GENERAL VIEW ... **Gt**

- **{[1840?]}**
- *General view of the world*
- ***Libraries*** : KMK
- ***Notes***

1. KMK reports that this is a damaged copy of MITCHELL, S.A. (1840a).

GEOGRAPHICAL ... ACCOUNT **NG**

- **1829 (01)**
- *A geographical and historical account of the great world*; to which is added, a voyage to its several islands; with a vocabulary of the language; and a map
- LONDON; James Ridgway, Piccadilly
- 21 cm.; pp. [4], 58
- ***Libraries*** **[IDT]**
- ***Notes***

1. A satire, not a special geography.

GEOGRAPHICAL COMPANION **Y OG**

- **{1802} (01)**
- *A geographical companion to Mrs Trimmer's Scripture*, Ancient, and abridged Histories; with prints, [calculated to render the study of history more interesting to children and serve as an easy introduction to the knowledge of the earth]. 3 pt.
- LONDON; J. Harris
- 24mo
- ***Libraries*** L
- ***Notes***

1. BL and NUC record further eds. (both London; J. Harris): 1811, 125 pp. (L); 1816, New, 92 pp. (: DLC, MH).

GEOGRAPHICAL DESCRIPTION ... (1608) **?**

- **{1608} (01)**
- *A geographical description of all the empires and kingdoms*, both of continent and islands, in this terrestrial globe
- Place/publisher n.r.
- ***Libraries*** n.r.
- ***Source*** Item 789 in E.G.R. Taylor (1934:228). Might be the (05) ed. of ABBOT (1599) with the information derived from some secondary source.

GEOGRAPHICAL DICTIONARY ... **Dy**

- **{1602} (01)**
- *A geographical dictionary*: in which are described the most eminent countreys, towns, ports, seas, streights, and rivers of the whole world: very useful for the understanding of all modern histories
- LONDON
- ***Libraries*** n.r.
- ***Source*** Cox (1938:337). In neither BL, NUC, nor Pollard and Redgrave (1946). Given the similarity in the printed form of the date of this ed. to that of the next, as well as the similarity in the titles, it seems likely that this ed. is a ghost; it is treated as such in THE CHRONOLOGY OF PUBLICATION.

- **{1662} (02)**
- *A geographical dictionary*. In which
- [LONDON]; By J.C. to be sold by Henry Brome
- 8vo
- ***Libraries*** O
- ***Source*** Wing (1948)

- **{1676}, 2nd (03)**
- [*A geographical dictionary*]
- [LONDON]; By A.M. for Henry Brome
- 12mo

- ***Libraries*** O
- ***Source*** Wing (1948)

- **1678, 3rd, corrected (04)**
- *A geographical dictionary*, in ... world. Very ...
- LONDON; printed by M.C. for Henry Brome at the Gun in S. Paul's Church-Yard, the West end
- 12mo, 14 cm.; pp. [4], 132, [7]
- ***Libraries*** **L** : OCl
- ***Notes***

1. BL catalogues this dictionary under Duval (Pierre), following information contained in 'The Stationer to the Reader' (i.e., the preface). NUC catalogues it by title.

- **{1681}, 4th, corrected (05)**
- *A geographical dictionary* ...
- LONDON; printed by S.R. for Henry Brome ...
- 14 cm.; 2 p. l., 171 p.
- ***Libraries*** : CLU, ICN
- ***Source*** NUC (NG 0126944)

- **{1687}, 5th, very much amen-[sic] and corrected (06)**
- *A geographical dictionary* ... world. Together with a new and accurate map of the world
- LONDON; printed for Charles Brome, and sold by Benj. Crayle at the Peacock and Bible at the West-end of St. Pauls
- 13 cm.; 2 p. l., 171 [4] p.
- ***Libraries*** : CtY, OCl
- ***Source*** NUC (NG 0126946), which notes that the price was one shilling

GEOGRAPHICAL EXERCISES ... **Y mSG? Am**

- **{1815} (01)**
- *Geographical exercises compiled for the use of the boarding school and select school in Wilmington*
- [WILMINGTON, DE]; Robert Porter
- 13 cm.; 1 p. l., 46 p.
- ***Libraries*** : NNC
- ***Source*** NUC (NG 0126951) and Elson (1964:396)
- ***Notes***

1. NUC records 7th ed., 1842, Wilmington, n.p. (PU). Elson (1964:396) cites (01).

GEOGRAPHICAL GUIDE **Y OG**

- **{1805} (01)**
- *The geographical guide*; a poetical nautical trip round the island of Great Britain ...
- LONDON; printed for J. Harris
- 17 cm.; pp. 69

▸ *Libraries* : : CaOTP
▸ *Source* St. John (1975, 2:804)
▸ *Notes*
1. The full title shows that this book deals only with the British Isles.

GEOGRAPHICAL MAGAZINE **A n.c.**
▸ **{1790} (01)**
▸ *The geographical magazine; or, the universe displayed.* Containing a geographical, historical, and topographical description of all the empires, kingdoms, states, and provinces, in the habitable part of the terraqueous globe. Their situation, extent, climate, soil, rivers, lakes, forests, mountains, minerals, animals, and vegetables. With a faithful and circumstantial account of the religions, laws' [sic] manners, customs, virtues, vices, amusements, and diversions, of the human beings who people the various regions of Europe, Asia, Africa, and America. In which will be interwoven the particulars of their respective forms of governments, their armies and navies, commerce, products, and manufactures. Drawn from unerring sources of information. Forming a complete system of geography. Embellished with maps, perspective views, representations of the manners and customs of the inhabitants, and whatever else may tend to illustrate or embellish a performance of so extensive and interesting a nature
▸ LONDON; printed for the proprietors, and sold by R. Butters, No. 79, Fleet-street; Symmonds, Pater-noster Row; and all other booksellers
▸ 3 vols.
▸ *Libraries* n.r.
▸ *Source* Fox and Stoddart (1975)
▸ *Notes*
1. 'The first *Geographical Magazine* was launched in 1790. The venture is today something of a bibliographical mystery. It does not seem to have been noticed or reviewed in the contemporary periodical press, while among all the major institutional libraries of Great Britain only one set survives' (Fox and Stoddart 1975:485). Fox and Stoddart also provide a photograph of the tp., which is the source of the information given above, and say that the first volume was devoted to Spain and Portugal, the second to France and Switzerland, and the third to Germany and central Europe. No more were published. They also give a short account of the rise of magazines in general.

   In stating that this publication was the first geographical magazine, they overlooked MARTYN (1782-3).

GEOGRAPHICAL QUESTIONS ... **Y mSG?**
▸ **{1802} (01)**
▸ *Geographical questions and answers*, with a brief chronology of ... England ... to which is prefixed a general statement of the powers of Europe ... for the instruction of young minds
▸ ETON, Berkshire
▸ 15 cm.; pp. 64
▸ *Libraries* : ICN

▸ *Source* NUC (NG 0127019)
▸ *Notes*
1. NUC records further eds. (both Eton, Berkshire): 1807, 66 pp. (CtY); 1812, 62 pp. (CtY).

GEOGRAPHICAL READER ?
▸ **{1804} New (01)**
▸ *Geographical reader*. New ed.
▸ LONDON; Isbister
▸ V. 3
▸ *Libraries* : PU

GEOGRAPHY AND ASTRONOMY ... **Y SG?**
▸ **{1800} (01)**
▸ *Geography and astronomy familiarized*, for youth of both sexes
▸ LONDON; printed for J. Wallis, by T. Gillet
▸ Pp. 64
▸ *Libraries* n.r.
▸ *Source* ESTC
▸ *Notes*
1. NUC records a further ed.: 1801, London, 64 pp. (DLC).

GEOGRAPHY AND ATLAS **Y mSG**
▸ **1876, new (01)**
▸ *Geography and atlas*. The royal school series. No. 1. For junior classes
▸ LONDON; T. Nelson and Sons, Paternoster Row; Edinburgh; and New York
▸ 18.5 cm.; pp. [7], 8-64
▸ *Libraries* : : **Capc**
▸ *Notes*
1. Information with respect to the earth as a whole, including elementary cartography, occupies pp. 7-25; condensed information of a utilitarian nature, some of it in semi-tabular form and related to places (chiefly towns and cities) is provided about the British Isles (pp. 26-44) and other countries and continents (pp. 45-64).
   A senior class-book is said to be in preparation.

GEOGRAPHY EMBELLISHED ... ?
▸ **{[1802]} (01)**
▸ *Geography embellished* with a variety of views from nature for the book-case of instruction and delight
▸ LONDON; J. Marshall
▸ 9.5 cm.; 56 p., 2 l.
▸ *Libraries* : DLC : CaOTP
▸ *Source* NUC (NG 0127319) and St. John (1975, 2:804)

▸ *Notes*
1. St. John (1975, 2:804) notes that the date is recorded on the frontispiece.

GEOGRAPHY FOR YOUTH ... **Y**

▸ **1787, 3rd (01)**
▸ *Geography for youth*, or, a plain and easy introduction to the science of geography, for the use of young gentlemen and ladies: containing an accurate description of the several parts of the known world. To which are added, geographical questions, and a table of the longitude and latitude of the most remarkable places on the terraqueous globe. Illustrated by twelve maps, on which are delineated the new discoveries made by Commodore Byron, and the Captains Wallis, Carteret, and Cook. The third edition, greatly enlarged, corrected, and improved
▸ LONDON; printed for W. Lowndes, No 77, Fleet-Street
▸ 12mo, 19 cm.; pp. [4], 260
▸ ***Libraries*** **L**
▸ *Notes*
1. A school textbook. The special geography occupies pp. 19-246, of which 105 pp. are devoted to Europe, 62 to America, and 10 to the new discoveries in the Pacific.

▸ **1790, 4th, considerably enlarged ... (02)**
▸ *Geography for youth* ... to which are subjected geographical questions and ... Illustrated by copper plates
▸ LONDON; printed ...
▸ 12mo, 19 cm.; pp. viii, 339, [1]
▸ ***Libraries*** **L**
▸ *Notes*
1. Spot-checking suggests that the new material is fairly well scattered throughout.

▸ **1797, anr ed., considerably ... (03)**
▸ *Geography for youth* ... plates. A new edition, considerably enlarged, corrected and improved
▸ LONDON; printed ... No 76, Fleet-Street
▸ 12mo, 18 cm.; pp. xii, 384, [4]
▸ ***Libraries*** **L** : DLC
▸ *Notes*
1. Spot-checking did not reveal the location of the changes suggested by the pagination.

GEOGRAPHY, IN EASY DIALOGUES **Y OG?**

▸ **{1816} (01)**
▸ *Geography, in easy dialogues*, for young children. By a lady
▸ LONDON; printed for N. Hailes
▸ 14 cm.; pp. 72
▸ ***Libraries*** : : CaOTP
▸ ***Source*** St. John (1975, 2:804)

▸ *Notes*
1. Probably limited to England.

GEOGRAPHY MADE FAMILIAR ...
See NEWBERY (1748).

GEOGRAPHY REFORMED ... **OG**
▸ **1739 (01)**
▸ *Geography reform'd*: or, a new system of general geography treated of in all its branches, mathematical, historical and technical, according to a more accurate and compleat analysis of the science, than has been given by former authors. In three parts
▸ LONDON; printed for Edward Cave, at St John's Gate
▸ 8vo, 19 cm.; pp. [10], 302, [13]
▸ ***Libraries*** **L**
▸ *Notes*
1. A methodological treatise. The full title of the issue of 1740 (02) gives a fair idea of its coverage. See also note to entry (03).

▸ **{1740} (02)**
▸ *Geography reformed* ... geography according to an accurate analysis of the science, augmented with several necessary branches omitted by former authors. In four parts. I. Of the nature and principles of geography ... II. Of mathematical geography ... III. Historical geography ... IV. Of technical geography ... The whole illustrated with notes and references to the principal geographers ... There is added a copious index of the terms contained in the work, answering the end of a dictionary of general geography
▸ LONDON; printed by E. Cave
▸ 5 p. l., 302, [13] p.
▸ ***Libraries*** E : OClW
▸ ***Source*** ESTC and NUC (NG 0127424)

▸ **1749, 2nd (03)**
▸ *Geography reformed* ...
▸ LONDON; E. Cave
▸ 8vo, 19 cm.; pp. [10], 302, [13]
▸ ***Libraries*** **L**
▸ *Notes*
1. Despite the differences in the titles, this ed. is all but identical to that of 1739. The last two paragraphs of the Preface of 1739 have been omitted. The body of the text is unchanged, as are the page decorations. It reads as a work by a scholar with an interest in the problems of obtaining accurate information about the countries of the world. He is more concerned with the construction of accurate maps (and globes) than with the descriptions that would accompany them. Sources are cited. Its objective is similar to that of [MEAD] (1717), but it is more scholarly than the earlier work, and a quick reading of the two discourages the conjecture that they might be by the same author. On the basis of interest and level of scholarship, one candidate

for the authorship is John Green, whom G.R. Crone has shown to be the author of a number of anonymous works in related fields published between 1736 and 1755. If Crone is correct, however, John Green and Bradock Mead are the same man. In that case, the difference in the level of scholarship must be attributed to the experiences of Mead's life. See [MEAD] (1717).

- **{1754}, 2nd (04)**
- [*Geography reformed*]
- LONDON
- 12mo
- ***Libraries*** GU, LcP
- ***Source*** ESTC

GERBIER, Balthazar **NG**

- **1649a (01)**
- *The first lectvre, of an introduction to cosmographie*, (which is a description of all the world) read publickly at Sr. Balthazar Gerbier his Academy, at Bednall-Greene
- LONDON; printed by Gartrude Dawson, and are to be sold by Hanna Allen at the Crown in Popes-Head Alley
- 4to, 19 cm.; pp. 13
- ***Libraries*** L, O, [IDT] : CtY
- ***Notes***

1. The part of Ptolemy's cosmography having to do with the outermost heavens and the stars. The only sign of the changes in astronomy that were to come is that it is acknowledged that people in Italy, using 'prospective glasses' (i.e., telescopes), had found that the heavens contained far more stars than had been known to the Ancients.

- **1649 (02)**
- *The first lecture of ... cosmographie*: being a description of all the world. Read publiquely at Sr. Balthazar Gerbiers Academy. Imprimatur, Hen: Scobell, Cleric: Parliamenti
- LONDON; printed at London for Robert Ibbotson dwelling in Smithfield neer Hosier Lane
- 4to, 19 cm.; pp. [4], 16
- ***Libraries*** L

GERBIER, Balthazar **NG**

- **1649b (01)**
- *The second lecture being an introduction to cosmographie*: read publiquely at Sr. Balthazar Gerbiers Academy. On Bednall Greene
- LONDON; printed for Robert Ibbitson in Smithfield near the Queens Head Tavern
- 4to, 19 cm.; pp. [4], 19
- ***Libraries*** L, O

▸ *Notes*
1. Deals with the motion of the sun, moon, and planets as understood in the ptolemaic model of the universe (the Copernican hypothesis is rejected explicitly); also deals with the four elements and the position of the earth in the universe.

GERBIER, Balthazar **OG**
▸ **1649c (01)**
▸ *The first lectvre, of geographie,* (which is a description of the terrestriall globe) read publickly at Sr. Balthazar Gerbier his Academy, at Bednall-Greene
▸ LONDON; printed by Gartrude Dawson, and are to be sold by Hanna Allen at the Crown in Popes-Head-Alley
▸ 4to, 19 cm.; pp. [2], 12
▸ ***Libraries*** **L** : CLU
▸ *Notes*
1. Concerned with the 'geometricall' part of geography, i.e., the astronomical division of the earth, plus measures of distance, and the jargon of geography (i.e., the technical terms describing portions of land and sea).

GIBBON, F.E. **Y SG?**
▸ **{1854} (01)**
▸ *Geography, catechism*
▸ [LONDON]; Relfe
▸ 18mo
▸ ***Libraries*** n.r.
▸ ***Source*** BCB, Index (1837-57:110)
▸ *Notes*
1. BL records an ed. in [1881], 106 pp., and NUC one in 1885 (PPL).

GIBBON, F.E. **?**
▸ **{1855} (01)**
▸ *Geography, etc.*
▸ [LONDON]; Relfe
▸ 18mo
▸ ***Libraries*** n.r.
▸ ***Source*** BCB, Index (1837-57:109)

GILBERT, James **Y SG?**
▸ **{1845} (01)**
▸ *Geography, schools*
▸ [LONDON]?; Groombridge
▸ 12mo
▸ ***Libraries*** n.r.
▸ ***Source*** BCB, Index (1837-57:110)

GILBERT, James **Y SG?**
- **{[1846]} (01)**
- *Gilbert's geography for family and schools*
- LONDON; Dover
- 12mo
- ***Libraries*** L

GILES, John Allen **Y SG?**
- **{[1858]}, 4th (01)**
- *First lessons in geography ...*
- LONDON; (Guildford, printed)
- 16mo
- ***Libraries*** L

GILES, John Allen **Y SG?**
- **{1851} (01)**
- *Geography in question and answer*, for the use of little children
- LONDON; Law
- 16mo
- ***Libraries*** L
- ***Source*** BL and BCB, Index (1837-57:109)

GILL, George **Y SG?**
- **{[1873]} (01)**
- *The Oxford & Cambridge geography*, etc.
- LONDON
- 8vo; pp. 164
- ***Libraries*** L
- ***Source*** BL, which notes: '[Gill's school series]'

GILL, George **Y**
- **{[1875-9]} (01)**
- *A second (– sixth) standard geography, etc.*
- LONDON
- 8vo, 5 pts.
- ***Libraries*** L
- ***Source*** BL, which notes: '[Gill's school series]'
- ***Notes***

1. According to Lochhead (1980:108), only that part [the 5th presumably] devoted to the sixth standard would have been special geography.

GILL, George **Y SG?**
- **{[1867]}, 3rd (01)**
- *A series of geographical lessons*. Third edition
- LONDON, Liverpool [printed]

- 16mo, 2 pts.
- ***Libraries*** L

GILL, George **Y SG?**
- **{[1868]} (01)**
- *A series of lessons in geography for home use*. Cheap edition especially prepared for elementary schools
- LONDON, Liverpool (printed)
- 12mo
- ***Libraries*** L
- ***Source*** BL, which notes: 'Title from wrapper'

GILL, George **Y SG?**
- **{1882} (01)**
- *The Whitehall geographical readers*
- LONDON; George Gill and Sons
- ***Libraries*** : : CaBVaU
- ***Notes***

1. The CaBVaU catalogue records the title as: *Gill's School Series: the Whitehall ...*

GLASGOW GEOGRAPHY ... **A**
- **{1822} (01)**
- *Glasgow geography*, containing a physical, political and statistical view of the various empires, kingdoms, states ... in the known world
- GLASGOW
- Vols
- ***Libraries*** n.r.
- ***Source*** James Thin (53-59 South Bridge, Edinburgh, Scotland), Catalogue 351, *General second-hand and antiquarian*, n.d. (ca. 1976), p. 19
- ***Notes***

1. See note to (02) below.

- **1825 (02)**
- *The Glasgow geography*, containing a physical, political and statistical view of the various empires, kingdoms, states, etc. etc. in the known world; including the most recent discoveries and latest political arrangements; with a philosophical view of universal history, and a compendium of astronomy, mineralogy, etc. To which is subjoined a historical appendix, comprising a concise narration of the wars consequent on the French Revolution, till the Treaty of Paris, 1815. Embellished with appropriate engravings, and illustrated by correct statistical tables and a complete atlas. By several literary gentlemen
- GLASGOW; printed and published by Khull, Blackie, & co., and A. Fullarton & co
- 22 cm., 5 vols.; Vol. I, pp. [6], vii, 86, 3-702; Vol. II, pp. [4], 879; Vol.

III, pp. [3], iv, 944; Vol. IV, pp. [3], iv, 863; Vol. V, pp. [4], 328; [3], iv-xxvi, 434

- *Libraries* : : **Capc**
- *Notes*

1. May belong to a sequence of publications that began in 1805 (SYSTEM OF GEOGRAPHY), and continued in 1829 (BELL, 1829-32). The opening 86 pp. of Vol. I are devoted to a conventional history of the world, from its creation as recorded in Genesis, to the fall of the Roman empire. Africa occupies pp. 1-227 (2nd ser.); this section opens with 26 pp. on the fauna of the continent; then come three paragraphs that present, in a microcosm, five of the salient characteristics of the intellectual character of the period. One is the use of knowledge derived from Classical authors. The other four – scepticism, gentility, ethnocentrism, and religious orthodoxy – are combined in the following passage: 'The external appearance [of negroes], can have escaped the attention of very few. The jet black colour, the soft skin, the woolly hair, the flat nose, the thick prominent lips, are marks too palpable not to be generally observed. A conformation in which they differ so much from the other inhabitants of the world has given rise to the question, whether the negroes are not a distinct and separate race of man? With regard to this question, we remark in the first place, that it is one which will not be readily entertained by any person acknowledging the divine authority of the scriptures; in the second place, it is a question, which, if considered apart from revelation, is extremely difficult in itself, and the decision of which belongs rather to the naturalist than to the geographer. What has chiefly contributed to render the negroes remarkable, is that commerce, carried on by Europeans, of which they from the principal commodity. Of this commerce it is unnecessary to say much in this place. Every one, whose judgment is unbiased by the most powerful of all prejudices, interest, intuitively perceives that it is cruel and unjust' (Vol. I, p. 27).

   The final 475 pp. of the volume are devoted to Asia, including the islands in the Pacific.

   Vol. II is devoted to the countries of northern Europe, including Russia and the British Empire (i.e., the United Kingdom), where Scotland is treated first. For this country some statistical information intended to summarize the character of the climate is presented – a product of the keeping of meteorological records that had begun in the previous century.

   Vol. III contains the balance of Europe. The section on Rome (pp. 233-41) contains some purple prose, the pleasure of whose composition must have been close to sinful for the, presumably, Calvinist author.

   Vol. IV and part of V are devoted to America (pp. 1-634 of IV to the North; pp. 634-863 of IV, and pp. 1-168 of V to the South). In the light of recent events, the following passage, which closes the two pp. devoted to the Falkland Islands, has more than average interest: 'The history of the ridiculous disputes between Great Britain and Spain concerning these miserable islands, is one of the numerous evidences of the necessity of the study of geography among statesmen, as nothing but a complete geographical ignorance concerning them, could have raised such an unnecessary alarm on both sides' (Vol. IV, p. 862). The events following the passage of 150 further years suggest that it is more than geographical ignorance that is at fault.

The first series of pp. in Vol. V include 39 on Iceland and the Faroe Islands; these bring the special geography to an end. Then come sections on astronomy and physical geography. The volume concludes with the second series of numbered pp. devoted to that historical appendix identified on the tp.

The scale of this work, its date, and its provenance all suggest that a comparison with BELL (1829-32) would be interesting.

GLEASON, Benjamin **Y Am**

- **{1814}, 2nd (01)**
- *Remembrancer. Geography, on a new and improved plan*, topographically demonstrated, with maps, charts, and globes, by delineation, reference, and instruction ... Lectures ...
- BOSTON; Munroe & Francis
- 19.5 cm.; pp. 148
- ***Libraries*** : DLC, InU, MB, MBAt, MH, MHi, MWA, NjR, NN
- ***Source*** NUC (NG 0251170)
- ***Notes***

1. Discussed in Sahli (1941, *passim*).

GLEIG, G[eorge] R[obert] **Y SG?**

- **{1854} (01)**
- *A child's first book of geography*
- LONDON
- ***Libraries*** n.r.
- ***Source*** Allford (1964:246). Taken to be one of *Gleig's School Series*, 58 vols., 1850-[75?], recorded in BL.

GOLDSMITH, J., Rev.
See PHILLIPS, Richard.

GOODRICH, Charles Augustus **Y Am**

- **{[c1826]} (01)**
- *Outlines of modern geography* on a new plan carefully adapted to youth, with numerous engravings of cities, manners, costumes & curiosities ...
- BRATTLEBOROUGH, [VT]; Holbrook
- Pp. 252
- ***Libraries*** : PU, RPB
- ***Source*** NUC (NG 0322472). See note to (02) below.
- ***Notes***

1. NUC records further eds.: 1826, Hartford [CT] (MH, PPeSchw); 1827, Boston, 252 pp. (CtW, MH, MiU, ODW); 1827, 4th, Boston, 252 pp. (ICU, MH, MiDW, NNC); and 1827, 6th, Boston (MH).

- **{1827} (02)**
- *Outlines of modern geography*, on a new plan ... youth. With ... costumes and curiosities. Accompanied by an atlas
- BRATTLEBOROUGH, VT; published by Holbrook & Fessenden. Stereotyped by James Conner, New-York
- 15.5 cm.; pp. [5]-252
- ***Libraries*** : MB, MH : **CaBVaU**
- ***Notes***

1. The content is dominated by information about North America, the United States in particular. The text proceeds by way of question and answer, but many of the toponyms that form part of the answers are omitted, obliging the students to consult an atlas in order to learn the missing facts.
2. NUC records further eds.: 1828, Brattleborough [VT] (MH); 1830, Boston, 252 pp. (MH, MHi, WaU); 1836, Boston (MH); 1837, 5th, Boston, 252 pp. (MiU); BL records ed. of 1837, Boston, 252 pp.
3. Elson (1964:398) cites one of the Boston eds. of 1827.

GOODRICH, Charles Augustus **A Am**

- **{1836} (01)**
- *The universal traveller*: designed to introduce readers at home to an acquaintance with the arts, customs, and manners, of the principal modern nations on the globe; ... Derived from the researches of recent travellers of acknowledged enterprise, intelligence and fidelity; and embodying a great amount of entertaining and instructive information
- HARTFORD, [CT]; Canfield and Robins
- 19.5 cm.; pp. xii, [13]-610
- ***Libraries*** : ICJ, MiU, ODW, PPG, ViU
- ***Source*** NUC (NG 0322516). See note to (02).
- ***Notes***

1. NUC records further eds.: 1836, 2nd, Hartford, 501 pp. (CtY, MH, MnHi, MWA, OClWHi, OU, PPFr, T); 1837, 3rd, Hartford, 504 pp. (ICRL, WaPS).

- **1838 (02)**
- *The universal traveller* ... globe. Embracing a view of their persons - characters - employments - amusements - religion - dress - habitations - modes of warfare - food - arts - agriculture - manufactures - superstitions - government - literature, etc. etc. Derived ... imbodying ... Second edition
- HARTFORD; Canfield & Robins
- 18.5; pp. [3], 4-504
- ***Libraries*** : FMU, Njk, **NN**, OrU, PU, ViU
- ***Notes***

1. A popular anthropology rather than a special geography. There is no astronomical section, nor anything on maps and globes; it is indifferent to location and physical geography. Provides, on a country-by-country basis, information on the topics listed on the tp. The U.S. is dealt with first, at

greatest length, and with marked self-satisfaction. Goodrich acknowledges Yale College library as the location of his many sources; also that 'for the articles on the United States, France, and Italy, he [the author] takes pleasure in acknowledging his obligations to a distinguished literary friend – *Rev. Royal Robbins*' (presumably ROBBINS 1830). He quotes fairly extensively, sometimes citing, and sometimes failing to cite, his sources.
2. The date is provided by NUC (NG 0322519).
3. NUC records further eds. (all Hartford, CT, from 1843): 1839, 2nd, Hartford, 504 pp. (MH, PPPCPh); 1839, New York, 504 pp. (NIC); 1842, 504 pp. (OCl); 1843, 504 pp. (DLC, ICJ, NIC, OClJC, OO); 1845, 504 pp. (ICJ, LU, OU); 1848, 540 pp. (PPT); 1849, 545 pp. (DLC, OClW); 1850 (CtY).

GOODRICH, Samuel Griswold (Parley, Peter) **Y Am**
- ▸ **{1831} (01)**
- ▸ *The child's book of American geography*; designed as an easy and entertaining work for the use of beginners. By the author of Peter Parley's tales
- ▸ BOSTON; Waitt & Dow
- ▸ 18 cm.; pp. 64
- ▸ ***Libraries*** : DLC, MB, MH
- ▸ ***Source*** NUC (NG 0324118)
- ▸ ***Notes***

1. Discussed in Brigham and Dodge (1933:17), where the title is given as *The Child's Own Book of Geography.*

- ▸ **1832 (02)**
- ▸ *The child's book of American geography* ... beginners. With sixty engravings
- ▸ BOSTON; Waitt & Dow
- ▸ 4to, 18 cm.; pp. [2]-64
- ▸ ***Libraries*** L : CtY, MH, NNC
- ▸ ***Notes***

1. Limited essentially to the United States. Refers to the 'eastern' continent, but provides only a map, three lines of text, and four illustrations.
2. Catalogued by BL under American Geography. NUC (NG 0324119) notes: 'On cover: By the author of Peter Parley's tales. First ed., 1831.' NUC also notes that the work was translated into Ojibwa (NG 0324005) and modern Greek (NG 0324132).
3. NUC records further ed.: 1837, 2nd, Boston, 64 pp. (CtY, DLC, ICU, IdU, MH, NjP, NNC, OWoC).

GOODRICH, Samuel Griswold **Y Am**
- ▸ **{1850a} (01)**
- ▸ *A comprehensive geography and history, ancient and modern*
- ▸ NEW YORK; Huntington and Savage
- ▸ 27.5 cm.; pp. 272
- ▸ ***Libraries*** : DLC, OO
- ▸ ***Source*** NUC (NG 0324169)

▸ *Notes*
1. NUC records further ed.: 1853, New York, 272 pp. (DLC, NNC).
2. Culler (1945, *passim*) discusses (01); Hauptman (1978:431) includes a reference to an ed. of 1855 as part of a general discussion, quotes directly from it (1978:432), and refers to its illustrations (1978:436); Elson (1964: 402) cites the ed. of 1853.

GOODRICH, Samuel Griswold (Parley, Peter) **Y**
▸ **{[186-?]}, new (01)**
▸ *Geography and atlas* ... New ed., carefully adapted to English schools and families
▸ LONDON; Cassell, Petter, and Galpin
▸ Pp. (4) [7]-88
▸ ***Libraries*** : MH
▸ ***Source*** NUC (NG 0324307), which notes: 'Cassell's educational works'

GOODRICH, Samuel Griswold (Parley, Peter) **mSG Am**
▸ **{[c1844]a} (01)**
▸ *Manners and customs of the principal nations of the globe.* By the author of Peter Parley's tales
▸ PHILADELPHIA; Thomas, Cowperthwaite & co
▸ 18 cm.; pp. iv, [5]-352.
▸ ***Libraries*** : CtY, PPWa, ViLxW
▸ ***Source*** NUC (NG 0324530), which notes: '[c1844]'
▸ ***Notes***
1. NUC records further eds.: 1845, Boston, 352 pp. (DLC, MB, MH, MWA, OU, ViU); 1845, New York, 352 pp. (MSaE, NcU, ViU; not at MtBc, NUC notwithstanding); and 1848, Boston (MB).

▸ **{1849} (02)**
▸ *Manners and customs* ...
▸ BOSTON; published by Rand and Mann, No. 3 Cornhill
▸ 18cm; pp. [7], iv, [5]-352
▸ ***Libraries*** : **MtBC**, Nh, NN
▸ ***Notes***
1. As its title implies, the book deals with a narrower range of topics than the norm for special geography. The author has a thesis, which is expressed in the second sentence of the Introduction thus: 'Though mankind, in all ages and countries, possess the same elements of character, these are modified or controlled by the potent influences of *climate, religion*, and *government*' (pp. 5-6; emphasis in the original). The ways in which these three influences bring about their effects are considered in the Introduction, but not in equal detail. 'In respect to religion, we may remark that not only the ceremonies of worship, but those which belong to births and burials, courtship and marriage, as well as many others, are shaped or modified by its potent influence' (p. 11). So much for that topic. Government receives three paragraphs: one is devoted to the restraints placed on individuals by despotic authority, one to the observation that the situation is different where 'gov-

ernment be free,' and the third begins as follows: 'There are other modes in which the power of government upon national character is made manifest, but we have not space to notice them here' (pp. 11-12). In contrast, 'the influence of climate' is the major preoccupation of pp. 6-11. Race is also referred to as having an influence on national character, but, in wording that succeeds in being both explicit and ambiguous, its role is played down.

Having stated his thesis, however, Goodrich scarcely mentions it again. Once the introduction is finished, it is left to the readers to decide for themselves how, and to what degree, the 'potent influences' make themselves felt in bringing about the customs and manners that are described.

The attitude presented in the text is that of a middle-class traveller of the 19th century, naively self-confident in the belief that American ways provide the norm for the planet. The account (pp. 38-40) of the table manners of English travellers observed while staying at an inn, for example, suggests, in both style and detail of observation, the outlook of a British missionary reporting to his supporters at home the behaviour of some heathen tribe in some 'remote' corner of the world.

Although youth may have been the intended audience, this is not stated to be the case in the Introduction, which can be read as being directed to a general audience.

The use of countries as the framework within which to give information, and the sequence in which they are presented, mark the book as falling partially within the genre of special geography. Almost three quarters of the text is devoted to Europe.

2. NUC records further eds.: 1855, Boston, 352 pp. (ViU); and 1856, Boston, 352 pp. (KMK).

GOODRICH, Samuel Griswold (Parley, Peter) **Y Am**

- **1845a (01)**
- *A national geography*, for schools; illustrated by 220 engravings, and 33 maps; with a globe on a new plan: by ... author of Peter Parley's tales
- NEW YORK; Huntington & Savage, 216 Pearl Street. C.A. Alvord, printer, corner of John and Dutch Streets
- 4to, 28.5 cm.; pp. [2]-108
- ***Libraries*** **L** : DLC, NcA, PP
- *Notes*

1. Deals with all the countries of the world. The bigotry and prejudice so abundantly present in GOODRICH, S.G. (1835), see below, are largely if not entirely absent.
2. NUC records further eds.: 1846, Cincinnati (MH); 1846, 4th, New York; Huntington and Savage [Culler (1945:308)]; 1848, improved, New York, 112 pp. (DLC); 1850, New York, 71 pp. (DLC); and 1852, New York (MH) [NG 0324583 notes: 'The new national geography'].

GOODRICH, Samuel Griswold **Dy Am**

- **{1832}a (01)**
- *A new universal pocket gazetteer.* Containing descriptions of the most remarkable empires, kingdoms, nations ... in the known world

- BOSTON; Carter, Hendee & Co.
- 16mo; pp. 297
- ***Libraries*** : CtY, MB
- ***Source*** NUC (NG 0324586)

- **{1832} (02)**
- *A new ... gazetteer*, containing ... kingdoms ... in the known world, with notices of manners, customs, religion ... and population
- BOSTON; W. Hyde
- Pp. 297
- ***Libraries*** : CtY, INS, MB, MH, MiU, MnU, NcD, OClWHi
- ***Source*** NUC (NG 0324587)

GOODRICH, Samuel Griswold (Parley, Peter) **Y Am**

- **{1860} (01)**
- *Parley's four quarters of the world*
- PHILADELPHIA; Desilver
- 4 vols. in 1
- ***Libraries*** : MH
- ***Source*** NUC (NG 0324294), which notes: 'Contents: The tales of Peter Parley about Europe. Rev. ed. The tales ... about Asia. Rev. ed. The ... Africa. Rev. ed. The ... American. Rev. ed.'

GOODRICH, Samuel Griswold (Parley, Peter) **Y Am**

- **{1844b} (01)**
- *Parley's geography for beginners, at home and school*
- NEW YORK; Huntington & Savage
- ***Libraries*** : KyBgW
- ***Source*** Sahli (1945:381) and NUC (NG 0324308)
- ***Notes***

1. According to NUC (NG 0324560), this is a subsequent ed. to *Peter Parley's method of telling about geography to children*. See GOODRICH, S.G. (1829) below.
2. Discussed in Nietz (1961:228), and Sahli (1932, *passim*).
3. NUC and BL record further eds. (all New York): 1845, 160 pp. (L : MH, MiD, NCU); 1846, 160 pp. (L : MU, MWelC, PU); 1847, 160 pp. (L : MH, NjP); 1848 (L : PPWa); 1849 (: MH); 1850, 2nd (L : MH) [Culler (1945:308)]; 1853 (L : MH); and 1854, 160 pp. (L : NNC).

GOODRICH, Samuel Griswold (Parley, Peter) **Y**

- **{[ca. 1840]} (01)**
- *Parley's tales about Europe, Asia, Africa, America, and Australia*. Ed. by the Rev. T. Wilson [pseud]
- LONDON; Darton and Clark
- 14.5 cm.; 1 p. l., [v]-viii p., 2 l., [vii]-viii, [9]-536 p.
- ***Libraries*** : DLC : CaOTP
- ***Source*** St. John (1975, 1:181)

▸ *Notes*

1. St. John (1975, 1:181-2) notes that: 'Goodrich wrote his *Tales of Peter Parley about America, Europe, Africa and Asia* in four separate volumes between 1827 and 1830. For this English edition they were edited and probably rewritten by Samuel Clark who used the pseudonym, the Rev. T. Wilson.'
2. BL records a title that differs only in the replacement of Australia by Oceania (published in 1892). At first sight the substitution of Oceania for Australia in the title does not justify treating it as a different work from (01). However, (01) shows prejudices similar to those in GOODRICH, S.G. (1835), below. Because both works are the product of London publishing houses, and because it is in them that the racial and national prejudices that disturb a reader of the late twentieth century are to be found, whereas there are few signs of such biases in those American eds. that I have examined, the possibility that they are a contribution of the British editor seems real. Not that the American versions are entirely free of prejudice; see, e.g., note to entry (02) of GOODRICH, S.G. (1829), which follows.
3. See GOODRICH, S.G. (1835), below, for a title that is similar except that it does not begin with *Parley's ...*

GOODRICH, Samuel Griswold (Parley, Peter) **Y Am**

▸ **1829 (01)**

▸ *Peter Parley's method of telling about geography to children*[: with nine maps and seventy-five engravings principally for the use of schools]

▸ HARTFORD; H. and F.J. Huntington

▸ 14.5 cm.; pp. viii, [9]-122

▸ ***Libraries*** : CtHT, DLC, NNC

▸ ***Source*** NUC (NG 0324552)

▸ *Notes*

1. The words in [ ] are derived from CIHM.
2. NUC records further eds.: 1830, Hartford, 114 pp. (DLC, MH : CaQQS); 1831, Boston (MH) [NG 0324554 notes: 'Cover reads: New York, Collins and Hannay']; 1831, Hartford, 114 pp. (CtY, MB, MH); and 1832, New York (MH).

▸ **1834 (02)**

▸ *... Parley's method of telling about geography ...*

▸ BOSTON; Carter, Hendee & co.

▸ 13 cm.; pp. [4], vi, 11-120

▸ ***Libraries*** : NN

▸ *Notes*

1. Information is provided in numbered paragraphs, the numbers being the key to corresponding questions. The author is identified in the statement of copyright; he seeks to create the impression that information will be presented in the form of travellers' tales; the text, however, is essentially factual, though somewhat colloquial in style. He also seeks to 'inculcate lessons of morality and religion' (p. vi). This, presumably, is the justification for several adverse judgements passed on people living in countries remote

from the U.S.: 'The changes that take place among such nations [those of Africa] are not very great. They have their wars, and sometimes a whole tribe is killed, or perhaps driven to some other region. Sometimes, too, a fatal disease comes among them, and sweeps off a whole people. But such things do not have a very extensive influence, and are soon forgotten' (p. 104). The final 11 pp. are a concentrated exposition of bigotry and prejudice that conclude with the following words: 'Let us fear to do wrong, because God can punish us. Let us love to do right, because God will reward us' (p. 120).

2. NUC records further eds.: 1834, Hartford, (CtY, MH); 1835, Hartford, 120 pp. (FMU) [NG 0324559 notes: 'c1829']; 1836, New York, 120 pp. (DLC, ICU); 1836, Philadelphia, (MH {badly mutilated}); 1837, New York, 120 pp. (MH, NjR); [1838], New York, 120 pp. (DLC) [NSG 0038410 notes, (1): 'c1829'; (2): 'Issued in covers printed from the same setting of type (except variant imprint statement) as those of the ed. published in New York by F.J. Huntington, 1838']; 1838, New York; F.J. Huntington, 120 pp. (ICU, MiU); 1838, Philadelphia, 120 pp. (CSt, CtY); 1839, New York, 120 pp. (CtY, MH, NN); 1839, Philadelphia, 120 pp. (PU); 1840, New York, 120 pp. (NN); 1841, New York, 120 pp. (NN); 1842, New York (MH); 1843, New York (MH, PU); and 1844, New York (CtY).

   NUC (NG 0324560) notes: 'Later edition, 1844, has title: Parley's geography for beginners, at home and school ...'
3. Nietz (1961:210) discusses (01); Hauptman (1978:430-1, n. 34, and 436) discusses (02).
4. A microform version of the ed. of 1830 is available from CIHM.

GOODRICH, Samuel Griswold **Am**

- **{1840} (01)**
- *A pictorial geography of the world*, comprising a system of universal geography, popular and scientific ... and illustrated by more than one thousand engravings ... With a copious index, answering the purpose of a gazetteer
- BOSTON; Otis, Broaders and company; New York; Tanner and Disturnell
- 27.5 cm.; pp. 1008
- ***Libraries*** : DLC, MBAt, MeB, MH, MWA, OO, PP, PPPD, ViU
- ***Source*** NUC (NG 0324637), which notes: 'Edited by T.S. Bradford. Issued in 10 parts.' Discussed in Nietz (1961:228)
- ***Notes***

1. NUC records further eds. (all Boston): 1840, 2nd, 493 pp. (CtY, ICJ, MH, OCl, OClWHi, PPAmP, ViU); 1840, 3rd, 1008 pp. (MH, PHi, PPAN, PPWa); 1841, 1008 pp. (MWA); and 1841, 4th, 1008 pp. (NIC, OClW).

- **{1841}, 8th (02)**
- *A pictorial geography* ... scientific, including a physical, political, and statistical account of the earth and its various divisions, with numerous sketches from recent travels ... 8th ed.
- BOSTON; C.D. Strong
- 2 vols.

- ***Libraries*** : AU, ICRL, MH
- ***Source*** NUC (NG 0324642)

- **{1841}, 9th (03)**
- *A pictorial geography* ... scientific ... illustrated by more than one thousand engravings ...
- BOSTON; C.D. Strong
- 26.5 cm., 2 vols.
- ***Libraries*** : CSmH, DLC, MWA, NjR, PV
- ***Source*** NUC (NG 0324643)

- **{1841}, 10th (04)**
- *A pictorial geography* ... scientific ... With a copious index, answering the purpose of a gazetteer. 10th ed.
- BOSTON; C.D. Strong
- 27 cm.; pp. 1008
- ***Libraries*** : IaU
- ***Notes***

1. NUC records further eds. (all Boston): 1842, 11th, 2 vols. (MnHi); 1845, 12th, 1008 pp. (MWA); 1845, 13th, 2 vols. (ICRL, MiU, PPL); 1847, 14th, 1008 pp. (INS); and 1849, 16th, 1003 pp. (INS, OCU, PV).

- **{1851}, 19th (05)**
- *A pictorial geography* ... scientific, including ... travels: and illustrated by more than one thousand engravings ... With ... 19th ed.
- BOSTON; C.D. Strong
- 26 cm.; pp. 1008
- ***Libraries*** : NNC
- ***Source*** NUC (NG 0324649)
- ***Notes***

1. NUC records a further ed: 1856, new, Boston, 2 vols. (MB, MH, MWA, NN).

GOODRICH, S[amuel] G[riswold] **Y Am**

- **{[1850]b} (01)**
- *A primer of geography*
- NEW YORK; Huntington & Savage
- 19 cm.; pp. 62
- ***Libraries*** : DLC
- ***Source*** NUC (NG 0324793)
- ***Notes***

1. Discussed in Culler (1945, *passim*); Hauptman draws on it three times (1978:436), and refers to it in general discussion (1978:431, 434).

GOODRICH, Samuel Griswold **Y Am**

- **{1830} (01)**
- *A system of school geography chiefly derived from Malte-brun*, and arranged according to the inductive plan of instruction

- BOSTON; Carter & Hendee, etc.
- ***Libraries*** : MH
- ***Source*** NUC (NG 0324897)
- ***Notes***

1. NUC (NG 0324896) records an ed. in 1826, but also refers to it as '12th ed. Hartford, F.J. Huntington; [etc.] 1836.' MiDSH report that the earlier date is an error; the correct date is 1836; see (05).

- **1830 (02)**
- *A system of school geography ...*
- HARTFORD; Collins & Hannay, New York. Carter & Hendee, Boston. Towar J. & D.M. Hogan, Philadelphia. Plaskitt & Co., Baltimore. A.S. Beckwith, Providence. S. Coleman, Portland. J. & J.W. Prentiss, Keene. J.H. Nash, Richmond, Va.. C.D. Bradford & Co., Cincinnati. J.N. Whiting, Columbus. W.W. Worsley, Louisville. Hastings & Tracy, Utica. Glazier & Co. Hallowell. P.A. Brinsmade, Augusta
- 12mo, 16 cm.; pp. viii, 9-320
- ***Libraries*** **L** : DLC, MH
- ***Notes***

1. Intended to teach by giving 'particulars first, and general views afterwards.' The child is to start by learning about its immediate locality, and then to build on that knowledge by learning, from the book, about more remote places. Many questions are provided in the text, most of which can only be answered by consulting maps, which are not provided in the book.

- **{1832}, 2nd (03)**
- *A system of school geography ...*
- HARTFORD; F.J. Huntington
- 12mo; pp. 288
- ***Libraries*** : AU, MB, MH, MnU, OClWHi
- ***Source*** NUC (NG 0324899)
- ***Notes***

1. NUC (NSG 0038433) records what is probably another copy of this ed. at NmU.
2. NUC records a further ed.: 1833, 6th, Boston (MH).

- **{1833}, 8th (04)**
- *A system of school geography ...*
- HARTFORD; F.J. Huntington, etc.
- ***Libraries*** : MH, ICU
- ***Source*** NUC (NG 0324901), which notes: 'Half-title: The Malte-Brun school geography'
- ***Notes***

1. NUC records further eds.: 1833, 8th, Raleigh, 288 pp. (NcU); 1834, 10th, Hartford (CtY, MH); 1835, 11th, Hartford, 288 pp. (CtY, ICRL, MH, NcU, NN, OClWHi); and 1836, 11th , Hartford, 288 pp. (DLC, NIC, OClW)

- **1836, 12th (05)**
- *A system of school geography ...*

- HARTFORD; F.J. Huntington [disintegration of the paper makes the list of joint publishers undecipherable]
- 16.5 cm.; pp. [7], 8-288 pp.
- ***Libraries*** : **NN**
- ***Notes***

1. The author addresses his readers, who are presumed to be senior pupils in school. Their attention is first drawn to the local scene; then to the local county; they are then introduced to maps; then to a relatively detailed survey of the U.S., and finally to the rest of the world. This is what is meant by the 'inductive method' of the title; see (01) above.

   The content may be derived from Malte-Brun, but the style is not. The latter takes the form of discursive remarks made by a traveller, e.g., in Scotland: 'As these mountains, lochs, and valleys, are woven into the poetry and romance of Scott, Burns, and others, we have read them from childhood, and associated them in our minds with all that is wild, pastoral and pleasing' (p. 133).
2. NUC records further eds.: 1837, 12th, New York, 288 pp. (DLC, InU, MiEM); 1838, 20th, New York (MH); 1839, 20th, New York, 288 pp. (DLC, MBAt, MH, OO); 1839, 27th, New York, 288 pp. (MH, NcD); and 1842, 29th, New York, 288 pp. (NcU).
3. Carpenter (1963:256-7) provides a general discussion; Nietz (1961:228) discusses (01), though wrongly dating it to 1831; Nietz (1961:217-21) and Sahli (1941, *passim*) discuss the 27th ed. of 1839.

GOODRICH, Samuel Griswold **A Am**

- **1832b (01)**
- *A system of universal geography, popular and scientific*, comprising a physical, political, and statistical account of the world and its various divisions; embracing numerous sketches from recent travels; and illustrated by engravings of manners, costumes, curiosities, cities, edifices, remarkable animals, fruits, trees, and plants
- BOSTON; Carter, Hendee & Co.
- 8vo, 24 cm.; pp. [6], 9-920, [2]
- ***Libraries*** **L** : MBAt, MH, MtBC, Nh, WaU
- ***Notes***

1. The original intention, according to the preface, 'had nothing more in view than a compilation from Malte Brun and Bell,' but after study this 'original work was undertaken.' In general, information about countries is provided under the separate headings of physical and political geography. The United States, dealt with on a state-by-state basis, occupies pp. 9-356.
2. A comparison with LAURIE (1842) might be interesting.

- **{1832} (02)**
- *A system of universal geography ...*
- CINCINNATI; Roffe & Young
- 24.5 cm.; 3 p. l., [9]-920, [2] p.
- ***Libraries*** : DLC, NIC
- ***Source*** NUC (NG 0324912)

- **{1832} (03)**
- *A system of universal geography ...*
- NEW YORK; Collins & Hannay
- 24.5 cm.; 4 p. l., [9]-920 p.
- ***Libraries*** : NcD, ViU
- ***Source*** NUC (NG 0324913)

- **{1832} (04)**
- *A system of universal geography ...*
- PHILADELPHIA; Key, Mielke & Biddle
- 24 cm.; pp. 920
- ***Libraries*** : MB, N, PPWa, PU
- ***Source*** NUC (NG 0324914)

- **{1833}, 2nd (05)**
- *A system of universal geography ...*
- BOSTON; C. Hendee & co. New York; Collins & Hannay. [etc.]
- 23.5 cm.; pp. xvi, [9]-975
- ***Libraries*** : CtY, MB, MH, NN, PPF, PU
- ***Source*** NUC (NG 0324915), which notes: '"A brief commercial dictionary": p. [956-65].'

GOODRICH, Samuel Griswold (Parley, Peter) **Y**

- **1835 (01)**
- *Tales about Europe, Asia, Africa, and America.* By Peter Parley, author of tales about natural history, etc. Embellished with 137 engravings
- LONDON; printed for Thomas Tegg and Son, Cheapside; Tegg, Wise, and Co., Dublin; Griffin and Co., Glasgow; and J. and S.A. Tegg, Sydney, Australia
- 16mo, 13 cm.; pp. [2], x, 506
- ***Libraries*** L
- ***Notes***

1. 'The design of this little work is to convey to young persons, under the guise of amusement, the first idea of geography and history' (p. [iii]). Its probable effect will surely have been to inculcate prejudices based on the crudest misunderstanding of racial and national character. A special sort of special geography. See note to GOODRICH, S.G. (ca. 1840), above.
2. NUC and BL record further eds.: 1839, 4th, London, 540 pp. (: IaU, NcD); 1842, 5th, London, 540 pp. (: NjP); 1846, 6th, London (: NIC); and 1854, London, 424 pp. (L).
3. See GOODRICH, S.G. (18–?), above, for a similar title, but with the heading *Parley's ...*

GOODRICH, Samuel Griswold (Parley, Peter) **Y Am**

- **{1845b} (01)**
- *The world and its inhabitants*. By the author of Peter Parley's tales
- BOSTON; Bradbury, Soden & co.
- 17.5 cm.; 1 p. l., iv, [5]-328 p.

- ***Libraries*** : DCL, MB, MH, MiU, OO, RPB, ViU
- ***Source*** NUC (NG 0325188), which notes that the work is pseudonymous. WaS has disposed of the copy it held.
- ***Notes***

1. NUC records further eds.: 1845, Boston, 328 pp. (CU-I); 1845, New York, 328 pp. (NcU, ViU); 1845, Philadelphia, 328 pp. (CtY, OClWHi, PCC, PPWa, ViLxW); 1848, Boston (MB); 1849, Boston (formerly held at Nh); 1855, Boston, 328 pp. (ViU); 1856, Boston (MH); 1860, Boston (MB); 1864, Boston, 328 pp. (OCl, ViU); and [n.d.], London, [n.p.] (InU).

GOODRICH, Samuel Griswold **? Am**

- **{1855} (01)**
- *The world as it is, and as it has been*; or a comprehensive geography and history, ancient and modern
- NEW YORK; J.H. Colton and company
- 27 cm.; pp. 272
- ***Libraries*** : DLC, FSaHi, ICRL, NcA, OClWHi, ViU
- ***Source*** NUC (NG 0325198)
- ***Notes***

1. NUC (NG 0325199) records a further ed. in 1858 (NIC), noting 'c1857.'

GORDON, George **mSG**

- **1726 (01)**
- *An introduction to geography, astronomy, and dialling*. Containing the most useful elements of the said sciences, adapted to the meanest capacity, by the description and uses of the terrestrial and celestial globes. With an introduction to chronology
- LONDON; printed and sold by J. Senex, at the Globe against St. Dunstan's church, Fleet-Street; G. Strahan, at the Royal Exchange; W. and J. Osborn and T. Longman, in Pater-noster-Row; C. King, in Westminster-hall; and by the author
- 8vo, 20 cm.; pp. [12], iv, [4], 188, 40
- ***Libraries*** **L**, SanU : FU, ICU, OkU
- ***Notes***

1. Primarily concerned with problems resulting from the sphericity of the earth, but includes a summary special geography (pp. 24-38), identifying the principal natural and political divisions of the globe.

- **1729, 2nd, in which ... (see below) (02)**
- *An introduction to geography ... dialling*: containing ... globes: with ... chronology. The second edition. In which, besides many other great additions, are about twenty paradoxes belonging to the globes, intirely new. As also, the construction and uses of refracting and reflecting telescopes, etc.
- LONDON; printed for A. Bettesworth, at the Red-Lyon in Pater-Noster-Row
- 8vo, 20 cm.; pp. xvi, 318, [2]
- ***Libraries*** **L** : CSt, WU

▸ *Notes*
1. The special geography has been expanded considerably (pp. 25-96); it now includes modest amounts of information about the countries whose locations and subdivisions have been identified.

▸ **1742, 3rd in which ... (as in 1729 to ... globes) (03)**
▸ *An introduction to geography ...*
▸ LONDON; printed for C. Hitch, at the Red-Lion in Pater-Noster-Row
▸ 8vo, 20 cm.; pp. xvi, 318, [2]
▸ ***Libraries*** L : NN
▸ *Notes*
1. Spot-checking and the pagination suggest no change since 1729.
2. NUC records a further ed.: 1765, 3rd, London, 318 pp. (OkU, PHi).

GORDON, James **A n.c.**
▸ **1790-4 (01)**
▸ *Terraquaea; or, a new system of geography and modern history*
▸ LONDON; printed for the author. Sold by C. Dilly; and in Dublin by William Porter, Skinner Row
▸ 8vo, 22 cm., 4 vols.; Vol. I, pp. [4], xxxii, 336. [Vol. I only]
▸ ***Libraries*** L (Vol. I only) : MoU (Vols. II-IV from later eds.)
▸ *Notes*
1. Starts with a general survey of the features of the earth's surface, including some systematic human geography. The section on special geography begins with Spain; Portugal and Switzerland follow. On the basis of the BL holding it seems that the rest of the world will be dealt with in later volumes. In fact, coverage is limited to Europe; see below.

▸ **1794, 2nd (02)**
▸ *Terraquaea: ...*
▸ DUBLIN; printed for the author, by William Porter, Skinner-Row
▸ 8vo, 21 cm., 4 vols.; Vol. I, pp. [2]., xxxiv, 335; Vol. II, pp. xxiv, 383; Vol. III, pp. [2], x, 371; Vol. IV, pp. [8], vi, 405
▸ ***Libraries*** L : DLC
▸ *Notes*
1. Vol. I is as before; Vol. II deals with Italy and Germany, and includes a list of subscribers: Vol. III covers Britain and Ireland (includes a list of subscribers at the front); Vol. IV deals with Iceland, Lappland, the Scandinavian countries, Belgium, and France (again there is a list of subscribers).

GORDON, Pat **Y**
▸ **1693 (01)**
▸ *Geography anatomized: or, a compleat geographical grammer.* Being a short and exact analysis of the whole body of modern geography, after a new, plain and easie method, whereby any person may in a short time attain to the knowledge of that most noble and useful science. Comprehending, a most compendious account of the continents, islands, peninsula's [sic], isthmus, promontories, mountains, oceans, seas, gulphs, straits, lakes,

rivers and chief towns of the whole earth. As also the divisions, subdivisions, situation, extent, air, soil, commodities, manners, government, religion in all countries of the world. To which is subjoin'd the present state of the European plantations in the East and West Indies, with a reasonable proposal for the propagation of the blessed gospel in all pagan countries. The whole work carefully performed according to the exactest and latest discoveries. Illustrated with divers maps

- LONDON; printed by J.R. for Robert Morden and Thomas Cockerid [sic], at the Atlas in Cornhil, and at the Three Leggs in the Poultrey
- 12mo, 17 cm.; pp. [16], 208, [4]
- *Libraries* EU, **L**, LRS, O : CtY, DLC, NNH
- *Notes*

1. This was the first special geography to be referred to by its author as a grammar. It was intended for the children of the gentry. The first 114 pp. are devoted to presenting information in tabular form. Potted descriptions of the countries follow on pp. 115-98. Spot comparisons show it to contain more information than EACHARD (1691), and the text reads much more smoothly than does that of the earlier work. The tables also give the impression of providing information in a clear fashion, though the clarity is, at least occasionally, achieved at the expense of distorting the spatial realities of the countries described.

- **1699, 2nd, much improv'd and enlarg'd (02)**
- *Geography ... or, the compleat ...* a new and curious method. Comprehending I. A general view of the terraqueous globe. Being a comprehensive system of the true fundamentals of geography; digested into various definitions, problems, theorems, and paradoxes: with a transient survey of the whole surface of the earthly ball, as it consists of land and water. II. A particular view of the terraqueous globe. Being a clear and pleasant prospect of all remarkable countries upon the face of the whole earth; shewing their situation, extent, division, subdivision, cities, chief towns, name, air, soil, commodities, rarities, archbishopricks, bishopricks, universities, manners, language, government, arms, religion. Collected from the latest authors, and illustrated with divers maps
- LONDON; printed for Robert Morden and Thomas Cockerill; at the Atlas in Cornhill, and in Amen-Corner
- 8vo, 18 cm.; pp. [18], 402, [1]
- *Libraries* **L**, NC : CtY
- *Notes*

1. Pp. 1-42 are wholly new. They contain: definitions of geographical terms, geographical problems (having to do with location on the globe, local time, difference in time, and other problems generally dependent on the sphericity of the earth and its astronomical relations with the sun), and geographical paradoxes; these last would probably be called riddles today. Some depend for their solution on the sphericity of the earth and its relations with the sun; most depend upon puns or other sleights of meaning. Then follow many tables listing the location of features of physical geography (islands, lakes, rivers, etc.). Pp. 59-402 contain information of the type found in the 1693 ed. but the tabular and textual information for any given country are

now grouped together. As the greater length suggests, much additional information has been provided.

- **{1702}, 3rd, corrected, and somewhat enlarg'd (03)**
- *Geography anatomized ... or, the geographical grammar.* Being a ...
- LONDON; R. Mordon, T. Cockerill, and R. Smith
- 12mo; 10 p. l., 428 p.
- ***Libraries*** : ICU, MWA, NN
- ***Source*** NUC (NG 0333561)

- **{1704}, 4th, corrected, and somewhat enlarg'd (04)**
- *Geography anatomized ...*
- LONDON; printed for S. and J. Sprint, John Nicholson, and S. Burroughs ... Andr. Bell ... and R. Smith ...
- 18 cm.; 12 p. l., 428 p. 1 l.
- ***Libraries*** : CLU, CtY, MH, NN
- ***Source*** NUC (NG 0333563)

- **{1708}, 5th, cor., and somewhat enl. (05)**
- *Geography anatomized ...*
- LONDON; printed for J. Nicholson, [etc.]
- Pp. 428
- ***Libraries*** : CU, ICU, InU, IU, MH, MiU, NNC, OClWHi, RPB
- ***Source*** NUC (NG 0333564)

- **{1711}, 6th, corr., and somewhat enl. (06)**
- *Geography anatomized ...*
- LONDON; printed for J. Nicholson, [etc.]
- Pp. 428
- ***Libraries*** LcU : IU, MH : CaBVaU
- ***Source*** ESTC and NUC (NG 0333565)

- **1716, 7th, corrected and somewhat enlarg'd (07)**
- *Geography anatomized ...* languages ...
- LONDON; printed for J. Nicholson, J. and B. Sprint, and S. Burroughs, in Little Britain; Andr. Bell, at the Cross-Keys and Bible in Cornhil, and R. Smith under the Royal Exchange
- 8vo, 18cm; pp. [24], 428, [4]
- ***Libraries*** L : DLC, IU, MWA : CaAEU, CaBViP
- ***Notes***

1. Since 1699 some of the geographical paradoxes have been replaced by others, the sequence has been changed, and six more have been added. In the section on countries, a few items of new information have been added, but most of the additional 26 pages of text are accounted for by a more generous spacing of the letterpress.

- **{1719}, 8th, cor., and somewhat enl. (08)**
- *Geography anatomized ...*
- LONDON; J. and B. Sprint [etc.]
- 18 cm.; 12 p. l., 428 p.

- ***Libraries*** : CtY, DLC, MH, NN
- ***Source*** NUC (NG 0333567)

- **1722, 9th, corrected and somewhat enlarg'd (09)**
- *Geography anatomized* ... enlarg'd. And a set of new maps. By Mr. Senex
- LONDON; printed by S. Palmer, for R. Knaplock, J. and B. Sprint, S. Burroughs, D. Midwinter, A. Bettesworth, R. Ford, A. Ward, and J. Clark
- 8vo, 20 cm.; pp. [24], 416
- ***Libraries*** L
- ***Notes***

1. The reduction in number of pages since 1716 is achieved by using pages left blank at the end of the sections on some countries in that edition.

- **1725, 10th, corrected and somewhat enlarged (10)**
- *Geography anatomized* ...
- LONDON; printed for R. Knaplock ... A. Bettesworth, J. Brotherton, R. Ford ...
- 8vo, 20 cm.; pp. [24], 416
- ***Libraries*** L : CtY, PPL
- ***Notes***

1. Seemingly another issue rather than a new edition.

- **1730, 12th, corrected, and somewhat enlarg'd (11)**
- *Geography anatomized* ...
- LONDON; printed for J. and J. Knapton, R. Knaplock, D. Midwinter, A. Bettesworth, B. Sprint, J. Osborn and T. Longman, J. Brotherton, R. Ford ... J. Clarke
- 8vo, 20 cm.; pp. [24], 416 pp.
- ***Libraries*** L : CSmH, MH, OCU, PPL, RPJCB
- ***Notes***

1. Again no sign of change.

- **{1733}, 12th (12)**
- *Geography anatomiz'd; or, the geographical grammar*; with maps by Senex
- LONDON
- 16mo
- ***Libraries*** : MBAt, PHi
- ***Source*** NUC (NG 0333572)

- **1733, 13th, corrected and somewhat enlarged (13)**
- *Geography anatomized* ... [as in 1702-30]
- LONDON; printed for J.J. and P. Knapton, R. Knaplock, D. Midwinter and A. Ward, A. Bettesworth and C. Hitch, B. Sprint ...
- 8vo, 20 cm.; pp. [24], 432
- ***Libraries*** L : DLC, PHi
- ***Notes***

1. The additional pp. 417-32 contain a gazetteer.

- **1735, 14th, corrected and somewhat enlarged (14)**
- *Geography anatomized* ...

- LONDON; printed for J.J. and P. Knapton, D. Midwinter, A. Bettesworth ... B. Sprint, A. Ward, S. Birt, T. Longman ...
- 8vo, 20 cm.; pp. [24], 432
- ***Libraries*** L
- ***Notes***

1. No sign of change.

- **1737, 15th, corrected, and somewhat enlarged (15)**
- *Geography anatomized ...*
- LONDON; printed for J. and P. Knapton ...
- 8vo, 20 cm.; pp. [24], 432
- ***Libraries*** L : FU, ICU, MB, NcD, TxU
- ***Notes***

1. Seemingly unchanged.

- **1740, 16th, corrected, and somewhat enlarged (16)**
- *Geography anatomized ...*
- LONDON; printed for D. Midwinter, A. Ward, J. and P. Knapton, S. Birt, T. Longman, J. Clarke (Exchange), C. Hitch, J. Brotherton, J. Hodges, T. Cooper, and the executrix of R. Ford
- 8vo, 20 cm.; pp. [24], 432
- ***Libraries*** L, LvU : TxU
- ***Notes***

1. As before.

- **{1741}, 17th, corrected, and somewhat enlarged (17)**
- *Geography anatomized ...*
- LONDON; D. Midwinter, A. Ward ...
- Pp. [22], 432
- ***Libraries*** : CU, ICN, MH, OGK, TxU, ViU : CaAEU
- ***Source*** NUC (NG 0333576)

- **{1744}, 18th, cor. and somewhat enl. (18)**
- *Geography anatomized ...*
- LONDON; printed for A. Ward, [etc.]
- Pp. 432
- ***Libraries*** : MoU, NIC, NN, PHi, RPJCB
- ***Source*** NUC (NG 0333578)

- **1747, 16th, corrected and somewhat enlarged (19)**
- *Geography anatomized ...*
- DUBLIN; printed by Oli. Nelson, in Skinner Row, for G. Risk, G. Ewing, W. Smith, and P. Crampton, in Dean Street, and G. Faulkner, in Essex Street, Booksellers
- 8vo, 21 cm.; pp. [16], 414
- ***Libraries*** L
- ***Notes***

1. Quick comparison suggests that the reduction in number of pp. compared with (16) was achieved by fitting more text to a page, not by reducing the number of words.

- **1749, 19th, corrected, and somewhat enlarged (20)**
- *Geography anatomized ...*
- LONDON; printed for J. and P. Knapton, J. Brotherton, J. Clarke, S. Birt, T. Longman, C. Hitch, R. Hett, J. Hodges, M. Cooper, J. Davidson, J. and J. Rivington, and J. Ward
- 8vo, 21 cm.; pp. [20], 432
- ***Libraries*** **L** : CSt, MBAt, MH, OOxM, PMA, RPJCB
- ***Notes***

1. No explanation for the diminution in the number of preliminary pages was sought.

- **1754, 20th, corrected and inlarged (21)**
- *Geography anatomized* ... religion. With a new set of maps ingraved on a larger scale by Emanuel Bowen, Geographer to his Majesty
- LONDON; printed ... S. Birt, D. Browne, T. Longman, C. Hitch, J. Hodges, J. Shuckburgh, J. and J. Rivington, J. Ward, J. Wren, and M. Cooper
- 8vo, 21 cm.; pp. [20], 416
- ***Libraries*** E, **L** : CtY, DLC, ICU, OO : CaBViPA
- ***Notes***

1. For the first time since 1702, the claim that this latest ed. is 'corrected and somewhat enlarged' is justified. A little extra information is given for several countries. Most of the saving in space needed to accommodate the additional information is achieved by the omission of 'An appendix comprehending a brief account of the European plantations in Asia, Africa, and America: as also some reasonable proposals for the propagation of the blessed gospel in all pagan countries,' which has been present since 1693 and unaltered since 1699.

- **{1760}, 20th, cor. and enl'd and ... maps by Mr. Senex (22)**
- *Geography anatomized ...*
- DUBLIN; Grierson
- Pp. 414
- ***Libraries*** : PP
- ***Notes***

1. Discussed in Downes (1971:382), and Warntz (1964, *passim*).
2. Microform versions of (06), (17), and (21) are available from CIHM.

GORDON, William **A**

- **1789 (01)**
- *A new geographical grammar and complete gazetteer*; upon an improved, enlarged, and scientific plan. Wherein those parts of astronomy which are connected with geography, the use of both globes, and the nature, use, and construction of maps, are minutely explained; and the continents of Europe, Asia, Africa, and America, with their islands, including all the valuable discoveries make in the most remarkable voyages and travels through both hemispheres, copiously described: together with the rise, progress, and present state of all the empires, kingdoms, and states in the known world; with respect to situation, extent, climate, soil, produce, government, power,

opulence, trade, manufactures, genius, language, manners, customs, religions, etc. Embellished and illustrated with a new set of maps and other engravings accommodated to this work. To which are added the elements of chronology. And a copious list of remarkable events, with their dates, from the earliest account of time to the present. By ... of the Mercantile Academy, and author of The Universal Accountant, etc. etc. etc.

- EDINBURGH; printed for Alexander Guthrie, No. 25, South Bridge Street
- 4to, 26.5 cm.; pp. [5], viii-xiii, [1], 615
- ***Libraries*** L
- ***Notes***

1. In Europe, it begins with Scotland. For the countries of Europe, it provides information at the level of county or province. Provides for each a table of cities with their latitude, longitude, and river on which located. Then comes a section headed air, soil, produce, trade, manufactures, etc. Ends with information about the country in general. Deals with countries in other continents in a similar way, although sometimes only the concluding, general section is retained. Provides occasional direct quotations from other works, with the source cited on some occasions. Ends with appendices listing rivers and mountains of all countries, the currencies of Europe, two chronologies, and an index (pp. 529-615).

GOUINLOCK, G. and J., comps. **Y Can**

- **{1845} (01)**
- *A system of general geography*; including outlines, or a first course for beginners, on an improved and easy plan, adapted to the interrogative or intellectual mode of tuition. The natural peculiarities, productions, manufactures, commerce, etc. of the different countries are pointed out; and a concise description given, of every town of importance in the world. Scripture geography. Introduction to astronomy and problems on maps and the globes. With very copious exercises throughout. Compiled and arranged for the use of schools in British America
- TORONTO; Hugh Scobie
- 18.5 cm.; pp. viii, 9-247
- ***Libraries*** : DLC, ICU : CaBVaU, CaOONL, CaOTP
- ***Source*** NUC (NG 0352941)
- ***Notes***

1. A microform version is available from CIHM.

GRANT, Horace **Y SG?**

- **{1861}, new (01)**
- *Geography for young children* ... New edition. With a biographical notice of the author, by E. Chadwick
- LONDON; Bell
- 16mo
- ***Libraries*** L
- ***Source*** BL and ECB, Index (1856-76:130)

GRAY, Thomas
▸ *The civil service geography* (see SPENCE 1867)

GREEN, John
See note to GEOGRAPHY REFORMED (03), and note to MEAD (1717).

GREEN, Samuel **Y Am**
▸ **{1818} (01)**
▸ *A geographical grammar*
▸ NEW LONDON, CT; Samuel Green
▸ ***Libraries*** : NNC
▸ ***Source*** Elson (1964:396); holding confirmed by NC

GREGORY, J. **Y**
▸ **{1739} (01)**
▸ *Manual of modern geography*
▸ ***Libraries*** n.r.
▸ ***Notes***
1. Known from a reference to it in the Preface of entry (02).

▸ **1740, 2nd ... (see below) (02)**
▸ *A manual of modern geography*, collected from about 20 of the best authors. Containing a short, but comprehensive and entertaining account of all the known world; the situation, extent, product, government, religion, customs, etc. of every country. To which is added, a more full and particular account of England, and the present royal family, privy council, parliament, civil government, army, navy, bishopricks, counties, chief towns, etc. The whole interspers'd with sketches of history, and curiosities, the like not to be met with in any small book ever publish'd. With an alphabetical index. The second edition, revised, and corrected; wherein is added, an account of all the royal families in Europe, etc.
▸ LONDON; printed for Richard Hett, at the Bible and Crowe, in the Poultry, and Jer. Roe, bookseller in Derby
▸ 16.5 cm.; pp. [2], xii, 189, [3]
▸ ***Libraries*** : : **CaAEU**, CaOLU
▸ ***Notes***
1. Acknowledges (p. vii), though disparagingly, that a similar work was published in 1737 (see LENGLET DU FRESNOY 1737). Claims (pp. viii-ix) to draw on Lock[e]'s *Thoughts Concerning Education* for his 'method of education.' Attributes (p. xi) the speedy appearance of the 2nd ed. to the good reception accorded to the 1st. Despite the disclaimers of the tp., it is no more biassed towards English topics than were many special geographies whose titles suggested an evenhanded treatment of the countries of the world. The additional material on Continental royal families that distinguishes this ed. from its predecessor is extremely brief.

- **{1748}, 3rd (03)**
- *A manual of modern geography* ... interspersed with sketches of history, and curiosities. Whereunto is added, a table of latitudes, and longitudes of some of the principal cities in the world. By Emmanuel Bowen
- LONDON; R. Hettml [sic]
- 16mo; xii, 200 pp. 3 l.
- ***Libraries*** : NN

- **{1760}, 4th, corrected etc. (04)**
- *A manual of modern geography*, containing a short ... account of all the known world ... 4th edition corrected etc. To which is added ... hydrography; or an account of water, the oceans, seas, gulphs, lakes, streights, and the most considerable rivers in all parts of the world. Whereunto is ... added latitudes and longitudes ... cities and towns in the world ... By ... Bowen, Geographer to his majesty
- LONDON; Dilly and Robinson
- ***Notes***

1. Discussed in Bowen (1981:151).
2. A microform version of (02) is available from CIHM.

GREIG, John **Y**

- **{1810} (01)**
- *The world displayed*; or, the characteristic features of nature and art exhibited ... intended for youth, etc.
- LONDON; C. Cradock & W. Joy, etc.
- 12mo; pp. vi, 664
- ***Libraries*** L : CtY
- ***Source*** BL and NUC (NG 0505118)

[GREY, Richard] **mSG**

- **1730 (01)**
- *Memoria technica: or, a new method of artificial memory*, applied to and exemplified in chronology, history, geography, astronomy. Also Jewish, Grecian and Roman coins, weights and measures, etc. With tables proper to the respective sciences; and memorial lines adapted to each table
- LONDON; printed for Charles King in Westminster-Hall
- 8vo, 18 cm.; pp. xviii, [6], 119
- ***Libraries*** **L**, SanU
- ***Notes***

1. Intended to help educated people memorize the superabundant facts that were a mark of their education. The section devoted to geography (i.e., to the names of places) occupies pp. 49-84; it verges on the driest of special geographies, being essentially a catalogue of place names listed by continent and country. In the preface, 'Dr. Wells Treatise of antient and present geography' is identified as the chief source (p. xi); see WELLS (1701).
2. BL identifies the author. The book clearly filled a need, for it went through many eds., as recorded in NUC: 1732, 2nd (CLU, CtY, DLC, InU, MoSU, OrU, TxU, ViU); 1737, 3rd (CLU, CtY, InU, IU, MoU, NB,

NcD, ViU); 1756, 4th (CtY, NIC, NjP); 1778, 5th (CtY, IU); 1781, 6th (CtY, IU); 1790, Wolverhampton (CtY, PPL); 1796, Dublin (ICN, MWA, OO, OOxM, PU, PV, RPB, ViW); 1799, (CtY, MBAt, MH, NcD); 1805, 8th (NcD); 1812, (MiU); 1819, Oxford (CtY, MeB, NN); 1821, Oxford (DLC); 1824, Oxford (CtY, DLC, ICJ, MoSU, TxU); 1831, Oxford (CtY); 1836, Oxford (DLC); 1838, Oxford (PU); 1840? London (DLC); 1846, Oxford (CtY, DLC); 1857, Oxford (DLC); 1872, Oxford (CtY); and 1880, Oxford (MB).

NUC also notes: 'Artificial memory: or, a selection of words from "Grey's Memoria technica" with numerous additions, arranged in the order of the events by a lady for the use of her pupils, 1827, London; W. Gilbert; vi, 31 p.' (ICU).

GRIFFEN, Joseph **Y Am**

- **{1833} (01)**
- *Elements of modern geography*; or, easy and systematic steps to the acquisition of geographical knowledge ...
- GLEN'S FALLS, [NY]; printed by A. Smith
- 15 cm.;pp. 227
- ***Libraries*** : DLC, N
- ***Source*** NUC (NG 0517325). See (02), which follows.

- **{1839}, 2nd, revised (02)**
- *Elements of modern geography*
- TRENTON; D.D. Clark
- ***Libraries*** n.r.
- ***Source*** Sahli (1941:381)
- ***Notes***

1. Discussed in Nietz (1961:217-9) and Sahli (1941, *passim*).

GROVE, George **Y**

- **{1875} (01)**
- *Geography* ... with maps, etc.
- [LONDON]
- 16mo
- ***Libraries*** L
- ***Notes***

1. BL gives a cross reference to Green, J.R. [ed.], *History Primers, etc.*, justifying London as the place of publication.
2. NUC records further eds.: (all New York, D. Appleton): 1877, 3 p. l., [5]-126 (: CU, DHEW, MA, NN, PHC, ViU); 1878, 139 pp. (: MH, MoU, NcD); 1882 (: MH); 1883, 139 pp. (: DHEW); 1884, 127 pp. (: CtY, IEN). BL and NUC also record eds. in 1889.
3. The author is better known as the editor of *A Dictionary of Music and Musicians*.

GROVES, E[dward]? ?
- **{1842} (01)**
- *Geography, modern*
- LONDON? Houlston
- 12mo
- ***Source*** BCB, Index (1837-57:110)

- **{1860}, new (02)**
- *A compendium of modern geography* ... New edition, considerably enlarged
- LONDON
- 8vo
- ***Libraries*** L
- ***Notes***

1. That this entry is a successor ed. to the previous one is a surmise.

GUMMERE, Samuel R. **Y Am**
- **{1816} (01)**
- *Elemental exercises in geography* [for the use of schools]
- PHILADELPHIA; Kimber & Sharless
- Pp. 162
- ***Libraries*** : CtY, PHC, PPT, PSC
- ***Source*** NUC (NG 0600479)
- ***Notes***

1. According to NUC, the title changed from *Elemental* to *Elementary* with the 2nd ed.
2. NUC records further eds.: 1817, 2nd (PP); 1820, 3rd, Philadelphia, 160 pp. (OClWHi, PHi, PSC); 1825, 5th, Philadelphia, 196 pp. (NNC, PHi); 1827, 6th, Philadelphia (PPeSchw); 1828, 6th, Philadelphia, 180 pp. (PSC, PU); 1830, 7th, Philadelphia, 180 pp. (PHC, PSC); 1834, 8th, Philadelphia, 180 pp. (PHC, PSC); and 1837, 9th, Philadelphia, 180 pp. (PSC, Vi).
3. Elson (1964:398) cites the 5th ed. of 1825.

GUTHRIE, William **A**
- **1807, 1st [12mo] (01)**
- *Guthrie's geographical grammar, in miniature*; containing the present state ... [as in GUTHRIE (1770), below, except as follows] ... Newtonian system, and ... face of nature, since ... nations; their ... military strength, orders of knighthoods, etc. IX ... London. To which are added ... [as in (04) of GUTHRIE (1770)] ... present time. The astronomical part by James Ferguson, F.R.S. To which have been added the late discoveries of Dr Herschell, and other eminent astronomers. Illustrated with a correct set of maps
- MONTROSE, [Scotland]; printed by D. Buchanan, for Wilson & Spence, Booksellers, York
- 12mo, 14 cm.; pp. [2], 358
- ***Libraries*** **L**

GUTHRIE, William **A**

- **1795 (01)**
- *Guthrie's universal geography* improved: being a new system of modern geography: or a geographical, historical, and commercial grammar; and present state of all the several kingdoms of the world. Containing I. An account of the figures, motions, and distances, of the planets, according to the Newtonian system, and the latest observations. II. A new general view of the earth, considered as a planet; with many useful geographical definitions and problems. III. A display of the grand divisions of the globe into land and water, continents and islands. IV. The situation, extent, &. &. &. &. of empires, kingdoms, provinces, and colonies. V. Their climates, air, soil, vegetable productions, metals, minerals, natural curiosities, seas, rivers, bays, capes, promontories, and lakes. VI. The birds and beasts peculiar to each country. VII. Observations and remarks on the changes that have been any where observed upon the face of nature, since the most early periods of history. VIII. The history and origin of nations; their various forms of government, religion, laws, revenues, taxes, naval and military strength. IX. The genius, manners, customs, and habits of the people in various parts of the globe. X. Their language, learning, arts, sciences, manufactures, and commerce. XI. The chief cities, structures, ruins, and artificial curiosities throughout the world. XII. The longitude, latitude, bearings, and distances of principal places from London. To which are added, I. A new and copious geographical index, with the names and places alphabetically arranged. II. A new genealogical table and account of all the sovereigns in the world. III. A chronological table of remarkable events, from the Creation to the present time. A new edition, enlarged, corrected and improved, with great additions; being enriched with the most recent discoveries of the latest voyagers and travellers. The historical part comprising the latest and most important occurrences and events, particularly the revolutions in France, Sweden, Poland, &c. The astronomical part, including the latest discoveries of James Ferguson, F.R.S. Dr. Herschell, and other eminent astronomers. Illustrated with a set of large and accurate maps
- LONDON; printed for the proprietors; and sold by all booksellers in town and country
- 25.5 cm.; pp. [3], vii, [9]-956
- ***Libraries*** : **CtY**, PPAN
- ***Notes***

1. Although the tp. asserts that this a subsequent, not a first, ed., there are reasons for not assigning this book to the main series of GUTHRIE (1770). The text is changed in rather the same sort of way that the tp. is; namely, the overall impression is the same, but the details differ. For example, the order in which the continents are dealt with has been changed. In addition, the style strikes a different note. There still are, at the beginning of each chapter, tables of the subdivisions of the countries, but thereafter there are no headings; instead the text is continuous. The balance of interest also seems to have changed; e.g., in the section on New York State, there has been a marked increase in the attention devoted to religion and education compared with the 1796 (27) ed. of GUTHRIE (1770).

   The fact that the book to which this is a note, i.e., a 4to ed., without the publisher, the printer, or any booksellers being identified was published in

London in the same year that a much revised ed. of the main series first appeared in the U.S. raises the possibility that this is a pirated version of the latter. See GUTHRIE (1794-5), below.

- **1820, 3rd American, with ... (as below) (02)**
- *A universal geography*; or, a view of the present state of the known world. Containing 1 ... [as in GUTHRIE (1770)] ... curiosities. To which are added, 1. A geographical index, with the names alphabetically arranged. 2. A chronological table of remarkable events, from the creation to the present time. 3. A list of men of learning and science. Originally compiled by William Guthrie, esq. The astronomical part by James Ferguson, F.R.S[.] To which have been added the late discoveries of Dr. Herschell, and other eminent astronomers. Accompanied with twenty-one correct maps. The third American edition, with extensive additions and alterations, by several American editors. In two volumes
- PHILADELPHIA; published by Benjamin Warner, No. 171, High Street. Also for sale at his shop in Richmond, [VA] and by Wm P. Bason, Charleston, [SC]
- 22 cm., 2 vols.; Vol. I, pp. [viii], 514; Vol. II, pp. [5], 10-640. (The MWA catalogue notes 'page 334 of Vol. 2 wrongly numbered 433.'
- ***Libraries*** : CtY, DLC, **MWA**, OClWHi, OO, PMA, PPL, ViU, WaSpG
- ***Notes***

1. The Advertisement (pp. v-vii) is, with the exception of one word, identical to that in the 1809 ed. of GUTHRIE (1794-5), which suggests that the publisher saw it as the 3rd ed. of that series, rather than of the *Universal Geography*.

GUTHRIE, William **A**

- **1770 (01)**
- *A new geographical, historical, and commercial grammar*; and present state of the several kingdoms of the world. Containing I. The figures, motions, and distances of the planets, according to the Newtonian system and the latest observations. II. A general view of the earth considered as a planet; with several useful geographical definitions and problems. III. The grand divisions of the globe into land and water, continents and islands. IV. The situation and extent of empires, kingdoms, states, provinces, and colonies. V. Their climate, air, soil, vegetable production, metals, minerals, natural curiosities, seas, rivers, bays, capes, promontories, and lakes. VI. The birds and beasts peculiar to each country. VII. Observations on the changes that have been any where observed upon the face of nature since the most early periods of history. VIII. The history and origin of nations: their forms of government, religion, laws, revenues, taxes, naval and military strength. IX. The genius, manners, customs, and habits of the people. X. Their language, learning, arts, sciences, manufactures, and commerce. XI. The chief cities, structures, ruins, and artificial curiosities. XII. The longitude, latitude, bearings, and distances of principal places from London. With a table of the coins of all nations, and their value in English money. Illustrated with a new and correct set of maps, engraved by Mr. Kitchen
- LONDON; J. Knox, at No. 148, near Somerset-House, in the Strand

- 4to, 21.5 cm.; pp. [2], iii-vii, [1], ii-xlvi, 656
- *Libraries* : CSmH, **CtY**, DLC, IaU, NcU, TxU : CaBVaU
- *Notes*

1. Watt notes: 'Of this popular Grammar, Knox, the bookseller is said to be the real compiler' (Watt 1824, 1:452). If this assertion is correct, Knox could have obtained most, if not all of his information from Guthrie, William, *A general history of the world ... Including all the empires, kingdoms, and states, their revolutions, forms of government, laws, religion, customs and manners, the progress of learning, arts, sciences, commerce and trade. Together with their chronology, antiquities, public buildings, and curiosities of nature and art ...*, London, J. Newbery (etc.), 12 vols., 8vo, 1764-7 (Watt 1824, 1:452, and NUC NG 0611901). The Knox in question is, presumably, the publisher.

   Intended for the general public, rather than for scholars or the aristocracy. The author admits, in the preface, to an ethnocentric bias in his history. There is an introduction to astronomy and physical geography, but it is brief. The Introduction also reflects the utilitarian spirit of the time with a section: 'Of the origin of nations, laws, government, and commerce.'

   Further general notes on this book are provided after entries (02) and (45), below. See also GUTHRIE (1795 and 1794-5).

- **1771, 2nd (02)**
- *A new geographical, historical, and commercial grammar ...* from London. XIII. A general index. With a table ... Illustrated with a set of large maps, engraved by Mr. Kitchen, geographer
- LONDON; printed for J. Knox ...
- 8vo, 20.5 cm., 2 vols.; Vol. I, pp. [3], iv-x, 11-484; Vol. II, pp. [2], 3-409, [1]
- *Libraries* **L** : **CtY**
- *Notes*

1. Vol. I is devoted to northern Europe, Scotland (61 pp.), England (225 pp., of which history occupies 87), Wales (7 pp.), Ireland (24 pp.), and other British islands (4 pp.). Vol. II deals with the rest of Europe (185 pp.), Asia (99 pp.), Africa (44 pp.), and America (155 pp.).

   To the Preface of the 1st ed. is added one paragraph: 'Tho' the book was chiefly intended for schools, and the more uninformed part of mankind, we have pleasure to find, by the rapidity of its sale, and the universal approbation it has met with, that it has attracted the notice of those who are best able to judge of the execution and ... has already found a place in the libraries of the learned. One advantage it certainly possesses ... throughout ... the author seems to have divested himself of political, religious, and national prejudices; and when he discovers any bias, it is always in favour of civil and religious liberty' (p. x). For a less favourable opinion of Guthrie, see note 3 to GUTHRIE (1794-5), below.

- **1771, 3rd, improved and ... (see below) (03)**
- *A new geographical, historical, and commercial grammar ...* geographer. The third edition, improved and enlarged; the astronomical part by James Ferguson, F.R.S.
- LONDON; printed for J. Knox ...

- 8vo, 21 cm.; pp. viii, [9]-728
- ***Libraries*** L, O : CtY, DLC, MBAt, PU, ViU
- ***Notes***

1. By the use of smaller type, the amount of text per page has been increased by about one quarter compared with the previous ed. This probably accounts for the change in pagination; spot-checking revealed no changes in the text. The table of coins referred to in the titles of 1770 and 1774 is included in this ed. even though it is not referred to in the tp.

- **1774, 4th, improved and enlarged (04)**
- *A new geographical, historical, and commercial grammar* ... places from London. To which are added, I. A geographical index, with the names of places alphabetically arranged. II. A table of coins of all nations, and their value in English money. III. A chronological table of remarkable events from the creation to the present time. Illustrated with a new and correct set of maps, engraved by Mr. Kitchen, geographer. The fourth edition, improved and enlarged; the astronomical part by James Ferguson, F.R.S.
- LONDON; printed for J. Knox ...
- 8vo, 22 cm.; pp. viii, [9]-728
- ***Libraries*** L : CtY, RPJCB
- ***Notes***

1. The pagination suggests no change since 1771.

- **{1774, 1776}, new ... (see below) (05)**
- *A new geographical, historical, and commercial grammar* ...
- LONDON; printed for J. Knox and sold by E. and C. Dilly ... and G. Robinson
- 22 cm., 2 vols.; Vol. 1, a new ed. improved and enl.; Vol. 2, 5th ed. improved and enl.
- ***Libraries*** : CtY, DLC, NNH : CaOTU
- ***Source*** NUC (NG 0611918), which notes that the date of Vol. 1 is 1776

- **1777, new, improved and enlarged (06)**
- *A new geographical, historical, and commercial grammar* ... [as in 1774 (04)] ... geographer. A new edition, improved ... F.R.S.
- LONDON; printed for John Knox, Edward and Charles Dilly and George Robinson
- 8vo, 22 cm.; pp. 766
- ***Libraries*** L, OSg : DLC
- ***Notes***

1. A little new material has been added on the British isles; otherwise the changes in pagination seem to be the result of a reorganization of the table of coins and the addition of a chronology of the world.

- **1779, 6th, improved and enlarged (07)**
- *A new geographical, historical, and commercial grammar* ... geographer. The sixth edition, improved ... F.R.S.
- LONDON; printed for Edward and Charles Dilly, in the Poultry; and George Robinson, Pater-Noster-Row.
- 8vo, 22 cm.; pp. 766

- *Libraries* L : MWA
- *Notes*

1. Pagination suggests no change since 1777.

- **{1780}, improved and enlarged (08)**
- *A new geographical, historical, and commercial grammar ...*
- DUBLIN; printed for J. Williams and J. Exshaw
- 22 cm.; pp. 8, [2], [9]-766
- *Libraries* : ViU
- *Source* NUC (NG 0611921)

- **{1780}, new, with great ... (see below) (09)**
- *A new system of modern geography*; or, a geographical, historical, and commercial grammar; and present state of the several kingdoms of the world ... The astronomical part by James Ferguson, F.R.S. A new edition, with great additions and improvements. Illustrated ...
- LONDON; C. Dilly [etc.]
- [4to], 28 cm.; pp. ix, [2], 795
- *Libraries* : DLC, NNH
- *Notes*

1. NUC (NG 0611966) looks on this title, presumably on the grounds that it is different, as a sign of a different book. However, an examination of the '3rd, 4to ed.' of 1786 - see (14) below - suggests that it is only the title that has changed. As a consequence, the eds. in the two series of titles are treated as a single series here.

- **{1782} 7th, improved and enlarged (10)**
- *A new geographical, historical, and commercial grammar ...*
- LONDON; printed for C. Dilly and C. Robinson
- Pp. 766
- *Libraries* : CtY, MH, MWA, PPL
- *Source* NUC (NG 0611922)

- **{1782} (11)**
- *A new system of modern geography* ... [as in 1780, 4to (09)]
- LONDON; C. Dilly
- 30 cm.; pp. ix, [2], 855, [24]
- *Libraries* : MiD, RPJCB : CaBVaU
- *Source* NUC (NG 0611967)
- *Notes*

1. Taken to be an ed. of GUTHRIE (1770) for the reasons given in (09), above.

- **{1783}, 8th, with great additions and improvements (12)**
- *A new geographical, historical, and commercial grammar ...*
- LONDON; printed for C. Dilly, and G. Robinson
- 8vo, 22 cm.; pp. 10, [2], [9]-848
- *Libraries* SanU : CtY, CU, MHi, NNC, RPB, ViU
- *Source* ESTC, CtY

- *Notes*
1. NUC (NG 0611924) records 848 (i.e., 852) pp.

- **1785, 9th, with great ... (see below) (13)**
- *A new geographical, historical, and commercial grammar* ... geographer. The ninth edition, with great additions and considerable improvements; the astronomical ... F.R.S.
- LONDON; printed for Charles Dilly, in the Poultry; and G.G.J. and J. Robinson, in Pater-Noster-Row
- 8vo, 21 cm.; pp. 928
- ***Libraries*** L, O : DLC, MWA, NcU, NhM, PPL, ViLxW
- *Notes*
1. In an 'advertisement' at the beginning, it is stated that material has been added to the descriptions 'of Russia, Poland, Sweden, Denmark, Spain, Switzerland, the two Sicilies, the East Indies, and other countries' (the last includes Iceland); and also to the historical section of 'each kingdom,' especially England and the newly formed United States of America.

- **1786, 3rd [4to], with great ... (see below) (14)**
- *A new system of modern geography*; or, a geographical ... [as in (02)] ... F.R.S. The third edition, with great additions and improvements, and a copious index. Illustrated with a set of large and accurate maps
- LONDON; printed for C. Dilly, in the Poultry; and G.G.J. and J. Robinson, Pater-Noster-Row
- 4to, 28 cm.; pp. [2], ix, [3], 950, [25]
- ***Libraries*** L
- *Notes*
1. The advertisement refers to changes 'since the last quarto edition.' Because of the use of wider margins and larger type, each page carries only some 5 per cent more text than the 8vo eds. of 1785 and 1787. As this ed. is longer than its smaller relatives, it must contain more material than they do.

- **1787, 10th, corrected (15)**
- *A new geographical, historical, and commercial grammar* ... Newtonian system, and the latest ... geographer. The tenth edition, corrected; the astronomical ... F.R.S.
- LONDON; printed for C. Dilly ...
- 8vo, 22 cm.; pp. 928
- ***Libraries*** L : CtY, CU, KyU, NcD, OFH, PPA, ViLxW, ViU : CaBVaU, CaQMBN
- *Notes*
1. The advertisement of the previous (i.e., 1785) ed. still present, proclaiming changes since 1783, but the pagination suggests that there have been none since 1785.

- **{1788}, 11th, cor. (16)**
- *A new geographical, historical, and commercial grammar* ...
- LONDON; printed for C. Dilly
- 8vo, 23 cm.; pp. iv, [5]-10, [2], 928

- ***Libraries*** SanU : CtY, MH, ViU
- ***Source*** ESTC and NUC (NG 0611930)

- **1788, 4th [4to], corrected (17)**
- *A new system of modern geography* ... [as in the 3rd 4to ed. of 1786]
- LONDON; printed for C. Dilly, in the Poultry; and G.G.J. and J. Robinson, Pater Noster Row
- 4to, 29 cm.; pp. xii, 869, [26]
- ***Libraries*** **OSg** : CtY, MiD, WaS
- ***Notes***

1. A comparison of the sections on America with those in the 8vo ed. of 1777 shows that they differ only in the omission or inclusion of occasional sentences and paragraphs.

- **{1789}, 10th, corrected (18)**
- *A new geographical, historical, and commercial grammar* ... considered as a planet: with several ... face of nature, since ... origin of nations[,] their forms ... strength, orders of knighthood, etc. IX ... The astronomical part by James Ferguson, F.R.S. Illustrated with a correct set of maps, engraved by Mr Kitchen, geographer. The tenth edition, corrected
- DUBLIN; printed for John Exshaw, No. 98, Grafton Street
- 28 cm.; pp. iv, 10, 934
- ***Libraries*** : RP
- ***Source*** Letter from RP

- **{1789}, considerably enl. and cor. (19)**
- *An improved system of modern geography*; or, a geographical ... grammar; containing the ancient and present state of all the empires, kingdoms, states, and republics in the known world ... Originally comp. by ...
- DUBLIN; printed by John Chambers
- 27 cm.; pp. xvi, 994, [14]
- ***Libraries*** : NPV : CaOHM

- **{1790}, 12th, corr. and improved (20)**
- *A new geographical, historical, and commercial grammar* ... considered as a planet; with ... origin of nations: their forms ... twelfth edition, corrected and improved
- LONDON; C. Dilly and G.G.J. Robinson
- 22 cm.; pp. 10, [2], 894
- ***Libraries*** O : CtY, ICN, LU, PSC, RP, TU, ViU
- ***Source*** ESTC, NUC (NG 0611934), and letter from CtY

- **{1792}, 13th, corr. (21)**
- *A new geographical, historical, and commercial grammar* ...
- LONDON; printed for C. Dilly, [etc.]
- 23 cm.; pp. [3], iv, [1], 6-10, [3], 2-928 (pp. 421 misnumbered as 521, 672 as 272, and 684 as 844)
- ***Libraries*** : CtY, DLC, MWA, OrP, PMA, RPJCB : **CaBViPA**

- **{1792}, 2nd [see note below] (22)**
- *A new geographical, commercial, and historical grammar* ... several empires and kingdoms of the world: containing I. The distances, figures, and revolutions of the celestial bodies ... coins of all the different nations ... remarkable events ... The whole executed on a plan similar to that of W. Guthrie, esq. by a society in Edinburgh; the astronomical part collected from the works of James Ferguson ... enriched with the late discoveries of Dr. Herschel, and other eminent astronomers; embellished with an elegant set of maps ... exhibiting more fully the new geographical discoveries than those to be met with in any former publication. - Second edition
- EDINBURGH; printed for Alexander Kincaid, and sold by all the booksellers
- 8vo, 22 cm., 2 vols.
- ***Libraries*** : CtY
- ***Source*** CtY
- ***Notes***

1. CtY treats this as an ed. of GUTHRIE (1770). However, the wording of the short version of the title they include in their catalogue suggests that it is better described as an ed. of KINCAID (1790), although that work may be a pirated version of GUTHRIE (1770). See also note 2 to (25), below.

- **{1792}, 5th [4to], corr., improved and greatly enl. (23)**
- *A new system of modern geography* ...
- LONDON; printed for C. Dilly ... and G.G.J. and J. Robinson
- 28 cm.; pp. ix, 1049
- ***Libraries*** : CtY, DFo, NIC, NjN : CaNBSaM
- ***Source*** NUC (NG 0611970)

- **{1794}, 13th, corrected and improved (24)**
- *A new geographical, historical, and commercial grammar* ...
- DUBLIN; printed for John Exshaw
- Pp. iv, [5]-934
- ***Libraries*** : NNC
- ***Source*** NUC (NG 0611937)
- ***Notes***

1. It is probable that a copy of this ed. is also held at NN (NG 0611971), which notes that the tp. is missing. SanU may also have a copy, though it seems more likely that they hold a copy of (25).

- **1794, 14th, corrected ... (as below) (25)**
- *A new geographical, historical, and commercial grammar* ... world. Containing I. The figures, motions, and distances of the planets, according to the Newtonian system ... military strength, orders of knighthood, etc. IX ... time. The astronomical ... Fergusgn [sic] F.R.S. To which have been added the late discoveries of Dr. Herschell, and other eminent astronomers. Illustrated with a correct set of maps engraved from the most recent observations and draughts of geographical travellers. The fourteenth edition, corrected and considerably enlarged
- LONDON; printed for Charles Dilly, in the Poultry; and G.G. and J. Robinson, in Pater-Noster-Row

- 8vo, 23 cm.; pp. vi, 7-965 [971 because of mispagination]
- *Libraries* L, O, SanU[?] : CtY, MeB
- *Notes*

1. The advertisement, dated 1793, outlines the additions that have been made; these include material on various countries in Europe and Asia, as well as on some of the states of the U.S. The layout of the tp. is slightly different from that of the 12th ed. of 1790, (21) above.
2. If Kincaid pirated the version of GUTHRIE (1770) printed in London (see KINCAID 1790), it seems that the London publishers returned the compliment with respect to the 'discoveries of Dr Herschell.' See (22), above.

- **1794-5, 1st American, corr., improved, and greatly enl.**

See GUTHRIE (1794-5) below.

- **1795, 15th, corrected, and greatly enlarged (26)**
- *A new geographical, historical, and commercial grammar ... travellers. The fifteenth edition ...*
- LONDON; printed for C. Dilly ...
- 8vo, 22 cm.; pp. [3], iv-xii, [4], [7]-12, 965
- *Libraries* L, SanU : **CtY**, DLC, RPJCB, ViU : CaQQS
- *Notes*

1. The advertisement of 1794 (dated 1793), is repeated, but the pagination, once the errors of 1794 have been allowed for, suggests no change since then.

- **1795, 6th [4to], corrected, and greatly enlarged (27)**
- *A new system of modern geography* ... [as in the 1794, 8vo] ... travellers. The sixth edition ...
- LONDON; printed for C. Dilly ... and G.G.J. and J. Robinson
- 4to, 29 cm.; pp. [12], 1098, [28]
- *Libraries* L, LvU : CtY, DLC, MH, NN, PP, ViW : CaOLU
- *Notes*

1. The advertisement identifies the additions made in both the 5th and the 6th 4to eds.

- **1796, 16th, corrected, and considerably enlarged (28)**
- *A new geographical, historical, and commercial grammar* ... travellers. The sixteenth edition ...
- LONDON; printed for Charles Dilly ...
- 8vo, 22cm; pp. [3], iv-xii, [4], 971 [965 because 773-8 repeated]
- *Libraries* : **CtY**, MH, MWA
- *Notes*

1. The Preface in this ed. occupies pp. [vii]-xii, whereas in the 15th (26) it occupied pp. [7]-12. The layout of the tp. of this ed. is slightly different from that of the 14th of 1794, i.e., (25), above.

- **{1798}, 17th, corrected and considerably enlarged (29)**
- *A new geographical, historical, and commercial grammar ...*
- LONDON; printed for Charles Dilly, in the Poultry; and G.G. and J. Robinson ...

- 21 cm.; pp. xii, [4], 1019
- ***Libraries*** O : CtY, MHi : CaBVaU
- ***Source*** ESTC and NUC (NG 0611943)

- **1799, 15th, corrected, and greatly enlarged (30)**
- *A new geographical, historical, and commercial grammar* ... travellers. The fifteenth edition ...
- MONTROSE, [Scotland]; printed for Da. Buchanan, & Jas. Morison, sold by them, J. Fairbairn, A. Constable, Edinburgh; W. Coke, Leith; J. Gillies, & J. Duncan, Glasgow; the booksellers of Dundee, Aberdeen, etc.
- 8vo, 21 cm.; pp. iv, [4], 1066
- ***Libraries*** L : MWA, NjN, NNC : CaOKU
- ***Notes***

1. Has a preface identifying the additions made since the previous ed.

- **{1799}, new (31)**
- *A new geographical, historical, and commercial grammar ...*
- MONTROSE, [Scotland]; printed by and for David Buchanan
- 22 cm.; pp. 1066
- ***Libraries*** : NIC
- ***Notes***

1. According to NIC, this ed. is distinct from that shown in (30) above.

- **1800, 18th, corrected and considerably enlarged (32)**
- *A new geographical, historical, and commercial grammar* ... [as in (28)] ... travellers. The eighteenth edition ...
- LONDON; printed for G.G. and J. Robinson, in the Poultry; and J. Mawman (successor to Mr. Dilly) in the Poultry, [and] by S. Hamilton, Falcon-Court, Fleet-Street
- 8vo, 22 cm.; pp. xii, [4], 1056
- ***Libraries*** L : NN, PU : CaQQS
- ***Notes***

1. An advertisement, dated 1800, describes the additions since 1798.

- **{1801}, 18th, corrected and greatly enlarged (33)**
- *A new geographical, historical, and commercial grammar ...*
- LONDON; published and sold by Vernor & Hood, D. Ogilvy & son [etc.]
- 21.5 cm.; pp. viii, 1076
- ***Libraries*** : MiU, PSt, ViU
- ***Source*** NUC (NG 0611950)

- **1801, 19th, corrected and considerably enlarged (34)**
- *A new geographical, historical, and commercial grammar* ... travellers. The nineteenth edition ...
- LONDON; printed for G.G. and J. Robinson and J. Mawman
- 8vo, 22 cm.; pp. xii, [14], 1056
- ***Libraries*** L : DLC, NcD
- ***Notes***

1. An advertisement states the changes made since 1800.

- **{1805}, 20th, corr., and considerably enl. (35)**
- *A new geographical, historical, and commercial grammar ...*
- LONDON; J. Walker, Wilkie, and Robinson
- Pp. 1034
- ***Libraries*** : ICarbS, KyU : CaQMNB
- ***Source*** NUC (NG 0611954)

- **{1806}, 20th, cor. and considerably enl. (36)**
- *A new geographical, historical, and commercial grammar ...*
- LONDON; printed for J. Walker, Wilkie and Robinson, Scatcherd and co. [etc.]
- 21.5 cm.; pp. xii, [4], 1036
- ***Libraries*** : CtY

- **{1807}, new and ... (see below) (37)**
- *A new geographical, historical, and commercial grammar* ... new and enlarged, brought down to the present time
- EDINBURGH; printed by J. Moir, for C. Mitchel & Co., Perth
- 8vo; pp. xii, 897
- ***Libraries*** : ICU, NN
- ***Source*** NUC (NG 0611956)

- **1808, 21st, corrected and considerably enlarged (38)**
- *A new geographical, historical, and commercial grammar* ... travellers. The twenty-first edition ...
- LONDON; printed for J. Johnson; F. and C. Rivington ...
- 8vo, 22 cm.; pp. xi, [4], 1036
- ***Libraries*** **L** : MiU
- ***Notes***

1. An advertisement states the changes made since 1805.

- **{1810}, 19th, enl. (39)**
- *A new geographical, historical, and commercial grammar ...*
- MONTROSE, [Scotland]; D. Buchanan
- Pp. vi, 973
- ***Libraries*** : MH, PU
- ***Source*** NUC (NG 0611961), and Blackwell's Rare Books Catalogue A75

- **{1810}, 22nd, corr., and greatly enl. (40)**
- *A new geographical, historical, and commercial grammar ...*
- LONDON; Vernor, Hood and Sharpe
- 22 cm.; 542+ pp.
- ***Libraries*** : DLC

- **1811, 7th [4to], corrected, and greatly enlarged (41)**
- *A system of modern geography*; or, a ... [as in (27)] ... astronomers
- LONDON; printed for F.C. and J. Rivington ...
- 4to, 27 cm.; pp. x, [2], 1115
- ***Libraries*** **L**

▸ ***Notes***
1. An undated advertisement admits, in effect, that little notice has been taken of newly formed countries in Europe, but claims that unspecified additions keep the book up to date.

▸ **{1812}, 22nd, corr. and considerably enl. (42)**
▸ *A new geographical, historical, and commercial grammar ...*
▸ LONDON; Rivington
▸ Pp. 1048
▸ ***Libraries*** : CtY, OOxM

▸ **{1815} (43)**
▸ *A new geographical, historical, and commercial grammar ...*
▸ PHILADELPHIA; Johnson and Warner
▸ 22 cm.
▸ ***Libraries*** : DLC, MH, PHi, PPA, PPAN, ViU
▸ ***Source*** NUC (NG 0611964)

▸ **1819, 23rd, studiously ... (as below) (44)**
▸ *A geographical, historical, and commercial grammar ...* curiosities: to which are added, I. a geographical index ... arranged, and their latitudes and longitudes; II. a table of coins, and their ... money; III. a chronological ... events, from ... time; and, IV. an obituary of eminent and illustrious persons, of every age and nation. The astronomical part by James Ferguson, F.R.S. Illustrated with a correct set of maps engraven from recent observations. The twenty-third edition, studiously revised and carefully corrected
▸ LONDON; printed for F.C. and J. Rivington; J. Cuthell; W. Clarke and sons; Cadell and Davies; Scatcherd and Letterman; Longman, Hurst, Rees, Orme, and Brown; Lackington and co.; Darton, Harvey, and co.; John Richardson; J. Booker; T. Boosey and sons; E. Williams; R. Scholey; J. Asperne; J. Mawman; J. Black and son; Baldwin, Cradock, and Joy; T. Tegg; Sherwood, Neely, and Jones; Ogle, Duncan, and co.; T. Hughes; G. and W.B. Whittaker; R. Reynolds; E. Edwards; Simpkin and Marshall; Wilson and sons, York; and Stirling and Slade, Edinburgh
▸ 8vo, 22 cm.; pp. xix, 923
▸ ***Libraries*** L, OSg
▸ ***Notes***
1. An advertisement, dated 1819, describes the additions made since the previous ed.

▸ **1827, 24th, studiously ... (45)**
▸ *A geographical, historical, and commercial grammar*: exhibiting the present state of the world; and containing I. An account of the planetary system. II. A ... planet; with geographical ... III. A history of the progress of geographical science; a description of the varieties of the human race; and a sketch of the various religions. IV. The grand divisions ... V. The situation and extent ... VI. Their climates ... vegetable and mineral productions, natural curiosities, seas, rivers, and lakes. VII. Observations ... XI. ...

curiosities. To which are added I. ... Illustrated with a new and correct ... maps. The twenty-fourth edition ...
- LONDON; printed for F. and J. Rivington ...
- 8vo, 24 cm.; pp. xiv, [2], 928
- *Libraries* L : DLC : CaOTU
- *Notes*

1. An advertisement, dated 1827, identifies the changes made since the previous ed.

- **1843 (46)**
- *A geographical, historical, and commercial ... grammar, in miniature ...* [as in (45)] ... F.R.S. A new edition, revised, greatly enlarged, and brought down to the end of 1842 by R.A. Davenport
- LONDON; printed for Thomas Tegg, 73, Cheapside; and R. Griffin & Co., Glasgow
- 12mo, 15 cm.; pp. iv, 824
- *Libraries* L, O
- *Notes*

1. The use of tiny print may allow the entire text of the 1827 ed. to be retained, and perhaps even enlarged.
2. By almost any measure, Guthrie's *Grammar* was the most popular special geography of them all. A handful of school textbooks published in the second half of the nineteenth century, for which new eds. were prepared every year, exceeded the number through which this book went in the seventy-three years that separated the first from the last, but it is unlikely that any other description of all the countries in the world sold more copies to willing buyers. Establishing a convincing explanation of the book's popularity would be an challenging exercise in historical cultural analysis.
3. Discussed in Downes (1971:382-3), East (1956), and Warntz (1964, *passim*). Dryer (1924:118) provides some information about the American ed. of 1815.
4. Microform versions of (01), (05), (15), (19), (21), (23), (26), (27), (29), (30), (32), (35), and (45) are available from CIHM.

GUTHRIE, William **A Am**

- **1794-5, 1st American ... (as below) (01)**
- *A new system of modern geography*: or, a geographical, historical, and commercial grammar; and present state of the several nations of the world. Containing, 1 ... [as in GUTHRIE 1770] ... 12. The longitude ... and distances from Philadelphia. To which are added, I. A new and copious geographical index, with the names and places alphabetically arranged. II. A new genealogical table and account of all the sovereigns in the world. III. A chronological table of remarkable events, from the Creation to the present time. IV. The late discoveries of Herschell, and other astronomers. A new edition, enlarged, corrected and improved, with great additions; being enriched with the most recent discoveries of the latest voyagers and travellers. The historical part comprising the latest and most important occurrences and events, particularly the revolutions in France, Sweden, Poland, etc. The astronomical parts corrected by Dr.Rittenhouse. Illustrated with a

set of large and accurate maps. The first American edition, corrected, improved, and greatly enlarged

- PHILADELPHIA; M. Carey
- 27 cm., 2 vols. (see below for pagination)
- ***Libraries*** : AU, CSt, **CtY**, DI-GS, DLC, FSaHi, GU-De, ICN, ICU, KyHi, KyLx, KyLxT, KyU, LNT, MB, MBAt, MdE, MeB, MH, MiU, MnHi, **MWA**, NcA, NcU, Nh, NIC, NN, NNNAM, NWM, OClW, OClWHi, OKentU, OO, PHC, PHi, PMA, PP, PPAmP, PPL, PU, PV, PWW, RPJCB, ScU, TNJ, TU, ViL
- ***Notes***

1. Beginning in 1780 the first sentence of this title was used to open the titles of the 4to eds. of GUTHRIE (1770); see (09) of that entry.
2. Pagination in the MWA copy is: Vol. I (1794), pp. [3], 4-572; Vol. II (1795), [3], iv-xi, [1], 704, 43. At CtY the pagination is: Vol. I [as at MWA]; Vol. II, pp. [3], iv, 702, 43. CtY has two copies of this ed., one of them bound in a single volume. The catalogue at MWA notes: 'page 435 of Vol. 1 wrongly numbered 439. Pages 237-240 of Vol. 2 wrongly numbered 187-190.'

   Vol. II, pp, v-xi of the MWA copy carries a list of subscribers, together with the towns where they live; pp. [iii]-iv notes: 'the work has been extended to fifty-six numbers, instead of forty-eight.'
3. Vol. I contains a 'Preface of the American Editor.' Among other things the editor comments at length on what he perceives as defects in G.'s book. Chief among these is G.'s tendency to praise the English and disparage everybody else. This provides an interesting contrast to the claim made in the paragraph added, in 1771, at the end of the Preface of the 1st ed. of GUTHRIE (1770); see the note to (02) of that entry, above. These remarks of the American editor conclude as follows: 'The alterations and additions in the present [edition], are so numerous, that it better deserves the title of an original work, than some untitled transcripts of Guthrie, which, under a different name, have been introduced to the world. The article on Sweden is entirely written new again; of Ireland and Scotland, we have already taken notice. In Russia, Holland, the Austrian Netherlands, Germany, and Prussia, we have made large additions. The literary articles of Spain and Italy, are much more complete than formerly; and we have, throughout, almost in every page, made a variety of corrections, which, though, separately taken, they are minute, yet all together, form a very considerable improvement. The work of Guthrie has the appearance of being written at various times, and by different hands; for passages are sometimes evidently thrust in, which are unconnected with the context. In point of common grammar, we have made corrections past number' (p. 10).

   Further extensive changes are noted in pp. [iii]-iv of Vol. II. The largest single set of changes is concerned with 'the united states of America [which] occupy from p. 243 to p. 580 inclusive in the present volume ... The Rev. Jedidiah Morse ... has furnished the principal part of it.'

   Occasional notes and references to sources are placed in footnotes.
4. One of the most widely held works in this guide, being rivalled only by MORSE (1789).

- **{1809}, 1st American [8vo], improved (02)**
- *A new geographical, historical, and commercial grammar* ... [as in (01) above] ... military strength, orders of knighthood, etc. 9. The genius ... distances ... from London. To which are added a geographical ... time. The astronomical parts by James Ferguson, F.R.S., to which have been added, the late discoveries of Dr. Herschell, and other eminent astronomers. Illustrated with twenty-five correct maps. The first American edition improved. In two volumes
- PHILADELPHIA; published by Johnson & Warner, and for sale at their book stores in Philadelphia; Richmond, Virginia; and Lexington, Kentucky
- 22 cm., 2 vols.; Vol. I, pp. [x], 556; Vol. II, pp. [4], 496. (The MWA catalogue notes 'page 213 wrongly numbered 313 and [some?] page numbers omitted in numbering of Vol. 1; page 117 wrongly numbered 711 in Vol. 2.'
- ***Libraries*** : CSt, DGU, InU, KyLoF, MH, **MWA**, NBu, NNC, OClW, PPAN, ViU
- ***Notes***

1. The Advertisement (pp. [iii-v]) describes the changes made from the previous ed. It is then stated: 'In publishing this first American edition of Guthrie's geography, no reasonable care or expense has been spared to render it the most perfect edition of the *octavo size* [emphasis added] that has yet appeared' (p. [v]). The emphasis is added because the distinction in sizes may be a clue to the differences between this ed. and that of 1794-5 [(01) above]. There are signs that the former is closer to the English editions than the latter; for example the account of the American War of Independence was omitted from the earlier ed., but is present in the latter.

   For what may be another ed. of this series, see the 1820 (02) ed. of GUTHRIE (1795), above.
2. Discussed in Wright (1959).

GUY, John **Y SG?**

- **{1840} (01)**
- *Geography for children* [on a perfectly easy plan adapted for the use of schools and private families]
- LONDON
- 12mo
- ***Libraries*** L
- ***Notes***

1. The full title is known from the 60th ed. of 1865.
2. BCB, Index (1837-57:130) records an ed. in 1848; CIHM records an ed. in 1852 (: : CaOLU); NUC records further eds.: 1865, 60th, London, 126 pp. (: CaOTP); 1872, 84th (ICU). St. John (1975, 2:805) records an ed.: 1865, 16th [sic], London, 126 pp. (: : CaOTP).
3. A microform version of the ed. of 1852 is available from CIHM.

GUY, John **Y**

- **[1854] (01)**
- *Mother's own catechism of geography*

- LONDON; T. Allman and Son 42, Holborn Hill. And sold by all booksellers
- 12mo, 14 cm.; pp. 71
- ***Libraries*** **L**
- ***Notes***

1. Extremely simple (e.g., the use of an * to indicate a footnote is explained). Often not much more than a list of places. Guidance on the pronunciation of names is given.

GUY, [Joseph] **Y SG?**

- **{1846} (01)**
- *[Guy's] First geography* [for the younger classes, with seven maps, and questions for examination at the bottom of each page]
- ? Cradoc
- 18mo
- ***Libraries*** n.r.
- ***Source*** BCB, Index (1837-57:110)
- ***Notes***

1. The full title is known from the ed. of 1858, itself known from an advertisement in the 1858 (25th) ed. of Joseph Guy, *School Geography* (i.e., GUY, Joseph, 1810).
2. ECB, Index (1856-76:130) records ed. of 1874.

GUY, [Joseph] **Y SG?**

- **{1864}, new (01)**
- *Geography, preparatory.* New edit
- [LONDON]; Simpkin
- ***Libraries*** n.r.
- ***Source*** ECB, Index (1856-76:131)
- ***Notes***

1. The book is attributed to Joseph rather than to John because of the publisher.

GUY, Joseph **Y**

- **{1810} (01)**
- *School geography*
- LONDON
- ***Libraries*** n.r.
- ***Source*** Peddie and Waddington (1914:227)

- **1811, 2nd (02)**
- *Guy's school geography*, on a new and easy plan; comprising not only a complete general description, but much topographical information, in a well digested order; exhibiting three distinct parts, and yet forming one connected whole. Expressly adapted to every age and capacity, and to every class of learners, both in ladies' and gentlemen's schools
- LONDON; printed for C. Cradock and W. Joy, Paternoster Row, C. Law,

Ave Maria Lane; and to be had of all booksellers, with full allowance to schools

- 12mo, 15 cm.; pp. viii, 170
- ***Libraries*** **L**
- ***Notes***

1. Retains the preface of the 1st ed. Primarily a special geography, with advice to the teacher on what should be learned by heart. There is a short section dealing with problems of location and time on the earth. Ends with questions suitable for setting to students using the book as a text.

- **1816, 5th, with considerable ... (see below) (03)**
- *Guy's school geography* ... capacity and ... learners both ... Fifth edition, with considerable improvements, including the new continental arrangements
- LONDON; printed for Baldwin, Cradock, and Joy, Paternoster Row; and C. Law, Ave Maria Lane
- 12mo, 15 cm.; pp. x, [2], 173, [7]
- ***Libraries*** **L**
- ***Notes***

1. There have been only modest changes made since 1811.
2. BL and NUC record further eds. (all London): 1817, 6th, 173 pp. (L); [1830], 11th, 182 pp. (**L** : DLC) [modest changes since 1817]; [1831?] 12th, 182 pp. (**L** : CtY) [1831 the year of publication according to NUC (NG 0615923), though with a ?. BL has two copies, one dated c1830 and the other c1835. The former carries a handwritten note on the flyleaf: 'David Gilmours Book, Glasgow, July 29, 1832']; 1840, 15th (: ICU); [CIHM records 1843, 16th (: : CaQQS)].

- **1855, new ... (see below) (04)**
- *Guy's school geography* ... corrected to the present time by John Tillotson
- LONDON
- 12 mo; 208 pp.
- ***Libraries***: **L**
- ***Notes.***

1. There are modest changes compared with the text of 1831.

- **1858, 25th, revised ... (see below) (05)**
- *Guy's school geography* ... revised, enlarged and thoroughly corrected
- LONDON
- 12mo; 244 pp.
- ***Libraries***: **L**
- ***Notes***

1. The table of contents suggests considerable changes since 1831, and some since 1855. A section on physical geography has been added; this is not mentioned on the tp. A new preface states that, as demand for the book has continued strong, the publishers 'have had the work thoroughly revised and considerably enlarged.'

- **1860, 26th, revised, etc. (06)**
- *Guy's school geography* ... revised, etc. [by W.C. Stafford]

- LONDON
- 12mo; 244 pp.
- ***Libraries***: **L**
- ***Notes***

1. Though not mentioned on the tp. until this ed., the chapter on physical geography shows no sign, on spot-checking, of differing from that of 1858. The pagination suggests that there have been no changes in any other part of the text either.
2. BL records further eds.: 1863, 27th; 1866, 28th; 1869, 29th; 1873, 30th; and 1882, 31st.
3. A microform version of the ed. of 1843 is available from CIHM.

GUY, Joseph [Jr] **Y**

- **{1852} (01)**
- *The illustrated London geography*. By ... of Magdalen Hall, Oxford
- LONDON; [printed by Levey, Robson and Franklyn], 227 Strand.
- 20.5 cm.; viii, 132 pp.
- ***Libraries*** : : CaOLU
- ***Notes***

1. Special geography occupies pp. 11-108.
2. The words in [ ] are derived from NUC (NG 0615936), which records further ed.: 1853, London, 140 pp. (MB, MH, NN).
3. A microform version of (01) is available from CIHM.

GUYOT, Arnold Henry **Gt**

- **{1866a}**
- *Common-school geography*
- NEW YORK; C. Scribner & company
- ***Libraries*** n.r.
- ***Source*** Elson (1964:403)
- ***Notes***

1. See the body of the title of the next entry, and my comments in the Introduction on the work of Ruth Elson (p. 27).

GUYOT, Arnold Henry **Y Am**

- **{1866b} (01)**
- ... *The earth and its inhabitants. Common-school geography* ...
- NEW YORK; C. Scribner & company.
- 31 cm.; 2 p. l., 147 pp.
- ***Libraries*** : DLC, GMW, MH, MiU, OO
- ***Source*** NUC (NG 0617232), which notes: '(Guyot's geographical series, no. II),' and also: 'Prepared with the co-operation of Mrs. Mary H. Smith'
- ***Notes***

1. NUC records further eds. (all New York): 1867, 154 pp. (MdBP, OClWHi, ViU); 1867, 147 pp. (MH, N); 1868, 154 pp. (MH, OkU); 1869, 160 pp. (FU, MH, ViU); 1870, 160 pp. (MH, MnHi, ViLxW); 1871, 160 pp. (ICJ, MH, OO, OrU); 1873 (MH); and 1875, 160 pp. (MH).

2. Carpenter (1963:258-60) provides a general discussion, as do Brigham and Dodge (1933:23), and Dryer (1924:122-3); Hauptman (1978:427-8) draws on it; Nietz (1961:230) discusses (01); Elson (1964:403-4) cites (01) and the ed. of 1874.

GUYOT, Arnold Henry **Y Am**
- **{[1867]} (01)**
- ... *The earth and its inhabitants. Intermediate geography*
- NEW YORK; Charles Scribner and company
- 29 cm.; pp. 87
- ***Libraries*** : NcA
- ***Source*** NUC (NG 0617241), which notes: '(Guyot's geographical series)'
- ***Notes***

1. NUC records further eds. (all New York): [1867], 118 pp. (MdBP, MH, MiEM, OClWHi, PP); 1868, 90 pp. (CtY, MH, OO); 1869, 96 pp. (DHEW, DLC, I, MH, OClW, OO, ViLxW [though some of these must hold the next ed. NG 0617244 notes: 'Prepared with the coöperation of Mrs Mary H. Smith']; 1870, 118 pp. (WHi, and some of DHEW, DLC, I, MH, OClW, OO, ViLxW); 1871, 118 pp. (NN); 1872, 118 pp. (MH, T); c.1873 (MH); c1874, 118 pp. (DLC, MH); and 1879, 118 pp. (DHEW, DLC).
2. Carpenter (1963:258-60) provides a general discussion; Nietz (1961:230) discusses (01); the ed. of 1869 is discussed in Culler (1945, *passim*); Elson (1964:403) cites the ed. of 1867.

GUYOT, Arnold Henry **Y Am**
- **{1868} (01)**
- ... *Elementary geography for primary classes*
- NEW YORK; C. Scribner & company. Boston, E.P. Dutton & co.
- 22 cm.; pp. 96
- ***Libraries*** : DLC, MH, MoU, OO, PU
- ***Source*** NUC (NG 0617296), which notes: 'Prepared with the cooperation of Mrs. Mary H. Smith'
- ***Notes***

1. According to Culler (1945:308) this is the 3rd ed.
2. NUC records further eds.: 1869 (MB), 1871 (DHEW, ViU), 1872 (DHEW), 1873 (CtY, DHEW), c1875 (ICU, MiU), c1879 (DLC, MH), and 188–? (ViU); [c1879], New York and Chicago (MH); 1879?, New York, 96 pp. (NcD); also an ed. in 1876 translated into Dakota (DLC, ICN, PU) [NG 0617335].
3. Carpenter (1963:258-60) provides a general discussion; Culler (1945, *passim*), and Nietz (1961:230) discuss (01).

GUYOT, Arnold Henry **Y Am**
- **{c1882a} (01)**
- *The Guyot geographical reader and primer*; a series of journeys around the world (based upon Guyot's introduction) with primary lessons
- NEW YORK, Cincinnati [etc.]; American Book company

- 19 cm.; pp. vi, 298
- ***Libraries*** : CtY, ICU, MiU
- ***Source*** NUC (NG 0617306)
- ***Notes***

1. The relationship of this book to GUYOT (1882b) is not known.

GUYOT, Arnold Henry **Y Am**

- **{[1874]} (01)**
- ... *Guyot's grammar-school geography*
- NEW YORK; Scribner, Armstrong and Co.
- 33 cm.; pp. [4], 134
- ***Libraries*** : DHEW, DLC, ICJ, OClWHi, OO
- ***Source*** NUC (NG 0617320), which notes that the date is that of the Preface, and also: '(Guyot's geographical series)'
- ***Notes***

1. *Pace* NUC, not on shelf list at NN.

- **{1882}, 3rd (02)**
- *Grammar-school geography*
- NEW YORK; Charles Scribner's Sons
- ***Libraries*** n.r.
- ***Source*** Culler (1945:308)
- ***Notes***

1. Carpenter (1963:258-60) provides a general discussion; Nietz (1961:230) discusses (01); Culler (1945, *passim*) discusses (02).

GUYOT, Arnold Henry **Y Am**

- **{[1875]} (01)**
- *Guyot's new intermediate geography*
- NEW YORK; Scribner, Armstrong & company. Chicago; Hadley brothers [etc.]
- 30 cm.; 2 p. 1., 98, [10] p
- ***Libraries*** : CtY, DLC, NN, OClWHi, OO
- ***Source*** NUC (NG 0617345), which notes: '(Guyot's geographical series)'
- ***Notes***

1. NUC records further eds. (all New York): [c1875], 115 pp. (MH) [NG 0617346, which notes: 'Massachusetts ed.']; [c1876] (MH) [NG 0617347, which notes: 'Half-title: Geography of Massachusetts']; and c1882, 100, 9 pp. (MH) [NG 0617347, which notes, 1: '[Massachusetts ed.]'; 2. '(Guyot's geographical series).'
2. Culler (1945, *passim*) discusses (01); Hauptman (1978:432) quotes from (01), and (1978:433-5) thrice refers to it in general discussion.

GUYOT, Arnold Henry **Y Am**

- **{1869} (01)**
- ... *Introduction to the study of geography*
- NEW YORK; C. Scribner and company. Boston; E.P. Dutton & co.

- 25 cm.; pp. iii, [1], 118
- ***Libraries*** : DLC, ICJ, MH
- ***Source*** NUC (NG 0617324)
- ***Notes***

1. The similarity of the titles and the sizes of the two books both suggest that this may be a continuation of *Primary; or, introduction to the study of geography* (see GUYOT 1866b, below).
2. For the 1873 ed. of this title, NUC notes: 'c.1868' (see (04), below).
3. NUC records further eds. (all New York): 1871 (DHEW, MH) [NG 0617325, which notes: 'Cover reads: New York, Scribner, Armstrong, and co., 1873']; 1872, 118 pp. (DHEW, MH); [1873?] (MH) [NG 0617327, which notes: 'c.1868']; 1879? (MH); and 1896? (DLC).
4. Carpenter (1963:258-60) provides a general discussion; Elson (1964:403) cites an ed. of 1868.

GUYOT, Arnold Henry **Y Am**

- **{1866b} (01)**
- ... *Primary; or, introduction to the study of geography*
- NEW YORK; C. Scribner and company. Cleveland; Ingham & Bragg, [etc.]
- 25 cm.; pp. iii, [1], 118
- ***Libraries*** : DLC, MH, PPiU
- ***Source*** NUC (NG 0617366), which notes: 'Prepared with the co-operation of Mrs. Mary H. Smith'
- ***Notes***

1. NUC records further issues/eds.: 1867 (MH, OO), and 1868 (MdBP, MH).
2. This title may be continued as *Introduction to the study of geography ...* (see note 1 in the previous entry, i.e., GUYOT 1869).
3. Discussed in Brigham and Dodge (1933:23), Culler (1945, *passim*), and Nietz (1961:198, 223?, 230); cited by Elson (1964:403).

GUYOT, Arnold Henry **Y Am**

- **{1882b} (01)**
- *Scribner's geographical reader and primer*; a series of journeys round the world (based upon Guyot's introduction) with primary lessons
- NEW YORK; C. Scribner's sons
- 19cm; pp. vi, 282
- ***Libraries*** : DLC
- ***Source*** NUC (NG 0617371)
- ***Notes***

1. The relationship of this book to GUYOT (1882a) is not known.
2. NUC records further eds.: 1882, New England ed., New York, 298 pp. (ICU, MB, MH); and [c1882], New York and Chicago, (MH).

H

H., T. **Y**

- **{1707} (01)**
- *A short way to know the world*: or the rudiments of geography: being a new familiar method of teaching youth the knowledge of the globe, and the four quarters of the world.
- LONDON; printed for Tho. Osborne
- 8vo; pp. [12], 222, [2]
- ***Libraries*** : DFo, ICU
- ***Source*** NUC (NS 0518866), which catalogues under Short and notes: 'A second edition, 1712, in the British Museum written by T.H.'

- **1712, 2nd, with additions (02)**
- *A short way to know the world ...*
- LONDON; printed for T. Osborne in Gray's-Inn, next the Walks
- 12mo, 16 cm.; pp. [24], 234, [18]
- ***Libraries*** L, O : CLU, CtY, DFo, MHi, NN, Vi : CaOTP
- ***Notes***

1. The first 19 pp. are devoted to general geography, the remainder to special, with Europe receiving 60 per cent of the space. The catechism technique of question and answer is used, for the first time in English; the German originals of HÜBNER (1738) did so as early as 1696.

   The author claims to be the first to give the length and breadth of every country in units of standard length (i.e., Italian miles). He also provides a comparative table of national units of length. In his section on general geography he mentions both the Ptolemaic and Copernican cosmologies, showing some preference for the latter. In the preface he mentions the geographies of Gordon, Morden, and Falconer, and also the work of Eachard (i.e., GORDON 1693; MORDEN 1680; the 'Introduction to geography' of *System of geography* 1701 [see note 1 of (02) of THESAURUS GEOGRAPHICUS]; and EACHARD 1691).

- **{1745} (03)**
- *A short way ... world*: or, a compendium of modern geography ... in two parts. I. General geography ... II. Special geography ... The whole improved with maps adapted to the work, and copper-plates
- LONDON; Printed for T. Osborne
- 17 cm.; pp. xvi, 381 (25)
- ***Libraries*** : CtY

HALE, Nathan **Y Am**

- **{1830} (01)**
- *An epitome of universal geography*, or a description of the various countries of the globe; with a view of their political condition at the present time. With sixty maps

- BOSTON; N. Hale; Richardson, Lord and Holbrook
- 19 cm.; iv, 404 p.
- ***Libraries*** : IU, MB, MH, MiU, MWA, NN, NNC, OO
- ***Source*** NUC (NH 0042669) and Elson (1964:399)

HALL, Mary L[ucy] **Y Am**

- **{1864} (01)**
- *Our world, or, first lessons in geography ...*
- BOSTON; Crosby & Nichols
- 19 cm.; pp. 177
- ***Libraries*** : DLC, OrU
- ***Source*** NUC (NH 0054399)

- **{1866} (02)**
- *Our world* ... geography for children
- BOSTON; S.F. Nichols
- ***Libraries*** : MH
- ***Notes***

1. NUC records further eds. (all Boston): 1867 (H); 1868 (MH); 1870 (MH); 1872, 116 pp. (DHEW, IU, MH); 1873 (MH); 1875 (MH, OO); 1876 (MH); 1880 (MH); 1881, 119 pp. (DHEW, MH, NN); 1882 (DHEW, OU); 1886 (MH); and in 1888 and 1891.
2. Discussed in Culler (1945, *passim*). Elson (1964:404) cites the ed. of 1872.

HALL, Mary L[ucy] **Y Am**

- **{1872} (01)**
- *Our world*. No. II. A second series of lessons in geography
- BOSTON; Ginn brothers
- 30.5 cm.; 2 p. l., 176 p.
- ***Libraries*** : DLC, OCl
- ***Source*** NUC (NH 0054414)
- ***Notes***

1. NUC records further issues/eds. (all Boston): 1873 (CtY, MH); 1875 (DLC, MB, MH); 1877, 176 pp. (MH, OClWHi, PP); 1878 (PP); 1882 (MH); and 1887 (MH).

HALL, S[amuel] R[ead] **Y SG? Am**

- **{1831} (01)**
- *The child's book of geography*
- SPRINGFIELD, [MA]; Merriam, Little & co. New York, Collins and Hannay; [etc.]
- 13 cm.; pp. 96
- ***Libraries*** : DLC

- **{1833}, 2nd (02)**
- *The child's book of geography*. With outlines of countries, cuts, and eight copper plates

- HARTFORD; D.F. Robinson & co.
- 13.5 cm.; pp. 96
- *Libraries* : NBuG

HAMEL, Nicolas **Y tr**

- **{1800} (01)**
- *The world in miniature*: containing a curious [and faithful] account of the [situation, extent, climate, productions, government, population, dress, manners, curiosities, etc. etc. of the] different countries of the world, [compiled from the best authorities;] with [proper] reference to the [most essential] rules of the French language and translations of the difficult words & idiomatical expressions; a book particularly useful to students in geography, history, or the French language. By ... rector of the town of l'Aigle, author of a French grammar, and several other school books]
- LONDON
- 12mo
- ***Libraries*** L
- ***Source*** BL, with supplementary information, placed in [ ], taken from BoC 7047
- ***Notes***

1. An ed. of 1803 is known from an advertisement in entry (02) of PHILLIPS, R. (1803). Further eds. are known: 1806, 2nd (MoSU); 1815, 4th ed. (L); 1820, 5th (: : CaOTP); 1841, New (L). According to BoC 7047, the collation of the 5th ed. is pp. [i-ii], iii-iv, [1], 2-12, [13], 14-251, [252].

HAMILTON, A.G. **mSG**

- **1839 (01)**
- *A new key to unlock every kingdom*, state, and province in the known world. Containing the length, breadth, and population of each kingdom, and their chief cities; produce, government, revenue, military and naval strength, arts, religion, etc. etc. A large distance table of England, Ireland, Scotland, and Wales, with the principal travelling stations of France and the Netherlands, and their distance from each other. Also, a description of the ocean, land, deserts, mountains, volcanoes, rivers, falls, rapids and lakes; zones and climates, days and nights, winds, tides, and minerals; geographical discoveries; a table of chronological events, from the creation to the present time; races of men; manners of nations; state of society; heights of mountains, largest rivers, etc. etc. By ... author of the Guide to the Globe, and Geographical Dictionary, etc.
- LONDON; sold by Adam and co.; Leeds, Mason and Scott; Manchester, Ainsworth and sons; Sheffield, Gilbert; Newcastle-on-Tyne, Pickersgill, and all booksellers
- 17.5 cm.; pp. [5], vi-viii, 100
- ***Libraries* GP**
- ***Notes***

1. Closer to a dictionary than a special geography, at least as far as descriptions of the countries of the world are concerned, for they are listed in alphabetical order. These descriptions are telegraphic in form, and occupy

pp. 26-51. The balance of the book is devoted to concise discussions of the topics referred to in the tp. The principal sources are identified. In the chronology, 4,000 B.C. is favoured as the date of the creation of the world.

HARPER'S INTRODUCTORY GEOGRAPHY **Y SG?**

- **{[1877]} (01)**
- *Harper's introductory geography*[; with maps and illustrations prepared expressly for this work by eminent American artists]
- NEW YORK; Harper Brothers
- ***Libraries*** n.r.
- ***Notes***

1. NUC (NH 0126867) notes that copyright was granted in 1877. The words in the title set in [ ] are taken from the ed. of 1880.
2. NUC records the following eds. (all New York): 1879 (MH); 1880 [c1877], 112 pp. (CU, MH, WaS, WaWW); 1881 (MH, MiU); 1883, 112 pp. (DLC); 1884, 112, 8 pp. (DLC) [NH 0126870 notes: '"Special geography of the Pacific states": 8 p. at end']; 1885, 114 pp. (DLC, Nh); 1886, 112, 8 pp. (DHEW, MH) [NH 0126872 notes: '"Special geography of the Pacific states": 8 p.']; and [1890, c1878]; [c1896]; [1905]; and [c1914].

HARPER'S SCHOOL GEOGRAPHY **Y SG?**

- **{1876} (01)**
- *Harper's school geography.* With maps and illustrations prepared expressly for this work by eminent American artists
- NEW YORK; Harper & brothers
- 32 × 25 cm.; 1 p. l., ii, 124 p.
- ***Libraries*** : DLC, OrU, PP
- ***Notes***

1. NUC records further eds. (all New York): 1876, Indiana ed. (DLC); 1876, Massachusetts ed. (DLC); 1876, New York ed. (DLC); 1876, Pennsylvania ed. (DLC); 1877 (DLC); 1877, Kansas ed. (DLC); 1877, Maine and Massachusetts ed. (DLC); 1877, Minnesota, Iowa, Nebraska, Kansas, and Missouri ed., 128 pp. (MoU); 1877, [156 pp.] (DLC); 1877, New Jersey, Pennsylvania, and Delaware ed., [146 pp.] (DLC, IEdS); 1877, North Central states ed., [153 pp.] (DHEW, DLC, FTaSU, MH, OO); 1877, Wisconsin and Michigan ed., 126 pp. (DLC); 1878, Iowa, Kansas, Minnesota, Nebraska, and Missouri ed., 156 pp. (DLC); 1878, 148 pp. (DLC, IEdS); 1878, New York ed., 126, 10 pp. (DHEW, DLC, MH, MoU, NN, OO); 1879, 128 pp. (DLC, MB, MH); 1880, 128 pp. (MH, OU); 1880, 128, 17 pp. (DLC); 1881, New Jersey, Pennsylvania, and Delaware ed., (MH); 1881, North Central states ed. (OkU); 1882, New England ed., 128, 26 pp. (CtY, N, PP, ViU); 1883, Minnesota, Iowa, Nebraska, Kansas, and Missouri ed., 155 pp. (DLC); c1885, North Central states ed. (MH, NN); c1885, Ohio, Indiana, and Illinois ed., 128, 20 pp. (OO); 1885, Pacific states ed., 128, 25, 4 pp. (DLC); c1885, 128, 17 pp. (WaU) [NH 0126952 notes: '"Special geography of the states of the Columbia and Alaska territory": 17 p. at end']; 1886, New Jersey, Pennsylvania, and Delaware ed. (MH); 1887, New England ed. (MH); and 1888; 1890; 1896; and 1903.

HARRIS, Alexander **A Am**

- **{1862} (01)**
- *A geographical handbook* ... With a copious index. Adapted as an aid to the student of history
- LANCASTER, PA; printed at the Daily Express office
- 21 cm.; 427 p., 1 l.
- ***Libraries*** : DLC, PPT
- ***Source*** NUC (NH 0129444), which classifies it as Geography, i.e., as a book intended for adults

- **{1864}, 2nd (02)**
- *A geographical handbook*: or, a description of the different countries ... also a concise outline of the principal cities, towns and villages of each, giving their kind of manufactures, population and notable events of history ... Adapted ...
- LANCASTER, PA; Pearsol & Geist, printers
- 21 cm.; iv, [5]-432 p.
- ***Libraries*** : DLC, KU, OU, PHi, PPT
- ***Source*** NUC (NH 0129445)

HARRIS, J. **?**

- **{1712} (01)**
- *Geographical dictionary*
- Place/publisher n.r.
- ***Libraries*** n.r.
- ***Source*** Allford (1964:239)

HARRIS, John **Y mSG**

- **1825?, 4th, with ... (see below) (01)**
- *The traveller*; or, an entertaining journey round the habitable globe; illustrated with plates, consisting of views of the principal capital cities of the world. Fourth edition, with numerous emendations and additions
- LONDON; John Harris, corner of St. Paul's Church-Yard
- 17 cm.; pp. [v], vi-viii, [3], 2-219
- ***Libraries*** : : **CaBVaU**
- ***Notes***

1. 1825 may be the date of the first ed.; it is taken from the illustration facing the tp.
2. Seeks to educate by entertaining; avoids the 'technicalities' of schoolbooks and 'premature constraints upon the memory' (p. v). 'To prove the inaptitude of young persons to study, and their dislike of whatever bears the appearance of a task, would be unnecessary; it is continually observed ... by every parent, and tutor ... the author has endeavoured ... to overcome the one by the stimulus of novelty, and to evade the other by ... giving to instruction the guise of casual remarks in the course of a pleasant excursion' (p. vi).

   Initially, the style is condescending and paternalistic; it soon becomes a recitation of facts; at times, especially when identifying the location and dimension of the counties of England, it is little more than a list in the form of a text. Europe receives the most attention, with the rest of the world being dealt with in some 70 pp. Once the UK has been left behind, it is the customs and manners of people that receive most attention - *National Geographic* without pictures.

HART, Joseph C. **Y mSG? Am**

- **{1827}, 2nd, improved (01)**
- *An abridgement of geographical exercises*, for practical examination on maps of geographical exercises ... Accompanied by an atlas of fourteen maps
- New York; R. Lockwood
- 15 cm.; pp. vi, [7]-108
- ***Libraries*** : DLC

- **{1827}, 4th, improved (02)**
- *An abridgment of geographical exercises* ... maps. Written for the Junior Department of the New York high schools, and adopted by the Public School Society. Accompanied ...
- NEW YORK; R. Lockwood
- ***Libraries*** : MH
- ***Notes***

1. NUC records further eds.: 1829, 6th (DHEW); and c1826, 14th (MiU).

HART, Joseph C. **Y mSG? Am**

- **{1824} (01)**
- *Geographical exercises*; containing 10,000 questions for practical examinations on the most important features of the maps of the world and the United States, by Melish, Lay's map of the state of New York, and maps of America, Europe, Asia, and Africa, by Arrowsmith ...
- NEW YORK; W.A. Mercein, printer
- 15 cm.; pp. vii, [9]-155
- ***Libraries*** : DLC
- ***Notes***

1. NUC records further eds. (both New York): 1825, 155 pp. (NN); and 1857, 'revised, enlarged, and improved, by ... Charles B. Stout, 141 pp. (DLC).
2. Brigham and Dodge (1933:14) discuss (01).

HART, Joseph C. **Y mSG Am**

- **{1851} (01)**
- *A popular system of practical geography*, for the use of schools, and the study of maps ...
- NEW YORK; Cady & Burgess
- 16.5 cm.; pp. viii, [9]-132
- ***Libraries*** : DLC

▸ *Notes*
1. NUC records a further ed.: 1854, New York, 134 pp. (DLC).

HARTLEY, John [the Rev., of Leeds] ?
▸ **{1816}, 3rd (01)**
▸ *Outlines of geography* ... Third edition
▸ LONDON; printed in Leeds
▸ 12mo
▸ ***Libraries*** L
▸ ***Notes***
1. See also, HARTLEY, John [Wesleyan minister].

▸ **{1821} (02)**
▸ *Geography, outlines*
▸ Place/publisher n.r.
▸ ***Libraries*** n.r.
▸ ***Source*** Peddie and Waddington (1914:227)

HARTLEY, John [Wesleyan minister] **Y SG?**
▸ **{1816}, 2nd (01)**
▸ *Geography for youth* ... Second edition
▸ LONDON
▸ 12mo
▸ ***Libraries*** L
▸ ***Notes***
1. According to BL, the author was a Wesleyan minister. This identification seems more authoritative than that of NUC; see note to (02), below.

▸ **{1828}, 5th (02)**
▸ *Geography for youth* adapted to the different classes of learners
▸ LONDON
▸ ***Libraries*** : OO
▸ ***Source*** NUC (NG 0149243), which speculates that the author might be the Rev. John Hartley of Leeds; see HARTLEY, John [the Rev., of Leeds], above.

HARWOOD, Thomas ?
▸ **{1804} (01)**
▸ *Geography*
▸ LONDON
▸ 12mo
▸ ***Source*** London Catalogue (1811) and Allibone (1854)

HATTON ?
▸ **{17–?} (01)**
▸ *Hatton's geography*

- LONDON
- 8vo
- *Libraries* n.r.
- *Source* London Catalogue (1786)

HEALE, Edmund Markham **Y SG?**

- **{1853} (01)**
- *A manual of geography compiled for the use of military students*
- LONDON; Whittaker
- 12mo
- *Libraries* L
- *Source* BL and BCB, Index (1837-57:110)

- **{[1863, (1862)]}, 3rd (02)**
- [*A manual of geography ...*]
- LONDON
- 12mo
- *Libraries* L

HENISCH, Georg **OG**

- **{[1591?]} (01)**
- *The principles of geometrie, astronomie, and geographie ...* Gathered out of the tables of the astronomical institutions of Georgius Henischius. By Francis Cooke ...
- LONDON; J. Windet, sold at the house of F. Cooke
- 8vo, 14.5 cm.; pp. [43] l.
- *Libraries* C : CSmH, CtY
- *Source* NUC (NH 0276786) and Pollard and Redgrave (1946)
- *Notes*

1. The second part of the title makes clear that this is not a special geography.

HERON, Robert (1796)
See NEW AND COMPLETE (1796).

HESLOP **?**

- **{1831} (01)**
- *Geographical exercises*
- Place/publisher n.r.
- *Libraries* n.r.
- *Source* Peddie and Waddington (1914:226)

HEWITT, J. **Y SG?**

- **{1864}, new (01)**
- *Geography, elements*. New edit
- ? Darton

- 18mo
- ***Libraries*** n.r.
- ***Source*** ECB, Index (1856-76:130)

HEYLYN, Peter (1652) **A**

- **{1642}**
- *Cosmographie in four bookes*. Containing the chorographie and historie of the whole world, and all the principall kingdoms, provinces, seas, and isles thereof
- London; printed for Henry Seile ...
- ***Libraries*** : CLU
- ***Source*** NUC (NH 0347275). According to CLU, this ed. is a ghost; see (01), below.

- **{1649}**
- *Cosmographia*
- ***Source*** E.G.R. Taylor (1934:294). In neither BL nor NUC. Taylor provides no information beyond the title. Almost certainly a ghost; see (01), below.

- **1652 (01)**
- *Cosmographie ... contayning ...*
- LONDON; printed for Henry Seile, and are to be sold at his shop over against Saint Dunstans Church in Fleetstreet (the individual books were printed in different presses)
- Folio, 35 cm.; 4 pts. in 1 vol. [Books 1-3 and the two parts of Book 4 each has its own tp. The wording given above is that of the tp. to Book 1]; pp. [14], 324, 278, 258, [2], 198, [17]. [According to (NH 0347275) the collation is: 'folio, 35 cm.; 4 pts in 1 vol; signatures: A-M^{6}, (n)-(p)^{6}, (q)^{4}, N-Z^{6}, Aa-Dd^{6}, Aa-Yy^{6}, Zz^{7}, Aaa-Xxx^{6}, Yyy^{2}, Aaaa-Zzzz^{4}, Aaaaa-Ccccc^{4}, Ddddd-Fffff^{2}. Numerous errors in paging.']
- ***Libraries*** BfQ, CMp, DC, **L**, LCP, LMT, O, [IDT] : CLU, CSmH, CtY, DLC, InU, MdAN, MiU, MWA, N, NN, NNUT, NPV, OPClWHi, OrU, PBL, PPAmP, RPJCB, ViU, WaSpG : CaBVaU, CaBVPA
- ***Notes***

1. A massive work much concerned with the division of territories, their boundaries, and their names, together with the histories of their rulers. It is the first large-scale special geography by an English author (see AVITY). Its frequent republication suggests that it was the standard authority in its field for two generations. Its popularity probably set the style for much that was to follow later.

   In an introductory address to the reader, H. identifies the point of view from which he writes, noting: 'I am an Englishman; and, which is somewhat more, a churchman.' By the latter term he meant that he was a cleric of the Church of England. In the first half of the seventeenth century, to occupy such a position would usually have entailed the adoption of a particular political outlook, which, in this case, it certainly did. He also reports that he wrote the book in part because he had much time on his hands (being then, as a consequence of his active support for the losing side in the

Civil War, without professional employment). In addition, he was unhappy about the errors he had made in writing his *Microcosmos* (HEYLYN 1621).

H. displays a scholar's knowledge of Classical and Biblical sources, doing so in prose that Billinge (1983) would surely categorize as 'mandarin' in style. The additional information he provides, relative to the earlier work, tends to belong to history rather than to geography.

2. The publication history of the work is complex. NUC records 33 issues or editions, but one of them is certainly a ghost, and several of the others probably are. The entries that follow are derived from NUC, guided by information obtained by examining copies held by L and LRGS.

- **1657, 2nd (02)**
- *Cosmographie* … books. Containing … principal …
- LONDON; printed for Henry Seile …
- Folio. The L copy is massive, with a page length of 45 cm.; other libraries report only 35 cm. As in (01) there are four books bound as a single work; as before, Books 1 to 3 have their own tps., as do both parts of Book 4, and also the appendix; this time, however, the pagination is continuous. Pp. [14], 1098 (with some mispaging), [4], 1089-95, [21]. Some libraries record the number of pp. as 1098, others as 1095; given the information recorded at L and confirmed by CLU (NH 0347287), it seems likely, however, that they all refer to the same ed.
- ***Libraries*** C, DC, GU, **L**, LC, LPO, NoU, OH : AU, CLU, CoU, CU, CtY, DFo, FSaHi, IaU, MdBJ, MH, MHi, MoSW, N, NcD, NIC, NN, NNHi, OrP, OU, PU, RP, RPJCB : CaBViPA
- ***Notes***

1. Because the order in which the material is presented has been much altered, it is difficult to say how much has been added, though it is clear that some has. The individual parts were again printed separately, with Book 1, which contains half of Europe, Book 3 (Asia), and Book 4 (Africa and America) all bearing the date 1656. NUC refers to this as the 2nd ed., presumably following the frontispiece that precedes the tp., where there is a statement to this effect.

In a minor quirk of presentation, mountain ranges that form the frontiers between countries are dealt with separately from the countries.

- **{[1660 (1662)]}, 2nd (03)**
- *Cosmographie* … foure bookes contayning …
- LONDON; Printed for P. Chetwind
- Folio, 35 cm.; 4 vols. in 5 [bound in 1 vol.]
- ***Libraries*** : KyU, MH, MHi
- ***Source*** NUC (NH 0347288 and 0347290). The date of publication is taken from an appendix. On the tp. the date appears as 1660.

- **{1665}, 3rd (04)**
- *Cosmographie … four books*. Containing …
- LONDON; printed for Philipp [sic] Chetwind
- 36 cm.; pp. 1095
- ***Libraries*** ShP : OHi, PPT
- ***Source*** NUC (NH 0347291)

- **1666, 3rd (05)**
- *Cosmographie*, in ... world, and all ... thereof. With an accurate and an approved index of all the kingdoms, provinces, countries, inhabitants, people, cities, mountains, rivers, seas, islands, forts, bayes, capes, forests, etc. of any remarque in the whole world; much wanted and desired in the former, and now annexed to this last impression, revised and corrected by the author himself immediately before his death
- LONDON; printed for Anne Seile, and are to be sold at her shop over against St. Dunstans Church in Fleet-Street
- Folio, 42 cm.; 4 pts. (pagination continuous): pp. [12], 1095 [some mispaging], [39]
- ***Libraries*** AbrwthU, C, E, EU, **L**, LDW, LIT, SkPt : CLSU, CSmH, CtY, DLC, RPJCB, ViHo, ViU, WU
- ***Notes***

1. The tps. of the separate books all bear the date 1665, as does the frontispiece; the latter also identifies this as the 3rd ed. On paging through, few changes from 1657 are apparent; most of those noticed involve only a single paragraph here and there; the longest of them involved the reorganization of the 7 pp. on Palestine. The maps of individual countries, present in (02), are omitted.

- **{1666 (Wing 1982), 1667 (NUC)} (06)**
- *Cosmographie ...*
- LONDON; printed for Philip Chetwind
- 34 cm.; 6 p. l., 1098, [2], 1 l., 1089-95, [40] p.
- ***Libraries*** DU : CtY, MHi
- ***Source*** Wing (1982) and NUC (NH 0347296)

- **1669, 3rd (07)**
- *Cosmographie* in ... forts bays ... world: ...
- LONDON; printed for Anne Seile: and are to be sold by George Sawbridg [sic], Thomas Williams, Henry Broom, Thomas Basset, and Richard Chiswell
- Folio, 36 cm.; 4 pts. [with separate pagination; see (01)]; pp. [13], 301 (some mispaging), [5], 226, [2], 230, [2], 162, [41]
- ***Libraries*** AbrwthWN, BlaR, Brf, C, **L**, LcP, LvU, NP, O : CLU, CU, DFo, FU, ICN, IU, MB, MiU, MnU, MWA, NIC, PPA, PU
- ***Notes***

1. Tps. of Books 2, 3, and 4 carry the date 1668. The frontispiece identifies this as the 3rd ed. The book also carries an imprimatur dated 1664. Although the pages are smaller than in (05), the lost space has been taken from the margins, and the amount of text per page has been increased. Because the spacing of the text has been reorganized, it is difficult to make precise comparisons, but spot-checking suggests that the contents are unchanged since 1666. NUC (NH 0347302) notes that in some copies the engraved title reads: '3rd edition ... 1667.' It also notes: 'A description of this edition may be found in an article "Philip Chetwind and the Allott copyrights", by Henry Farr, in the Library, 4th ser., v. 15, p. 152-153.'

- **1670 (08)**
- *Cosmography*, in ... geography ... history ... forts, bays ... world; ... revised, corrected and inlarged by ...
- LONDON; printed for Philip Chetwind
- Folio, 35 cm.; 4 pts. (with continuous pagination); pp. [12], 1095 (with some mispaging), [41]
- ***Libraries*** CGc, **L**, NP, OMe, WinStG : DLC, LU, MWA, NjP, NN, PMA, PPL
- ***Notes***

1. Tps. of Books 1, 2, and 3 carry the date 1665; Book 4 is dated 1662. The layout of the pages is in a style similar to that of the eds. published by Henry Seile (i.e., 1642, 1652, and 1657). Spot-checking revealed no difference from the text of (07), and thus also of (05).

- **1674, revised and corrected ... (09)**
- *Cosmography* in ... world; much wanted ...
- LONDON; printed for Anne Seile and Philip Chetwind
- Folio, 36 cm.; 4 pts. (with separate pagination); pp. [13], 303, [3], 226 (the pagination of Book 1 continues to 443, which is followed by 139), [2], 230, [2], 162, [41]
- ***Libraries*** AbrwthWN, CJ, HeP, **L**, OChCh, [IDT] : CLU, CU, InU, LU, MH, MiU, MWA, NcD, NjP, OGK, RPJCB, TxU, ViU : CaBViPA
- ***Notes***

1. The tps. of the separate books carry the date 1673, but the frontispiece is that of (07), and like that ed., this one carries the imprimatur of 1664. Leafing through suggests that, apart from the major mispagination in Book 2 and a few very slight changes, this ed. is identical to that of (07) (in some cases mispaginations in the earlier version are repeated in this one). Whether Anne Seile and Philip Chetwind were in competition earlier or not, they have now joined forces.

   Wing (1982) distinguish between two issues of this ed.; they have different tps.

- **1677 (10)**
- *Cosmography* ... world: and ...
- LONDON; printed by S.C. for P. Chetwind and A. Seile, and are to be sold by T. Basset, J. Wright, R. Chiswel, and T. Sawbridge
- Folio, 34 cm.; 4 pts. [pagination, including errors, as in 1674)
- ***Libraries*** C, DC, **L**, LLI, O, [ICD] : CLU, CSmH, CtY, CU, DFo, DLC, FU, IaU, LU, MeB, MiU, NcD, NjP, NN, NNG, PPL, ScU, WaSpG
- ***Notes***

1. NUC (NH 0347328) notes: 'A reprint of the 6th edition, being the 3rd edition with a new title-page and "an alphabetical and exact table ..." added to it; other issues have dates 1673 and 1674, cf Sabin, Bibl. amer.'

- **1682 (11)**
- *Cosmography* ...
- LONDON; printed for P.C.T. Passenger at the Three Bibles of London-Bridge, B. Tooke at the Ship in St. Paul's Church-Yard, and T. Sawbridge at the Three Fowers de Luces in Little-Britain

- Folio, 35 cm.; 4 pts. (paginated separately); pp. [14], 303, [3], 226, [2], 230, [2], 162 (misnumbered 562), [39]
- ***Libraries*** C, CEm, LinC, **L**, LRGS, O, [IDT] : CtY, ICN, MnU, NBu, NNC, OClW, PU : CaOTU

- **1703 (12)**
- *Cosmography* ... isles thereof. Improv'd with an historical continuation to the present times by Edmund Bohun, esq: with a large and more accurate index, than was in any of the former editions, of all the kingdoms ... world: revised and cleared from a multitude of mistakes, which had crept into former impressions. And five new-engrav'd maps, according to the best and most exact projection
- LONDON; for Edw. Brewster, Ric. Chiswell, Benj. Tooke, Tho. Hodgkin, and Tho. Bennet
- Folio, 41 cm.; 4 pts. (with continuous pagination); pp. [11], 1132, [36]
- ***Libraries*** **L**, O, OSg, SanU : CLU, CtY, DI-GS, DLC, ICarbS, IEG, MChB, NjP, PPL
- ***Notes***

1. Separate tps. for the books are retained despite the continuous pagination; the 2nd and 3rd are dated 1701, and the fourth 1702. Bohun made relatively few changes to the first two books. His chief additions come in the 3rd (Asia), with some made also to America, though nothing like the 100 additional pp. that is suggested by mispagination.
2. Discussed in Baker (1928:269-70 and 1935:131-2), Bowen (1981:92-3), Gilbert (1972:47-50), and Taylor, E.G.R. (1934:139-43).
3. See note 1 of (03) of THESAURUS GEOGRAPHICUS.

HEYLYN, Peter **A**

- **1621 (01)**
- *Microcosmos*, or a little description of the great world. A treatise historicall, geographicall, politicall, theologicall
- OXFORD; printed by Iohn Lichfield, and Iames Short, printers to the famous universitie
- 4to, 20 cm.; pp. [16], 420, [3]
- ***Libraries*** AbrwthWN, C, **L**, Lam, O : CSmH, CtY, DFo, NN, RPJCB, TxU
- ***Notes***

1. A scholarly work. The preface contains a statement of objectives and methods. Sources are cited in marginal notes. In the first two eds., the author is identified only as H., P.
2. Facsimile ed. (Johnson/TOT).

- **1625, augmented and revised (02)**
- *Microcosmos* [in Greek script]. A little description of the great world
- OXFORD; printed by John Lichfield and William Turner, and are to be sold by W. Turner and T. Huggins
- 4to, 20 cm.; pp. [16], 814
- ***Libraries*** C, **L**, O : CSmH, CtY, DFo, DLC, MWA, NcD, NIC, OWoC : CaBViPA

- *Notes*
1. The preface is shorter than in 1621, and is now chiefly a justification for the many additions and corrections. The spelling has been modified considerably. The growth in bulk seems to have been fairly well distributed, with roughly twice as much being written about most countries.

- **1627, 3rd, revised (03)**
- *Microcosmos ...*
- OXFORD; printed by J. L[ichfield] and W. T[urner] for William Turner and Thomas Huggins
- 4to, 20 cm.; pp. [20], 810
- ***Libraries*** BfQ, C, DU, **L**, O, [IDT] : CSmH, CtY, DFO, DLC, ICN, LNT, MoU, PHi, RPJCB, TU
- *Notes*
1. Sources are no longer cited. Most changes are ones of detail, with information given about more places, and some information eliminated. The text on some countries is almost unaltered (e.g., Spain). In the preface to his *Cosmography*, HEYLYN (1642) stated that he had not been responsible for the revisions to the 3rd and later eds. of *Microcosmos*.

- **1629, 4th (04)**
- *Microcosmos ...*
- OXFORD; printed by W. T[urner] for William Turner and Thomas Huggins
- 4to, 19 cm.; pp. [20], 809 (with some misnumbering)
- ***Libraries*** **L**, **O**, OMa : CSmH, CSt, CtY, DFo, GEU, ICN, MH, MiU, NcU, NjP, NN, NNC, PU, TxU : CaBVaU
- *Notes*
1. Type reset and the errata of the previous ed. incorporated in the text; otherwise the text is unchanged apart from the occasional rewording of a sentence.

- **1631, 5th (05)**
- *Microcosmos ...*
- OXFORD; printed by Will. Turner
- 4to, 19 cm.; pp. [20], 809 (with some misnumbering)
- ***Libraries*** CEm, Lam, LU, **O**, WarSE : CtY, DFo, DLC, ICN, MH, MiU, MWA, NcU, OCl, OClWHi
- *Notes*
1. Apart from the very occasional change of a word, the text is as in (04).

- **1633, 6th (06)**
- *Microcosmos ...*
- OXFORD; printed for William Turner and Robert Allott
- 4to, 19 cm.; pp. [20], 809 (with some misnumbering)
- ***Libraries*** AbrwthWN, **L**, LvU, LU, LUU, O, WigMD : CSmH, CtY, DFo, DLC, InU, IU, MB, Mi, NN, OU
- *Notes*
1. Apart from some changes in spelling, the text is the same as (05).

- **1636, 7th (07)**
- *Microcosmos ...*
- OXFORD; printed for William Turner, and are to be sold at the black Beare in Pauls Church-Yard
- 4to, 20 cm.; pp. [20], 809 (with some misnumbering)
- ***Libraries*** BrF, C, EU, **L**, LvU, LU, NcU, O : CSmH, CU-A, DFo, DLC, MnU, NjP, NN, NNUT
- ***Notes***

1. The misnumbering survives a resetting of the type. Text unchanged.

- **1639, 8th (08)**
- *Microcosmos ...*
- OXFORD; printed by William Turner
- 4to, 20 cm.; pp. [20], 809 (with some misnumbering)
- ***Libraries*** **L**, O : CLU, CSmH, CtY, DFo, DLC, IU, MChB, NN
- ***Notes***

1. The misnumbering survives yet another setting of the type.

- **{1639} (09)**
- *Microcosmos ...*
- ***Libraries*** LDW, LUU, O : CSmH
- ***Source*** Pollard and Redgrave (1946), which notes that the date is misprinted as 1939, and that tables and indices are bound at the end
- ***Notes***

1. Discussed in Baker (1928:269-70, and 1935:131-2), Bowen (1981:70-71), and Taylor, E.G.R. (1934:108).

HIGDEN, Ranulf **mSG**

- **{1482} (01)**
- *Polychronicon*
- [WESTMINSTER]; William Caxton
- Folio, 28 cm.; pp. [21], ccccxxviii, [1] l.
- ***Libraries*** BlaR, C, GU, L, MRu, NP : CSt, CtY, DFo, DLC, ICN, MoSW, NcGW, NN, NSyU, PBL, RPB
- ***Notes***

1. 'The *Polychronicon* of Ranulph Higden ... contains a *Mappa Mundi* [i.e., a verbal description, not a map, of the known world], compiled from a host of sources, running from Pliny, Isidore and Macrobius down to Giraldus Cambrensis' (Taylor, E.G.R. 1930:4).

   The bibliographic information provided above is taken from NUC (NG 0358170), which notes: 'Books 1-7 edited by William Caxton from John Trevisa's translation of Higden's Latin text; book 8 bringing the history to 1460 was compiled by Caxton. The first book comprises a brief description of the world, and a more particular account of Great Britain.' According to Cox (1938), Trevisa made the translation in 1387.

- **{1495} (02)**
- *Polichronicon*
- [WESTMINSTER; Wynken de Worde]

- Folio, 26 cm.; pp. 50 p. l., cccxlvi f., 1 l.
- ***Libraries*** C, E, L, MRu : CtY, DLC, InU, MiU, N, NN, PPL, PPRF, PPT, TxDaM
- ***Source*** NUC (NH 0358172)

- **{1527} (03)**
- *Polichronicon*
- [SOUTHWERKE; printed by P. Treveris for J. Reynes]
- Folio, 30 cm.; pp. 49 p. l., 346 numb. l.
- ***Libraries*** BlaR, C, CheR, L, LU, O, PrR : CLU, CSmH, CtY, DFo, DLC, ICN, InU, MH, MWiW, NjP, NN, NNPM, RPB
- ***Source*** NUC (NH 0358178 and 0358176), which notes: 'This is a reprint of the edition printed by Wynken de Worde in 1495. The only variation appears to be the introduction of a few woodcuts and the omission of the date of Wynken de Worde's edition at the end.'

HILEY, Richard **Y SG?**

- **{1848a} (01)**
- *The child's first geography, etc.*
- LONDON; Longmans.
- 12mo.
- ***Libraries*** L
- ***Source*** BL and BCB, Index (1837-57:110)
- ***Notes***

1. It seems likely that there was only one author named Hiley who wrote special geography, though identified variously as R. and Richard by different authorities.

HILEY, R. **? n.c.**

- **{1872a} (01)**
- *Compendium of Asiatic, African, American, and Australian geography*: with historical notices of the principal countries. Forming a sequel to Hiley's 'Compendium of European Geography,' etc.
- LONDON
- 8vo
- ***Libraries*** L
- ***Source*** Allford (1964:248)
- ***Notes***

1. Together with the next entry, it does deal with the whole world.

HILEY, R. **? n.c.**

- **{1872b} (01)**
- *Compendium of European geography and history*
- LONDON
- ***Libraries*** : NB

HILEY, R. **Y SG?**
- **{1848b} (01)**
- *Elementary geography*
- LONDON
- ***Libraries*** n.r.
- ***Source*** Allford (1964:245)
- ***Notes***

1. ECB, Index (1856-76:130) records an ed. in 1872.

HILEY, Richard **Y SG?**
- **{1843} (01)**
- *Progressive geography*, adapted to ... junior classes
- LONDON
- 12mo
- ***Libraries*** L
- ***Notes***

1. BL records a further ed.: 1848, 2nd.

HINCKS, T[homas] D[ix] **?**
- **{1842} (01)**
- *Geography, modern*
- [LONDON]? Groombridge
- 18mo
- ***Libraries*** n.r.
- ***Source*** BCB, Index (1837-57:110)

HINCKS, Thomas Dix **Y SG?**
- **{1819}, 8th (01)**
- *A short and easy introduction to geography*. Eighth edition, with ... alterations
- CORK
- 12mo
- ***Libraries*** L

HISTORICAL POCKET LIBRARY **Y**
- **1790 (01)**
- *The historical pocket library*; or, biographical vade-mecum. Six volumes. Consisting of I. The heathen mythology. II. Ancient history. III. The Roman history. IV. The history of England. V. Geography. VI. Natural history. The whole forming a moral and comprehensive system of historical information for the amusement and instruction of the young nobility of both sexes
- LONDON, [Bath, printed]; George Ridley
- 16mo
- ***Libraries*** **L**

▸ *Notes*
1. Vol. V is a special geography. Catalogued by BL under Pocket.
2. NUC records a further ed.: 1818-19, Boston, 6 vols. in 3 (InU, MWA, N, NIC).

HISTORY OF ALL NATIONS **Y**
▸ **{1777} (01)**
▸ *The history of all nations.* Giving a brief and entertaining account of the situation, customs, manners, genius, temper, religious and other ceremonies, trade, manufactures, arts and sciences, government, policy, laws, cities, rivers, mountains, and the most material natural curiosities in every country throughout the known world. Interspersed with a variety of elegant cuts, representing the habits of the people of the different kingdoms. To which is added, a short and easy introduction to geography, with a cut, describing the several circles of the sphere; and an another exhibiting the order and course of the planets round the sun
▸ EDINBURGH; printed by David Paterson
▸ 13 cm.; pp. xii, 204
▸ ***Libraries*** : : CaOTP
▸ ***Source*** St. John (1975, 2:806)

▸ **{1800}, new and enlarged (02)**
▸ *The history of all nations ...* Interspersed with upwards of twenty elegant cuts ... sphere. Together with a new and accurate map of the world. Designed for the use of schools. A new and enlarged edition
▸ LONDON; printed for G.G. and J. Robinson; and S. Hodgson, Newcastle
▸ 15.5 cm.; pp. xv, 222
▸ ***Libraries*** : : CaOTP
▸ ***Source*** St. John (1975 1:183-4)

HODGINS, John George **Y Can**
▸ **1863 (01)**
▸ *Easy lessons in general geography*, with maps and illustrations: being introductory to 'Lovell's General Geography' By ... LL.B., F.R.G.S., author of 'Geography and history of the British colonies,' 'Lovell's general geography,' etc.
▸ MONTREAL; printed and published by John Lovell, St Nicholas Street; and sold by R. & A. Miller. Toronto: R. & A. Miller, 62 King Street East
▸ 20.5 cm; pp. [1-3], 4-22, [23], 24-34, [35], 36, [37], 38-80 [BoC]
▸ ***Libraries*** : MH : CaOOA, CaOTP
▸ ***Source*** BoC 9112 and CIHM
▸ ***Notes***
1. Special geography occupies pp. 23-80.
2. BoC 9112 notes: 'At the head of title: Lovell's series of school-books. Maps drawn by F.N. Boxer, A. & C.E. Montreal [sic], A.E. Graham, Toronto and T.C. Scoble, Toronto C.W. [sic] and engraved by Fisk & Russell, New York and Schonberg & Co. New York. Illus. signed by Hunter, Jackson and J.H. Walker.'

3. CIHM records an ed. in 1872, Montreal (: : CaOLU). There were further eds.: 1874, Montreal (L : : CaBVaU), and 1876, Montreal (: : CaBVaU).
4. Microform versions of (01) and the ed. of 1872 are available from CIHM.

[HODGINS, John George?] **Y Can**

- **1880 (01)**
- *Lovell's advanced geography for the use of schools and colleges*: with maps, illustrations, statistical tables
- MONTREAL; John Lovell & Son
- Pp. 148, [3]
- ***Libraries*** : : CaBVaU, CaOONL
- ***Notes***

1. That H. might be the author of this book, as he is of others published by Lovell, is speculation on my part.
2. Makes extensive use of tiny print. Provides much detail on Canada. Price: $1.50.
3. A microform version is available from CIHM.

[HODGINS, John George?] **Y Can**

- **1877 (01)**
- *Lovell's first steps in general geography*, [with maps and illustrations; being introductory to Lovell's 'Easy lessons in general geography']
- MONTREAL; John Lovell & Son
- Pp. [3], 4-50
- ***Libraries*** : : CAOTU
- ***Notes***

1. That H. might be the author of this book, as he is of others published by Lovell, is speculation on my part.
2. The text consists wholly of questions and answers. Special geography occupies pp. 17-50.
3. A microform version is available from CIHM.

HODGINS, John George **Y Can**

- **{1861} (01)**
- *Lovell's general geography, for the use of schools*, with numerous maps, illustrations and brief tabular views
- LONDON; Sampson Low, son & co., etc.
- ***Libraries*** : MH
- ***Source*** NUC (NH 0422614), which notes: 'Lovell's series of school-books'

- **1861 (02)**
- *Lovell's general geography ...*
- MONTREAL; printed and published by J. Lovell; Toronto: R. & A. Miller
- Pp. [3]-72
- ***Libraries*** : : **CaOLU**, CaOTP
- ***Source*** CIHM. See note to (03) and (04), below.

▸ *Notes*
1. Special geography of a highly condensed type occupies pp. 36-72. See also the note to (04), below. BoC 8925 records a variant ed. of the same year and publishers, but with a different collation.

▸ **{1861} (03)**
▸ *Lovell's general geography ... views*. By ... LL.B., author of 'Geography and history of the British colonies.'
▸ MONTREAL; printed and published by John Lovell, St. Nicholas Street; and sold by R. & A. Miller. Toronto: R. & A. Miller, 62 King Street East
▸ 28.5 cm; pp. 1-100
▸ ***Libraries*** : : CaOTP
▸ ***Source*** BoC 8925
▸ *Notes*
1. There is no obvious way of reconciling the information given in the sources with respect to the Canadian eds. of 1861.
2. ECB, Index (1856-76:130) records an ed. in 1862, Montreal, and there was another in 1866, Montreal, 100 pp. (: : CaNSWA).

▸ **1868, revised ... (see below) (04)**
▸ *Lovell's series of school-books*. Lovell's general geography, for the use of schools; with ... views. By ... LL.B., F.R.G.S. Revised edition: with entirely new maps and additional illustrations. Authorized by the Council of Public Instruction for Ontario
▸ MONTREAL; printed and published by John Lovell, St. Nicholas Street; and for sale at the bookstores
▸ 4to, 28 cm.; pp. [5], 6-104
▸ ***Libraries*** : **NN** : **CaBVaU**
▸ *Notes*
1. Intended for Canadian schools; explicitly reduces the volume of information found in British and American texts about their respective countries; in its place gives a disproportionate amount of information about the British colonies in North America. The Preface is dated 1867, presumably the date of revision. Information is provided in numbered paragraphs, with questions at the foot of the page, keyed to the numbers. The tone is utilitarian, the product of a commercial and administrative outlook.
2. BL and NUC records further eds. (all Montreal): 1869, 104 pp. (: NIC); 1874 (: : CaBVaU); 1875 (L); 1876 (: : CaBVaU).

▸ **1879 (05)**
▸ *Lovell's intermediate geography*, with maps and illustrations; being introductory to Lovell's Advanced Geography
▸ MONTREAL; John Lovell
▸ Pp. 104
▸ ***Libraries*** : : CaBVaU, CaOONL
▸ *Notes*
1. The change in title is treated as being simply that. Though a detailed comparison of the text was not made, the content is similar in tone to that of (04). Beyond that, the pagination is similar, perhaps identical, and the dates of publication proceed in sequence.

2. CIHM records an ed. in 1888. Microform versions of (02), (05), and the ed. of 1888 are available from CIHM.

HOLLAND, [John] Y

- **{1816}, 5th (01)**
- *A system of ancient and modern geography*, with a series of geographical examinations. The fifth edition, corrected and much improved
- Place/publisher n.r.
- ***Libraries*** n.r.
- ***Source*** Advertised in the 5th (i.e., 1816) ed. of GUY (1810), where the author is identified as Mr. Holland of Bolton. BL identifies John Holland (HOLLAND 1798) as being 'of Bolton.'

HOLLAND, John Y

- **1798 (01)**
- *A system of geography*, with a series of geographical examinations
- MANCHESTER; printed at the office of W. and T. Cowdroy, Market-Street-Lane
- 12mo, 18 cm.; pp. vii, [1], 66, [1]
- ***Libraries*** L
- ***Notes***

1. A rather slight special geography, being little more than list of places, with their locations, plus occasional supplementary information.
2. BL records a further ed.: 1802, Manchester, 65 pp.

HOLLAND, John ?

- **{1828}, 7th, corrected and much improved (01)**
- *A system of modern and ancient geography* ... Seventh edition ...
- LONDON; Baldwin & Cradock
- 12mo; pp. viii, 172
- ***Libraries*** L
- ***Notes***

1. Only an examination of the texts would show whether this work and HOLLAND (1816) are the same, or whether the reversal in the order of the adjectives in the title is a sign that they are different.

HOLMES, Jean/John **A Can tr**

- **{1836} (01)**
- *Abridgment of modern geography* ... of sacred geography ... part 1 ... first edition
- QUEBEC
- 191 pp.
- ***Libraries*** n.r.
- ***Source*** Hamelin (1960:349)
- ***Notes***

1. Born of Protestant stock in Vermont, Holmes migrated to Canada and

immersed himself wholeheartedly in the French-Catholic culture of Lower Canada, becoming both a priest and a successful academic. Hence the two versions of his forename.
2. Discussed in Savard (1962a, 1962b, 1962c), and Wade (1954).

HOLMES, John **Y**
- **{1751} (01)**
- *The grammarian's geography and astronomy ancient and modern*, exemplified in the use of the globes terraqueous and coelestial. To which all the terms of art, parts of the globe, and problems thereon to be performed, with the use of maps, are so plainly and methodically consider'd and treated of, as scarce ever to be forgot when once taught and shown by the diligent tutor. In two parts. Particularly adapted to the capacities of the young gentlemen studying the Classicks; as well as, useful and entertaining to all others, who, not having had opportunity of acquainting themselves with mathematical calculations, are yet desirous of some knowledge of the earth and heavens. The geographical part comprehending the ancient and modern names, situation, government, religion, bounds, dimensions, length, and breadth of most places in the world; including land and water, namely continents, islands, peninsulas, isthmus's [sic], promontories, capes, coasts, mountains, with oceans, seas, lakes, straits, gulfs, rivers, and countries, kingdoms, cities and towns; with the latitude and longitude of the most principal parts, and their bearing and distances from London; in view of the Latin and Greek Classicks, Homer, Virgil, Herodotus, Justin, Xenophon, Caesar, Plutarch, Livy, Thucydides, Sallust, Dionysius Periegetes, Pausanias, Josephus, Eusebius, Silius Italicus, Lucian, Florus, Nepos, Eutropius, Quintus Curtius, and the rest: with the adventures, voyages, and travels of Ulysses, Aeneas, Nebuchadnezzar, Cyrus, Alexander the Great, Hannibal, Julius Caesar, our Saviour Jesus Christ, St. Paul, the rest of the apostles, and many others in both sacred and profane history. The astronomical part containing a description of the laws, order, number, names, distances, magnitude, motions, and appearances of the heavenly bodies, sun, moon, and planets, with the problems belonging thereto; an account of the several systems of the universe, and a defence of the true solar one; the history of the rise, progress and present perfection of astronomy; the Classical stories relating to the planets signs and constellations delineated on the celestial globe. With a dramatic epilogue called Coelum Reformatum. The whole illustrated with necessary maps and schemes neatly engraved on copper. By ... master of the publick grammar school, in Holt, Norfolk
- LONDON; printed for W. Strahan; and sold by C. Hitch and L. Hawes, in Pater-noster-row
- 8vo, 18 cm.; pp. [22], 276, [30]
- ***Libraries*** **L** : CSmH, KyU, NN
- *Notes*

1. Unlike most other authors of special geography, Holmes seems to be as much at home with astronomy and the Classics as he is with geography. The Dedication is written in Latin, suggesting fluency in the use of that language. As a teacher, he is interested in the problem of the great quantity of information

that must be learned by heart. To help cope with this problem he resorts to mnemonics that involve cards, taking advantage of their four suits and thirteen values; he also uses versification, which may be his own. He provides 'annotations' to the text at the foot of virtually every page. It is a discursive book written by a lively mind, but even he cannot avoid passages that are prose in form but lists in substance.

HOPWOOD, Henry **Y SG?**
- **{1845} (01)**
- *An introduction to the study of modern geography.* With a chapter on the geography of the Christian church
- CAMBRIDGE, Rugeley, London
- 12mo
- ***Libraries*** L
- ***Source*** BL, where, from cross references, it can be learned that this book is Vol. 14 of *The Juvenile Englishman's Library*, 21 vols., 1844-9 ed. successively by F.E. Paget and J.F. Russell
- ***Notes***

1. BCB, Index (1837-57:110), records an ed. in 1846, published by Masters.

HOPWOOD, Henry **Y SG?**
- **{1856}, 3rd (01)**
- *School geography*, with a chapter on the ecclesiastical geography of the British empire ... 3rd edition
- LONDON
- 18mo
- ***Libraries*** L

HORN, Joseph Stephenson **Y SG?**
- **{[1870]}a (01)**
- *H[orn]'s geography*
- [LONDON]
- 8vo
- ***Libraries*** L
- ***Source*** BL, which notes: 'Title from wrapper'
- ***Notes***

1. BL records further eds.: 1870, [Manchester], 48 pp.; and also 1898, 58th and 59th; 1928, 74th; and 1932, 75th.

HORN, Joseph Stephenson **Y**
- **{[1875]} (01)**
- *J. Horn's new codes series. Grammar and geography.* Standard 2-6
- MANCHESTER; [John Heywood?]
- 16mo
- ***Libraries*** L

▸ ***Notes***
1. According to Lochhead (1980:108) only Standard VI would be special geography.

HORN, Joseph Stephenson **Y SG?**
▸ **{[1869]} (01)**
▸ *The scholar's geography, etc.*
▸ MANCHESTER
▸ 16mo, 2 pts.
▸ ***Libraries*** L
▸ ***Source*** BL, which notes: 'Pt 2 has a separate titlepage'
▸ ***Notes***
1. ECB, Index (1856-76:131) records an ed. of Pt. 2 in 1873, [Manchester]. BL records further eds.: 1893, 11th; 1894, 12th; and 1898, 13th.

HORN, Joseph Stephenson **Y SG?**
▸ **{[1870]}b (01)**
▸ *Horn's twopenny geography*; containing a series of lessons on the geography of the British empire and the five continents
▸ MANCHESTER
▸ 12mo
▸ ***Libraries*** L
▸ ***Source*** BL, which notes: 'Title from wrapper'
▸ ***Notes***
1. See HORNIUS (1786*).

HORNIUS [HORN] **?**
▸ **{1786}* (01)**
▸ *Hornius's geography*
▸ LONDON
▸ 4to
▸ ***Libraries*** n.r.
▸ ***Source*** General Catalogue (1786)

HORT, William Jillard **?**
▸ **{1816} (01)**
▸ *The new geography*[; or an introduction to modern geography ... containing the arrangements concluded by the Congress at Vienna]
▸ Place/publisher n.r.
▸ ***Libraries*** n.r.
▸ ***Source*** Peddie and Waddington (1914:227)
▸ ***Notes***
1. The words in the title set in [ ] are taken from the BL entry for the ed. of 1827.
2. BL records an ed.: 1827, Bristol.

HUBBARD, John **Y Am**

- **{1803} (01)**
- *The rudiments of geography*, being a concise description of the various kingdoms, states, empires, countries, and islands in the world. Together with their latitudes, longitudes, extent, [etc.] ... With an introduction explaining the astronomical part of geography. To which is added, a chronological table of the most important events which have happened from the creation of the world to the present day
- WALPOLE, NH; Thomas & Thomas
- 17.5 cm.; 2 p. l., [9]-232 p.
- ***Libraries*** : MH, NNC
- ***Source*** NUC (NH 0573710)
- ***Notes***

1. NUC records further eds. (all Walpole unless otherwise noted): 1805, 2nd, 240 pp. (MB, NN, NNC, NSyU, PPM); 1807, 3rd, 240 pp. (DLC, OO); 1808, 4th, also Troy, NY, 240 pp. (CSmH, DLC, MWA, NBuG, PU); 1811, 5th, 16mo, 240 pp. (MiU, Nh, NN) [NH 0573715 notes: 'Last page wrongly numbered 402']; 1814, Barnard, VT, 219 pp. (AU, CSt, DLC, MeB, MH, MiU, MWA, TxU, ViU).

- **1814, 6th, revised and corrected (02)**
- *The rudiments of geography*: being ... world: together ... extent, boundaries, rivers, lakes, air, climate, soil, produce, manufactures, chief towns, population, religion, and learning. With an introduction ... added, an enlarged chronological ... Sixth edition ...
- BARNARD, VT; printed by J.H. Carpenter for J. Dix
- 12mo.; 240 pp
- ***Libraries*** : CSt, CtW, DNLM, ICU, InU, IU, MiU, MtHI, NcD, NjP, NN, NNC, NSyU, TxU, VtMiS, VtU
- ***Notes***

1. The additional wording in the title was supplied by VtMiS.
2. Cited by Elson (1964:394).

HÜBNER, Johann **Y tr**

- **1738 (01)**
- *An introduction to geography, by way of question and answer.* Particularly design'd for the use of schools: giving a general description of all remarkable countries in the world; of their situation, extent, division, cities, rivers, soil, commodities, rarities, archbishopricks, bishopricks, universities, manners, government, religion, etc. To which is prefix'd, an explanation of the sphere, or of all the terms any ways necessary for the right understanding of the terraqueous globe. With the addition of a short dictionary of the most common names of antient geography. Together with an index of the principal places, rivers, etc. mentioned in this book. Translated and improved from the last edition of Mr. Hubner's Geography, written originally in German
- LONDON; printed for T. Cox, at the Lamb, under the Royal Exchange
- 12mo, 16 cm.; pp. [6], 283 (168-69 entered twice), [25]
- ***Libraries*** O

▸ *Notes*

1. It is possible that the original version of this work [*Kürze Fragen aus der alten und neuen Geographie*], published in 1696, introduced to geography the technique of setting out information in the form of question and answer. It seems reasonable to assume that the use of this form must have reinforced strongly the tendency of teachers to require their pupils to learn so-called facts by heart.

▸ **1742, 2nd (02)**

▸ *A new and easy introduction to the study of geography, by way of question and answer.* Principally design'd for the use of schools: in two parts. Containing I. An explication of the sphere; or of all such terms as are any ways requisite for the right understanding of the terraqueous globe. II. A general description of all the most remarkable countries throughout the world; of their respective situations, extents, divisions, cities, rivers, soils, commodities, curiosities, archbishopricks, bishopricks, universities, customs, forms of government, and religion, etc. To which is added a compleat set of maps. Likewise a compendious dictionary of the most common names of ancient geography, explain'd by those which they now bear; as also, an alphabetical index of the principal places that are mentioned throughout the work. Written originally in High Dutch by the late celebrated Mr Hubner, and now faithfully translated with additions and improvements. The second edition, carefully revis'd and corrected by J. Cowley, geographer to his Majesty

▸ LONDON; printed for T. Cox, at the Lamb, under the Royal-Exchange; and J. Hodges, at the Looking-Glass, over against S. Magnus Church, London-Bridge

▸ 12mo, 17 cm.; pp. vi, 271, [23]

▸ *Libraries* BrP, **OSg** : DLC

▸ *Notes*

1. BL and NUC record further eds.: 1746, 3rd, 271, pp. (L : MH, RPJCB); 1753, 4th, London, 271 pp. (: ICU, ViU); 1777, 7th, London, 270 pp. (**L** : CtY) [pagination suggests few changes since 1742, and spot-checking revealed none].
2. Discussed in Bowen (1981:153-4).

HÜBNER, Johann **Gt**

▸ **{1746}, 3rd, revised ... (see below)**

▸ *Introduction to the study of geography, by way of question and answer.* In schools. [With] a compleat set of maps. Likewise, a compendious dictionary. Written originally in High Dutch. 3rd ed., revised and corrected by J. Cowley

▸ LONDON; printed for T. Cox, etc.

▸ 12mo

▸ *Libraries* : MH, RPJCB

▸ *Notes*

1. The change in title from HÜBNER (1738) is slight. In addition, RPJCB states that an examination of the book shows it to be the 1746 ed. of HÜBNER (1738). It is thus reasonable to think that this book (i.e., HÜBNER 1746) is a ghost.

HUGHES, Edward **Y SG?**
- **{1851} (01)**
- *Geography for elementary schools*. First course
- LONDON
- 12mo
- ***Libraries*** L
- ***Notes***

1. BL records an ed. in 1873, London.

HUGHES, Edward
- ***Notes***

1. H. was responsible for the 1851 ed. of PHILLIPS, R. (1803), described as being 'revised, corrected and greatly enlarged' by him.

HUGHES, Samuel, and Jas. E. DENNIS, compilers **Y Can**
- **c1883 (01)**
- *A primer of map geography*, for pupils preparing for promotion examinations[,] pupils preparing for entrance examinations[,] pupils preparing for intermediate examinations[,] students preparing for teachers' certificates[,] and all official examinations. With recent departmental examination papers from the Provinces of Ontario, Manitoba, and Nova Scotia. Compiled by ... First English Master, Toronto Collegiate Institute, and ... Head Master, Woodstock Model School
- TORONTO and Winnipeg; W.J. Gage and company
- 19 cm
- ***Libraries*** : : **CaBVaU**
- ***Notes***

1. Information is provided largely in the form of tables 'to prevent the waste of time in poring over a prosy text-book' (Preface).

HUGHES, William **Y SG?**
- **{1850} (01)**
- *A child's first book of geography* [arranged in a series of easy reading lessons]
- LONDON
- 18mo
- ***Libraries*** L
- ***Notes***

1. Identified in BL as one of Gleig's School Series. The Gleig in question was George Robert (1796-1888).
2. BCB, Index (1837-57:110) records an ed. in 1854, London (Vaughan 1985).

HUGHES, William **Y**
- **{1859} (01)**
- *A class book of modern geography*; with examination questions
- LONDON, Liverpool (printed)

- 8vo
- *Libraries* L
- *Notes*

1. Discussed in Vaughan (1985). See note to (02).

- **1865, 4th (02)**
- *A class-book of modern geography*, with examination questions. By ... F.R.G.S., author of 'A Manual of Geography,' 'A Class-Book of Physical Geography,' etc.
- LONDON; George Philip and son, 32, Fleet Street: Liverpool: Caxton Buildings, South John Street, and 51, South Castle Street
- 16.5 cm.; pp. [iii], iv-vii, [2], 2-362, [1], 2-16
- *Libraries* : : **CaBVaU**
- *Notes*

1. The information provided about places is primarily that which can be obtained from maps of small, or, in the case of the United Kingdom, intermediate scale, supplemented with information about climate, terrain, economic activity, and size of population, plus a miscellany of historical and ethnographic facts whose content varies from country to country.

   The current methodology of both geography and pedagogy is succinctly summarized as follows: 'The "Examination questions" appended to each section ... are mainly designed for oral use; and have been framed under the conviction that frequent oral examination – pursued with constant reference to maps ... is the soundest test of a learner's advancement in geographical knowledge. But the greater number of them are equally adapted for the purpose of written examination. The use which the author ... recommend[s] ... of these questions [is]: – that after *each* lesson gone through by the teacher and professedly learned by the pupils, the real amount of knowledge gained by the learner should be tested by oral examination ... In addition, the author would, at not unfrequent intervals, apply the test of written examination. The added labour imposed upon the teacher ... will be amply recompensed by the certain test of progress thus allowed' (pp. [iii]-iv).

   The final series of numbered pages carry advertisements for atlases and maps prepared by W.H. and published by George Philip and son.
2. BL and NUC record further eds.: 1873, 8th, London, 362 pp. (: CU); 1881, rev. by J.F. Williams, London, Liverpool (printed), 380 pp. (L); 1885, London, 403 pp. (L); 1887, London, 403 pp. (L); and 1892, 1893, and 1898.

HUGHES, William **Y**

- **{1860} (01)**
- Geography, elementary class book [*Elementary class-book of modern geography*]
- [LONDON]; Philip
- *Libraries* n.r.
- *Source* ECB, Index (1856-76:130)
- *Notes*

1. The words in the title set in [ ] are taken from BL entries for later eds.
2. BL and NUC record further eds.: 1861, London, Liverpool (printed), 3 pt.

(L); 1882, new, revised by J.F. William, London, Liverpool (printed), 182 pp. (L); 1885, London (: MH). BL also records eds. in 1888, 1892, 1907, 1921, and 1926. NUC also records eds. in 1903 and 1910.

HUGHES, William, ed. **mSG**

- **{1858} (01)**
- *The family atlas*. Philips' family atlas of physical, general, and classical geography
- LONDON; George Philip & son
- Imp. 4to
- ***Libraries*** n.r.
- ***Notes***

1. 'A series of 52 maps ... accompanied by illustrative letterpress, describing the natural features, climate, productions and political divisions of each country, with its statistics, brought down to the latest period, and a copious consulting index' (Advertisement at the back of (02) of HUGHES 1859). There was at least one subsequent ed.

HUGHES, W[illiam] **Y SG?**

- **{1851a} (01)**
- *Geography for beginners*
- LONDON; Longman
- 18mo
- ***Libraries*** n.r.
- ***Source*** BCB, Index (1837-57:110)

HUGHES, William **OG**

- **{1870} (01)**
- *Geography: what it is, and how to teach it*, etc.
- LONDON
- 8vo
- ***Libraries*** L
- ***Notes***

1. Discussed in Vaughan (1985).

HUGHES, William **Y SG?**

- **{1886} (01)**
- *The intermediate class-book of modern geography* ... Abridged from Professor Hughes class-book of modern geography, by J.F. Williams
- LONDON; G. Philip & Son
- 8vo; pp. xii, 244
- ***Libraries*** L

HUGHES, William **Y**

- **1851b (01)**

- *A manual of European geography*: embracing the physical, industrial, and descriptive geography of the various countries of Europe. For the use of schools and colleges
- EDINBURGH; Adam and Charles Black. London; W. Hughes, Aldine Chambers, Paternoster Row; and all booksellers
- 17 cm.; pp. [1], ii-xi, 339
- ***Libraries*** [IDT]
- ***Source*** Based on a course of lectures given by Hughes, apparently with some success (p. xi). Intended to be read rather than consulted, though it does have tabular material. Subsequently bound and issued as the first part of HUGHES (1852a).
- ***Notes***

1. Discussed in Vaughan (1985).

HUGHES, William **Y n.c.**

- **1852a (01)**
- *A manual of geography*. Part II. Containing the geography of Asia, Africa, America, Australasia, and Polynesia
- LONDON; Longman, Brown, Green, & Longman
- 17 cm.; pp. [2], 321-661
- ***Libraries*** [IDT]

HUGHES, William **Y n.c.**

- **1852b (01)**
- *A manual of geography, physical, industrial, and political*
- LONDON; Longman, Brown, Green, & Longman
- 17 cm.; pp. [3], iv-xvi
- ***Libraries*** [IDT]
- ***Source*** Bound at the end of HUGHES (1852a)

- **{1855, 1856}, new (02)**
- *A manual of geography, physical, industrial, and political ...* New ed.
- LONDON
- 8vo, 2 pts.
- ***Libraries*** L
- ***Notes***

1. BL records further eds.: 1861, 1877, 1878, 1880, and 1889.

HUGHES, William **A**

- **1868, new ... (see below) (01)**
- *The modern atlas of the earth*, with an introduction to physical and historical geography, and an alphabetical index of the latitude and longitude of 70,000 places. New and revised edition, by ... F.R.G.S., Professor of Geography at King's College, London
- LONDON; Frederick Warne and co., Bedford Street, Covent Garden. New York: Scribner, Welford, and co.
- 36.5 cm.; pp. [8], [i]-xvi, [3]-232, 56

- ▸ *Libraries* : **MB**
- ▸ *Notes*

1. Contains abundant text, which flows smoothly, without headings. Deals primarily with elements of physical geography, especially terrain and climate; references to vegetation are couched in general terms. Varying amounts of information about the people of countries, and about the activities and customs are included. There is sometimes even a little history.

HUGHES, William **A**

- ▸ **1856 (01)**
- ▸ *The treasury of geography*, physical, historical, descriptive and political; containing a succinct account of every country in the world: preceded by an introductory outline of the history of geography; a familiar inquiry into the varieties of race and language exhibited by different nations; and a view of the relations of geography to astronomy and the physical sciences. Designed and commenced by the late Samuel Maunder; continued and completed by ... F.R.G.S., late Professor of Geography in the College for Civil Engineers; author of a 'Manual of Geography,' etc.
- ▸ LONDON; Longman, Brown, Green, Longmans, & Roberts
- ▸ 12mo, 16.5 cm.; pp. [v], vi-x, 924
- ▸ *Libraries* L : **MB**, DCL, ICRL, NjR, PPiU, PPL, PV
- ▸ *Notes*

1. 'The introductory portion of the volume, as far as p. 20, is from Mr. Maunder's own pen. The account of Asiatic Turkey (pp. 400-465) embodies material which had been collected by Mr. Maunder ... For all the remainder ... Mr. Hughes alone is responsible' (p. [v]). Describes countries with the use of numbered sections, e.g., I. extent, natural features, climate, and productions; II. topography, which includes the description of some cities; III. political geography, which includes economic activity. The details of organization are not consistent throughout the book.

   Sources are occasionally cited in footnotes. There are occasional extensive quotations from other authors. The index is unusually long, occupying pp. 877-924.
2. BL and NUC record further eds.: 1860, 924 pp. (L : DLC, N); 1866, 924 pp. (L : PV); 1869, 924 pp. (L : I); 1872, 892 pp. (L : DLC); 1875, 892 pp. (L : INS); and 1878, 892 pp. (L : NN).

HUNTINGTON, Nathaniel Gilbert **Y SG? Am**

- ▸ **{1836} (01)**
- ▸ *Introduction to modern geography for beginners and common schools* preparatory to the use of his and other larger works upon the subject. Accompanied by a new and corresponding atlas
- ▸ HARTFORD; Reed & Barber
- ▸ Pp. 166
- ▸ *Libraries* : OClWHi, OO, PPiU
- ▸ *Source* NUC (NH 0627535)
- ▸ *Notes*

1. NUC records a further ed.: 1838, 166 pp. (AU, CtY, MH).

HUNTINGTON, Nathaniel Gilbert **Y Am**

- **{1833} (01)**
- *A system of modern geography* ... designed to answer the two-fold purpose of a correct guide to the student, and of a geographical reading book ... illustrated by a variety of cuts and tables, and an atlas
- HARTFORD; E. Huntington & co.
- 16 cm.; pp. viii, [9]-304
- ***Libraries*** : DLC
- ***Notes***

1. BL and NUC record further eds.: 1834, Hartford, 304 pp. (: MH, NN); 1835, Hartford, 306 pp. (L : CoU, DHEW, DLC, IU, KyHi, MiU, NN) [according to Sahli (1941:382) this is the 2nd ed.]; 1836, Hartford, 306 pp. (: DLC, OkU, PPL); and 1836? Raleigh, NC, 306 pp. (: DHEW, DLC, NcD, NjR, NN, OClWHi, PPL).
2. Discussed in Nietz (1961:196-7, 200, 214-15, 217) [N.B., it is a presumption, based on the date, that it is this work of Huntington's to which N. is referring], and Sahli (1941, *passim*); cited by Elson (1964:400).

HUSSEY, G. **mSG**

- **1670 (01)**
- *Memorabilia mundi: or, choice memories, of the history and description of the world*
- LONDON; printed for the author, and are to be sold by F. Smith at the Elephant Castle without Temple-Bar
- 12mo, 14 cm.; pp. [16] 185, [5]
- ***Libraries*** **L**, O
- ***Notes***

1. Close to special geography in spirit, as it describes all continents and many countries, but without much order. Much space is given to England, where several individual counties are described.

HUTCHINSON, J[ames] **Y SG?**

- **{1849} (01)**
- *Geography, easy lessons*
- ? Wright
- 18mo
- ***Libraries*** n.r.
- ***Source*** BCB, Index (1837-57:110)
- ***Notes***

1. NUC (NH 0638439) records a further ed.: 1874, 79th, London, 110 pp. (NN), and identifies this Hutchinson as James.

I

IMPERIAL CYCLOPAEDIA **Dy n.c.**

- **{[1853]} (01)**
- *The imperial cyclopaedia.* Literary division. The cyclopaedia of geography, history, etc. Part 1. A-Agent [no more published]
- LONDON
- 8vo
- ***Libraries*** L

INTRODUCTION TO GEOGRAPHY ... (1735) **Y mSG?**

- **{1735}, 2nd (01)**
- *An introduction to geography and astronomy*: containing chiefly, the description and use of the terrestrial and celestial globes. With many new and useful particulars not yet made publick in any other treatise on this subject. For the benefit of learners. The 2d. ed.
- LONDON; printed for C. Price and sold by J. Atkinson etc.
- 15.5 cm.; pp. vii, 210 [5]
- ***Libraries*** : NNC, NPV
- ***Source*** NUC (NI 0141933)

INTRODUCTION TO GEOGRAPHY ... (1768) **Y SG? els**

- **1768 (01)**
- *An introduction to geography*; to which is added a short explanation of the use of the artificial terrestrial globe and of the astronomical system. For the use of schools. With permission of superiors
- BRUGES; Joseph van Praet
- 8vo, 18 cm.; pp. 108
- ***Libraries*** L, O
- ***Notes***

1. The text consists of questions and answers suitable for school children. The special geography is placed before the general.

INTRODUCTION TO GEOGRAPHY ... (17–?) **Y**

- **{[17–]} (01)**
- *An introduction to geography*, for the use of Mrs. Davis's little society
- [LONDON]
- 13.5 cm.; pp. 186
- ***Libraries*** : CLU, ICU
- ***Source*** NUC (NI 0141934)

INTRODUCTION TO GEOGRAPHY ... (1802) **?**

- **{1802} (01)**

- *Introduction to geography*
- Place/publisher n.r.
- ***Libraries*** n.r.
- ***Source*** Peddie and Waddington (1914:226)
- ***Notes***

1. May be a subsequent edition of the preceding.

IRVING, Christopher **Y**

- **{[1820]} (01)**
- *A catechism of general geography*: containing the situation, extent, mountains, lakes, rivers, religion, government, etc. of every country in the world ...
- LONDON; Romsey
- 12mo; 14 cm.; pp. iv, [5]-80
- ***Libraries*** L : CtY
- ***Source*** BL and NUC (NI 0159114)
- ***Notes***

1. The title shows that the meaning given to the terms special and general geography by Varen are no longer current.
2. BL records 12th ed. in 1867, rewritten and arranged by J.P. Bidlake, etc.

J

J., G. **A**

- **1718 (01)**
- *Geography epitomiz'd: or, the London gazetteer*. Being a geographical and historical treatise of Europe, Asia, Africa, and America. Wherein the several empires, kingdoms, principalities, states, provinces, islands, counties, bishopricks, and chief cities of the whole world are particularly describ'd; their extents, situations, etc. Together with a concise account of the inhabitants, their behaviour, manners, politicks, religion, etc. The rivers of note, productions of soil, rarities of nature, and riches of the respective countries: likewise the crown-revenues, ways of government, forces, antiquity, etc. of every state: with a particular description of King George's dominions in Germany. To which are added, an introduction to geography, and knowledge of the globe, and three tables. The first of distances from London, to the most considerable cities and market-towns in England and Wales, answering to a map of roads; the second of all the cities and towns in Great-Britain; and the third of foreign towns
- LONDON; printed for Charles Rivington, at the Bible and Crown in St. Paul's church-yard; Jer. Batley and Tho. Warner in Pater-Noster-Row; and J. Sackford in Lincolns Inn Square
- 8vo, 16 cm.; pp. [12], 239, [23]
- ***Libraries*** L : CLU, NN
- ***Notes***

1. Gives locations of countries and lists their areal subunits; names the chief products and identifies the character of the people or some fact known about

the country. Reminiscent of ABBOT (1599) and HEYLYN (1621). England and Wales, which are dealt with county by county, occupy pp. 19-71.
2. Preface signed by G.J. NUC lists under *Geography epitomiz'd* (NG 0127324).

JAMES, Thomas **Y SG?**
- **{1854}* (01)**
- *Compendium of geography for Rugby school*
- Place/publisher n.r.
- ***Libraries*** n.r.
- ***Source*** Allibone (1854)

JAMIESON, Alexander **Y**
- **{[1820?]} (01)**
- *A grammar of universal geography*, and of elementary astronomy, for the use of schools and private instruction
- LONDON?
- ***Source*** St. John (1975, 1:184)
- ***Notes***
1. St. John (1975, 1:184) records the 2nd ed., noting that the Preface is dated 1820.

- **{1823}, 2nd (02)**
- *A grammar of universal geography* ... instruction. With a new set of maps
- LONDON; printed for N. Hales
- 14.5 cm.; pp. viii, 252
- ***Libraries*** : : CaOTP
- ***Source*** St. John (1975, 1:184)

JAMIESON, Alexander **?**
- **{1831} (01)**
- *Modern Geography*
- Place/publisher n.r.
- ***Libraries*** n.r.
- ***Source*** Peddie and Waddington (1914:227)

JAUFFRET, Louis François **Y tr**
- **{1804} (01)**
- *The travels of Rolando*; containing, in a supposed tour round the world, authentic descriptions of the geography, natural history, manners and antiquities of various countries. Translated from the French of L.J. Jauffret ... [by Lucy Aikin]
- LONDON; printed for Richard Phillips, by J. Taylor, and sold by Tabart and co.; and all booksellers
- 15 cm., 4 vols.
- ***Libraries*** : : CaOTP

- ▸ ***Source*** St. John (1975, 1:185), which provides the name of the translator
- ▸ ***Notes***

1. NUC records further eds.: 1813, 3rd (MH); 1823, 2 vols. (IEN, MMeT).
2. For notes on the book, see the note to (02), below.

- ▸ **1853, newly corrected ... (see below) (02)**
- ▸ *Travels of Rolando; or, a tour round the world.* Translated by Miss Aikin. Newly corrected and revised by Cecil Hartley, A.M., author of the 'Circle of the Sciences,' etc. Illustrated by William Harvey
- ▸ LONDON; George Routledge & Co., Farringdon Street
- ▸ 17 cm.; pp. [3], iv-xii, 506 [2]
- ▸ ***Libraries*** : **MH**
- ▸ ***Notes***

1. Intended for youth, to teach them knowledge of countries by means of a story about the fictional travels of a young man. These take him through most of Africa and southern Asia. In an effort to provide information about all the countries of these regions, including those which the protagonist did not visit, space is made for digressions about such places.
2. NUC records a further ed.: 1854, 4th, 506 pp. (MH, ViU).

JEFFERYS **Gt?**

- ▸ **{1786}* (01)**
- ▸ *Geography improved*
- ▸ LONDON
- ▸ Folio
- ▸ ***Libraries*** n.r.
- ▸ ***Source*** General Catalogue (1786). Possibly a ghost; see JEFFERYS (1767).

JEFFERYS, Thomas **Y**

- ▸ **{1767} (01)**
- ▸ *The study of geography improved*; designed for the use of schools, as well as private tutors: being a more certain and expeditious method of conveying the knowledge of that science, and fixing it in the memory of young persons, than any hitherto made public
- ▸ LONDON
- ▸ Folio
- ▸ ***Libraries*** n.r.
- ▸ ***Source*** Watt (1824, 2) where listed under Jefferies

JOHNS, B[ennett] G[eorge] **Y SG?**

- ▸ **{1864}, new (01)**
- ▸ *Geography, elements.* n[ew] e[d.]
- ▸ Lockw. (Darton's School Library)
- ▸ 18mo
- ▸ ***Libraries*** n.r.

- ***Source*** ECB, Index (1856-76:130)
- ***Notes***

1. BL records further eds.: 1881, London, 186 pp.; and 1892.

JOHNS, Charles Alexander **Y SG?**

- **{1872} (01)**
- *The child's first book of geography*
- LONDON, [Oxford, printed]; Bell
- 16mo.
- ***Libraries*** L
- ***Source*** BL and ECB, Index (1856-76:130)

JOHNSON, R. **Dy**

- **1776 (01)**
- *The new gazetteer: or geographical companion.* Containing a general and concise account, alphabetically arranged, of all the empires, kingdoms, states, provinces, cities, towns, seas, harbours, bays, rivers, lakes, and mountains, in the known world: and more particularly in Great-Britain, Ireland, and America. The whole intended as a useful pocket vade-mecum for the readers of public news-papers, and for young students in geography
- LONDON; printed for Edward and Charles Dilly in the Poultry, and R. Baldwin in Pater-noster Row
- 11 cm.; pp. [144]
- ***Libraries*** : **MH**

JOHNSON, W.R. **Y mSG?**

- **{1809} (01)**
- *Goldsmith's grammar of geography, rendered into easy verse.* Describing the situation, manners, and produce of all nations. For the use of young persons
- LONDON
- Pp. 208
- ***Libraries*** : IU

JOHNSTON, Alexander Keith **Dy**

- **{1851} (01)**
- *Dictionary of geography,* descriptive, physical, statistical, and historical, forming a complete general gazetteer of the world
- LONDON; Longman, Brown, Green, and Longman
- 22 cm.; 1 p. l., [v]-viii, 1432 p.
- ***Libraries*** : CtY, DLC, MB, OClWHi, ViU
- ***Source*** NUC (NJ 0132806)
- ***Notes***

1. NUC records further eds.: 1852, 1432 pp. (MB, NIC, NN, PPT); and 1862, 1352 pp. (OrU : CaBVaU).

JOHNSTON, Alexander Keith **Dy**

- **{1877}, new, thoroughly rev. (01)**
- *A general dictionary of geography*, descriptive, physical, statistical, historical; forming a complete gazetteer ... New ed., thoroughly rev.
- LONDON; Longmans, Green, and co.
- 24.5 cm.; 4 p. l., 1513 p.
- ***Libraries*** : DLC, DN, DN-Ob, MB, MH
- ***Source*** NUC (NJ 0132810)
- ***Notes***

1. The addition of *general* to the title may indicate that this is a separate work from JOHNSTON (1851), rather than simply a new ed.
2. NUC records a further ed.: 1882, 1513 pp. (MH, NIC).

JOHNSTON, Alexander Keith **Y**

- **1886, revised ... (see below) (01)**
- *An intermediate physical and descriptive geography* abridged from the physical, historical, and descriptive geography by the late ... Revised and corrected to date. For the use of schools
- LONDON; Edward Stanford, 55, Charing Cross, S.W.
- 8vo, 19 cm.; pp. xi, [1], 283
- ***Libraries*** **L**
- ***Notes***

1. Opens with 40 pp. of systematic geography, summarizing the physical and human elements; the balance is special geography.

JOHNSTON, Alexander Keith **A**

- **{1880} (01)**
- ... *A physical, historical, political, and descriptive geography*
- LONDON; E. Stanford
- 20.5 cm.; xiii p., 1 l., 487 p.
- ***Libraries*** : DLC, MB, MChB, OU
- ***Source*** NUC (NJ 0132925), which notes: '(The London geographical series).' See note to (03), below.

- **{1881}, 2nd (02)**
- *A physical ... and descriptive geography*
- Place/publisher n.r.
- ***Libraries*** : DI-GS

- **1885, 3rd, revised ... (See below) (03)**
- *A physical ... and descriptive geography*. Revised by E.g., Ravenstein
- LONDON; Edward Stanford, 55, Charing Cross, S.W.
- 8vo, 20 cm.; pp. xi, [2], 490, [6]
- ***Libraries*** **L** : CtY
- ***Notes***

1. Part of the London geographical series. 154 pp. of systematic physical and human geography; the balance is special. Europe and Asia have ca. 100 pp. devoted to each; Africa and America, ca. 50 pp. each, and Australia just

over 40 pp. This is, presumably, the source of from which JOHNSTON (1886*) and (1881*) were derived.
2. NUC records further eds.: 1890, 4th; 1896, 5th; 1908.

JOHNSTON, Alexander Keith **Y**
- **{1881} (01)**
- ... *A school physical and descriptive geography*
- LONDON; E. Stanford
- Pp. 407
- ***Libraries*** : MiD
- ***Notes***
1. BL and NUC record a further ed.: 1884, 3rd, 407 pp. (L : DLC, NBuG, NN).
2. One of the London geographical series. More detailed than JOHNSTON (1886), above.

JOHNSTON, Robert **Y SG?**
- **{1877} (01)**
- *The competitive elementary geography,* etc.
- LONDON
- 12mo.
- ***Libraries*** L
- ***Source*** BL, which records 2nd ed. in 1879

JOHNSTON, Robert **Y SG?**
- **{1872} (01)**
- *The competitive geography*
- LONDON; Longman, Green and Co.
- 8vo; pp. iv, 512
- ***Libraries*** L : MiD
- ***Source*** BL and NUC (NJ 0135464)
- ***Notes***
1. BL and NUC record further eds.: 1874, 2nd, 520 pp. (L); 1877, 3rd, 523 pp. (L); 1879, 4th, 521 pp. (L); 1880, 5th, recommended by the commissioners of intermediate education in Ireland, 521 pp. (L : NNU-W); and 1894 (L).
2. NUC (NJ 0135465) notes: '(Johnston's civil service series).'

JOHONNOT, James, ed. **Y Am**
- **{1882) (01)**
- *A geographical reader*. Comp. and arranged by ...
- NEW YORK; D. Appleton and company
- 19 cm.; xiv, 418 p.
- ***Libraries*** : DLC, MB, MiU, OCU, OCX, OrP
- ***Source*** NUC (NJ 0137312)

- ***Notes***

1. NUC records further eds. (all New York, D. Appleton): 1883 (MH); 1884 (VIU); 1885 (OClWHi); 1887 (MH); and also 1890.

JONAS, Evan **Gt?**

- **{1811}* (01)**
- *Geographical Grammar*
- LONDON
- 8vo, 2 vols.
- ***Source*** London Catalogue (1811). Probably a ghost: see JONES, Evan (1775)*.

JONES, Edward **Gt**

- **{1773}**
- *Young geographer & astronomer's best companion*
- LONDON
- 12mo
- ***Source*** Allibone (1854), which also records '2nd ed., 1792.' Probably a ghost; see JONES, Evan (1773).

JONES, Evan **A**

- **{1775}* (01)**
- *A new and universal geographical grammar*: or, a complete system of geography ... A new edition, corrected. In two volumes
- LONDON; printed for G. Robinson and S. Bladon
- ***Libraries*** SwnU (Vol. I only)
- ***Notes***

1. SwnU reports that the author describes himself, on the tp., as 'Teacher of classics and geography, at Bromley in Kent.' Explicitly intended for an adult audience.

- **{1786} (02)**
- *Geographical grammar*
- LONDON
- 8vo, 2 vols.
- ***Libraries*** n.r.
- ***Source*** General Catalogue (1786)

JONES, Evan **Y**

- **1773 (01)**
- *The young geographer and astronomer's best companion*. Containing I. The elements of modern geography, in which, besides many other useful articles, the latitude and longitude of a great variety of places are given from the latest observations. II. A comprehensive system of ancient geography, both sacred and profane, particularly adapted to the illustration of the Classic authors, and of the historical parts of the Bible. III. The description and use of the

celestial and terrestrial globes, in which particular attention has been given to the regular disposition, and most convenient solution of a numerous collection of problems, which are succeeded by several ingenious and entertaining paradoxes for the exercise of the learner. - Also the principles of dialling, as it is performed and illustrated by the globes; - and the construction, and use of the different kinds of maps. IV. The elements of astronomy, in which, besides a large account of the solar system, and of the various motions, revolutions, etc. of the planets and comets, are given the theory of the four seasons, the harvest-moons, eclipses, tides and several other phenomena too numerous to be here mentioned. To this part is added a copious appendix, containing the elements of chronology, a science very intimately connected with that of astronomy. The whole is illustrated with the necessary engravings. And, though principally intended for the use of schools, may serve as a convenient memorandum-book for those gentlemen and ladies who have already been instructed in the sciences above-mentioned
- ▸ LONDON; printed for R. Baldwin, No. 47, in Paternoster-Row
- ▸ 8vo, 17 cm.; pp. [1], x, [2], 261, [3]
- ▸ ***Libraries*** **L** : MH : CaOTP
- ▸ ***Notes***

1. The special geography (pp. 5-40) contains little beyond tables of the subdivisions of countries, with their chief cities and their locations. Acknowledges that the table of latitude and longitude was taken from the 'ingenious Mr. Donne.'
2. If this is the work recorded in Allibone (1854), i.e., JONES, Edward (see above), then there was a 2nd ed. in 1792.

JUVENILE RAMBLER **Y mSG**
- ▸ **{1838} (01)**
- ▸ *The juvenile rambler*; or, sketches and anecdotes of the people of various countries, with views of the principal cities of the world
- ▸ LONDON; John Harris
- ▸ 13 cm.; pp. 225
- ▸ ***Libraries*** : : CaOTP
- ▸ ***Source*** St. John (1975, 1:185)

K

KEITH, R.M. **?**
- ▸ **{1847} (01)**
- ▸ *Geography, system*
- ▸ [LONDON]; Longman
- ▸ 12mo
- ▸ ***Libraries*** n.r.
- ▸ ***Source*** BCB, Index (1837-57:110)

KEITH, Thomas ?
- **{1835}, new (01)**
- *Elements of geography*. New ed.
- LONDON
- ***Libraries*** n.r.
- ***Source*** Allibone (1854)
- ***Notes***

1. It seems likely that the Keith whom different sources identify variously as T. and as Thomas is the same man.

KEITH, T. ?
- **{1809} (01)**
- *Geography*
- Place/publisher n.r.
- ***Libraries*** n.r.
- ***Source*** Peddie and Waddington (1914:226). This could be a subsequent ed. of KEITH, T. (1787), which follows.

KEITH, Thomas **Y**
- **{1787} (01)**
- *A short and easy introduction to the science of geography*, etc.
- LONDON
- 12mo
- ***Libraries*** L
- ***Notes***

1. For the full title and information on the content, see (02), below.
2. BL records 5th ed. in 1805.

- **1812, 7th, corrected and improved (02)**
- *A short and easy introduction to the science of geography*: containing an accurate description of the situation, extent, boundaries, divisions, chief cities, etc. of the several empires, kingdoms, states, and countries, in the known world: with the nature, use, and construction of maps. The seventh edition ...
- LONDON; printed for C. Law, Scatcherd and Letterman, Longman, Hurst, Rees, Orme and Brown, and Gale and Curtis; by Law and Gilbert, St.John's-Square, Clerkenwell
- 17.5 cm.; pp. [3], iv-viii, 181, [3]
- ***Libraries*** : : **CaAEU**
- ***Notes***

1. Keith's view of geography is shown by his choice, as a motto for the tp., of the following quotation from the writings of John Locke: 'The learning of the figure of the globe, the situation and boundaries of the four parts of the world, and that of particular kingdoms and countries, being only an exercise for the eyes and memory, a child with pleasure will learn and retain them.'

   The principal sources from which information has been drawn are identified in the Preface. It is also made clear there that the book is intended primarily for use in schools. The British Isles are dealt with first, and in greatest detail. However, thanks to the availability of the works of Jedidiah MORSE, each of the newly united states receives separate treatment. In places the text approaches a list in the form of prose. 'The nature and use of maps' occupies pp. 145-67; much of this space is devoted to the exposition of problems concerned with the establishment of location on the surface of the earth, and of the differences in time between such places. Keith's background in mathematics is visible in pp. 168-77, which are devoted to 'The geometrical construction of maps.'

KEITH, Thomas **Y SG?**
▸ **{1826} (01)**
▸ *A system of geography, for the use of schools*, etc.
▸ LONDON
▸ 12mo
▸ ***Libraries*** L
▸ ***Notes***
1. Allibone (1854) records an ed. in 1847.

KELLY, Christopher **A**
▸ **1814, 1817 (01)**
▸ *A new and complete system of universal geography*; or, an authentic history and interesting description of the whole world, and its inhabitants: comprehending a copious and entertaining account of all the empires, kingdoms, states, republics, and colonies, of Asia, Africa, America, and Europe; as consisting of oceans, continents, islands, promontories, capes, bays, peninsulas, isthmuses, gulphs, lakes, rivers, canals, harbours, mountains, volcanoes, deserts, etc. with their respective situations, extent, latitude, longitude, boundaries, climate, air, soil, metals, minerals, vegetable productions, and every curiosity, natural and artificial, worthy of notice throughout the face of nature, both by land and water. Likewise an exact account of the population, military and civil governments, revenues, laws, trade, commerce, arts, sciences, manufactures, agriculture, religion, customs, language, manners, genius, habits, ceremonies, amusements, and general character and pursuits of the various nations of the earth. Including, details of the most remarkable battles, sieges, sea-fights, and revolutions, that have taken place in different parts of the habitable globe. With faithful accounts of all the new discoveries, that have been made by the most celebrated navigators of various nations, from Columbus, the first discoverer of America, to the death of the renowned Captain Cook, and those who have succeeded him, to the present time. Also, interesting extracts from the most authentic narratives of modern travellers, and the missionaries who have been sent out by different societies; forming a complete collection of voyages and travels, illustrative of the present state of the known world. To which will be subjoined, a useful compendium of astronomy, with remarks on the use of globes, etc. The whole concluding with a copious index, upon a plan entirely new, and designed to form a general gazetteer of the world. Embellished with numerous engravings, executed by artists of the first ability, and illustrated by correct statistical

tables, and a new set of accurate maps, forming an ornamental and complete atlas

- LONDON; printed for Thomas Kelly, No. 53, Paternoster-Row, by Rider and Weed, Little Britain
- 4to, 27 cm., 2 vols.; Vol. I (1814), pp. [7], ii-xx, 774; Vol. II (1817), pp. [5], 2-1105, [1]
- ***Libraries*** **L**
- ***Notes***

1. Aimed at the popular rather than the academic public ('the lady's library, the tradesman's parlour, and the peaceful retirement of the sequestered cottage' [Preface], and subsequent references to 'the historian ... politician ... military and naval officers ... merchant and trader ... moralist,' all of whom, it is said, will find the book of interest). Vol. I contains Asia, Africa, and America; Vol. II deals with Europe, approximately one-third of the space being devoted to the British Isles. The section on Europe contains much recent history. Fairly extensive use of direct quotation is made throughout both volumes. In some instances the quotations run to several pages in length.
2. Downes notes that K. still uses 'climate' in the original Greek sense. He also asserts that 'Kelly was the first of our geographers to use the Scottish censuses of Webster (1755) and Sinclair (1799) and the first government census (1801)' (Downes 1971:384).

- **1818 (02)**
- *A new ... universal geography ...*
- LONDON; printed ... by Weed and Rider
- 4to, 27 cm., 2 vols.; Vol. I, pp. [7], ii-xx, 774; Vol. II, pp. [5], 2-1105, [1]
- ***Libraries*** O, **OSg** : CaQQS
- ***Notes***

1. The copy held at CaQQS has Vol. I, 1818, and Vol. II, 1817.

- **{1819-22} (03)**
- *A new ... universal geography* ... of the whole world and its inhabitants. With faithful accounts of all the new discoveries ... to the present time. Also a useful compendium of astronomy, with remarks on the use of globes, etc. ...
- LONDON; T. Kelly.
- 4to, 28 cm., 2 vols.
- ***Libraries*** : DLC
- ***Source*** NUC and James Thin, Edinburgh, Catalogue 345, *General second-hand and antiquarian*, n.d., c1975, item 1794

- **1826-8 (04)**
- *A new ... universal geography ...*
- LONDON; printed for Thomas Kelly, No. 53, Paternoster-Row, by Rider and Weed, Little Britain [Vol. I]
- 28 cm., 2 vols.; Vol. I, pp. [7], ii-xx, [1], 6-774; Vol. II, pp. [5], 2-1105, [1]
- ***Libraries*** : MB : **CaBViPA**
- ***Notes***

1. Vol. I of the CaBViPA holding is the first ed.; the date of Vol. I of this ed.

is not known with certainty, but is presumed to be 1826 as recorded by MB. The section devoted to England, which is in Vol. II, ends (pp. 990-4) with a 'report of a committee of the House of Commons, presented on 19 Feb.' 1817. That, combined with the lack of any change in the pagination of Vol. II when compared with the 1st ed., suggests that the text is unchanged since then.

- **1830-2 (05)**
- *A new ... universal geography ...*
- LONDON; printed for Thomas Kelly, No. 17 Paternoster-Row, by J. Rider, Little Britain
- 27 cm., 2 vols.; Vol. I (1830), pp. [6], ii-xx, [5], 6-774; Vol. II (1832), pp. [2], 2-1105, [1]
- ***Libraries*** : **CtY**
- ***Notes***

1. The fact that the pagination is almost unchanged since the 1st ed., together with the fact that the history of France is brought down to the peace treaty ending the Napoleonic war, while that of England ends in 1816, makes it probable that no significant changes have been made since the first ed.
2. See the next entry for a further ed. with a change in title.
3. Discussed in Downes (1971:384).
4. A microform version of (02) is available from CIHM.

KELLY, Christopher

- **{1850} (01)**
- *An authentic history and entertaining description of the world and its inhabitants*: comprehending an account of the four ancient divisions of the world as well as of that new general division, which includes Australasia and Polynesia, under the modern designation of Oceania, forming a complete universal geography, embodying the discoveries that have been made by the most celebrated navigators and travellers, from Columbus to the present time. To which is prefixed an introductory chapter, showing the connexion of geography with astronomy, and containing a compendium of the latter science. A new edition, edited by Thomas Bartlett, including the discoveries of Ross, Parry, Franklin, Clapperton, Denham, etc.
- LONDON; T. Kelly
- 29 cm., 2 vols.
- ***Libraries*** : N
- ***Notes***

1. N confirms that, despite the change in title, this is a later ed. of KELLY (1814, 1817).

KELSALL, Charles (Mela Britannicus) **OG**

- **{1822} (01)**
- *Remarks touching geography* [especially that of the British isles; comprising strictures on the hierarchy of Great Britain]
- Place/publisher n.r.
- ***Libraries*** n.r.

- ***Source*** Peddie and Waddington (1914:227)
- ***Notes***

1. BL and NUC (NK 0085444) record a further ed.: 1825 (L : MH). The full title suggests strongly that the work is marginal to this guide.

KENNION, C[harlotte] **Y SG?**

- **{1825} (01)**
- *Modern Geography*
- Place/publisher n.r.
- 12mo
- ***Libraries*** n.r.
- ***Source*** Peddie and Waddington (1914:227)
- ***Notes***

1. The Christian name is taken from BL, which records 2nd ed. in 1845.

KENNY, William David

- *Kenny's Goldsmith's grammar of geography* ... (see (03) of PHILLIPS 1803)

KENNY, William Stopford

See note 2 to (02) of PHILLIPS (1803) for an ed. 'corrected and improved' by W.S. Kenny.

KENNY, William Stopford **Y**

- **{1856} (01)**
- *Kenny's school geography*; or, earth and heaven ... [with exercises for examination at the end of each division. To which is added, a treatise on astronomy, with problems on the terrestrial and celestial globes]
- LONDON
- 12mo
- ***Libraries*** L
- ***Notes***

1. The words in the title set in [ ] are taken from NUC, which records: 1866, 5th (CtY).

KILBOURN, John **Y SG? Am**

- **{1813} (01)**
- *A compedious system of universal geography*: designed for the use of schools and youth in general
- WORTHINGTON; Putnam & Israel, printers, Zanesville
- 19 cm.; pp. xiii, [14]-266
- ***Libraries*** : CSmH, NN, PHi, OFH
- ***Source*** NUC (NK 0133841)

KINCAID, Alexander **A**

- **1790 (01)**
- *A new geographical, commercial, and historical grammar*; and present state of the several empires and kingdoms of the world. Containing I. The distances, figures, and revolutions of the celestial bodies, as demonstrated by Sir Isaac Newton, and observed by the latest astronomers. II. A description of the earth, considered as one of the revolving bodies in the general system; with several definitions and problems necessary for understanding the science of Geography. III. An account of the great divisions of the surface of the globe into continents, islands, oceans, seas, etc. IV. An accurate and particular description of all the empires, kingdoms, and states in the world, as situated with regard to one another. V. The climate and soil, with its productions, whether vegetable or mineral; the natural curiosities; a particular description of the sea-coasts, including all the bays, capes, promontories, and adjacent rocks; with an account of the most remarkable lakes and rivers. VI. An account of the animals, whether birds, beasts, fishes, or insects, found in each country. VII. Observations on the various parts of the globe in which history makes mention of any remarkable change having naturally taken place. VIII. An history of all nations from their origin; laws, government, religion, manners, and customs; distinctions of rank, revenues, taxes, strength by sea and land. IX. The genius and external appearance of the people. X. An account of their learning, arts, language, manufactures, and commerce. XI. The chief cities, artificial curiosities, antiquities, etc. To which is added, I. A geographical index, with the names of places, alphabetically arranged. II. A table of the coins of all the different nations, with their value in Sterling money. III. A chronological table of remarkable events, from the earliest accounts to the present time. The whole executed on a plan similar to that of W. Guthrie, esq; by a Society in Edinburgh. The astronomical part collected from the works of James Ferguson, F.R.S. Enriched with the late discoveries of Dr. Herschel, and other eminent astronomers. Embellished with an elegant set of maps, engraved on purpose, more numerous, accurate, and exhibiting more fully the new geographical discoveries than those to be met with in any former publication
- EDINBURGH; printed for Alexander Kincaid, and sold by all the booksellers in town and country
- 8vo, 22 cm., 2 vols.; Vol. I, pp. [6], 576; Vol. II, pp. [8], 598, [2]
- ***Libraries*** **L** : NN, PPT
- ***Notes***

1. The dedication, Preface, and table of contents of the BL copy are bound at the beginning of Vol. II. Vol. I is devoted primarily to Europe, and, within Europe to the British isles. Within that section, Scotland receives a relatively great amount of space. Much history is included. New discoveries, chiefly in the Pacific, are dealt with separately.

   The title is nearly identical in form to that of GUTHRIE (1770), although the wording differs in detail throughout. CtY treats later eds. of this work (i.e., Kincaid) as eds. of Guthrie; see notes to (22) and (25) of GUTHRIE (1770).

KNIGHT, Charles **Dy**
- **1854-5 (01)**
- *The English cyclopaedia.* A new dictionary of universal knowledge. Conducted by ... [Assisted by A. Ramsay and J. Thorne]
- LONDON; Bradbury and Evans
- 4to, 29 cm., 25 vols.
- ***Libraries*** L : CtY, CU, DLC, DSI, ICN, ICRL, IU, LU, MB, MeB, MH, MiU, NB, NjP, NN, NNC, NRU, OClWHi, OrP, OU, PP, PPiPT, PPL, PV, RPB, T, WaS
- ***Notes***

1. It is a dictionary; the format is double columns, and there are at least 1,000 numbered columns in each of Vols. I to III; the pages are unnumbered. As the print is small, the volume of information provided is large, reaching down the urban hierarchy to the village level in the case of the UK.

   Sources are occasionally cited, both in the body of the text and at the end of individual articles. Vols. I to IV, plus the supplement are devoted to geography, and they are 'based upon The Penny cyclopaedia of the Society for the diffusion of useful knowledge' (Preface).

KRISHNAMOHANA, Vandipadhyaya **A n.c. els**
- **{1848} (01)**
- *Geography.* Part I. Containing a description of Asia and Europe. Compiled [by K.V.] from Murray's Encyclopaedia of geography, Malte-Brun's Geography, and other works
- CALCUTTA
- 12mo
- ***Libraries*** L
- ***Source*** BL, which notes: 'No. VIII of the "Encyclopaedia Bengalensis, edited by ..."'

L

LAMBERT, Claude François **mSG tr**
- **[1750?] (01)**
- *Curious observations upon the manners*, customs, usages, different languages, government, mythology, chronology, antient and modern geography, ceremonies, religion, mechanics, astronomy, medicine, physics, natural history, commerce, arts, and sciences, of the several nations of Asia, Africa, and America. Translated from the French of M. L'abbé Lambert
- LONDON; printed for G. Woodfall, at the King's-Arms, Charing-Cross; W. Russel, at Horace's-Head, without Temple-Bar; and W. Meyer, in May's-Buildings, St. Martin's-Lane
- 8vo, 20 cm., 2 vols.; Vol. I, pp. [2], iv, 411; Vol. II, pp. [2], 404, [16]
- ***Libraries*** L : CSmH, CU, MiD, NhD, NN, PPL, RPJCB, Vi

▸ *Notes*
1. Not systematically concerned with location, and countries not dealt with in contiguous sequence; thus not special geography. However, most, and probably all, the information it contains could be found in that class of work. On the whole it belongs to the tradition of anthropology rather than geography.
2. NUC also records what appears to be a separate series of eds., based on a different translation, with holdings as follows: 1750 (CtY, DLC, ICN, InU, IU, MiU, MnHi, MnU, NB, NhD, NIC, NP, PHi); 1754 (MnHi); and 1760 (ICN).

LAMBERT, Johann Heinrich **OG tr**
▸ **{1800} (01)**
▸ *The system of the world*. Translated from the French, by James Jacque, esq.
▸ LONDON; Vernor and Hood; [etc.]
▸ 8vo, 17 cm.
▸ ***Libraries*** L : AU, CtY, DLC, MB, NjP, ODW
▸ ***Source*** BL and NUC (NL 0050545), which notes: 'The original work was first published (Augsburg, 1761), under title "Cosmographische Briefe über die Enrichtung des Weltbaues". The present translation is from the abridged edition, Systeme du monde, of "Lettres cosmologiques sur l'organisation de l'univers".' Thus, not a special geography.

LANDMAN, George **Dy**
▸ **1840, new ... (see below) (01)**
▸ *A universal gazetteer*; or, geographical dictionary. Founded on the works of Brookes and Walker ... A new edition, revised with upwards of eighteen hundred additional names of places
▸ LONDON; Longman, Orme and co., W.T. Clarke; T. Cadell; J.M. Richardson; Baldwin and Cradock; J.G.F. and F. Rivington; A.K. Newman and co.; Hamilton and co.; Whittaker and co.; Allen and co.; Sherwood and co.; Duncan and Malcolm; Simpkin, Marshall and co.; J. Souter; Cowie and co.; J. Dowling; Smith, Elder, and co.; Houlston and Stoneman; T. Bumpus; J. Templeman; Capes and co.; E. Hodgson; S. Hodgson; R. Mackie; J. Wacey; W. Edwards; H. Washbourne; Harvey and Darton; J. Fraser; J. Chidley; L.A. Lewis; W. Morrison; H. Bickers; J. Snow; C. Dolman; Haywood and Moore; L. Booth. - Liverpool: G. and J. Robinson. - York: Wilson and sons. - Cambridge: J. and J. Deighton. - Edinburgh: Sterling and co.
▸ 21.5 cm.; pp. [3], iv, 773.
▸ ***Libraries*** [IDT]
▸ *Notes*
1. A brief Preface is devoted largely to responses to criticisms levelled against previous eds. to the effect they lacked accurate information on the distances between places. Neither objective nor justification for publication given.

LAURIE, J. Werner L., *Mrs* **Y SG?**
▸ **{1869a} (01)**

- *Maxwell's first lessons in geography* for the young. By the author of Home and Its Duties, etc.
- EDINBURGH
- 16mo
- ***Libraries*** L

LAURIE, J. Werner, *Mrs* Y

- **{1869b} (01)**
- *Maxwell's general geography*
- EDINBURGH; Thomas Laurie. London; Simpkin, Marshall & Co.; and Hamilton, Adams & Co.
- 8vo, 17 cm.; pp. [2], 165
- ***Libraries* L**
- ***Notes***

1. Very similar in content, organization, and level to LAURIE (1864), which follows.

LAURIE, James Stuart, ed. Y

- **1864 (01)**
- *Manual of elementary geography*
- LONDON; Thos. Murby, 32, Bouverie-Street, Fleet-Street, E.C.; and Simpkin, Marshall, & Co., E.C.
- 8vo, 17 cm.; pp. [6], 118
- ***Libraries* L**
- ***Notes***

1. Provides information about the countries of the world that school children are supposed to learn by heart. On the cover the book is identified as *The Standard Manual of Geography* (see below). It is BL that identifies Laurie as the editor of the work.
2. ECB, Index (1856-76:130) records an ed. in 1868. BL records 5th ed. in 1870, 152 pp.

- **1877, anr ed. (02)**
- *The standard manual of geography*
- LONDON; John Marshall & Co., 42 Paternoster Row, E.C.; Simpkin & Co.; Hamilton, Adams & Co.; Whittaker & Co., E.C. Manchester; John Heywood. And all booksellers
- 8vo, 16 cm., 5 pts. (available separately)
- ***Libraries* L**
- ***Notes***

1. Though the pagination is continuous, there are four parts, each with its own tp. as follows: Pt. I, British Empire (contains 32 pp.); Pt. II, Europe (pp. 33-63); Pt. III, Asia, Africa, America, etc. (pp. 65-96); Pt. IV, America and Oceania (pp. 97-120). However, despite the claim on its tp., Pt. III deals only with Africa and Asia. BL (which provides the date) is thus justified in stating that there are only four parts. However, the cover of Pt. IV (in their separate versions) states, 'complete in 5 parts.'

   Bound with Pts. I to IV of the BL copy is *Laurie's Outline of Sacred*

Geography. Palestine and Asia Minor. [Published by] The Central School Depot, 22 Paternoster Row, E.C.; Simpkin & Co.; Hamilton, Adams & Co.; Kent & Co., E.C.; The Scholastic Trading Companies; and all booksellers. [pagination] 121-34. Also bound with this work is *Laurie's Examination Questions in Geography. Designed to accompany 'Laurie's Complete Manual of Geography'; and for examinations in general* [Published by] The Central ... Despite the change in title, this work is essentially the same work as *The Manual of Elementary Geography.*

LAURIE, James Stuart, ed. **Y**

- **{[1868]} (01)**
- *The sixpenny geography* ... Part I. Definitions – the British Empire. Part II. Europe. Part III. Asia, Africa, America, and Oceania
- LONDON; The Central School-Depot, and General Printing Agency, 22 Paternoster Row. John Marshall & Co., E.C.; Simpkin & Co., E.C.; Hamilton, Adams & Co., E.C.
- 8vo, 16 cm.; pp. 96
- ***Libraries*** L
- ***Notes***

1. BL records 8th ed. in 1877.

LAURIE, James **A**

- **1842 (01)**
- *System of universal geography, founded on the works of Malte-Brun and Balbi*; embracing a historical sketch of the progress of geographical discovery, the principles of mathematical and physical geography, and a complete description, from the most recent sources, of the political and social condition of all the countries in the world; with numerous statistical tables, and an alphabetical index of 12,000 names
- EDINBURGH; Adam and Charles Black; and Longman, Brown, Green, & Longman; London
- 8vo, 22 cm.; pp. xxiii, [1], 1063
- ***Libraries*** EU, **L**
- ***Notes***

1. Because extensive use is made of small print, the amount of information contained is very large. Begins with mathematical, physical, and political geography (pp. 29-136); aside from the index, descriptive geography occupies the balance (Europe, pp. 137-624; Asia, 625-800; Africa, 801-71; America, 872-966; Oceania, 967-1018). Appears comparable to GOODRICH (1832). Was used as his principal source by BOHN (1861).
2. Allford (1964:245) states that H.G. Bohn was the editor. BCB, Index (1837-57:110) records another ed. in 1844.

- **1849, new, revised and corrected throughout (02)**
- *System ... geography; founded* ... embracing the history of geographical discovery ... sources, of all the countries in the world. With alphabetical index of 13,500 names

- EDINBURGH and London; Adam and Charles Black; and Longman, Brown, Green and Longmans
- 8vo, 23 cm.; pp. xxiii, [1], 1067, [1]
- ***Libraries*** **L**, [IDT]
- ***Source*** BL, which records another ed. in 1851

LAVALLEE, Th[eophile Sebastien] **mSG tr**

- **1868 (01)**
- *Physical, historical, and military geography*: from the French of Th. Lavellée ... Edited, with additions and corrections, by Captain Lendy, F.G.S, F.L.S., etc. Director of the Practical Military College at Sunbury
- LONDON
- 18 cm.; pp. [5], vi-xii, [5], 665, 4
- ***Libraries*** **GP**
- ***Notes***

1. Belongs to the school of *reine Geographie*. In this school, river basins are used as the units into which the surface of the earth is divided for descriptive purposes, rather than political units (Tatham 1957). According to Lendy, the work of Lavallée was made known in England in 1850 by the translation of Jackson, *Military Topography of Continental Europe*.

LAWSON, William **Y SG?**

- **{1875} (01)**
- *Class book of geography* ... With ... coloured maps, etc.
- EDINBURGH
- 16mo
- ***Libraries*** L

LAWSON, William **Y SG?**

- **{1882} (01)**
- *A geographical first book* ... With diagrams and coloured maps
- EDINBURGH; Oliver & Boyd
- 12mo; pp. 36
- ***Libraries*** L

LAWSON, William **Y SG?**

- **{1871, etc.} (01)**
- *Geographical primer* ... Adapted to the New Code, Standard IV. - 1872
- EDINBURGH; Oliver and Boyd
- 16mo
- ***Libraries*** L
- ***Source*** BL, which notes: 'Oliver & Boyd's New Code Class-Books,' and records further eds. in 1875; [1880], Edinburgh, 36 pp.; [1887], Edinburgh, 36 pp.; and 1889, 1891, and 1897

LAWSON, William **?**

- **{1865}a (01)**
- *Geography, the soldier's*
- ? Philip
- ***Libraries*** n.r.
- ***Source*** ECB, Index (1856-76:131)

LAWSON, William **Y SG?**

- **{1873, etc.} (01)**
- *Manual of modern geography: physical, political, and commercial* ... With coloured maps and illustrations
- [GLASGOW]; W. Collins
- 8vo and folio
- ***Libraries*** L
- ***Source*** BL, which notes: 'Collins' School Series,' and records 6th ed. in 1890

LAWSON, William **Y SG?**

- **{1864} (01)**
- *Outlines of geography for schools and colleges*
- LONDON, Liverpool [printed]; Philips
- 8vo
- ***Libraries*** L
- ***Source*** BL, which notes: 'Part of "Philips' Standard Series of Educational Works".' It records another ed. in 1875, London, Liverpool [printed].

LAWSON, William **Y SG?**

- **{1865}b (01)**
- *The young scholar's geography*
- LONDON, Liverpool [printed]
- 8vo
- ***Libraries*** L

LE BLANC, Vincent **mSG tr**

- **1660 (01)**
- *The world surveyed*: or, the famous voyages and travailes of Vincent Le Blanc, or, White, of Marseilles: who from the age of fourteen years, to threescore and eighteen, travelled through most parts of the world. Viz. The East and West Indies, Persia, Pegu, the kingdoms of Fez and Morocco, Guinny, and through all Africa. From the Cape of Good Hope, into Alexandria, by the territories of Monomotapa, of Preste John and Egypt, into the Mediterranean isles, and through the principal provinces of Europe. Containing a more exact description of several parts of the world, then hath hitherto been done by any other author. The whole work enriched, with many authentic histories. Originally written in French, and faithfully rendered into English by F[rancis] B[rooke], gent.

- LONDON; printed for John Starkey at the Mitre, near the Middle-Temple Gate in Fleet-street
- Folio, 27.5 cm.; pp. [10], 407, [13]
- ***Libraries*** L : CSmH, **CtY**, DLC, FJ, LU, MiU, MnU, MWiW, N, NIC, NjP, NN, OC, OCl, PBL, RPJCB, TU, WU
- ***Notes***

1. Though not mentioned on the tp., the narrator visited the American continent. He does provide information about the location and the boundaries of the countries he visits, though in general terms. The book is closer to special geography than to the literature of travel.

LEECH, William **Y**

- **{1801}, 5th, enlarged ... (see below) (01)**
- *Elements of geography*, [for the use of schools. The fifth edition, enlarged, improved and corrected]
- DUBLIN; printed by P. Wogan, No. 23, Old Bridge
- Pp. [3], 2-72
- ***Libraries*** : : CaQQS
- ***Notes***

1. Special geography of a highly condensed form occupies pp. 36-72.
2. A microform version is available from CIHM.

LENGLET DU FRESNOY, Pierre Nicolas **Y mSG tr**

- **{1742} (01)**
- *Geographia antiqua et nova*: or a system of antient and modern geography, with a sett of maps engraven from Cellarius's. Designed for the use of schools, and of gentlemen, who make the antient writers their delight or study. Translated from the French of Mr. L'abbé du Fresnoy, with great additions and improvements, from Ptolemy, Strabo, Cellarius, etc. To which is added a large index
- LONDON; printed for John and Paul Knapton, at the Crown in Ludgate-Street
- 4to, 26 cm.; pp. vi, [6], 157, [36]
- ***Libraries*** L : CLU, ICU, NBuG, PPL
- ***Notes***

1. BL records the author's names as shown above; NUC lists him as Nicolas Lenglet Dufresnoy.
2. Essentially a gazetteer (and atlas) of peoples and of the places where they lived, together with other toponyms. The primary focus consists of the countries described in Classical literature. Being laid out in the form of text rather than of tables, it has the appearance of special geography.

- **{1747} (02)**
- *Geographia antiqua et nova* ... geography, translated from the French with great additions and improvements, from Ptolemy, Strabo, Cellarius, etc. with 33 maps engraved on copper
- LONDON
- 4to

▸ *Libraries* n.r.
▸ *Notes*
1. Cox (1938:352), which is the source, gives the author's name as: Fresnoy, Languet du.

▸ **1768, 2nd, see below (03)**
▸ *Geographia antiqua et nova* ... index. The second edition. The whole together with the maps, corrected and improved
▸ LONDON; printed for Robert Horsfield, at No. 22, in Ludgate-Street
▸ 26.5 cm.; pp. [ii], iii-iv, [7], 2-105, [38]
▸ *Libraries* CaKU, **L** : DLC, NPV, ViU : **CaBvaU**
▸ *Notes*
1. A marked increase in the number of words per page allows what is probably much the same text as in 1742 to be carried on many fewer pages.

LENGLET DU FRESNOY, Pierre Nicolas **Y tr**
▸ **{1737} (01)**
▸ *The geography of children*: or, a short and easy method of teaching or learning geography ... Divided into lessons, by way of question and answer. With a small neat map of the world prefix'd, and also a list of the maps necessary for children. Translated from the French of Abbot Lenglet Dufresnoy, just published in Paris; with the addition of a more particular account of Great Britain and Ireland
▸ LONDON; printed for Edward Littleton; and John Hawkins
▸ 12mo, 15 cm.; pp. vii, 129
▸ *Libraries* L : : CaOTP
▸ *Source* BL and NUC (NL 0249073), which records only 29 pp. CaOTP confirms that this is a misprint.
▸ *Notes*
1. Deals with every continent; much concerned with size and relative location, thus a special geography.
2. BL records the author's names as shown above; NUC lists him as Nicolas Lenglet Dufresnoy.
3. According to St. John (1975, 1:186), the first French ed. was published in Paris in 1736.

▸ **{1738}, 2nd, corr. (02)**
▸ *The geography of children* ... geography. Designed principally for the use of schools ... Divided ...
▸ [LONDON]; printed for E. Littleton
▸ *Libraries* : CtY, IU
▸ *Source* NUC (NL 0249074)

▸ **{1776}, 10th (03)**
▸ *Geography for children* ... Lenglet du Fresnoy, and now greatly augmented and improved ... The 10th ed. To which is prefixed, a method of learning geography without a master, for ... grown persons
▸ LONDON; G. Keith
▸ 17 cm.; pp. x, [2], 148

- ▸ ***Libraries*** : DLC
- ▸ ***Source*** NUC (NL 0249057)
- ▸ ***Notes***

1. NUC treats the geography *for* children, i.e., (01), as a different title from the geography *of* children, i.e., (03). Bowen (1981:150), however, ignores the minor variations in the title, and confirms what the numbering of the editions already suggests, namely, that only one work is involved.
2. NUC records 14th ed. in 1783, London (MH).

- ▸ **1787, 15th (04)**
- ▸ *Geography for children* ... geography: designed ... schools. Whereby even children may in a short time know the use of the terrestrial globe and geographical maps, and all the considerable countries in the world; their situation, boundaries, extent, divisions, islands, rivers, lakes, chief cities, government and religion. Divided into lessons, in the form of question and answer: with a new general map of the world, and also a list of maps necessary for children. Translated from ... improved throughout the whole. 15th ed. To which is prefixed ... for the use of such grown persons as have neglected this useful study in their youth. And a table of the latitude and longitude of the most remarkable places mentioned in this work. As also a print of the orrery
- ▸ LONDON; printed for J. Johnson, No. 72, St. Paul's Church-Yard, and E. Newberry, at the corner of Ludgate Street
- ▸ 12mo, 16 cm.; pp. x, [2], 151, [6]
- ▸ ***Libraries*** **L** : MH

- ▸ **1791, 16th (05)**
- ▸ *Geography for children* ... map of the world, the spheres, and also ... work
- ▸ LONDON; printed ...
- ▸ 12mo, 18 cm.; pp. xii, 154 (121-22 and 143-44 switched), [2]
- ▸ ***Libraries*** **L** : CLU, MH
- ▸ ***Notes***

1. More words per page than in 1787. Some questions and answers are new, and some are rephrased.
2. ESTC records 17th ed. in 1793 (EU). NUC records further eds.: 1795, 18th, London, 143 pp. (CLU); and 1795, 19th, London (MH). St. John records an ed.: 1797, 19th, London, 143 pp. (: : CaOTP).

- ▸ **1798, 13th [sic] (06)**
- ▸ *Geography for youth*, tr. from the French, and augmented.
- ▸ Philadelphia
- ▸ 156 pp.
- ▸ ***Source*** : PHi
- ▸ ***Notes***

1. Inferred to belong to this main entry, despite the difference in the short title, on the basis of the general congruity of other bibliographic details, and in particular the fact that this is identified as the 13th ed.
2. NUC records a further ed.: 1799, 20th, London, 152 pp. (MH, OClWHi).

- **1800, 22nd [sic] (07)**
- *Geography for children*; or ... situations ...
- SHREWSBURY, [England]; printed for Sandford and Maddocks
- 12mo, 17 cm.; pp. xii, 154
- *Libraries* L
- *Notes*

1. Pagination and spot-checking both suggest no changes since 1791.

- **1804, 30th (08)**
- *Geography for children*: or a ... answer; with ... Fresnoy. Comprising a short account of the recent changes which have taken place in various kingdoms and states, to the present time. To which is prefixed ...
- TAUNTON, [England]; for Crosby and Co. Stationer's Court, London, and for the principal booksellers in town and country
- 12mo, 17 cm.; pp. v, [1], 126, [1]
- *Libraries* L
- *Notes*

1. Smaller type allows more words per page than in 1800. Takes note of at least some of the changes in boundaries that followed the French Revolution and subsequent wars.
2. St. John and NUC record further eds.: 1805, 21st (CaOTP); 1806, 22nd, London, 137 pp. (NN).

- **{1806}, 20th English and 1st Kentucky (09)**
- *Geography for children* ... improved by a teacher of Kentucky ...
- LONDON; S. Johnston and T. Newbury [sic]. Kentucky; reprinted and published by Joseph Charles.
- 17.5 cm.; pp. xii, 156
- *Libraries* : PU

- **{1806}, 16th (10)**
- *Geography for youth*; or, a short and easy method of teaching and learning geography ... Divided into lessons by way of question and answer ... Translated from the French of Abbé Lenglet du Fresnoy. An[sic] now greatly augmented and improved throughout the whole. 16th ed. To which is prefixed, a method of learning geography without a master ... and to this edition is now added, a table of the latitude and longitude ... also, a print of the orrery
- DUBLIN; printed for P. Wogen
- 18 cm.; pp. v, [1], [7]-202
- *Libraries* : CtY
- *Notes*

1. NUC records 23rd ed. in 1809, London, 137 pp. (NN)

- **1816, 26th (11)**
- *Geography for children*: or a short and easy method of teaching or learning geography. Designed principally for the use of schools: whereby children may, in a short time, know the use of the terrestrial globe and maps; and be able to find all the considerable countries in the world, and point out their situation, boundaries, extent, divisions, etc. etc. Divided ... world, and other plates. Translated from the French of Abbot Lenglet du Fresnoy ... [as in

1804] ... states. To which is added a table of the latitude and longitude of the most remarkable places mentioned in this work: and preceded by a method ... [as in 1787] ... youth
- LONDON; printed for F.C. and J. Rivington; Scatcherd and Letterman; G. Wilkie; Darton, Harvey, and Co.; Longman, Hurst, Rees, Orme, and Co.; John Richardson: Law and Whittaker; J. Mawman: J. Harris, Baldwin, Cradock, and Joy
- 12mo, 18 cm.; pp. v, [1], 137, [1]
- ***Libraries*** **L**
- ***Notes***

1. NUC records 30th ed. in 1825, London, 137 pp. (NN). BCB, Index (1837-57:110) records an ed. in 1852.
2. See GEOGRAPHY FOR YOUTH (1787).

LENGLET DU FRESNOY, Nicolas **NG tr**
- **{1730} (01)**
- *A new method of studying history, geography, & chronology.* With a catalogue of the chief historians of all nations, the best editions of their works, & characters if them. Written originally in French by M. Languet du Fresnoy ... And now made English with variety of improvements & corrections. To which is added a dissertation by Count Scipio Maffei of Verona, concerning the use of inscriptions & medals, by way of parallel ... By Richard Rawlinson ...
- LONDON; printed for C. Davis
- 20 cm., 2 vols.
- ***Libraries*** O : CtY, MH, MiU, NjP, PPL
- ***Source*** NUC (NL 0249114)
- ***Notes***

1. For other books by the author, whom NUC spells as above, see LENGLET DU FRESNOY, Pierre Nicolas.

LESSONS ... **Y mSG**
- **1791 (01)**
- *Lessons in geography; with an introduction to the use of the globes*
- LONDON; printed for T. Cadell, in the Strand
- 8vo, 20.5 cm.; pp. [2], 107
- ***Libraries*** **L**
- ***Notes***

1. The special geography, which is little more than a sequence of place names, plus the kings and queens of England listed in chronological order, occupies pp. 8-9. The remainder of the book expounds on the graticule and related phenomena. There are two sets of questions for the teacher to put to students.

LEYBOURN, William **OG**
- **1675 (01)**
- *An introduction to astronomy and geography*: being a plain and easie treatise of the globes. In VII parts. Containing I. The definitions of the lines, circles,

etc. upon the globe or sphere; and of several terms of art. II. The problems in astronomy methodically digested, with variety of examples. III. The several affections of triangles, and their solution upon the globe; with the variety of problems which every case contains. IV. The whole art of dyalling demonstrated and performed two several ways. V. The erection of an astrological figure of the heavens, according to the several ways of the ancient and modern astrologers. VI and VII. The explanation and uses of the terrestrial globe, with a brief geographical and hydrographical description of the earth and water

▸ LONDON; printed by J.C. for Robert Morden and William Berry; at the Atlas neer the Royal Exchange in Cornhil, and between York-house and the New Exchange in the Strand, London
▸ 8vo, 17 cm.; pp. [8], 234
▸ ***Libraries*** CT, ExC, **L**, LDW : CLU, CtY, DLC, IU, MiU, MWiW, NjP, NN, PPAmP
▸ ***Notes***

1. Wholly mathematical.
2. NUC records a further ed.: 1702, London, 184, 87 pp. (MiU).

LINDSAY, G. ?

▸ **{1833} (01)**
▸ *Concise summary of geography*
▸ Place/publisher n.r.
▸ ***Libraries*** n.r.
▸ ***Source*** Peddie and Waddington (1914:226)

LITTLE TRAVELLER **Y**

▸ **{ca. 1825} (01)**
▸ *The little traveller*; or, a sketch of the various nations of the world: representing the costumes, and describing the manners and peculiarities of the inhabitants. Embellished with fifteen beautifully coloured engravings. By J. Steerwell, jun. R.N.
▸ LONDON; published by A.K. Newman, & co.
▸ 16.5 cm.; pp. 34
▸ ***Libraries*** : : CaOTP
▸ ***Source*** St. John (1975, 2:808)
▸ ***Notes***

1. St. John (1975, 2:808) notes: 'Young James Steerwell describes his travels to "every quarter of the world" to his four cousins.' However, she catalogues the book under its title, creating the impression that Steerwell is a pseudonym. That interpretation is corroborated by the fact that it does not appear as the name of any author in NUC, and is listed by BL only as a pseudonym.
2 The same source also notes: 'An American edition was also published in Baltimore ... about 1825 ... The coloured engravings with a severely abridged text are repeated in *Travels over the land and over the sea*, (London, Darton & Clark, ca. 1840).'

LIZARS, Daniel **A**

- **1831 (01)**
- *The Edinburgh geographical and historical atlas*, comprehending a sketch of the history of geography; a view of the principles of mathematical, physical, civil, and political geography; an account of the geography, statistics, and history of each continent, state, and kingdom delineated. And a tabular view of the principal mountain chains of the world. Engraved on sixty-nine copper plates, and compiled from materials drawn from the newest and most authentic sources
- EDINBURGH; J. Hamilton, successor to D. Lizars. London; Whitaker, Treacher & co., [etc.]
- Folio, 38 cm.; pp. [4], 288, 16
- ***Libraries*** **L** (tp. torn) : DLC, NN, OCl
- ***Notes***

1. Catalogued by both BL and NUC under Lizars, yet NUC notes that Daniel Lizars died in 1812. Moreover, the Map division of BL records D. Hamilton as the compiler.
2. The text belongs to the tradition of special geography, with a marked bias towards history in general and towards British history in particular.

LIZAR'S ... ATLAS OF MODERN GEOGRAPHY ...
See WILLOX (1852).

LLOYD, Evan **Y**

- **1797 (01)**
- *A plain system of geography*; connected with a variety of astronomical observations, familiarly discussed in a conversation between a father and his son: containing an account of the figure, motion, and dimensions of the earth;– a view of the solar system; the motions, distances, etc. of the planets; a survey of the fixed stars;– an account of the circles belonging to the sphere, and of the different seasons, as arising from the earth's annual motion;– the nature and use of the terrestrial globe, with some necessary directions relative to maps;– the grand divisions of the earth into land and water, continents and islands;– the situation and extent of the several kingdoms, provinces, states, and empires; their soil, produce, governments, customs, manners, religion, etc. By ... schoolmaster. Illustrated with copperplates and maps
- EDINBURGH; printed for the author, by Mundell and Son, Royal Bank Close
- 17.5 cm.; pp. [5], vi-xxiv, 204
- ***Notes***

1. Special geography occupies pp. 73-202. Never was there such a docile and obliging son! The information provided by the father is less systematically presented than that in most special geographies.
2. Watt (1824, 2) records an ed. in 1798.

LOCKHART, John **Y SG?**

- **{1862} (01)**

- *Catechism of geography for the use of junior pupils in mixed schools*
- EDINBURGH; Longmans
- 16mo
- ***Libraries*** L
- ***Source*** ECB, Index (1856-76:130), and BL, which notes: 'Forming part of "Black's School Series"'

LONDON GAZETTEER ... **Dy**
- **1825 (01)**
- *The London [general] gazetteer, or geographical dictionary*: containing a description of the various countries, kingdoms, states, cities, towns, etc. of the [known] world; an account of the government, customs, and religion, of the inhabitants; the boundaries and natural productions of each country, etc., forming a complete body of geography, physical, political, statistical, and commercial
- LONDON; printed for W. Baynes and Son, Paternoster Row. Sold by Simpkin and Marshall ... & H.S. Baynes, Edinburgh
- 8vo, 3 vols.; pp. iv, xxi, [1], 712; 744; 676
- ***Libraries*** : MB, NN
- ***Notes***

1. Information provided by MB that their copy lacks the tp. of Vol. 1. The title as given above is taken from the tps. of Vols. 2 and 3. The words in [ ] are missing in the seemingly full title shown in Blackwell's Catalogue A1049, *Antiquarian Books on Travel and Topography*, 1976. The title invites comparison with CRUTTWELL (1798). See EDINBURGH GAZETTEER.

LORIOT, L. **Y tr**
- **1797 (01)**
- *A new short and easy method of geography*, French and English: chiefly calculated for the use of schools, and such persons as are desirous to learn by themselves that useful science, after the maps lately published by M. Robert Wilkinson; which would be best for those who study without a master
- READING, [England]; printed for the author, and sold by Smart and Cowslade; and Messrs Robinsons, Pater-Noster-Row, London
- 17 cm.; pp. [2], xii, [3], vi-viii, [1], 10-14, [1], 134
- ***Libraries*** L
- ***Notes***

1. Special geography by means of question and answer. English and French texts on facing pp. See GARNIER (1748).

LUFFMAN, John **?**
- **{1803} (01)**
- *Geographical principles*
- LONDON; Booth
- ***Libraries*** n.r.
- ***Notes***

1. 'In London Catalogue of Books, 1800-1822' (note to NUC NL 0550232).

NUC has an entry because DLC has nine maps that are marked 'Engraved for Luffmans's Geographical principles,' and 'Published may 1, 1803, by J. Luffman, no. 28, Little Bell alley, Coleman street, London.'

LYON, Sarah M. **Y OG Am**
- **{1848} (01)**
- *The musical geography*: a new and natural arrangement of the names of all the physical features of the globe ... Designed to be used with Mitchell's school atlas, or Pelton's outlines
- TROY [NY]; Young & Hartt
- 15 cm.; pp. 108.
- ***Libraries*** : DLC
- ***Source*** NUC (NL 0587469)

M

M., E.S. **Y SG?**
- **{1835} (01)**
- *Geographical text-book*
- ? Souter
- 12mo
- ***Libraries*** n.r.
- ***Source*** Peddie and Waddington (1914:226)

MacDOUGAL, Thomas Saint Clair **Y SG?**
- **{1835} (01)**
- *Descriptive outlines of modern geography* and a short account of Palestine ... With references to blank maps
- LONDON
- 12mo
- ***Libraries*** L
- ***Notes***

1. BCB, Index (1837-57:110) records an ed. in 1853; BL and NUC record: 1857, 12th, 179 pp. (L : DLC).

MacDOUGAL, Thomas Saint Clair **?**
- **{1839} (01)**
- *Geography*
- 12mo
- ***Source*** BCB, Index (1837-57:109)

MacFAIT, Ebenezer **OG**
- **1780 (01)**

- *A new system of general geography*, in which the principles of that science are explained; with a view of the solar system, and of the seasons of the year all over the globe; together with the most essential parts of the natural history of the earth. Part I
- EDINBURGH; printed for the author, and sold by J. Balfour[,] W. Creech, and other booksellers
- 8vo, 22 cm.; pp. viii, 356
- ***Libraries*** L, SanU : AzU, OkU
- ***Notes***

1. A general geography, in the sense of Varen. It is included here because such works were, as the author noted (p. 111), extremely scarce at that time. The word *climate* is used in the modern sense, a usage that was only becoming established at about that time (Bowen 1981:161, 220). Considerable space is devoted to the climates of northern countries. There is no sign that the second part was ever published.

MACHAN, R. **?**

- **{1841} (01)**
- *Geographical clock and companion*
- ? Hamilton
- ***Libraries*** n.r.
- ***Source*** BCB, Index (1837-57:109)

MacKAY, Alexander **Y SG?**

- **{1864} (01)**
- *Elements of modern geography for the use of junior classes*
- EDINBURGH; W. Blackwood and sons
- 8vo; pp. vii, 297
- ***Libraries*** L : ICN
- ***Source*** BL and NUC (NM 0060604)
- ***Notes***

1. BL and NUC record further eds. (all Edinburgh and London, except as noted): 1867, 4th (L); 1872, 11th, Edinburgh, 297 pp. (: ViU); 1872, 12th, Edinburgh (L : PPD); and 1885, rev., 297 pp. (L).

MacKAY, Alexander **Y SG?**

- **{1867} (01)**
- *First steps in geography*
- EDINBURGH; Blackwood
- 16mo; pp. 56
- ***Libraries*** : MB, PU
- ***Source*** NUC (NM 0060609)
- ***Notes***

1. The date is taken from ECB, Index (1856-76:130).
2. BL records a further ed.: [1869], Edinburgh.

MacKAY, Alexander **Y mSG?**
- **{1874} (01)**
- *The intermediate geography, physical, industrial and commercial.* [With extended section on the Australian colonies, etc.]
- EDINBURGH; W. Blackwood and Sons
- 8vo; pp. vi, 202
- ***Libraries*** L
- ***Notes***

1. BL records further eds.: 1882, 8th, Edinburgh, 232 pp.; and 1885, 10th, Edinburgh, 232 pp. BL also records an 18th ed. in 1894, and notes 'One of "Blackwood's class books, geographical series".' ECB, Index (1890-7:376) records 19th ed. in 1897.

MacKAY, Alexander **A**
- **1861 (01)**
- *Manual of modern geography, mathematical, physical and political on a new plan*, embracing a complete development of the river systems of the globe
- EDINBURGH and London; William Blackwood and Sons
- 8vo; pp. xvi, 695, [1], 8
- ***Libraries*** **L** : DS, MB, MH, MWC
- ***Notes***

1. A typical detailed nineteenth-century special geography; similar to HUGHES (1875). The rivers of individual countries are listed in tabular form, with information provided on their length, the area of their basins, and the capitals of the states and provinces located in each basin.
2. BL and NUC record further eds.: 1870, 1871, 2 pts., Edinburgh (L); 1873, Edinburgh (printed), London (L); 1876, Edinburgh, 676 pp. (L : NN); 1881, Edinburgh, 676 pp. (L : MiU); and 1885, Edinburgh, 676 pp. (L).

MacKAY, Alexander **Gt**
- {1885}
- *Modern geography*
- ***Libraries*** : DN
- ***Source*** NUC (NM 0060613)
- ***Notes***

1. This is probably a ghost. DN reports, first, that they 'are at a loss to account for the citation published in the *National Union Catalog* (i.e., NM 0060613)'; second, that while their own files suggest that the book they once held, and to which the NUC entry probably refers, was probably entry (06) of MacKay (1861), above, that book can no longer be found.

MacKAY, Alexander **Y SG?**
- **{1865} (01)**
- *Outlines of modern geography.* A book for beginners
- LONDON, Edinburgh (printed)
- 8vo.
- ***Libraries*** L

MacKAY, Alexander **Y SG?**
- **{1873} (01)**
- *A rhyming geography for little boys and girls*, etc.
- EDINBURGH; Bell and Bradfute
- 8vo; pp. vi, 154
- ***Libraries*** L
- ***Notes***

1. BL also records a further ed.: 1876, London.

MacLACHLAN, John **Y SG?**
- **{1863} (01)**
- *A first geography*
- EDINBURGH
- 18mo
- ***Libraries*** L
- ***Notes***

1. BL notes: 'Part of "M'Lachlan's Elementary Series",' and supplies the date.
2. ECB, Index (1856-76:130) records an ed. in 1864.

MAINWARING, Thomas **Y SG?**
- **{1808} (01)**
- *The elements of universal geography*; adapted to maps, with a selection of the most useful problems on the terrestrial and celestial globes, for the use of schools and private tutors
- Place/publisher n.r.
- 12mo
- ***Libraries*** n.r.
- ***Source*** Watt (1824, 2)

MAIR, John **A**
- **1762 (01)**
- *A brief survey of the terraqueous globe*. Containing, I. The description and use of globes. II. The construction and use of maps. III. Geography; or, a short view of the earth's surface, considered as inhabited by various nations: exhibiting, I. Their situation, extent, and boundaries. II. Their divisions, subdivisions, and chief towns. III. Their mountains of note, and principal rivers. IV. Their soil, produce, and commerce. V. Their government, revenues, land and sea forces. VI. Their religion, language, and literature
- EDINBURGH; printed for A. Kincaid and J. Bell and W. Gray, Edinburgh[;] and R. Morrison and J. Bisset, Perth
- 12mo, 16 cm.; pp. viii, 251
- ***Libraries*** L, SanU : CtY
- ***Notes***

1. The special geography occupies pp. 61-251, with Europe receiving relatively much attention, and Asia and Africa the least.

- **{1775} (02)**
- *A brief survey of the terraqueous globe*: containing the description ... globes, the construction ... maps, etc. Illus. with maps of the ancient and modern worlds
- Place/publisher n.r.
- F'cap, 8vo
- ***Libraries*** n.r.
- ***Source*** James Thin, Catalogue 331, *General second-hand and antiquarian*; item 2031

- **{[1780?]}, republished, with ... (see below) (03)**
- *A brief survey* ... III. Geography; or, a short view of the ancient and modern state of the several kingdoms of the world ... Written originally by John Mair, A.M. and now republished, with great additions, amendments, and improvements, illustrated with maps ... engraved by T. Kitchen
- EDINBURGH; printed for Bell and Bradfute, and Wm. Creech
- vi p., 1 l., 381 p.
- ***Libraries*** : NBu

- **{1789} (04)**
- *A brief survey of the terraqueous globe ...*
- EDINBURGH; Bell & Bradfute; Wm C[reech]
- 12mo; pp. vi, 381
- ***Libraries*** L

MAIR, Robert Henry (1868)
See (02) of BUTLER, Charles (1846).

MAJOR, Henry **Y**
- **{[1875]} (01)**
- *Major's new code geography.* Standard II-VI
- NOTTINGHAM
- 16mo, 5 pts.
- ***Libraries*** L
- ***Notes***

1. According to Lochhead (1980:108) only Standard VI would be special geography.

MALKIN, T. **?**
- **{1835} (01)**
- *Elements of geography*
- ***Libraries*** n.r.
- ***Source*** Peddie and Waddington (1914:226)

MALTE-BRUN, Conrad
- *A description of all parts of the world* ... (see (06) of MALTE-BRUN 1828)

MALTE-BRUN, Conrad **A Am tr**

- **{1828} (01)**
- *System of geography*. By M. Malte-Brun ... With additions and corrections, by James G. Percival. Embellished with a complete atlas, and a series of beautiful engravings
- NEW YORK and Boston; Stereotyped by James Conner, New York; S. Walker, Boston
- 4to, 3 vols.
- ***Libraries*** : ICRL, MChB, NjR, NN, ODW, OO
- ***Source*** NUC (NM 0156423), which notes: 'Translation of the author's Precis de la geographie universelle, Paris, 1810-29'
- ***Notes***

1. NjR reports that the work consists of 3 vols., with pages of maps and text interspersed.

- **1834, with additions and corrections (02)**
- *A system of universal geography*, or a description of all the parts of the world, on a new plan, according to the great natural divisions of the globe; accompanied with analytical, synoptical, and elementary tables. By ... editor of the 'Annales des voyages,' etc. With additions ... engravings. In three volumes
- BOSTON; printed and published by Samuel Walker. Published also in Philadelphia, by W. Moorhead, 124 Christian Street, and A.W. Rushton, 97 Wood Street;- in New York, by J. Moorhead, 357 Broome Street, and William Burnett, 17 Ann Street;- in Baltimore, by R. Reid, 48 Market Space, and D. Sullivan, Lexington Street;- and by their agents and the periodical booksellers in the principal cities and towns in the Union
- 28.5 cm., 3 vols.; Vol. I, pp. [8], [3]-640; Vol. II, pp. [8], xx, [3]-680; Vol. III, pp. [8], xv, [1], 681-1394
- ***Libraries*** L : DLC, IU, MB, **MWA**, OCl, OU, PPAmP, PHC, PPL, PPLT, PSC, PU, PV
- ***Notes***

1. On the verso of the tp., 'Entered according to Act of Congress, in the year 1834.'
2. NUC (NM 0156427) notes: 'Memoir on the life and writings of Malte-Brun. By M. J.J.N. Huot: V. 1. vii p. following Contents.'

- **{1836} (03)**
- *A system of universal geography* ..
- ***Libraries*** : DLC

- **1844 (04)**
- *A system of universal geography* ...
- BOSTON; printed and published by Samuel Walker
- 28.5 cm., 3 vols.; Vol. I, pp. [8], [3]-640; Vol. II, pp. [8], xx, [3]-680; Vol. III, pp. [8], xv, [1], 681-1394, [1]
- ***Libraries*** : AU, **MH**, OrP
- ***Notes***

1. In this edition the copyright statement is dated 1828. Judging by the size of the pages and of the type, this series could have held as much text as the six-

volume version published in Philadelphia in 1827-32, which was not copyrighted; see (05) of MALTE-BRUN (1822-33), below.

- **{1847} (05)**
- *A system of universal geography ...*
- BOSTON; printed and pub. by Samuel Walker
- ***Libraries*** : DN, ICRL, NIC
- ***Source*** NUC (NM 0156430)

- **1859, with additions ... (see below) (06)**
- *A description of all the parts of the world*, according ... globe; with analytical, synoptical and elementary tables; or, universal geography. By ... Percival. A new edition: containing recent geographical discoveries, changes in political geography, and other valuable additions; compiled from the late French editions of Malte-Brun, by MM. Huot and Lavallée, and other late authorities by W.A. Crafts. Beautifully illustrated with steel engravings and fine colored maps. In three volumes
- BOSTON; published by Samuel Walker, Jr.
- 4to, 3 vols.; Vol. I, pp. [7], 4-640; Vol. II, pp. [3], 4-680; Vol. III, pp. [2], 64 (issued in parts)
- ***Libraries*** : MH [incomplete?] : **CaBVaU**
- ***Notes***

1. NUC (NM 0156365) treats this as a distinct work rather than as a subsequent ed. of (01) to (05). The treatment here is based on the information contained on the tp. and the Preface, where it is stated that the first American ed., issued under the supervision of James G. Percival, is long out of print. NUC notes: 'Title taken from Vol. II.'

   Vol. I is devoted to physical geography, and the special geography of Asia and Oceania; Vol. II contains the special geography of Africa, America, and part of Europe; Vol. III contains the remainder of Europe.

- **{1863} (07)**
- *A description of all parts of the world* ... Crafts. Beautifully illustrated with steel engravings and fine colored maps
- BOSTON; S. Walker
- 31 cm., 3 vols.
- ***Libraries*** : CoU, DLC, NN, ViU
- ***Source*** NUC (NM 0156366)

- **{1865}, new ... (08)**
- *Universal geography*: being a description ... Lavallée, and the most recent American authorities. Beautifully ...
- BOSTON; S. Walker & co.
- 29.5 cm., 3 vols.
- ***Libraries*** : DLC : CaBVaU
- ***Source*** NUC (NM 0156451)
- ***Notes***

1. Despite the change in title, this is treated as the last ed. in the series published by Walker in Boston.

MALTE-BRUN, Conrad **A tr**

- **1822-33 (01)**
- *Universal geography*, or a description of all the parts of the world, on a new plan, according to the great natural divisions of the globe; accompanied with analytical, synoptical, and elementary tables. Improved by the addition of the most recent information, derived from various sources
- EDINBURGH
- 8vo, 21 cm., 9 vols.
- ***Libraries*** L, **OSg** : IU, KyLoU, PU
- ***Notes***

1. The pagination and titles of the individual volumes follow.
   Vol. I (1822). Containing the theory, or mathematical, physical, and political principles of geography. Printed for Adam Black; and Longman, Hurst, Rees, Orme and Black. Pp. iii-xx, 659.
   Vol. II (1822). Containing the description of Asia, with the exception of India. Printed for Adam ... Pp. [2], xiv, 640.
   Vol. III (1822). Containing the description of India and Oceania. Printed for Adam ... Pp. iii-xvi, 649.
   Vol. IV (1823). Containing the description of Africa and adjacent islands. Printed for Adam ... Orme, Brown, and Green. Pp. iii-xx, 503.
   Vol. V (1825). Containing the description of America and adjacent islands. Printed for Adam ... Pp. iii-xxiv, 613.
   Vol. VI (1827). Containing the description of Europe [see note below]. Printed for Adam Black; and Longman, Rees, Orme ... Pp. iii-xxii, 700.
   Vol. VII (1829). Containing the description of Prussia, Germany, Switzerland, and Italy. Printed ... Pp. iii-xxvii, [1], 744.
   Vol. VIII (1831). Containing the description of Spain, Portugal, France, Norway, Sweden, Denmark, Belgium, and Holland. Printed ... Pp. iii-xxii, [1], 675, [1].
   Vol. IX (1833). Containing the description of Great Britain and Ireland. Printed for Adam and Charles Black; and Longman ... Brown, Green, and Longman. Pp. iii-xix, iv-xxiv, 995.
2. Vol. IX contains a preface giving a history of the publication of the English ed., and a 'Memoir on the life and writings of Malte-Brun' by Mr J.J.N. Huot (v-xxiv, 2nd series). Pp. 625-995 are an index to the entire work.
   The subtitle of Vol. VI states that the volume contains a description of Europe; this is correct in the sense that it does have a long section on the physical geography of that continent. It also contains descriptions of Turkey in Europe, Hungary, Russia, and Poland.
3. Among the works listed in this guide, this one probably shares with that of Varen the highest esteem among present-day historians of geography.
4. Catalogued by BL under Bruun, Malthe Conrad.

- **1824-31 (02)**
- *Universal geography*; or, a description of all the parts of the world ... accompanied with analytical, synoptical, and elementary tables
- BOSTON; Wells and Lilly
- 21 cm., 8 vols.; Vol. I (1824), pp. [5], vi-xliii, [1], 635; Vol. II (1824), pp. [5], vi-xxx, 640; Vol. III (1825), pp. [5], vi-xxxi, [1], 649, [3]; Vol. IV (1825), pp. [5], vi-xxiii, 503; Vol. V (1826), pp. [5], vi-xxvii, [1], 615; Vol.

VI (1828), pp. [3], iv-xxiv, 700; Vol. VII (1829), pp. [3], iv-xxviii, 774; Vol. VIII (1831), pp. [3], iv-xxx, 820, lii
- ***Libraries*** : CSmH, CtY, DGU, DLC, DNLM, ICJ, IU, MBCo, MeB, **MH**, N, NjNbS, NNC, NWM, PPL, ViU, WaWW
- ***Notes***
1. According to NUC (NM 0156439), the entire edition was published in 1824; the dates given here are based on the copies held by MH.
   MH has two copies, one of 24.5 cm, with a deckle edge; it is incomplete, lacking Vol. VIII. Vol. VIII of the other copy shows that it was published by 'Lilly and Waite (late Wells and Lilly).

- **{1826-31} (03)**
- *Universal geography ...*
- BOSTON; Wells and Lilly
- 23 cm., 8 vols.
- ***Libraries*** : CSt, MiD, MoU, NjR, ViU
- ***Source*** NUC (NM 0156440), which notes: 'Vols. 3, 4, and 5 have note on t.-p: "Likewise additional matter, not contained in the European ed., and corrections".'

- **{1827-9} (04)**
- *Universal geography ..*
- PHILADELPHIA; Vols. 1-3 (1827), A. Finley; Vols. 4-6 (1829), J. Laval and S.F. Bradford
- 24 cm., 6 vols. (Vols. 5 & 6 paged continuously)
- ***Libraries*** : DLC, ICJ, IdU, IU, KyLxT, MiU, MWA, NcU, Nh, OClWHi, ODW, OFH, PP
- ***Source*** NUC (NM 0156441)

- **{1827-32} (05)**
- *Universal geography ...*
- PHILADELPHIA; published by Anthony Finley, north east corner of Fourth and Chesnut Streets. William Brown, printer
- 24 cm., 6 vols.; Vol. I (1827), pp. [3], iv-xxii, [2], 503; Vol. II (1827), pp. [3], iv-xv, [2], 529; Vol. III (1827), pp. [3], iv-xii, 439; Vol. IV (1829), pp. [3], iv-x, 450; Vol. V (1832), pp. [2], 427; Vol. VI (1832), pp. [2], [427]-823, [2], iv-xvii, [1], xxvi
- ***Libraries*** : **MH**, NcD, NcU, NN, OC, PP, PSt, ViU
- ***Source*** NUC (NM 0156443)
- ***Notes***
1. NUC records further eds.: 1829, Philadelphia, [several] vols. (KyU, NN, ViU); 1832, Philadelphia, 5 vols. (AU, DLC, NcD, PPDrop, TU); 1834, 2nd [Edinburgh], Edinburgh, 9 vols. (DLC).
2. BCB, Index (1837-57:109) records an ed. in 1851 as *Geography, Malte-Brun's*, [Edin.]? Bohn.

MANGNALL, Richmal **Y SG?**
- **{1815} (01)**
- *A compendium of geography*[, with geographic exercises,] for the use of

schools, [private families and all those who require knowledge of this important science ...]
- LONDON
- 12mo
- ***Libraries*** L
- ***Notes***

1. The words in the title set in [ ] are taken from the 3rd ed.
2. BL and NUC record further eds.: 1822, 2nd (L); 1829, 3rd, 592 pp. (: CLSU). BCB, Index (1837-57:110) records an ed. in 1845.

MANN, Herman **Y Am**

- **{1818} (01)**
- *The material creation*: being a compendious system of universal geography and popular astronomy. Vol. I - devoted to geography ...
- DEDHAM, MA; H. & W.H. Mann
- xii, 348 p.
- ***Libraries*** : CtY, MeB, NNC, NSyU, PHi
- ***Source*** NUC (NM 0176025)

- **{1818} (02)**
- *The material creation*: being a compendious system of universal geography and popular astronomy - particularly designed for American schools, academies, etc.
- DEDHAM, MA; H. & W.H. Mann
- 19 cm.
- ***Libraries*** : IU
- ***Source*** NUC (NM 0176026)
- ***Notes***

1. See VINSON (1818).
2. Cited in Elson (1964:396).

MANNING, Edward **Y Can**

- **{1864} (01)**
- *A catechism of geography*; being an easy introduction to the knowledge of the world, and its inhabitants: the whole of which may be committed to memory at an early age, and designed for pupils in infant preparatory schools. Adopted for the schools of New Brunswick, and brought down to the present state of geographical knowledge, by Edward Manning, English and mathematical master of the Saint John Grammar School
- SAINT JOHN, NB; published by J. & A. McMillan
- 16 cm; pp. [3], 4-80
- ***Libraries*** : : CaNBSM, CaOTP
- ***Source*** BoC 9223
- ***Notes***

1. Alston and Evans (1986) quote from the Preface: 'The want of a manual of geography for junior classes in which British America should occupy the largest space, has long been felt by many. Lovell's recent book is an important step in the right direction; but it is only suitable for senior

pupils, and does scant justice to the Lower Provinces. And his smaller manual is both expensive and inferior. The low price of the present book has of course prevented the insertion of maps.'
2. A microform version is available from CIHM.
3. See PINNOCK (1821, 1831a, 1831b).

MARKWELL, John **Y SG?**
- **{1872} (01)**
- *A junior geography on the principles of comparison and contrast*; with numerous exercises
- LONDON
- 8vo
- ***Libraries*** L
- ***Notes***
1. ECB, Index (1874-80:67) has an ambiguous entry in which one, and perhaps more, eds. are recorded in 1875-9. BL notes: 'See Morell (J.D.) Dr. Morell's English Series.'

MARKWELL, John **Y SG?**
- **{1874} (01)**
- ... *A senior geography, on the principles of comparison and contrast*. With four hundred exercises
- LONDON; W. Stewart & Co.; [etc. etc.]
- 18.5 cm.; pp. xii, [5]-320
- ***Libraries*** : NNC
- ***Notes***
1. BL records a further ed., [1876], and notes: 'See Morell (J.D.) Dr. Morell's Secondary Series.' See note to MORELL (1875).

MARKWELL, John **?**
- **{1875} (01)**
- *Geography, short*
- [LONDON]; Longmans
- ***Libraries*** L
- ***Source*** ECB, Index (1874-80:67)

MARTIN, Benjamin **OG**
- **1743 (01)**
- *A course of lectures in natural and experimental philosophy, geography and astronomy*: in which the properties, affections, and phaenomena of natural bodies hitherto discover'd, are exhibited and explain'd on the principles of the Newtonian philosophy, under the following heads, viz. Physics ... Mechanics ... Hydrostatics ... Hydraulics ... Pneumatics ... Phonics ... Light and colours ... Optics ... Astronomy ... Geography; or the theory of the earth, and the use of the terrestrial globe, explain'd ...
- READING, Berkshire; printed ... by J. Newberry ...

- 4to, 27 cm.; pp. [12], 125, [5]
- ***Libraries*** **L** : CtY, KU, WU
- ***Notes***

1. Includes no descriptions of any part of the earth, even at the level of the continents. The title shows the way that, at the time, geography overlapped the fields of physics and astronomy.

MARTIN, Benjamin [?] **Dy?**

- **{1760} (01)**
- *Gazetteer of the known world*
- LONDON
- ***Libraries*** n.r.
- ***Source*** Cox (1938:353); in neither BL nor NUC

MARTIN, Benjamin **OG**

- **{1758} (01)**
- *New principles of geography and navigation*, in two parts ..
- LONDON
- 50 cm.; pp. 98
- ***Libraries*** : RPB
- ***Source*** NUC (NM 0258636)
- ***Notes***

1. Judging by the other titles attributed to this author in NUC, this is likely to be mathematical geography, as that term was understood in the eighteenth century, rather than special geography.

MARTIN, Robert Montgomery, ed. **At**

- **{[1851?]} (01)**
- *The illustrated atlas and modern history of the world*, geographical, political, commercial, and statistical
- LONDON and New York; J. & F. Tallis
- [164] p. incl. 79 maps
- ***Libraries*** : OHi
- ***Notes***

1. NUC lists six entries with this title, with editions attributed to 1851 and 1857. Judging by the bibliographical information provided, the work is essentially an atlas. It was sometimes issued in two volumes, and two libraries (CtY, MB) are listed as holding only Vol. 2, whose title begins, *Index gazetteer of the world*. The work also seems to have been issued with a title that begins, *Tallis's illustrated atlas* (NM 0265462). The libraries holding some version of the work are (CSmH, CtY, DLC, ICJ, ICU, IU, MdBP, MH, N, NcD, NN, NPV, OHI, PPF, TxU : CaOLU).

MARTYN, William Frederick **A**

- **1782-3 (1784?) (01)**
- *The geographical magazine*; or, a new, copious, compleat, and universal sys-

tem of geography. Containing an accurate and entertaining account and description of the several continents, islands, peninsulas, isthmuses, capes, promontories, lakes, rivers, seas, oceans, gulphs, and bays, of Asia, Africa, Europe, and America; divided into empires, kingdoms, states, and colonies. With the climate, situation, extent and boundaries of each; and their several provinces, districts, capitals, cities, universities, towns, and villages. Also, the various forms of government, laws, religions, revenues, and naval and military powers, of the different countries; with all the castles, fortifications, sea ports, harbours, docks, arsenals, aqueducts, roads, public edifices, palaces, churches, mosques, temples, ruins, antiquities, natural and artificial curiosities, mountains, volcanoes, mines, metals, minerals, fossils, gums, trees, plants, fruits, flowers, herbs, and other vegetable productions. Their literature, arts, sciences, trade, manufactures, and commerce. The customs, manners, genius, dispositions, habits, amusements, and civil and religious ceremonies of the inhabitants; with the titles and distinctions of honour peculiar to each country. And pleasing and interesting descriptions of the infinite variety of birds, beasts, fishes, amphibious animals, reptiles, and insects. Likewise, an exact account of the coins, weights and measures, of the various countries; with tables reducing them to the value and standard of Great Britain; a geographical index, containing the names of places, alphabetically arranged; a biographical list of learned, eminent, and ingenious men, of every age and country; and a chronological table of remarkable events from the creation of the world. With a concise history of each country, from the earliest periods; comprehending an interesting and entertaining compendium of ancient and modern universal history. To which are prefixed an introductory treatise on the sciences of geography and astronomy; and their relation to each other: with the figures, motions, and distances of the planets, agreeable to the Newtonian system, and the observations, discoveries, and improvements, of Dr. Halley, Mr. Ferguson, and others. And a new and familiar guide to the use of the celestial and terrestrial globes. By ... assisted by the voluntary communications of several gentlemen of distinction in the different countries

- ▸ LONDON; printed for Harrison and Co. No. 18, Paternoster-Row
- ▸ 4to, 26 cm., 2 vols.; Vol. I (1782 [according to the tp., but see below]); pp. [6], vi-lxiv, 772; Vol. II (1783 [according to the tp., but see below]); pp. [4], iv-vi, 608
- ▸ *Libraries* CfU, **L**, O : CP, CSmH, CtY, CU, DLC, ICN, InU, LU, MdBP, MiD, MnU, NcD, NjR, NN, OU, PU, WaU : **CaAEU**, CaBViPA
- ▸ *Notes*

1. An unsystematic collection of facts that does not live up to the claims of the title. Much information is given on the customs and manners of some people, but there being no index, this can be found only by reading the whole. There is no uniformity in treatment or sequence of topics from country to country. It is unusual in putting Asia and Africa before Europe, and for giving more information about at least the larger Asian countries than about those of Europe.

   There may have been more than one issue. In the case of the BL copy, Vol. I has plates dated 1783, while Vol. II has plates dated 1784. The CaAEU copy has plates dated 1784 in Vol. I and 1785 in Vol. II.

According to Watt (1824), the author was not Martyn but William Fordyce Mavor.
2. For another seemingly similar venture, see GEOGRAPHICAL MAGAZINE (1790), including the note to that entry.

MATHER, Joseph H. **Y Am**
- **{1849}, revised and improved (01)**
- *Manual of geography*, embracing the key to Mitchell's series of outline maps, revised and improved
- UTICA, [NY]; Hawley
- Pp. 178
- ***Libraries*** : MH
- ***Source*** NUC (NM 0329860). NRHi no longer hold a copy. According to MH, this is a school textbook
- ***Notes***

1. NUC records further eds.: 1850, Hartford, 178 pp. (IU); and 1851, Hartford (MH).

MAUNDER, Samuel **Dy**
- **{[1845?]} (01)**
- *The little gazetteer*; or, geographical dictionary in miniature: describing the situation, extent, and other topographical features, with the commerce, manufactures, productions and general statistics, of every country in the world; to which are added, a comprehensive population table, and a list of the cities, boroughs, etc. of England and Wales, corrected up to the present time
- LONDON; printed for the proprietor
- 8.5 cm.; 3 p. l., [v]-xvi, 838 p.
- ***Libraries*** : DLC, IU, MH
- ***Source*** NUC (NM 0350270), which notes: 'added title page engraved: the little gazetteer, or, universal geographical dictionary, in miniature. Running title - the miniature universal gazetteer.'

MAUNDER, Samuel
- *The treasury of geography*. See HUGHES, W. (1856).

MAURY, Matthew Fontaine **Y Am**
- **{1881} (01)**
- ... *Elementary geography*. Designed for the primary and intermediate classes. Revised and abridged from the 'First lessons' [i.e., MAURY 1868a] and 'world we live in' [.e. MAURY 1868b]
- NEW YORK; University Publishing Company
- 24 cm.; pp. 104
- ***Libraries*** : CSmH, DLC
- ***Source*** NUC (NM 0360893), which notes: '(Maury's geographical series)'
- ***Notes***

1. According to the Preface of the ed. of 1892, the man who did the revising

and abridging was Mytton Maury. According to the same source, the work to which this is a note was designed as a companion to MAURY (1870).

On the basis of the ed. of 1892, the book presents the then conventional division of humanity into five races and four levels of development. Paragraphs are numbered, and have questions associated with them.

2. NUC records further eds. as follows: 1882, 104 pp. (ViU); 1886, 104 pp. (ViU); and also 1892, 1895, 1898, 1899, c1921, c1930, and c1931.
3. Carpenter (1963:262-3), and Dryer (1924:123) provide general discussion; Hauptman (1978:433) quotes from the ed. of 1892, and refers to it in a general discussion (1978:433-4).

MAURY, Matthew Fontaine **Y Am**

- **{1868}a (01)**
- *First lessons in geography*
- NEW YORK; Richardson & co. New Orleans; D.H. Maury
- 8vo; pp. 62.
- ***Libraries*** : DLC
- ***Source*** NUC (NM 0360913), which notes: '(Maury's geographical series).' Discussed in Brigham and Dodge (1933:24).
- ***Notes***

1. NUC records further eds.: 1871, 62 pp. (DLC); 1872, 62 pp. (ViU); 1876, 62 pp. (ViU); and 1878, 62 pp. (ViU).
2. Carpenter (1963:262-3) provides a general discussion.

MAURY, Matthew Fontaine **Y Am**

- **{[c1870]} (01)**
- ... *Manual of geography*: a complete treatise on mathematical, civil, and physical geography
- NEW YORK, Baltimore; University publishing company
- 31.5 cm.; pp. 162
- ***Libraries*** : DLC, TxU
- ***Source*** NUC (NM 0360936), which notes: '(Maury's geographical series)'
- ***Notes***

1. The following notes are based on the ed. of 1902 (NN). Still contains the Preface of the 1st ed., which presents the need for textbooks that can 'redeem the most delightful of subjects from the bondage of dry statistics ... and ... the drudgery of vague generalizations.' Divides humanity into the then conventional five races and four levels of development.
2. NUC records further eds.: 1880, New York, 136 pp. (DLC) [according to Culler (1945:309) this is the 2nd ed.]; [Elson (1964:404) records an ed. in 1881]; 1882, New York, 136 pp. (MH); 1885, New York, 136 pp. (OO); and 1887, New York, 136, 10 pp. (DLC) [NM 0360941 notes: 'Revised by Mytton Maury. "Extra pages containing special geography of Virginia": 10 p. at end']; and also 1891, 1892, 1895, 1898, 1899 (three times), and 1903.
3. Carpenter (1963:262-3), and Dryer (1924:123) provide general discussion; Brigham and Dodge (1933:24) discuss (01); Elson (1964:404) cites the ed. of 1881.

MAURY, Matthew Fontaine **Y Am**

- **{[1880?]}, revised (01)**
- *New complete geography*; revised
- NEW YORK; American Book co.
- ***Libraries*** : DLC, PP
- ***Source*** NUC (NM 0360970), which notes: 'c1880-1915'

MAURY, Matthew Fontaine **Y Am**

- **{1868}b (01)**
- ... *The world we live in*. [Intermediate]
- NEW YORK; Richardson and company. New Orleans; D.H. Maury, [etc.]
- 26.5 cm.; pp. 104
- ***Libraries*** : DLC
- ***Source*** NUC (NM 0361095), which notes: 'Maury's geographical series.' Discussed in Brigham and Dodge (1933:24), and Culler (1945, *passim*).
- ***Notes***

1. NUC records further eds.: 1871, 104 pp. (DLC, NcD); 1877, 104 pp. (ViU).

MAXWELL.
See LAURIE, J. Werner (1869a, 1869b).

McCORMICK, Henry **Y OG Am**

- **1885 (01)**
- *Practical work in geography*, for the use of teachers and advanced pupils. Being a guide for the young teacher in teaching preparatory, elementary and advanced work in geography, showing what to teach and what to omit from the text books. By ... Ph.D., Professor of History and Geography, in the Illinois State Normal University, Normal, Illinois
- CHICAGO; A. Flanagan, publisher
- 20 cm.; 324 pp
- ***Libraries*** : OrU
- ***Notes***

1. M. notes in his Preface that the greatest problem in *teaching* [my emphasis] geography is to distinguish between the facts that should be taught and those that should be omitted. In the body of the book, the distinction between the preparatory stage and those that follow is categorical, whereas that between the elementary and the advanced stages is one of degree.

   Whether knowingly or not, M. recommends that pupils should be introduced to the basic concepts of geography along the lines advocated by Pestalozzi and his followers for a century. In the preparatory stage, the basic concepts of scale, direction, distance, and relative location are presented at ever-increasing scales beginning with the classroom, and expanding via the schoolyard to the immediate neighbourhood. At the next, i.e., the elementary stage, there are two major additions: the shape and size of the earth, and the activities associated with the human occupation of places. These require a greater degree of abstraction, though McCormick provides advice on how they can be introduced with visual aids.

At the elementary stage the pupils are also provided with factual information about places, e.g., 'Honolulu is the capital and principal city [of the Sandwich Islands]' (p. 116; this happens to be the last fact provided in Part II of the book). Part III ends with the following information: '*Honolulu*, the capital of the Hawaiian Kingdom, is situated on the south side of the island of Oahu, 2,100 miles from San Francisco and 3,440 miles from Yokahama, Japan. The city has considerable commerce, and is an important port of call for vessels trading between the United States and the countries of Eastern Asia' (p. 300). In short, the 'advanced' level differs from the elementary level only in matters of fact, which, for some critics, raises the issue of its role in the curriculum.

McCULLOCH, John Ramsey **Dy**

- **{1841} (01)**
- *A dictionary, geographical, statistical, and historical,* of the various countries, places, and principal natural objects in the world. Illustrated with maps
- LONDON; Longman, Orme, Brown, Green, and Longmans
- 8vo, 23 cm., 2 vols.; Vol. I, pp. [5], vi-viii, 1020; Vol. II [1842?], pp. [5], [2]-948
- ***Libraries*** EU, L : MH, MnHi, MWA, NjNbS, NjP, NWM, OU, PPL, ViW : CaAEU, CaSSU
- ***Notes***

1. The Preface states: 'The quantity of letter-press contained in [this work] is fully equal to three and a half times the quantity contained in the last edition of *Pinkerton's Geography*, in two large volumes quarto!' The print is small, tiny in places; on pages where the latter occurs there are 1,900 to 2,000 words per page, so the total number of words in the complete work is probably more than 3 million.
2. NUC (NM 0030838) records an ed. in 1842-4, 2 vols. (KU, MWA, ODW); BL records 'New (supplement, etc)' in 1846-9, and 1849, 2 vols.

- **{1851}, new and ... (see below) (02)**
- *A dictionary ... world.* Illustrated with maps. A new and improved ed., with a supplement
- LONDON; Longman, Brown, Green, and Longmans
- 22 cm., 2 vols.
- ***Libraries*** L : DN, ICJ, MU, NNUT, NWM, PPC, ViU
- ***Source*** BL and NUC (NM 0030839)
- ***Notes***

1. NUC records a further ed. with a supplement in 1854, 2 vols. (DLC, KyU); BL and NUC record a further ed. in 1866, 'New, carefully rev. with the statistical information brought up to the latest returns by Frederick Martin,' 4 vols. (L : CtY, DLC, DN, DN-Ob, IaU, ICJ, MdBP, MnHi, NjP, NN : CaBViPA, CaOONL).
2. Microform versions of (01) and the ed. of 1866 are available from CIHM.

McCULLOCH, John Ramsey **Dy Am**

- **{1843-4} (01)**
- *McCulloch's universal gazetteer.* A dictionary ... world ... In which the articles relating to the United States have been greatly multiplied and extended, and adapted to the present condition of the country, and to the wants of its citizens. By Daniel Haskel ... Illustrated with seven large maps ...
- NEW YORK; Harper & brothers.
- 24.5 cm., 2 vols.
- ***Libraries*** : DLC, MH, NcU, NN, OO, OU, PP
- ***Source*** NUC (NM 0030902)
- ***Notes***

1. NUC records further eds. (all New York, 2 vols.): 1844-5 (DLC, MH, MiU, OCl, OClWHi, ViU, WaU); 1844-6 (Nh) {NM 0030904 notes: '[c1844]'}; 1845-6 (OClWHi); 1846 (Vi) [NM 0030906 notes: 'c1844']; 1847-8 (DLC, ICJ, MH, MnHi, NjR, PHi, T, ViU); 1848-51 (MB); 1849 (: : CaBViP); 1851 (MnHi, NN); 1852 (DCL, MH); and 1858 (OCl).

McNALLY, Francis **Y Am**

- **1855 (01)**
- ... *An improved system of geography.* Designed for schools, academies and seminaries
- NEW YORK; A.S. Barnes & co., 51 & 53 John Street. Philadelphia; J.H. Simon. Baltimore; J.W. Bond & co. Cincinnati; H.W. Derby. Chicago; D.B. Cook & co. St. Louis; L. & A. Carr. Mobile; Strickland & co. New Orleans; H.D. M'Ginness
- 30.5 cm.; pp. 93
- ***Libraries*** : DLC, OClWHi, OO, PP, PPeSchw
- ***Notes***

1. The tp. of the PPeSchw copy shows that the words elided from the beginning of the title read: 'National geographical series, No. 3.' Supplementary information obtained from NUC (NM 0084645).
2. NUC also records *System of geography*, 1855, NM 0084659, and *McNally's system of geography* NM 0084660, 1870, both at PPeSchw. That library, however, admits to holding only one title by McNally, namely, the one to which this is a note. The tp. verso of this book also states that the 'National geographical series' is made up of three titles: *Monteith's first lessons in geography* (i.e., MONTEITH 1855), *Monteith's manual of geography* (i.e., MONTEITH 1853), and *McNally's complete school geography.* For want of an alternative, the last title is taken to be this work, i.e., *An improved system of geography.*
3. This series was taken, in 1875, as the basis for Sadlier's series of schoolbooks; see CATHOLIC TEACHER.
4. NUC records further eds. (all New York except as noted): 1856, 100 pp. (MH); 1857, 99 pp. (CtY, MH); 1859, 100 pp. (MH) [According to Culler (1945:309) this is the 2nd ed.]; 1862, 110 pp. (MH, MiU) [NM 0084649 notes: '(National geographical series, no. 4)']; 1863, 110 pp. (DCL, OClWHi); 1864, 110 pp. (MB, MH, NN, OClWHi, PP); 1866, New York, Chicago, and New Orleans, 110, 12 pp. (CtY, OkU, MH, NBuHi) [NM

0084652) notes that the copyright was issued in this year]; 1867, 110 pp. (DLC, OO, PP, PSt) [NM 0084653 notes: '(National geographical series, no. 5)']; 1868, 110 pp. (MH, MnHi); 1873, 110 pp. (FTaSU, NN) {NM 0084655 notes: '[c1866]'}; 1874, 110, 12 pp. (Wa); 1875, New York, [etc.], 110, 12 pp. (ICJ, NNC) [NM 0084657 notes: '(National geographical series, revised edition)']; 1876, New York, Chicago (etc.), 110 p., 12 pp. (DHEW, NN, OO) [According to Culler (1945:309) this is the 2nd ed.].

NUC notes an ed at OKentU [NSM 0005187]. The entry gives the date as [1855], but the other information in the entry suggests that this refers to the date of the first ed.; see (01) above. The relevant information runs as follows: 'Rev. and enl. ed.; New York; A.S. Barnes & Burr; 31 cm.; 110 p.; "National geographical series, no. 4".' All this suggests a publication date in the years 1862-6.

5. Discussed in Culler (1945, *passim*); Elson (1964:404) cites the ed. of 1875.

McNALLY, F[rancis] **Y Am**

- **{1882} (01)**
- *McNally's system of geography for schools, academies and seminaries*. Rev. by James Monteith and S.T. Frost; and including Frost's 'Geography outside of textbooks.'
- NEW YORK and Chicago; A.S. Barnes and company
- 31.5 cm.; 1 p. l., 141 p.
- ***Libraries*** : DLC, LU, OO
- ***Source*** NUC (NM 0084661)

MEABY, George **Y SG?**

- **{1860}, 2nd (01)**
- *An initiatory geography in question and answer* ... Second edition
- LONDON; Relfe Bros
- 12mo
- ***Libraries*** L
- ***Notes***

1. ECB, Index (1856-76:130) records an ed. in 1862.

[MEAD, Bradock] **OG**

- **1717 (01)**
- *The construction of maps and globes*. In two parts. First, contains the various ways of projecting maps, exhibited in fifteen different methods, with their uses. Second, treats of making divers sorts of globes, both as to the geometrical and mechanical work. Illustrated with eighteen copper plates. To which is added, an appendix, wherein the present state of geography is consider'd. Being a seasonable enquiry into maps, books of geography and travel. Intermix'd with some necessary cautions, helps, and directions for future map-makers, geographers, and travellers
- LONDON; printed for T. Horne, J. Knapton, R. Knaplock, J. Wyat, T. Varnam and J. Osborn[,] D. Midwinter, R. Robinson, W. Taylor, J.

Bowyer, H. Clements, W. Mears, R. Gosling, W. Innys, J. Browne and W. Churchill

- 8vo, 19 cm.; pp. [32], 216
- ***Libraries*** **L** : AzU, CLU
- ***Notes***

1. Primarily a methodological work. As the title implies, it is concerned chiefly with maps, and with the problems involved in producing accurate ones. The author also has some comments on the texts that accompany the maps in special geographies. He mentions with qualified praise the geographies of Eachard, Mordern, Gordon, and Heylyn (i.e., EACHARD 1691, MORDEN 1680, GORDON 1693, and HEYLYN 1652). He is critical of *A Short Way to Know the World* (i.e., H., T. 1707), and the ATLAS GEOGRAPHUS (1711-17), of which he says, 'there are very good materials, but they want form' (p. 163). He notes that it was published in monthly installments (to which he attributes some of the disorder). His highest praise is reserved for 'that which goes by the name of Moll's Geography' (p. 164). This might be either (02) or (03) of THESAURUS GEOGRAPHICUS.

   BL identifies the author and adds that, subsequently, he changed his name to John Green. The identification was made by G.R. Crone (1949, 1951). See GEOGRAPHY REFORMED (1739).

MELA BRITANNICUS
See KELSALL (1822).

MELA, Pomponius **A tr**

- **{1585} (01)**
- *The worke of Pomponius Mela.* The cosmographer, concerning the situation of the world, wherein euery parte, is deuided by it selfe in most perfect manner, as appeareth in the table at the end of the booke. A booke right pleasant and profitable for all sortes of men: but speciallie for gentlemen, marchants, mariners, and trauellers, translated out of Latine by Arthur Golding gentleman
- LONDON; printed for Thomas Hacket, and are to be sold at his shop in Lumbert Street, under the signe of the Popes head
- 4to, 19 cm.; pp. [6], 93, [2]
- ***Libraries*** **L** : CSmH, CU, DFo, DLC
- ***Notes***

1. Mela's is the earliest Latin geography extant, dating from 43 A.D. Thomson (1948:225-6) says: 'It is a bad work ... He uses obsolete sources ... quite uncritically.' Despite these shortcoming, because he wrote in Latin and not Greek, Mela was influential in the early Middle Ages. Though his influence declined with the recovery of the more scholarly Greek authors in the later Middle Ages and the Renaissance, he shared in the generally high regard with which all the Classical authors were held (Wright 1925).
2. See note to (02), below.

- **1590 (02)**
- *The rare and singular worke of Pomponius Mela,* that excellent and worthy

cosmographer, of the situation of the world, most orderly prepared, and deuided euery parte by it selfe; with the longitude and latitude of euerie kingdom, regent, prouince, riuer, mountaines, cities and countries. Whereunto is added, that learned worke of Iulius Solinus Polyhistor, with a necessarie table for thys booke; right pleasant and profitable for gentlemen, marchaunts, mariners, and trauellers. Translated into Englishe by Arthur Golding gentlemen

- LONDON; printed for Thomas Hacket, and are to be solde at his shoppe in Lumbertstreete, under the signe of the Popes head
- 4to, 18 cm., 2 pts.; Pt 1, pp. [16], 124; Pt 2, pp. [228]
- ***Libraries*** C, L, O : CSmH, DFo, DLC, MWA, NNC
- ***Notes***

1. Part 1 carries the translation of Mela. The text has been reset since 1585, and there are some changes of spelling. The major difference, however, begins on p. 94. In 1585 the equivalent page came at the end of the text; now it is the start of a 30-page chorographical summary of the four continents. In Europe several countries are described individually, a body of text being followed by a list of the principal places with their longitude and latitude. For the other three continents, apart from an opening 1- to 2-page description of the continent as a whole, there is little text, but numerous locations. This whole section may have been inspired by the Fifth Book of CUNINGHAM (1559), which it resembles in organization.

   For Pt. 2, see SOLINUS (1587) (02).
2. Discussed in *Encyclopaedia Britannica*, 13th ed. (cited by Cox 1938:335).

MELISH, John **Y Am**

- **{1818} (01)**
- *A geographical description of the world*, intended as an accompaniment to the map of the world on Mercator's projection
- PHILADELPHIA; J. Melish and S. Harrison
- 21.5 cm.; 7, [ix]-xiii, [2], [9]-280 p.
- ***Libraries*** : AU, DLC, MiDSH, OKentU, ViW
- ***Source*** NUC (NM 0421851)
- ***Notes***

1. Classified as a textbook by NUC, which also records: 1822, Philadelphia, the author, new, greatly improved, 289 pp. (MoSW).

MERCATOR, Gerard **Gt**

- **{1632}**
- *Mercator's Atlas in Latin and English*
- ***Source*** E.G.R. Taylor (1934:272); in neither BL nor NUC; nor in Pollard and Redgrave (1976, 2). Given the attention that has been paid to the history of cartography, in particular its major figures, it seems probable that this is a ghost.

MERCATOR, Gerard [and Jodocus HONDIUS] **A tr els**

- **1636 (01)**

- *Atlas or a geographicke description of the regions, countries and kingdomes of the world*, through Europe, Asia, Africa, and America, represented by new and exact maps. Translated by Henry Hexham, Quarter-maister to the regiment of Colonell Goring
- AMSTERDAM; printed by Henry Hondius and Iohn Iohnson
- Folio, 48 cm., 2 vols.; Vol. I, pp. [20], 216, [3]; Vol. II, pp. [2], 216-462 [i.e., 246], [3]
- ***Libraries*** CEm, **L**, Lam, OE, OStJ : CSmH, DLC, ICN, IU, MB, MiU, MN, NjP, OClWHi, RPJCB, ViU
- ***Notes***

1. The maps were originally printed for a French edition of 1633. The tp. carries the original French wording, which was then covered by paste-on slips with the information of the English ed. (NUC [NM 0458270]).

   The wording on most of the maps is in Latin. Begins as in MERCATOR (1635; see below) with the nature of the universe and the creation of the world. There are fewer pages of text per map than in MERCATOR (1635), but because the pages are larger, it is possible that the text is not much reduced in length. Europe occupies the whole of Vol. I and pp. 216-400 of Vol. II.
2. NUC (NM 0458260) records an ed. of 1630, held by ICN, but Pollard and Redgrave (1976) do not acknowledge an ed. of so early a date; but see more below.

   Pollard and Redgrave (1976) add Hondius as the second author, and add the following information: 'Fr. title and imprint: Vol. 1: "*chez H. Hondius*, 1633" or "1636"; Vol. 2: "*sumptibus & typis H. Hondij*, 1633" or "1636". In some copies [OMan : ICN, MH], one or both of the imprint slips are missing. In the Latin "Editio decima" of 1630 and 1631 is a map by T. Pont: "A new description of the shyres Lothian and Linlitquo. *J. Hondius cœlavit sumptibus A. Hart*, [1630]" In later eds. including [this entry] the imprint is altered to: "*H. Hondius excudit*".'

   Pollard and Redgrave (1976) also record: '[Anr. issue, w. imprint:] *Amsterdam, chez H. Hondius (sumptibus & typis H. Hondi*), 1638. O, DU (frag.), MC (Vol. 2 only), SkPt (Vol. 2 only) : PU (Vol. 2 only). Apparently issued without English imprint slips.

- **{1641} (02)**
- *Atlas, or a geographicke description ...*
- Place/publisher n.r.
- 2 vols.
- ***Libraries*** : RPJCB
- ***Notes***

1. A facsimile ed. of (01) publised by TOT, 1968. The original is described by them as the only ed. of the Mercator-Hondius atlas with an English text.

MERCATOR, Gerard **A tr**

- **1635 (01)**
- *Historia mundi: or Mercator's atlas*. Containing his cosmographicall description of the fabricke and figure of the world. Lately rectified in divers places, as also beautified and enlarged with new mappes and tables; by the studious

industry of Ivdocvs Hondy. Englished by W[ye] S[altonstall] generosus, & Coll. Regin. Oxoniae
- LONDON; printed by T. Cotes, for Michael Sparke
- Folio, 29 cm.; pp. [25], 58, 930 (some misnumbering), [32]
- ***Libraries*** See notes below.
- ***Notes***

1. This is a special geography with many maps, rather than an atlas with text (see ORTELIUS 1606). The opening 49 pp. are concerned with the nature of the universe and the creation of the world (Genesis being the authority). Classical mythology also makes an appearance with a reference to 'the Race of Atlas' (p. 6, 1st series). In the main body of the text there are usually four pp. of text for each map. European countries and districts receive the bulk of the space (pp. 33-813, 2nd series). The final 32 pp. carry an index organized by countries.

   Pollard and Redgrave (1976) treat this as a variant of an ed. printed '*T. Cotes f. M. Sparke a. S. Cartwright.*' They record only three copies: **L** : CSmH, DFO. NUC (NM 0458352) treats it as the earliest ed, although at the time of publication no American library held a copy (MiU had a copy of the BL holding on film).

   Pollard and Redgrave (1976) record copies of their base version at C, E, L, O [IDT] : CtY, ICN, NN.

   NUC records two variants of (01); they differ from it in having been printed for Samuel Cartwright as well as Michael Sparke (specialists will note, however, that on a cartouche at the foot of the frontispiece of the BL copy, the names of both Michael Sparke and Samuell Cartwright appear). Copies are held by: CSmH, CtY (which holds both variants), DFo, DLC, ICN, InU, IU, NN, NRU, PPiU, RPJCB.

- **1635 (1637?), anr issue? [2nd ed?] (02)**
- *Historia mundi ...*
- LONDON; printed by T. Cotes ...
- Folio, 29 cm.
- ***Libraries*** AbrwthWN, **L**, [IDM] : CtY, IaU, ICN, InU, MiU, MWA, NcD, NcU, NN, NRU,TxU, ViU
- ***Notes***

1. NUC (NM 0458354) notes: 'Added tp. engraved, reading "second edytion" and dated 1637.' BL regards this as 1635. Spot checks suggest that the text and maps are identical to those of (01).

- **{1637}, 2nd (03)**
- *Historia mvndi or Mercators atlas.* Containing ... world. Latelij ... places as ... beutified ... mapps and tables bij ... industrie of Iodocvs Hondy ...
- LONDON; for Michael Sparke, and are to be sowld in greene Arbowre
- 4to; 3 p. l., [1038] p.
- ***Libraries*** : CSmH, DLC, RPJCB
- ***Source*** NUC (NM 0458355), which notes: 'Contains also the title-page of the English edition of 1635, upon which the imprint reads: London, printed by T. Cotes, for Michael Sparke and Samuel Cartwright, 1635. The publication changed hands two years after its first issue, and the engraved title was re-

engraved, having Sparke's name alone, with "Second edytion" added, and the date changed.'

- **{1639}, 2nd, i.e., 3rd (see below) (04)**
- *Historia mundi* ... world latelij ... places, as ... tables by ...
- LONDON; printed for Michael ...
- 30.5 cm.; 12 p. l., 56 p., 1 l., 930, [32] p.
- ***Libraries*** : MiU
- ***Source*** NUC (NM 0458361), which notes: 'This is actually the third edition; the second edition was 1637.'
- ***Libraries*** : RPJCB
- ***Notes***

1. As one of the major figures in the history of cartography, Mercator is dealt with in detail in all the standard histories of that subject (e.g., Bagrow 1964; Crone 1978), as well as being mentioned in histories of geography.

[MERITON, George] **Y**

- **1671 (01)**
- *A geographical description of the world.* With a brief account of the all several empires, dominions, and parts thereof. As also of the natures of the people, the customs, manners, and commodities of the several countreys. With a description of the principal cities in each dominion. Together with a short direction for travellers
- LONDON; William Leake, at the Crown in Fleet-Street, between the two Temple-Gates
- 12mo, 15 cm.; pp. [17], 352
- ***Libraries*** C, **L** (lacks tp.), O : CtY, IU, MiU, RPJCB
- ***Notes***

1. Contains a Preface stating that the book is an 'epitome,' presumably of one of the larger works of the time, prepared by a bookseller. Some passages in the text (e.g., the opening section on Europe) are strongly reminiscent of HEYLYN (1621), so entries (08), (09), and (10) of HEYLYN (1652) are the obvious candidates for being that 'larger work.' The section on America is very short (14 pp.), and only Mexico and Peru receive individual attention.
   It is intended for young people, and may be the first work addressed to that audience.
2. NUC (NM 0467550) records an additional [6] pp. at the end.

- **{1673}, 2nd (02)**
- [*A geographical description of the world*]
- ***Libraries*** n.r.
- ***Source*** Cox (1935:74)

- **1674, 2nd, enlarged ... (as below) (03)**
- *A geographical description of the world* ... travellers. The second edition enlarged and amended, with an addition of several islands countries, and places, not extant in the former impression
- LONDON; printed for William Leake, and John Leake, at the Crown ...
- 12mo, 14.5 cm.; pp. [25], 440, [4]

▸ ***Libraries*** **L** : CtY, IU, MdBJ, MH, MiU, NcU, NN
▸ ***Notes***
1. The author identifies himself in this ed. The amount of additional material is greater than the increase in the number of pages suggests because individual pages carry more text than in 1671. Eleven pp. on the world in general have been added at the beginning. Though material has been added to many countries, England receives by far the greatest amount, a description of every county having been added. The amount of text devoted to Wales and to Scotland has also been increased considerably.
2. This item appears in NUC as NM 0467552. According to NUC (NM 0467581), NN holds a book whose title is *Of the world and first of the same in general. The world is divided into four parts; Europe, Asia, Africa, America*. According to the information recorded in that NUC entry, the publisher, year of publication, and pagination are identical to those of the work to which this is a note. NN reports that the title *Of the world and first* 'was once part of the Library's original Lennox Collection but has since been lost or withdrawn.' Because no other copy of this book is known, its status is uncertain; however, it seems likely that it is an issue of the 1674 ed. of MERITON (1671) (i.e., the title to which this is a note) with a variant tp.

▸ **1679, 3rd, enlarged ... (see below) (04)**
▸ *A geographical description of the world* ... people; the ... travellers. The third edition, enlarged, and amended, with an addition of several islands, countries, and places, not extant in the former impressions
▸ LONDON; printed for William Leake ...
▸ 12mo, 15 cm; pp. [25], 388 [i.e., 384 (NM 0467554)]
▸ ***Libraries*** C, **L** : NN, TxU
▸ ***Notes***
1. The reduction in the number of pp. has been balanced by an increase in the amount of text per page. Spot-checking suggests that the text is unchanged from that of 1674.

MERITON, George **Gt**
▸ **{1674}**
▸ *Of the world and first of the same in general.* The world is divided into four parts; Europe, Asia, Africa, America
▸ ***Notes***
1. See note to (03) of MERITON (1671).

MIDDLETON, Charles Theodore, and others **A**
▸ **1777-8 (01)**
▸ *A new and complete system of geography*. Containing a full, accurate, authentic and interesting account and description of Europe, Asia, Africa, and America; as consisting of continents, islands, oceans, seas, rivers, lakes, promontories, capes, bays, peninsulas, isthmuses, gulphs, etc. and divided into empires, kingdoms, states, and republics. Together with their limits, boundaries, climate, soil, natural and artificial curiosities and productions, religion, laws, government, revenues, forces, antiquities, etc. Also the prov-

inces, cities, towns, villages, forts, castles, harbours, seaports, aqueducts, mountains, mines, minerals, fossils, roads, public and private edifices, universities, etc. contained in each: and all that is interesting relative to the customs, manners, genius, tempers, habits, amusements, ceremonies, commerce, arts, sciences, manufactures, and language of the inhabitants. With an accurate and lively description of the various kinds of birds, beasts, reptiles, fishes, amphibious creatures, insects, etc. including the essence of the most remarkable voyages and travels that have been performed by the navigators and travellers of different countries, particularly the late discoveries in the South Seas, and the voyages towards the North Pole; with every curiosity that hath hitherto appeared in any language respecting the different parts of the universe. Likewise many curious and interesting circumstances concerning various places, communicated by several gentlemen to the authors of this work. Also a compendious history of every empire, kingdom, state, etc. with the various revolutions they have undergone. To which will be added, a new and easy guide to geography, the use of globes, etc. with an account of the rise and progress of navigation, its improvements and utility to mankind. The whole embellished and enriched with upwards of one hundred and twenty most elegant and superb copper plates, engraved in such a manner as to do infinite honour to the respective artists by whom they are executed. These embellishments consist of views, maps, land and water perspectives; birds, beasts, fishes, etc. as also the various dresses of the inhabitants of different countries, with their strange ceremonies, customs, amusements, etc. etc. By ... assisted by several gentlemen eminent for their knowledge of geography

- ▸ LONDON; printed for J. Cooke, at Shakespeare's-Head, No. 17, in Pater-Noster-Row
- ▸ Folio, 37 cm., 2 vols.; Vol. I (1777), pp. [1], xxviii, 5-540; Vol. II (1778), pp. 546, [6]
- ▸ ***Libraries*** BikP, C, **L**, LeP : DLC, I, MB, NPV, OClWHi, PU, TxU : CaBVaU, CaOWaU
- ▸ ***Notes***

1. One of a number of large special geographies seemingly inspired by the discoveries made by Captain Cook (and just the first two voyages in the case of this one). Compared with MILLAR (1782), it is more orderly and there is less emphasis on the bizarre. It is unusual in that Vol. I is devoted to Asia and Africa, with each receiving approximately half the space. Vol. II deals with Europe and America, the latter getting less than one-fifth of the text devoted to it.
2. For a near contemporary work of comparable size see MILLAR (1782).

- ▸ **1778-9 (02)**
- ▸ *A new and complete system of geography* containing ... productions religion ...
- ▸ London; printed for J. Cooke ...
- ▸ Folio, 37 cm., 2 vols.; Vol. I, pp. [1], xxviii, 5-540; Vol. II, pp. 546, [10]
- ▸ ***Libraries*** **OSg** : CoGrS, DLC, MoSU, MnU, NIC
- ▸ ***Notes***

1. NUC (NM 0561266) also records an undated ed. (MiU).
2. A microform version of (01) is available from CIHM.

MIEGE, Guy **Y mSG**

- **1682 (01)**
- *A new cosmography, or survey of the whole world*; in six ingenious and comprehensive discourses. With a previous discourse, being a new project for bringing up young men to learning. Humbly dedicated to the honourable Henry Lyttleton, esq.
- LONDON; printed for Thomas Basset, at the George in Fleet-Street near St. Dunstans Church
- 8vo, 16 cm.; pp. [4], 146 [actually 148 as 127-128 are repeated]
- ***Libraries*** **L**, LPO : MH, RPJCB
- *Notes*

1. Takes the form of a dialogue between Philalethes and Sophronius. The former knows geography and is eager to see it taught to youth. In response to questions from Sophronius, he describes what is to be taught. Of the six discourses on geography, only one is concerned with the countries of the world; it forms a skeletal special geography.
2. Cox (1938:345) records the date of publication as 1683.

MILLAR, George Henry **A**

- **1782 (01)**
- *The new and universal system of geography*: being a complete history and description of the whole world. Containing a particular, full, accurate, circumstantial, and entertaining account, including the antient and present state, of all the various countries of Europe, Asia, Africa, and America, as divided into empires, kingdoms, states, republics, and colonies, and as subdivided into continents, islands, provinces, peninsulas, isthmuses, seas, oceans, gulphs, straits, rivers, harbours, deserts, lakes, promontories, capes, bays, districts, governments, principalities, etc. etc. together with their situations, extent, boundaries, limits, climate, soil, natural and artificial curiosities and productions; laws, religions, revolutions, conquests and treaties, antiquities, revenues, naval and military force, etc. likewise all the cities, capital towns, villages, their distance and bearing, universities, fortifications, castles, forts, sea-ports, mountains, volcanos, metals, aqueducts, docks, arsenals, minerals, fossils, ruins, palaces, temples, churches, structures, edifices, public and private buildings, roads, etc. contained in each part. Also a useful and entertaining historical and descriptive relation of all their customs, manners, genius, trade, commerce, agriculture, learning, policy, arts, sciences, manufactures, tempers, dispositions, amusements, habits, stature, shape, colours, virtues, vices, riches, or poverty, entertainments, language, and singular ceremonies at births, marriages, and funerals, titles of distinction, etc. of the different inhabitants: And a genuine history of all sorts of birds, beasts, fishes, reptiles, insects, vegetable productions, flowers, herbs, fruits, plants, gums, etc. found in the various regions. Including all the valuable discoveries made in the most remarkable voyages and travels to the different parts of the world from the earliest times to the present year 1782; particularly all the modern discoveries in the southern and northern hemispheres, etc. by Captain Cook, Lord Musgrave, Wallis, Carteret, Falconer, Byron, Anson, Forrest, Wraxall, Hanway, Clerke, Furneaux, Bougainville, Ives, Banks, Coxe, Dillon, Barretti, Sharp, Thickness, Algoretti, Drummond, Bruce, Carver, Suck-

ling, Chandler, Johnson, Twiss, Osbeck, Thompson, Solander, Dr. Cooke, Forster, Parkinson, Burnaby, Irwin, etc. etc. Comprising not only all the late discoveries in the Fox, and various other islands in the south seas, and towards the North Pole, but also those made in the Japanese Ocean, in the new northern archipelago, in North America, the West Indies, and those made by order of the Empress of Russia in the Red Sea, the Indian seas, eastern ocean, etc. etc. Also a great variety of curious particulars communicated to the author of this new work, by military and naval commanders, captains of ships, noblemen, private gentlemen, ingenious travelers, etc. and every curiosity extracted from various languages relative to the different parts of the universe. The whole being brought down to the present time, and forming the most extensive and original production on the subject ever published, wherein a great variety of improvements are included, not to be found in any other work of the kind. To which will be added, a new, complete, and easy introduction to geography and astronomy; giving an useful and entertaining explanation of the principles and terms of both sciences, their relation to each other, the figure, motion, etc. of the earth, planets, etc. latitude, longitude, use of maps, the compass, the nature of winds, constant and variable, of earthquakes, comets, meteors, thunder, lightening, air, and other authentic particulars, etc. By ... assisted by several gentlemen, celebrated for their knowledge in the science of geography, particularly William Langford, esq; who accompanied Capt. Cook in making new discoveries. Calculated to convey useful and entertaining knowledge to all ranks and degrees of people, for as a celebrated author justly observes, 'There is not a son or daughter of Adam but has some concern in geography.' The whole embellished with upwards of one hundred and twenty capital engravings, being the most elegant set of copper plates ever published in a work of this kind, and consisting of beautiful views, land and water prospects, dresses of the various inhabitants of different countries, their singular ceremonies, amusements, customs, etc. Also necessary maps, charts, draughts and plans, with the different objects in natural history, such as birds, beasts, fishes, amphibious animals, etc. These elegant embellishments are finished in the best manner by the celebrated Messrs. Pollard, Taylor, Rennoldson, Smith, Walker, Collyer, Wooding, Page, Sherwin, Grignon, Golder, and others, from curious original design and capital paintings made by Mr. Hamilton, Mr. West, Mr. Dodd, Samuel Wale, esq, etc. The necessary maps, etc. (absolutely forming of themselves the most valuable and complete Atlas ever delivered with any similar work) are all newly drawn and engraved according to the latest discoveries by Mr. Kitchen, Geographer and Hydrographer to his Majesty, Mr. Bowen, Mr. Conder, Mr. Lodge, etc.

- ▸ LONDON; printed for Alex, Hogg, 16, Paternoster Row
- ▸ Folio, 38 cm., 2 vols.; Vol. I, pp. [1], viii, 5-452; Vol. II, pp. [7], 457-821
- ▸ ***Libraries*** AbU, BrP, CfU, L, SwnU : CLU, CtY, DLC, ICN, KU, NBu, NN, NNH, ViU, WaS : CaQQS
- ▸ ***Notes***

1. A forerunner of the coffee-table book, with something to amaze or please on every page. Unsystematic despite the use of section headings. Does not live up to the extravagant claims of the title.
2. Price, when bound in one volume, £2-8-0.

3. For contemporary works of comparable size, see BANKES and others (1788?), and MIDDLETON (1777-8).

- **1783 (02)**
- *The new and universal system of geography ...*
- LONDON; printed for Alex. Hogg ...
- Folio, 39 cm., bound in 1 vol.; pp. [1], viii, 5-812, [8]
- ***Libraries*** **O** : CtY, NIC
- ***Notes***

1. Contains a list of 776 subscribers who bought the first edition.
2. A microform version of (01) is available from CIHM.

MILLER, Ebenezer **mSG**

- **1834 (01)**
- *A companion to the atlas*; or, a series of geographical tables on a new plan, forming a complete system of geography
- LONDON; Jackson and Walford, 18, St. Paul's Church Yard; J. Lindsay, St. Andrew's Street, Edinburgh; Stanfield, Wakefield; Baines and Newsome, Leeds; and Leader, Sheffield
- Folio, 34.5 cm.; pp. [82]
- ***Libraries*** **L**
- ***Notes***

1. There are 18 tables, each occupying a pair of facing pages; they are separated from the tables before and after by a pair of facing blank pages. It is intended as a reference work. Despite the accuracy of the title, the tables contain as much text as some books whose titles suggest they should consist of prose rather than of lists of facts. The price was 9/-, which the author acknowledges to be high.
2. NUC records a further ed. in 1838 (NN, PPF); and BCB, Index (1837-57: 109) one in 1839.

MILLER, Frederick **Y SG?**

- **{1875} (01)**
- *Elementary geography for elementary schools*. An introduction to physical and political geography
- LONDON, Malvern Link [printed]
- 8vo
- ***Libraries*** L

[MILLS, Alfred] **Y SG? Am**

- **{[1840?]} (01)**
- *The elements of geography made easy*. Embellished with neat ... engravings. Designed to render a general knowledge of the elements of geography and maps ...
- PHILADELPHIA; Morgan & Yeager
- 16mo; 12 l.

▸ *Libraries* : NN
▸ *Source* NUC (NM 0597783)

MILNER, [T.?] **Y SG?**
▸ **{1851}, 5th, enlarged, etc. (01)**
▸ *Milner's elements of geography for the use of schools.* To which is added, a concise account of the counties of England, by R. Scott
▸ LONDON; H. Harris
▸ 18mo; pp. 166.
▸ *Libraries* L
▸ *Notes*
1. It is likely that there is only one author of books related to special geography called Milner, though variously identified as T. and as Thomas.

MILNER, Thomas **A**
▸ **{[1863]} (01)**
▸ *The gallery of geography, a pictorial and descriptive tour of the world*
▸ LONDON & Edinburgh
▸ 8vo
▸ *Libraries* L
▸ *Notes*
1. BL and NUC record further eds.: 1864, London and Edinburgh, 950 pp. (: CtY, DLC, DN-Ob, NN, OO, PPL) [NM 0602278 notes that the date is taken from the preface]; 1864, Philadelphia, 2 vols. (: MH); 1869, Edinburgh (: I); [1872], New, with additions brought down to the beginning of the year 1772, etc., Glasgow, 8vo (L); 1872, New, Glasgow, 4to., 2 vols. (: NN). ECB, Index (1856-76:130) records an ed. in 1873, New, [Glasgow], 2 vols. BL and NUC record a further ed. in 1884, London, 1198 pp. (L : CtY).
2. See note to entry (02) below.

▸ **[1892?], new (02)**
▸ *The gallery of geography ...*
▸ LONDON; William Mackenzie, 69 Ludgate Hill, E.C.
▸ 25.5 cm, 8vo, 6 pts.; Part I, pp. [4], 200; Part II, pp. [2], 201-400 (no more seen)
▸ *Libraries* GP
▸ *Notes*
1. Opens with a 'prospectus' that shows the work to have been sold by subscription. The tone is jingoistic, and the general approach is that of a coffee-table book. Pt. I is devoted to exploration and discovery (160 pp.) and physical geography (40 pp.). Pt. II continues the section on physical geography; it ends with six paragraphs on 'man' (pp. 309-10). Europe follows, starting with England and Wales. For them information is given on a county-by-county basis, starting with features of physical geography, followed by a miscellaneous collection of antiquarian and topographical information. A list of 'important epochs' closes with the Parish Councils Act of 1891.

MILNER, Thomas **A**

- **1850 (01)**
- *A universal geography, in four parts: historical, mathematical, physical, and political.* By the Rev. ... M.A., F.R.G.S. Illustrated by ten maps, with diagrams and sections
- LONDON; The Religious Tract Society; instituted 1790. Sold at the Depository, 56, Pasternoster Row, and 65, St. Paul's Churchyard; and by the booksellers
- 12mo, 18.5 cm.; pp. [3], iv-xxiv, 527
- *Libraries* L : **MH**
- *Notes*

1. The date is taken from the end of the Preface. There it is also stated that the book is 'intended for the general home reader; for the emigrant who may wish to take ... a cheap and portable compendium of information relative to countries ... and for the use of colleges and schools ...' Divided into four parts. Pt. I (pp. 1-56) is devoted to 'historical geography,' and would today be called the history of geography; Pt. II (pp. 57-89) is devoted to 'mathematical geography,' i.e., chiefly the sphericity of the earth and problems of establishing location, and topics related to cartography and divisions of time; Pt. III (pp. 90-204) deals with physical geography, including within that topic a section devoted to the human race as a species within the animal kingdom; Pt. IV is 'political geography,' i.e., special geography.

- **{1876}, revised ... (see below) (02)**
- *A universal geography* ... Revised and brought down to the present time by Keith Johnston
- LONDON; Religious Tract Society
- 8vo
- *Libraries* L : DN
- *Source* BL and NUC (NM 0602311)

MIRROUR ... **mSG tr**

- **{1481} (01)**
- [An untitled work; the text begins: 'Here begynneth the table of the rubrices of this presente volume named *The Mirrour of the World* or thymage of the same ...]
- LONDON; William Caxton
- Folio; pp. 100 l., the first blank, without tp. or pagination. Sig a-m, in eights, n. four leaves. 29 lines to a page. Duff 401
- *Libraries* C (imp), L, MRu, O, WINStG : CSmH, DLC, MnU
- *Source* BL. British holdings from Pollard and Redgrave (1976), American from NUC (NI 0036446), which catalogues under Image du monde
- *Notes*

1. The NUC entry notes: 'Tr. by William Caxton from a prose version of the Image du monde (or Livre de clergie) attributed by some authorities to Goussuin, by others to Gautier, of Metz. It is derived from various Latin sources, chiefly the Imago mundi, probably comp. by Honorius Solitarius. Cf. George Sarton. Introd. to the hist. of science, Washington, 1931, v. 2, p. 591.'

NUC (NI 0036443) records an ed. published by the Early English Text Society (edited by O.H. Prior) in 1913 (for 1912). The entry provides the following information: '"Caxton's Mirrour ... the first work printed in England with illustrations ... was translated in 1480 ..." ... Translated from a prose version, sometimes known as the Livre de clergie, of the Image du monde, a compilation from various Latin sources, including the Imago mundi formerly attributed to Honorius Augustodunensis and now to Honorius Solitarius of Inciusus. The compilation is attributed by some authorities ... [as above]. The date of composition of the prose version used by Caxton is placed not far from the year 1246 ... generally assigned to the original rhymed version ...'

Cox (1935:69) draws on Prior's Introduction to add that the poem consisted of 6594 rhymed couplets, and further that 'its most interesting feature is its descriptions of strange countries, peoples, and animals, which are frequently mentioned in medieval literature.'

- **{1490} (02)**
- [Untitled as before] Here begynneth ... *myrrour of the world* ... same, etc.
- WESTMINSTER; Caxton
- Folio, 88 leaves; sig. a-l^8. 31 lines to a page
- ***Libraries*** C, Gu, L, MRu, OE : DLC, NN
- ***Source*** BL, with British holdings from Pollard and Redgrave (1976), American from NUC (NI 0036448)

- **{[1527?]} (03)**
- *The myrrour: & descrypcyon of the worlde* with many meruaylles. & the vii. scyences as gramayre[,] rethorike wyth the arte of memorye[,] logyke[,] geometrye[,] wyth the standarde of mesure & weyght and the knowledge how a man sholde mesure londe[,] borde[,] & tymber[,] and then arametryke wyth the maner of accou[n]tes and rekenynges by cyfres[,] and then misyke and astrnomye[,] with many other profytable and plesant comodytes
- [LONDON]; enprynted by me Laurnece Andrewe dwellynge in flete strete at the synge of the golde crosse by flete brydge ...
- Folio, 28 cm.; pp. [184]
- ***Libraries*** C, CMp, L, Lu, MRu : CSmH, MB
- ***Source*** BL, Pollard and Redgrave (1976), and NUC (NI 0036462), which shows the date as [1529?]. Pollard and Redgrave list under Vincentius, Bellovacensis.

- **{[1529?]}, anr issue (04)**
- *The myrrour & dyscrypcyon of the worlde* with many meruaylles. As gramayre ...
- [LONDON]; enprynted ...
- Folio
- ***Libraries*** : DFo
- ***Notes***

1. As noted above, Oliver H. Prior edited an ed. for the Early English Text Society in 1913 (Kraus reprint, 1987); it was republished by Oxford University Press in 1966.

MISSIONARY GEOGRAPHY ... **NG**

- **1825 (01)**
- *Missionary geography*; or, the progress of religion traced round the world. By an Irish clergyman, author of simple memorials
- LONDON; published by James Nisbet, 21 Berners Street, Oxford Street
- 12mo, 14 cm.; pp. [3], ii, 86
- ***Libraries*** **L**
- ***Notes***

1. Not a special geography, although it gives odds and ends of information about many places. There was a 2nd ed. in 1827.

MITCHELL, A. **Gt?**

- **{1853} (01)**
- *Geography, modern*
- ? Low
- 4to
- ***Libraries*** n.r.
- ***Source*** BCB, Index (1837-57:110)
- ***Notes***

1. Carpenter (1963:264), speaking of Samuel Augustus Mitchell, says that, 'he often dropped his first name.' See note to MITCHELL, S.A. (1846).

MITCHELL, Samuel Augustus **A Am**

- **{1837} (01)**
- *An accompaniment to Mitchell's map of the world, on Mercator's projection*; containing an index to the various countries, cities, towns, islands, etc., represented on the map ... also, a general description of the five great divisions of the globe, America, Europe, Africa, Asia, and Oceanica, with their several empires, kingdoms, states, territories, etc.
- PHILADELPHIA; Hinman and Dutton
- 23 cm.; pp. x, 11-572
- ***Libraries*** : AU, DLC, KyU, MU, OClWHi, OU, PP
- ***Source*** NUC (NM 0652958). The copy held by T is the issue of 1847. See note to (02) below.
- ***Notes***

1. BL and NUC record further eds. (all Philadelphia): 1838, Hinman and Dutton, 572 pp. (L : CtY, DLC, MdBP, NN); 1839, R.L. Barnes, 572 pp. (: MH, ViU); 1840, R.L. Barnes, 572 pp. (: DI-GS, DLC, OClWHi); 1841, R.L. Barnes, 572 pp. (: DN, ViU); 1842, R.L. Barnes, 572 pp. (: MH); 1843, [Philadelphia], S.A. Mitchell, 572 pp. (: CSmH, PP, PPA, PPLT, PV, ViLxW); 1844, Mitchell, 572 pp. (: MH, OClWHi); 1845, S.A. Mitchell, 572 pp. (: NcA, PPFr); and 1846, S.A. Mitchell, 572 pp. (: DLC).
2. Beginning in 1839, M. wrote a series of schoolbooks whose content was based, presumably, on this one (see below: MITCHELL, S.A. 1839, 1840b, 1840c, 1849); he also wrote a special geography intended for adult readers (MITCHELL, S.A. 1840a). Two decades later he produced a revised series of the textbooks (see below: MITCHELL, S.A. 1860*, 1860a, 1860b, 1865).

▸ **{1847} (02)**
▸ *Accompaniment to Mitchell's map of the world* ... [as in 1837] represented on the map, and so connected therewith, that the position of any place exhibited on it may be readily ascertained: also, a general description ...
▸ PHILADELPHIA; published by S. Augustus Mitchell, N.E. Corner of Market and Seventh Sts.
▸ 23cm
▸ ***Libraries*** : CtY, PU, T
▸ ***Source*** NUC (NM 0652968)
▸ ***Notes***
1. A facsimile of the tp. of the copy held by T shows that it is this ed., not 1837. The date of copyright is 1843.
2. 'The object of the following Accompaniment is ... to ... present in connexion with the Map, a distinct view of the principal geographical features of the world ... It is believed that the leading features in the general account given of each of the great divisions of the earth and their respective subdivisions, will be found sufficiently clear ... as to give ... a general idea of the present geography of the world' (Preface).
3. NUC records a further ed.: 1848, Philadelphia, S.A. Mitchell, 572 pp. (CtY, DLC).

MITCHELL, Samuel Augustus **Gt**
▸ **(1851)**
▸ *Easy introduction to the study of geography ...*
▸ ***Notes***
1. NUC (NM 0653080), which is one of two entries in NUC listing this title, is: *Easy introduction to the study of geography ... for instruction of children in schools and families*; Phila.; Cowperthwait, 176 pp. (OO). NUC notes: '(Mitchell's Primary geography ed. 2).' OO reports, however, that the book they have catalogued as *Easy introduction* has as its head title *Mitchell's primary geography*. NUC (NM 0653081), which is the other NUC entry, records an ed. of the *Easy introduction* in 1866. However, DLC, which is where this book is located, reports that the full title shows that the book is also a subsequent ed. of MITCHELL (1840c).
2. Elson (1964:402), Hauptman (1978, n. 61), and Sahli (1941, *passim*) all cite or discuss a book with this title; however, the information provided in the preceding notes makes it seem probable that the book in question is that listed below as MITCHELL, S.A. (1840c), i.e., the *Primary geography*; see note to (02) of that entry.

MITCHELL, Samuel Augustus **Y Am**
▸ **{1860}a (01)**
▸ *First lessons in geography [for young children.* Designed as an introduction to the author's Primary Geography]
▸ PHILADELPHIA; E.H. Butler & co.
▸ 19.5 cm.; iv p., 1 l., [7]-67 p.
▸ ***Libraries*** : DLC

- *Source* NUC (NM 0653083), which notes: 'Mitchell's new series of geographies'
- *Notes*

1. The words in the title set in [ ] are taken from the NUC entry for the ed. of 1864.
2. NUC records a further ed.: 1864, Philadelphia, (MH, OO). Culler (1945: 309) records a further ed.: 1865, 2nd, Philadelphia, E.H. Butler [Culler (1945, *passim*). NUC records further eds. (all Philadelphia): 1869 (MH); 1875, 72 pp. (CtY); 1884 (DHEW); [c1885] (MH); 1886 (DLC); yet further eds. in [1890?, c1885]; and 1918.

MITCHELL, Samuel Augustus **A Am**

- **{c1840}a (01)**
- *A general view of the world*, comprising a physical, political, and statistical account of its grand divisions ... with their empires, kingdoms, republics, principalities, etc; exhibiting the history of geographical science and the progress of discovery to the present time ... Illustrated by upwards of nine hundred engravings ...
- [PHILADELPHIA]; Thomas, Cowperthwait
- Pp. viii, 9-828
- *Libraries* : MiD
- *Source* NUC (NM 0653093)
- *Notes*

1. For information on the contents and their organization, see (02) below.
2. NUC records further eds. (Philadelphia except as noted): 1842, 612 pp. (DLC, MiU, NBu, ODW, PPLas); 1844, 612 pp. (OClW); 1845, 828 pp. (CLU, NIC, ViLxW); [1845], Richmond, VA, 828 pp. (NcD); 1846, 816 pp. (IEdS, KyU, ViU); 1848, 828 pp. (IU, NcA, OrPS); and 1850, 828 pp. (ViU).

- **1859 (02)**
- *A general view of the world ... divisions*, America, Europe, Asia, Africa, and Oceania, with their ... present time. A description of the chief mountains, rivers, lakes, plains, etc. of every section; the principal beasts, birds, fishes and reptiles; agricultural and mineral production; commerce, manufactures, education, government, arts, science, and literature. An account of the manners, customs, and condition of the inhabitants of each country together with a description of the chief cities and towns. Illustrated ... engravings of the principal vegetable productions, animals, noted edifices, curiosities, and races of men, in every region of the globe. The whole concluded by a general statistical survey of the various nations in the world, embracing the population of the United States according to the Census of 1840
- PHILADELPHIA; H. Cowperthwait & co.
- 8vo, 22 cm.; pp. [3], iv-vii, 9-828
- *Libraries* : **MB**
- *Notes*

1. Pure special geography. In the descriptions of the countries, modest amounts of information are generally provided about terrain, rivers, resources, and related economic activity, and then about religion, ethnic groups or language,

form of government, military strength, and cities. The order in which this information is given is not fixed, and varies in range of topic from country to country. M. usually finds some favourable point to make about the inhabitants of the various countries, and in general the tone foreshadows that of *National Geographic.*

The hope that even 'juvenile readers' will be able to understand the information presented in the book leaves the impression that the audience is expected to be chiefly adults.

MITCHELL, Samuel Augustus **Y Am**

- **{1840}b (01)**
- *Mitchell's geographical reader*: a system of modern geography, comprising a description of the world ...
- PHILADELPHIA; Thomas, Cowperthwait & co.
- 20 cm.; pp. viii, 9-600
- ***Libraries*** L (mislaid) : DLC, MB, MH, MoSW, NNC, OCU, PU, ViW
- ***Source*** BL and NUC (NM 0653115)
- ***Notes***

1. An advertisement in the 1849 ed. of MITCHELL (1840c), i.e., Mitchell's *Primary geography,* states that this work is a supplement to Mitchell's *School Geography* (presumably MITCHELL 1839); see note to entry (01) of Mitchell (1839).
2. Cited in Elson (1964:401).

MITCHELL, Samuel Augustus **Gt?**

- **{1854} (01)**
- *Mitchell's intermediate geography*
- PHILADELPHIA
- ***Source*** NUC (NM 0653118)
- ***Notes***

1. It is probable that this title belongs to the next entry.

MITCHELL, Samuel Augustus **Y Am**

- **{1849} (01)**
- *Mitchell's intermediate or secondary geography.* A system of modern geography, comprising a description of the present state of the world ...
- PHILADELPHIA; Thomas, Cowperthwait & co.
- 31 cm.; pp. 80
- ***Libraries*** : DLC, OClWHi, OO, PHi, PP
- ***Source*** NUC (NM 0653119)
- ***Notes***

1. NUC records further eds. (all Philadelphia): 1850, 80 pp. (DLC); 1852, Rev., 80 pp. (DLC, MB, MH, MOU) [NM 0653432 lists the title as: '... A system of modern geography ...' It is in a note to the entry that the fact that this is *M's intermediate ... geography* is revealed]; 1854 (*); 1855, Rev., 78 pp. (NcA); 1856, Rev., 86 pp. (NcU, ODW) [NM 0653122 notes: '"Geogra-

phy of North Carolina" by C.H. Wiley, p. 83-86']; and 1860, Rev., 82 pp. (OClWHi).
2. Carpenter (1963:264) provides a general discussion.

MITCHELL, Samuel Augustus **Y Am**

- **{1860}* (01)**
- [*Mitchell's new intermediate geography*]. A system of modern geography ...
- PHILADELPHIA; E.H. Butler & co.
- 31.5 cm.; pp. 102
- ***Libraries*** : CU, DLC, MH, OClWHi, OO
- ***Source*** NUC (NM 0653433), which notes '(Mitchell's new intermediate geography).' On the basis of this note, this work is interpreted as being the earliest copy of Mitchell's *New intermediate geography* yet located. According to NM 0653244, the 1st ed. was published in 1859.

- **{1865}, rev. (02)**
- [*New intermediate geography*]. A system of modern geography designed for the use of schools and academies
- PHILADELPHIA; E.H. Butler
- Pp. 103
- ***Libraries*** : DHEW, PPiU
- ***Source*** NUC (NM 0653454)
- ***Notes***

1. Discussed in Culler (1945, *passim*).

- **{1869}, rev. (03)**
- [*New intermediate geography*]. A system ... academies; illustrated by twenty-three copper-plate maps drawn and engraved expressly for this work from the latest authorities; and embellished with numerous engravings
- PHILADELPHIA; E.H. Butler & co
- 31 cm.; pp. 110
- ***Libraries*** : DHEW, DLC, MH
- ***Source*** NUC (NM 0653455)
- ***Notes***

1. NUC records further eds. (Philadelphia; E.H. Butler until 1872, then J.H. Butler until 1880, T.H. Butler in 1881, and then E.H. Butler again): 1871, Rev., 110 pp. (MnHi); 1872, Rev., 114 pp. (DLC, MH, PPT); 1875, Rev., 120 pp. (DLC) [NM 0653436 notes: '"A geography of the state of Connecticut ... By Horace Day": p. 115-120']; 1876, Rev., 114, 12 pp. (DLC, PPeSchw) [NM 0653437 notes: '"A geography of Pennsylvania ... By J.P. Wickersham": 12 p. at end']; 1877, Rev., 114, 12 pp. (DLC, PPeSchw) [NM 0653437 notes: '"A geography of Pennsylvania ... By J.P. Wickersham": 12 p. at end.' {A rare instance of two items occurring on the same NUC entry}]; 1877, Rev., 114, 8 pp. (DLC) [NM 0653438 notes: '"A geography of Rhode Island ... By Levi W. Russell": 8 p. at end']; 1877, Rev., 114, 8 pp. (DHEW, DLC) [NM 0653439 notes: '"A geography of West Virginia ... By W.K. Pendleton": 8 p. at end']; 1877, Rev., 114, 9 pp. (DLC) [NM 0653440 notes: '"A geography of Indiana ... By J.T. Scovell": 9 p. following p. 114']; 1877, Rev., 114, 12 pp. (DLC) [NM 0653441 notes: '"A

geography of Ohio ... By D.F. DeWolf": 12 p. at end']; 1877, Rev., 114, 12 pp. (DHEW) [NM 0653442 notes: '"A geography of Pennsylvania ... By J.P. Wickersham": 12 p. at end']; 1878, Rev., 114, 8 pp. (DLC) [NM 0653443 notes: '"A geography of Kansas ... By John A. Anderson ... and W.H. Rossington": 8 p. at end']; 1878, Rev., 114, 12 pp. (DHEW, DLC) [NM 0653444 notes: '"A geography of Illinois ... By J.M. Gregory": 10 p. at end']; 1878, Rev., 114, 12 pp. (DLC) [NM 0653445 notes: '"A geography of Virginia ... By John J. Lafferty": 12 p. at end']; 1879, Rev., 114, 12 pp. (DHEW, DLC) [NM 0653446 notes: '"A geography of Tennessee ... By J. Berrien Lindsey": 12 p. at end {in NM 0654327 NUC records the name as Lindsley}']; 1879, Rev., 126 pp. (DHEW, DLC, MH); 1880, Rev., 126, 12 pp. (PBa, ViU) [NM 0653457 notes: 'A geography of Virginia ... By John J. Lafferty, with a special t.-p. dated c1878: 12 p. at end']; 1881, Rev., 126 pp. (PPAN, PPeSchw, ViU); 1883, Rev., 126, 8 pp. (DLC) [NM 0653447 notes: '"A geography of Arkansas ... By James Mitchell": 8 p. at end']; 1886, c1881, New, 126 pp. (: : CaBVaU); with still further eds. in 1891, 1892, and 1895.

MITCHELL, Samuel Augustus **Y Am**

- **{1860}b (01)**
- *[Mitchell's] new primary geography ...*
- PHILADELPHIA; E.H. Butler & co.
- 22.5 cm.; pp. 95
- ***Libraries*** : DLC, ICU
- ***Source*** NUC (NM 0653273)
- ***Notes***

1. The word in the title set in [ ] is taken from the NUC entry for the ed. of 1865.

- **{1864} (02)**
- *The new primary geography*. Designed as an introduction to the author's New Intermediate Geography
- PHILADELPHIA; E.H. Butler & co.
- ***Libraries*** : MH, OO
- ***Source*** NUC (NM 0653274), which notes: 'At head of title: Mitchell's new primary geography, the second book of the series'

- **{1865} (03)**
- *Mitchell's new primary geography ...* The new primary geography: illustrated by twenty colored maps and embellished with a hundred engravings. Designed as ...
- PHILADELPHIA; E.H. Butler & co.
- 22 cm.; pp. 95
- ***Libraries*** : NNC
- ***Notes***

1. NUC records further eds. (Philadelphia; E.H. Butler until 1871, then J.H. Butler until 1879, T.H. Butler in 1881, and then E.H. Butler again): 1866, 96 pp. (ViU); 1868, 100 pp. (PU); 1871 (MH); 1874, 100 pp. (T); 1876, 100, [6] pp. (DLC) {NM 0653280 notes: '[Delaware ed.]'}; 1876, 100, [8]

pp. (DHEW) [NM 0653281 notes: 'A geography of Pennsylvania ... By J.P. Wickersham: 8 p']; 1877, 100 pp. (DLC, PP, WaS); 1878, 100, 8 pp. (DHEW, DLC, MH) {NM 0653283 notes: '[Kentucky ed.]'}; 1879, 114 pp. (DLC, NIC, T); 1881 (MH); and 1884, 114 pp. (DHEW); 1885, 114 pp. (DLC); and still further eds. in 1895 and 1913.
2. Culler discusses the ed. of 1866 (Culler, *passim*); Elson (1964:403) cites (03).

MITCHELL, Samuel Augustus **Y Am**
▸ **{1865} (01)**
▸ *Mitchell's new school geography*. Fourth book of the series. A system of modern geography, physical, political, and descriptive; accompanied by a new atlas of forty-four copperplate maps[, and illustrated by two hundred engravings]
▸ PHILADELPHIA; E.H. Butler & co.
▸ 18 cm.; pp 456
▸ ***Libraries*** : DLC :CaQQS
▸ ***Notes***
1. The words in the title set in [ ] are taken from the NUC entry for 1873.
2. NUC records further eds. (all Philadelphia): 1866 (PPL) [NM 0653448, supplemented by PPL letter noting that 'our cataloguer of the time assigned the date [1886] when we acquired the book, newly purchased, early in 1867']; 1867, 456 pp. (DLC, PPeSchw, ViU); 1871, 456 pp. (MnHi, NcU, ViU); 1873, 456 pp. (KAS, OClStM, PPFr); 1874, 456 pp. (MH, PPeSchw : CaQQS); 1876, 456 pp. (MB); and 1882 (MH, PPeSchw).
3. Judging by information provided on a preliminary page included in the 1869 ed. of MITCHELL (1839), i.e., *The School Geography*, a series of schoolbooks, identified by the inclusion of *New* in their titles, was introduced in 1866 [sic] to replace the 'old' series. Notwithstanding that innovation, the 'old' series continued to be reprinted, with the years of 'publication' included, well into the 1870s.

MITCHELL, Samuel Augustus **Y Am**
▸ **1840c (01)**
▸ *Mitchell's primary geography*. An easy introduction to the study of geography; designed for the instruction of children in schools and families
▸ PHILADELPHIA; Thomas, Cowperthwait & co.
▸ 14.5 cm.; pp. [3], iv, [3], 8-176
▸ ***Libraries*** : CtHT, MH, **NN**, PP, PU
▸ ***Notes***
1. Intended as a textbook for children in primary schools. The text is cast in the form of numbered lessons, which contain numbered statements. With these are associated questions and answers.

A close reading of the text yields some insights into the attitudes of the times. 'The introduction of moral and religious sentiments into books designed for the instruction of young persons, is calculated to improve the heart, and elevate and expand the youthful mind; accordingly, wherever the subject has admitted it, such observations have been made as tend to illustrate the excellence of the Christian religion, the advantages of correct moral principles, and the superiority of enlightened institutions' (p. iv). Three examples will show how this statement was translated into practice.

First: 'In the United States there is no nobility; every citizen is equal in civil and political rights. The son of the poorest man in the country, if he attends to his learning, and possesses abilities, may become President' (p. 38). No conflict is perceived with the following statement, which occurs in the section devoted to American indians. '8. The missionaries are good men, who teach savage nations to read the Bible, worship the true God, and to live like Christians. 9. The first settlers to the Western States were exposed to many dangers; their houses were often attacked, and their wives and children murdered by the Indians; these outrages, however, now take place less frequently, and the savages are every year becoming less numerous and formidable' (p. 51). On a related note, it is striking to see that, in the illustration 'Pocahontas saving the life of Captain Smith' (p. 71), Pocahontas is shown as having a white skin.

Second: 'The history of Asia begins with the creation of the world. This event, which took place near six thousand years ago, is recorded in the Holy Scriptures' (p. 130). '8. The Redeemer was constantly employed in performing acts of charity and mercy; yet he was barbarously crucified by the Jews: they were, in consequence, driven from their own country, and their chief city, was taken and destroyed by the Romans' (p. 132).

Third: 'Lesson 76. Religion. 1. Religion is that worship and homage which all men owe to God, as their creator and preserver. Religion is of two kinds, true and false. 2. True religion consists in worshipping God in the manner taught in Holy Scripture ... 7. Christ came to redeem mankind from sin and misery. He taught them [not *us*!] to worship God, in spirit and in truth ... 10. The Christian nations are the most enlightened and powerful in the world. Their inhabitants are the only people that enjoy the blessings of free government, and regard women as rational beings' (pp. 160-2).

Given the extracts just quoted, the following may surprise some people: 'Lesson 77. Astronomy ... 6. The immense distances at which the celestial bodies are placed from us, is the reason why they appear so small. If it were possible for us to go near them, we should, no doubt, find that they contain land and water, and are fitted for the habitation of rational and intelligent beings' (p. 163).

2. NUC (NM 0653342) gives the size as 15.5 cm and the pagination as iv, 5-176 p.

- **{1843} (02)**
- [*Mitchell's primary geography.*] An easy introduction to the study of geography
- PHILADELPHIA; Thomas, Cowperthwait & Co.
- ***Libraries*** n.r.
- ***Source*** Sahli (1941:382), and Culler (1945:309). As noted above, in Note 2 to MITCHELL (1851), i.e., the *Easy introduction to the study of geography*, neither Sahli nor Culler record Mitchell's *Primary geography* among the textbooks they discuss; it is very probable, however, that they did examine it, but referred to it by the second phrase in its title.

▸ *Notes*
1. BL and NUC record further eds. (all Philadelphia, Thomas, Cowperthwait): 1845, 176 pp. (: MB, MH); 1846, Rev., 178 pp. (L : MH); 1847 (: MH); and 1848, 178 pp. (: MH, NNC, OCIW).

▸ **1849, 2nd, revised (03)**
▸ *Mitchell's primary geography ...*
▸ PHILADELPHIA; Thomas, Cowperthwait & co.
▸ 8vo, 17 cm.; pp. v, 6-176, [2]
▸ ***Libraries*** **L** : DLC, MB, NIC
▸ *Notes*
1. Though physically enlarged compared with the first ed., there are very few changes in the text.
2. NUC records further eds. (both Philadelphia, Thomas, Cowperthwait): 1849, 2nd, rev., 178 pp. (CtY); and 1850, 2nd, rev., 178 pp. (CtY, MH, NNC).

▸ **{1851}, 2nd, rev. (04)**
▸ *Mitchell's primary geography* ... geography ... Illustrated ...
▸ PHILADELPHIA; Thomas, Cowperthwait & co.
▸ 17.5 cm.; pp. v, 6-176
▸ ***Libraries*** : DLC, MB, MH, OO
▸ ***Source*** NUC (NM 0653349)
▸ *Notes*
1. BL and NUC record further eds. (all Philadelphia): 1852, 3rd, rev., 176 pp. (: DLC, ICU, MH); 1853, 3rd, rev., 176 pp. (: MH, NNC, OO, PU); 1854, 3rd, rev., 176 pp. (: NN); 1855, 4th, rev., 144 pp. (: DLC, ICU, MH); 1856, 4th, rev., 144 pp. (**L** : MH) [The text is somewhat reduced in length since 1849; the section on astronomy has been replaced by a table showing the pronunciation of selected place names]; 1858, 4th, rev. (: MH) [According to Culler (1945:309), the 1858 ed. of the *Primary geography* was the 2nd, and was published by E.H. Butler & Co.]; 1859, 4th, rev., 144 pp. (: NNC); 1860, 4th, rev., 144 pp. (: T); 1863, 144 pp. (: ViU); 1866, 144 pp. (: DLC); and 1876, 100, 8 pp. (: DLC). NUC also records an ed. in 1917.
2. Hauptman (1978:429-30) discusses (01); he also refers to it in general discussion (1978:431, 433-4, and n. 61), and quotes from it (1978:436); both Sahli (1941, *passim*) and Culler (1945, *passim*) discuss the ed. of 1843. The fact that Culler (1945) examined both the 1843 and 1858 eds. without ever seeming to notice the difference in the titles by which he referred to them would probably not surprise anyone who has examined his thesis in a critical way. Elson (1964:402) cites the eds. of 1850 and 1853.

MITCHELL, Samuel Augustus **Y Am**
▸ **{1839} (01)**
▸ *Mitchell's school geography.* A system of modern geography, comprising a description of the present state of the world, and its five great divisions, America, Europe, Asia, Africa, and Oceanica, with their several empires, kingdoms, states, territories, etc ... [see below for full title which varied somewhat from ed. to ed.] Illustrated by an atlas of sixteen maps ...
▸ PHILADELPHIA; Thomas, Cowperthwait & co.

- 16.5 cm.; pp. 329
- *Libraries* : DLC, MH
- *Source* NUC (NM 0653398)
- *Notes*

1. A conventional special geography. On the basis of an examination of the ed. of 1845 [see (02) below], it was found to be similar in organization to MITCHELL (1840c; i.e., the *Primary geography*); thus it proceeds by way of question and answer. It seems likely, however, that it includes more and possibly a wider range of facts. More specifically, pp. 52-76 carry questions about maps, and pp. 77-336 are devoted to 'descriptive [i.e., special] geography.' A bias in terms of interest is visible in the fact that Asia, Africa, and Oceania are dealt with in pp. 276-336.

   It was intended as a textbook for elementary students, and is justified by the errors found in contemporary texts. The opening section (pp. 9-51) deals with 'geographical definitions,' i.e., the systematic exposition of physical and human geography. Five races of human beings are identified, and five stages of society: savage, barbarous, half-civilized, civilized, and enlightened (most other texts that were examined list only four, omitting the half-civilized).

   Despite the general similarity to the *Primary geography*, there are differences in tone, some of which are quite surprising. The topic of religion, for example, is presented in a relatively neutral way, for though religions are still divided into true and false, Christianity is not explicitly referred to as true. There is a brief reference to the creation of 'our first parents' (p. 276), but not to that of the world; there is also a reference to the life of 'our Saviour,' but no explicit link is made between his death and the fate of the Jews (see MITCHELL 1840c, above, note 1). Beyond that, in the section on America the references to Indians are less brutally prejudiced than those in MITCHELL (1840c).

   According to an advertisement in entry (03) of MITCHELL (1840c; i.e., the *Primary geography*), this work, i.e., the *School geography*, was intended as a sequel to the *Primary geography*. The same advertisement stated that MITCHELL (1840b; i.e., the *Geographical reader*) was a supplement to the *School geography*.
2. The book was originally accompanied by an atlas, which in this and succeeding eds. was referred to on the tp., as above. In the first ed., the atlas contained 16 maps. In many of the NUC entries that are, for the most part, the basis of those that follow, it is noted that the library that holds the book does not have the accompanying atlas. The atlas was expanded to 18 maps by 1843, 28 by 1846, 32 by 1852, and 34 by 1869.
3. NUC records further eds. (all Philadelphia, Thomas, Cowperthwait): 1840, 336 pp. (NcD) [Sahli (1941, *passim*) and Culler (1945, *passim*)]; 1841, 336 pp. (MB, MH); 1842, 336 pp. CtY, MH); 1843, Rev., 336 pp. (MB, MH, MHi, MsSM, NcA, NNC, OCl, PPStCh : CaQQS); and 1844 (MB, MH).

- **1845, revised (02)**
- *Mitchell's school geography* ... territories, etc. The whole embellished by numerous engravings of various interesting objects of nature and art; together with representations of remarkable and noted events. Simplified and adapted to the capacity of youth. Illustrated ... eighteen maps, drawn and engraven to accompany the work

- PHILADELPHIA; Thomas, Cowperthwait & co.
- 12mo, 17 cm.; pp. vii, 8-336
- *Libraries* L : IEdS, MH, NN, NNC, PPeSchw
- *Notes*

1. An advertisement preceding the preface, dated 1842, states that the work has been revised on the basis of the sixth (1840) Census of the United States. Revisions are planned at five-year intervals.

- **{1846}, 2nd, rev. (03)**
- *Mitchell's school geography* ... territories, etc. Embellished by numerous engravings. Adapted to the capacity of youth. Illustrated ... twenty-eight maps ...
- PHILADELPHIA; Thomas, Cowperthwait & co.
- 17 cm.; pp. 236 [sic]
- *Libraries* : CtY, DLC, NNC, ScCleU
- *Source* NUC (NM 0653405). The number of pp. is a misprint; according to DLC, it should read 336.
- *Notes*

1. NUC records further eds. (Philadelphia, and, if recorded, 336 pp.; Thomas, Cowperthwait until 1853, and E.H. Butler from 1858 on): 1847, 2nd, rev. (CtY, MH, N, NBuG); 1848, 2nd, rev. (MH, NcD, NNC, OU); 1849, 2nd, rev. (CtY, DLC, ICU, MH, Nh, OrCS, PHi); 1850, 3rd, rev. (DLC, ICU, MB, MH, NIC); 1851, 3rd, rev. (MB, MH, Nh, NN, NNC, OClWHi, PPeSchw); 1852, 4th, rev. (MH, NIC, PPT : CaAEU); 1853, 4th, rev. (CtY, ICN, MH); 1854, 4th, rev., Cowperthwait, Desilver & Butler (CtY, CU, IdU, MH, NcU, NN, NNC, PHi, PU, T, WaPS); 1855, Cowperthwait, Desilver & Butler, 4th, rev. (DHEW, MH, NBuG, NNC, OU); 1856, 4th, rev., H. Cowperthwait (NNC, OCl, OClWHi, PPeSchw, ViU, Wa : CaBVaU); 1857, 4th, rev., H. Cowperthwait (MH, NNC, PHi, PPeSchw); 1858 (MH, MoU, NIC, PPeSchw); 1859 pp. n.r. (PPeSchw); 1860, New rev. (CtY, DLC, MH, OO, ViU, WaU); 1861, New rev. (MiU, NNC, PHi, PP : CaBVaU); 1862, 38th (NBuG, NNC); 1863, 38th (DHEW, ICRL); 1864, 38th, pp. n.r. (MH, PPeSchw, WaS); 1866, Rev., pp. n.r. (CtHT, MU, NcA); 1868 (NN); 1869, Rev., pp. n.r. (NcD, ViU : CaQQS); 1870, Rev. (CtY, MH, OClStM); and 1873 (PU).
2. NUC (NM 0653397) lists an ed. that is held by ODW. That library reports that, apart from the words *A system of modern geography* and *school geography*, the information on the tp. is indecipherable. The fact that the book has 336 pp. confirms the assignment of the book to this main entry.
3. {1850} 3rd, rev., *A system of modern geography. [Mitchell's school geography]*. Philadelphia; Thomas, Cowperthwait, 400 pp. NUC (NSM 0040336). This is a ghost. MoU reports that an examination of the book shows it to have only 336 pp. It is thus the same as the work recorded as published in 1850, above.
4. Carpenter (1963:263-6) provides a general discussion, as do Brigham and Dodge (1933:17-18); Sahli (1941, *passim*) and Culler (1945, *passim*) discuss the ed. of 1840.
5. Microform versions of (01) and of the ed. of 1869 are available from CIHM.

MITCHELL, Samuel Augustus **Gt**
- **{1846}**
- *System of modern geography ...*
- ***Source*** NUC (NM 0653429)
- ***Notes***

1. DLC states that, on the basis of 'our brief examination,' this work is the *School geography*. It seems that all entries in NUC (NM 0653429-63, and NSM 0040333-7) that appear as referring to a work or works whose title(s) start(s) with *A system of modern geography* are better identified as either the *Intermediate geography* (MITCHELL 1849), the *New intermediate geography* (MITCHELL 1860*), the *New school geography* (MITCHELL 1865), or the *School geography* (MITCHELL 1839).

   DLC is not the only library to have copies of Mitchell's books appear in two different guises in NUC. MH reports that the *System of modern geography*, 1847, that appears in NM 0653449 is the same book as that entered as *Mitchell's school geography* in NM O653407; similarly, the *System of modern geography*, 1851, that appears in NM 0653450 is the same as the *School geography* that appears in NM 0653411.
2. Elson (1964:401-3) cites eds. of 1843, 1845, 1846, 1850, 1851, 1854, 1857, and 1862.

MODERN GEOGRAPHY ... **A SG and DY? Am**
- **{1811} (01)**
- *Modern geography: and a compendious general gazetteer*, illustrated with numerous maps, plates, etc. Containing a description of the empires, kingdoms, states, and colonies; with the oceans, seas, rivers, harbours, and isles, in the several quaters of the world, with a succinct history of the war, from the commencement of the French revolution until the year 1811. By a Society ...
- BERWICK; printed by W. Lochhead
- 22 cm.; 3 vols.
- ***Libraries*** : DLC

MODERN GEOGRAPHY ... **Y**
- **1852, 2nd, rev. (01)**
- *Modern geography simplified*: to which are appended, brief notices of European discovery, with select sketches of the ruins of ancient cities. Second edition, revised
- LONDON; Sims & M'Intyre.
- 12mo.
- ***Libraries*** L
- ***Notes***

1. A very slight work, designed for young children, and written in a condescending style.

MODERN SCHOOL GEOGRAPHY ... **Y Can**
- **{1865} (01)**

- ▸ Campbell's British American series of school-books. *Modern school geography and atlas*. Prepared for the use of schools in the British Provinces
- ▸ MONTREAL and Toronto; James Campbell, Publisher. St John, N.B.: J. & A. M'Millan & Co. Halifax, N.S.: A. & W. Mackinlay. Charlottetown, P.E.I. Laird & Harvey. St John's Newfoundland. J. Graham.
- ▸ Pp. [5], 2-? [mutilated]
- ▸ *Libraries* : : CaBVaU, CaQQS
- ▸ *Notes*

1. Information about the physical geography of places is given first, followed by their human geography, with a bias towards information of a commercial or civil-administrative nature. Relatively detailed with respect to Canada and the United States.

- ▸ **1876, 5th (02)**
- ▸ Canadian series of school-books. *Modern ... atlas*. Authorized by the Councils of Public Instruction for the Provinces of Ontario, Quebec, New Brunswick, and Prince Edward Island. Fifth edition
- ▸ TORONTO; J. Campbell
- ▸ Pp. [7], 8-67
- ▸ *Libraries* : : CaOTER
- ▸ *Notes*

1. CIHM records 1876, Toronto (: : CaOTER).
2. Microform versions of (01) and the ed. of 1876 are available from CIHM.

MOIR, George **Y At Can**

- ▸ **[1878] (01)**
- ▸ *Exercises in map geography for junior pupils*
- ▸ TORONTO; Copp, Clark
- ▸ Pp. [5], 6-43
- ▸ *Libraries* n.r.
- ▸ *Source* CIHM
- ▸ *Notes*

1. The date is taken from a statement in the Preface of the ed. of 1884; see note 3.
2. The verbal information is restricted to the names of places shown on the maps on facing pages, hence the decision to classify it as an atlas.
3. CIHM record an ed.: 1884, 7th, corrected and enlarged.

MOLL, Herman

- ▸ *A system of geography* ... See (02) of THESAURUS GEOGRAPHICUS.

MOLL, Herman

- ▸ *Present state of all nations* ... See PRESENT STATE.

MONTEITH, James **Y Am**

- ▸ **{1885}a (01)**

- ▸ *Barnes' complete geography*
- ▸ NEW YORK and Chicago; A.S. Barnes and company
- ▸ 31.5 cm.; pp. 139
- ▸ *Libraries* : DHEW, DLC, MiU, OO, PP
- ▸ *Source* NUC (NM 0723548)
- ▸ *Notes*

1. NUC records three variants of this ed.; they are distinguished by their provision of a supplementary section devoted to a particular state or region (all New York): ed. for New York, 141, 11, 8 pp. (ICRL, NN); ed. for Utah, 141, 6 pp. (DHEW); and Pennsylvania ed., 138 pp. (InStme).

- ▸ **{1885-96} (02)**
- ▸ *Barnes' complete geography*
- ▸ NEW YORK; Barnes
- ▸ Pp. 141, 14
- ▸ *Libraries* : NcRS

- ▸ **{1886} (03)**
- ▸ *Barnes' complete geography*
- ▸ NEW YORK and Chicago; A.S. Barnes and company
- ▸ 31.5 cm.; pp. 140
- ▸ *Libraries* : DLC, OClJC, OO
- ▸ *Source* NUC (NM 0723552)
- ▸ *Notes*

1. NUC records a variant ed.: [c1866], Pacific coast ed., 140, 21 pp. (CU). NUC records further variant eds.: 1890 (New England ed.), 1891 (Ed. for Idaho, Montana, and Wyoming), 1893 (Ed. for Utah), 1895 (Montana ed.), 1896 (Ed. for Kentucky and Tennessee), 1896 (New York), 1896 (Ed. for Texas and Oklahoma), 1896, 98 (Ed. for Missouri and Iowa), and c1914.

MONTEITH, James **Y Am**

- ▸ **{[1885]}b (01)**
- ▸ *Barnes' elementary geography*
- ▸ NEW YORK and Chicago; A.S. Barnes & company
- ▸ 24.5 cm.; pp. 92
- ▸ *Libraries* : DHEW, DLC, MiU
- ▸ *Source* NUC (NM 0723563)
- ▸ *Notes*

1. NUC records further eds.: 1896 and c1914.
2. The relationship of this book to MONTEITH (1874) is not known.

MONTEITH, James **Y Am**

- ▸ **{1874} (01)**
- ▸ *Elementary geography*. Taught by means of pictures, maps, charts ...
- ▸ NEW YORK; A.S. Barnes & Co.
- ▸ Sq. 8vo; pp. 80.
- ▸ *Libraries* : DLC
- ▸ *Source* NUC (NM 0723595)

▸ *Notes*
1. NUC records further eds.: 1874 (NN) [NM 0723594 notes: '(The independent course)']; 1875 (DHEW, MH, OO); 1879? c1874 (MH); and 1881, c1874 (MH).

▸ **{[1880]}, new, with improvements (02)**
▸ ... *Elementary geography* ... charts, diagrams, map drawing and blackboard exercises
▸ NEW YORK, Chicago (etc.); A.S. Barnes & company
▸ 23.5 cm.; pp. 80
▸ *Libraries* : DLC, MoU
▸ *Source* NUC (NM 0723598)

▸ **{[1883]}, ed. containing ... (see below) (03)**
▸ ... *Elementary geography* ... (Edition containing new preliminary lessons, a new system of review, new maps and illustrations)
▸ NEW YORK and Chicago; A.S. Barnes & company
▸ 25 cm.; pp. 80
▸ *Libraries* : DHEW, DLC, MoU
▸ *Source* NUC (NM 0723600)
▸ *Notes*
1. NUC records a further issue of this ed.: 1885? (MH), and a further ed.(?): in 189–.
2. The relationship of this book to MONTEITH ([1885]b), i.e., *Barnes' elementary geography*, above, is not known.
3. Culler (1945, *passim*) discusses (03).

MONTEITH, James **Y Am**
▸ **{1855} (01)**
▸ ... *First lessons in geography* ...
▸ NEW YORK; A.S. Barnes & co. St. Louis; Carr & Buchanan; [etc.]
▸ 17 cm.; 1 p. l., 62 p.
▸ *Libraries* : DLC
▸ *Source* NUC (NM 0723603)
▸ *Notes*
1. This book formed No. 1 in the 'National Geographical Series' published by A.S. Barnes & Co. MONTEITH (1853), below, formed No. 2, and McNALLY (1855) No. 3.
2. NUC records further issues of this ed.: 1856 (OClWHi); 1857 (OO, PP); 1858 (CtY, MH, OClWHi, OO); and 1859 (MoU).

▸ **{[c1862]} (02)**
▸ *First lessons in geography*[, on the plan of object teaching. Designed for beginners]
▸ NEW YORK and Chicago; A.S. Barnes and company
▸ 17.5 cm.; pp. 68
▸ *Libraries* : CU, ICU
▸ *Source* NUC (NM 0723608)

▸ *Notes*
1. According to Culler (1945:309) this is the 2nd ed. The words in [ ] are taken from the ed. of 1866.
2. NUC records further issues of this ed.: 1863 (DLC, MiU, OFH, OO); 1866 (MH); 1867 (MH, OClWHi); 1868 (MH); 1869 (MB, MoU); 1873 (MiU, MoU); and 1874 (DLC, OClWHi).

▸ **n.d., c1862 and 1873 (03)**
▸ *First lessons in geography*: [... as (02) ... beginners. By ... author of a series of geographies, maps, and globes]
▸ NEW YORK, Chicago and New Orleans; A.S. Barnes & Company
▸ 17 cm.; [5], 8-68 pp.
▸ ***Libraries*** : : **CaAEU**
▸ *Notes*
1. 'The plan of object teaching, by which the mind receives impressions through the medium of the eye, is here so combined with the map exercises, that a child just able to read is at once interested and instructed' ([3]).
   The text, which is at the elementary level, is in part descriptive, and in part takes the form of question and answer. Strongly pro-American and anti-Indian prejudices are present.
2. NUC records further eds. in 1889 and 1926.
3. Nietz (1961:231) discusses (01), and Culler (1945, *passim*) discusses (02).

MONTEITH, James **Y Am**
▸ **{1857} (01)**
▸ ... *Introduction to the manual of geography* ...
▸ NEW YORK; A.S. Barnes & co.
▸ 22 cm.; pp. 61
▸ ***Libraries*** : DLC
▸ ***Source*** NUC (NM 0723628), which notes: '(National geographical series, no. 2)'
▸ *Notes*
1. NUC records a further issue: 1858 (MH).

▸ **{[c1867]} (02)**
▸ ... *Introduction to the manual of geography*. Designed for junior classes in public and private schools
▸ NEW YORK, Chicago (etc.); A.S. Barnes & company
▸ 22 cm.; pp. 70
▸ ***Libraries*** : DHEW
▸ ***Source*** NUC (NM 0723630), which notes: 'On cover: edition for the Pacific coast'
▸ *Notes*
1. NUC records further issues (without state variations): 1868 (NIC, NN); 1870 (MoU); 1871 (OClWHi); and 1876 (NcU).

▸ **{[c1883]} (03)**
▸ *Introduction to the manual of geography* ...
▸ NEW YORK; American Book Co.

- 4to
- *Libraries* : MH
- *Source* NUC (NM 0723635)
- *Notes*

1. NUC also records an ed. in 1895.
2. Nietz (1961:231) discusses (01).

MONTEITH, James **Y Am**

- **{1856}, revised (01)**
- *Manual of geography, combined with history and astronomy*
- NEW YORK; A.S. Barnes & Co.
- 23.5 cm.; pp. 112
- *Libraries* : DLC, OCl
- *Source* NUC (NM 0723637), which notes: 1. '(National geographical series, no. 2).' 2. 'On cover: Monteith's and McNally's System.'
- *Notes*

1. The fact that the earliest known ed. of this title is said on the tp. to be *revised* suggests that this title is a continuation of MONTEITH (1853; i.e., the *Youth's manual*).
2. NUC records further issues of this ed.: 1856 (ICU); 1857 (MH); 1859 (MH); 1861 (NjP); 1863 (DHEW, Or) [According to Culler (1945:309) this is the 2nd ed.]; 1864 (MH, OO); 1868 (PHi); and 1869 (DLC, OClWHi).

- **{1870} (02)**
- *Manual of geography ... astronomy: designed for the intermediate classes in public and private schools*
- NEW YORK; A.S. Barnes & co.
- Sq. 8vo; pp. 124
- *Libraries* : NN
- *Notes*

1. NUC records later issues of this ed.: 1871 (MH); 1873 (DHEW); 1875 (OClWHi); 1876 (CU); 1877 (MH); 1881 (MH, PP); and 1896.
2. Nietz (1961:231) discusses (01); Hauptman (1978:431, 434-5) includes a reference to (02) in general discussions; Culler (1945, *passim*) discusses the ed. of 1863.

MONTEITH, James **Y Am**

- **{1872} (01)**
- *Monteith's comprehensive geography*[: local, physical, descriptive, historical, mathematical, comparative, topical, and ancient]. With map-drawing and relief maps
- NEW YORK AND Chicago; A.S. Barnes & company
- 31 cm.; 96 p., 1 l.
- *Libraries* DL
- *Notes*

1. The words in the title set in [ ] are derived from the NUC entry for 1873 [c1872].
2. See CATHOLIC TEACHER (1875).

3. NUC records a variant 'ed. for the Pacific coast,' New York, 104, (1), 12 pp. (DHEW), and then further eds. (all New York; Barnes): 1873 [c1872], 100 pp. (DHEW, NBuG); 1873 (MH); 1875 (MH, OO); 1876 (DLC, PPL); [1878] Maine ed., 104, 6 pp. (DLC) [NUC records other state variations of this ed. as follows: Massachusetts (DLC), New Hampshire (DLC, NcD), Vermont (DLC), Virginia (DLC, ViU); [1880], 104 pp. (DLC); [1882] Iowa ed., 104, 10 pp. (DLC) [NUC records other state variations of this ed. as follows: New York (MH), Pacific coast (DHEW, OrHi), Tennessee (T). It also records yet later eds.: 1889 (New), 1889 (New Jersey), 1890 (New England), c1896, 1900, and 1900 (Missouri and Iowa).
4. Culler (1945, *passim*) and Nietz (1961:200, 201-2) discuss (01).

MONTEITH, James **Y Am**

▸ **{[1884]} (01)**
▸ *Monteith's primary geography ...*
▸ NEW YORK and Chicago; A.S. Barnes & co.
▸ 25 cm.; 1 p. l., 3-22 numb. l.
▸ ***Libraries*** : DLC
▸ ***Source*** NUC (NM 0723671), which supplies the date

MONTEITH, James **Y Am**

▸ **{1853} (01)**
▸ *Youth's manual of geography*, combined with history and astronomy ...
▸ NEW YORK; A.S. Barnes & co. Cincinnati; H.W. Derby; [etc.]
▸ 8vo, 22 cm.; pp. 171
▸ ***Libraries*** L : DLC
▸ ***Source*** BL and NUC (NM 0723682)
▸ ***Notes***

1. This was the first of a series of school textbooks in geography that were published by A.S. Barnes (often in association with other publishers in cities other than New York) in sequences that lasted, in one case, until 1926. Monteith wrote most of them, but see also McNALLY (1855, 1882). Those discussed in secondary sources were all special geographies; as there is nothing in the titles of the others to suggest that they were significantly different, I have assumed that they all belong to the genre.
2. Judging by the 11th ed. of 1856, this work carries the head title 'National geographical series. No. 2.'
3. Uses an extreme form of question and answer; e.g., in the section on Europe, the text reads: 'Straits. Gibraltar? Dover?' The implication is that the student is looking at a map, and must find these places on it. Even more extreme are pp. 169-71, where 136 towns, 14 lakes, 75 rivers, 31 gulfs or bays, 16 seas, 3 channels, 49 islands, 23 mountains, 27 capes, and 10 peninsulas are named, all of whose locations a student was expected to be able to identify.
4. NUC records further eds. (all New York): 1854, 3rd (MH); 1856, 11th, 172 pp. (KU, NN, OO); 1859 (MH).
5. See CATHOLIC TEACHER.

MOORE, Henry **A**

- **1813? (01)**
- *A new and comprehensive system of universal geography*. Being a general and complete description of the several empires, kingdoms, states, and colonies, in Europe, Asia, Africa, and America; with the oceans, seas, and islands, in every part of the world. Containing I. The Newtonian system of the universe, together with the latest observations relative to the earth, considered as a planet, with useful geographical definitions and problems. II. The origin and history of nations, their forms of government, religion, population, laws, revenues, taxes, naval and military strength, orders of knighthood, etc. III. Their climate, soil, natural history, productions, curiosities, seas, rivers, lakes, etc. etc. IV. The manners, customs, amusements, persons, and habits of the people; their language, learning, manufactures, and commerce. V. The chief cities, towns, edifices, ruins, etc. etc. VI. The longitude, latitude, bearings, and distances of the principal places from London. To which are added, a copious geographical index, alphabetically arranged; tables of the coins of different nations, with their values in English money; and a chronology of all the remarkable events from the creation of the world to the present time. The whole digested upon a systematic plan, from the latest and best authorities, with notes, historical, critical, and explanatory. Illustrated by a complete atlas, and embellished with numerous engravings
- LONDON; printed by Macdonald and Son, 46, Cloth Fair, West Smithfield, for A. Whellier, No. 3, Paternoster Row; and sold by all the booksellers in the United Kingdom
- 4to, 27 cm.; pp. vi, xxvi, 1180
- ***Libraries*** **OSg** : WHi
- ***Notes***

1. On the basis of internal evidence, the book was published after 1812 and before the defeat of Napoleon in 1815. Biased towards Europe, and within Europe towards the British Isles, with history making up a substantial part of the bulk of the latter. It makes a serious attempt to live up to the claims of the tp.

MOORE, Jonas **mSG**

- **1681 (01)**
- *A new geography, with maps to each country, and tables of longitude and latitude*
- LONDON; printed for Robert Scot, at the Princes Arms in Little Britain
- 23.5 cm, 4to; pp. [4], 56, 32
- ***Libraries*** **EU** : DFo, DLC, PPA
- ***Notes***

1. This work is Pt. 2 of Vol. II of Moore, J., *A new systeme of mathematics*. An opening advertisement states that Moore died while working on this section of the complete work. It was an atlas and gazetteer, accompanied by a text in which the principal towns and rivers of politically defined territories are identified. Europe is dealt with in the first set of numbered pp.; the rest of the world is disposed of in the second set.
2. NUC (NM 0749182) records the pagination as: Pt. 1, [4] 56 pp.; pt. 2, 32 pp.

MORDEN, Robert **A**

- **1680 (01)**
- *Geography rectified; or, a description of the world,* in all its kingdoms, provinces, countries, islands, cities, towns, seas, rivers, bayes, capes, ports; their antient and present names, inhabitants, situations, histories, customs, governments, etc. As also their commodities, coins, weights, and measures, compared with those at London. Illustrated with above sixty new maps. The whole work performed according to the more accurate discoveries of modern authors
- LONDON; printed for Robert Morden and Thomas Cockeril. At the Atlas in Cornhill, and at the Three Legs in the Poultrey over against the Stocks-Market
- 4to, 21 cm.; pp. [13], 418
- ***Libraries*** **L** : CLU, CtY, DLC, MiU, MoSW, RPJCB, ViU
- *Notes*

1. Identifies its principal sources. Reasonably well-balanced in terms of information about different continents and countries, though Europe is favoured with a more detailed treatment. There is a section on coins, weights, and measures (pp. 157-96), and on jewels and exotic commodities (pp. 197-204).

- **1688, 2nd, enlarged ... (see below) (02)**
- *Geography rectified* ... ancient London. Illustrated with seventy six maps. The whole ... accurate observations and discoveries ... The second edition, enlarged with above thirty sheets more in the description and about twenty new maps
- LONDON; Printed for ... Cockerill, at ...
- 4to, 20 cm.; pp. [16], 596
- ***Libraries*** C, **L** (lacks pp. 305-20), LeP : DGU, DLC, ICN, ICU, LNT, MdBJ, MoSW, NcD, NhD, NNC, NNUT, PBL, PU, RPJCB, WHi
- *Notes*

1. The Preface reports that the first ed. was received favourably at Oxford and Cambridge. The order of countries has been changed to follow a more strictly contiguous sequence; this makes a quick comparison of material difficult. In general the additional material is fairly well distributed, though Europe receives proportionately the most coverage.

- **1693, 3rd, enlarged ... (see below) (03)**
- *Geography rectified* ... London. Illustrated with seventy eight maps. The third edition, enlarged, to which is added a compleat geographical index to the whole, alphabetically digested
- LONDON; printed for ...
- 4to, 20 cm.; pp. [10], 626, [72]
- ***Libraries*** C, **L**, LDW, LcU, O : CSmH, DLC, MA, MB, NN, NNE, NPV, PPAmP, PPiU, PPL, PU, RPJCB, ViU
- *Notes*

1. The order of countries is generally as before. The descriptions of some are unchanged, while others receive significant additions, e.g., Ireland and Lesser Tartary [i.e., the Ukraine]. The section on exotic commodities has been

omitted. Little if any material has been added to the descriptions of Asia and America, and not much to that of Africa. A long index of place names has been added.

- **{1697} (04)**
- *Geography rectified ...*
- LONDON; printed for R. Morden and T. Cockerill
- [6] p. l., 574 [2], 575-626, [27] p.
- *Libraries* : MiD

- **1700, 4th, enlarged ... (05)**
- *Geography rectified ...*
- LONDON; printed for R. Morden and T. Cockerill, and are to be sold by M. Fabian in Mercers Chappel-Porch in Cheapside, and Ralph Smith at the Bible under the Exchange in Cornhill
- 4to, 20 cm.; pp. [10], 626, [74]
- *Libraries* **L** : CtY, DLC, ICU, KU, MB, MBAt, MH, MWA, RPJCB : CaBViPA
- *Notes*

1. Spot-checking suggests that this is a reissue of 1693.
2. A microform version of (05) is available from CIHM.

MORDEN, Robert **OG**

- **{1702} (01)**
- *An introduction to astronomy, geography, navigation, and other mathematical sciences* made easie by the description and uses of the coelestial and terrestrial globes ...
- LONDON; printed for R. Morden and R. Smith
- 19 cm.; 12 p. l., 184, 87 p.
- *Libraries* L : DN, ICU, NN, ViNeM
- *Source* BL and NUC (NM 0762629)
- *Notes*

1. It could be taken to be a special geography only if the title were greatly abbreviated.

MORELL, [John Daniel?] **Y SG?**

- **{1875-9} (01)**
- *Geography, first steps*
- [LONDON]; Chambers
- *Libraries* n.r.
- *Source* ECB, Index (1874-80:67)

MORELL, [John Daniel?] **Y SG?**

- **{1875} (01)**
- *Geography, primary series*
- [LONDON]; Stewart
- *Libraries* n.r.

- *Source* ECB, Index (1874-80:67)
- *Notes*

1. BL catalogues the 1876 ed. of MARKWELL (1874) under Morell, John Daniel as: *Dr. Morell's Secondary Series. A senior Geography.*
   Note also that the English Catalogue, Index for 1875 bound with ECB, Index (1856-76), identifies *Morell* as *Dr. Morell* on one of the two occasions when he appears in it.

MORERI, Louis **Dy tr**

- **1694 (01)**
- *The great historical, geographical and poetical dictionary*; being a curious miscellany of sacred and prophane history. Containing, in short, the lives and most remarkable actions of the patriarchs, judges, and kings of the Jews; of the apostles, fathers, and doctors of the church; of popes, cardinals, bishops, etc. Of heresiarchs and schismaticks, with an account of their principal doctrines; of emperors, kings, illustrious princes, and great captains; of ancient and modern authors; of philosophers, inventors of arts, and all those who have recommended themselves to the world, by their valour, virtue, learning or some notable circumstances of their lives[.] Together with the establishment and progress both of religious and military orders, and the lives of their founders. The genealogy of several illustrious families in Europe. The fabulous history of the heathen gods and heroes. The description of empires, kingdoms, common-wealths, provinces, cities, towns, islands, mountains, rivers, and other considerable places, both of ancient and modern geography; wherein is observed the situation, extent and quality of the country; the religion, government, morals and customs of the inhabitants; the sects of Christians, Jews, Heathens and Mahometans. The principal terms of arts and sciences; the publick and solemn actions, as festivals, plays, etc. The statutes and laws; and withall, the history of general and particular councils, under the names of the places where they have been celebrated. The whole being full of remarks and curious enquiries, for the illustration of several difficulties in theology, history, chronology and geography. Collected from the best historians, chronologers, and lexicographers; as Calvisius, Helvicus, Isaacson, Marsham, Baudrand, Hoffman, Lloyd, Chevreau, and others; but more especially out of Lewis Moreri, D.D. his Sixth edition corrected and enlarged by Monsieur Le Clerk; in two volumes in folio. Now done into English. To which are added, by way of supplement, intermix'd throughout the alphabet, the lives, most remarkable actions, and writings of the illustrious families of our English, Scotch and Irish nobility, gentry, eminent clergy, and most famous men of all arts and sciences; as also, an exact description of these kingdoms; with the most considerable occurrences that have happened to this present time. By several learned men. Wherein are inserted the last five years historical and geographical collections of Edmond Bohun, esq; designed at first for his own geographical dictionary, and never extant till in this work
- LONDON; printed for Henry Rhodes, near Bride-Lane in Fleetstreet, Luke Meredith, at the Star in St. Paul's Church-Yard; John Harris, at the Harrow in the Poultry; and Thomas Newborough, at the Golden-Ball in St. Paul's Church-Yard

- Folio, 39 cm., 2 vols.; Vol. I [letters A-G], pp. [519, of which 504 are text]; Vol. II [letters H-Z], pp. [706, of which 704 are text]
- *Libraries* AbrwthU, ByseC, C, ChC, EU, LeP, **L** (wants page containing last part of Hannibal to first part of Harold, King of England), LILS, LvU, NP, O, SsaC : CtY, CU-S, DFo, DLC, MH, MnU, NBU, NcD, OC, OClW, OU, TxDaM, TxHU, TrU, ViU
- *Notes*

1. Has a place in this guide because the entries for geographical units are comparable in length with those found in, e.g., MORDEN (1680). The first ed. of the French original was published in 1673 (article on Moreri in this, 1688, ed.).
2. BL catalogues under Bohun, and under Dictionary.

- **1701, 2nd, revised (02)**
- *The great ... geographical, genealogical and poetical dictionary*; being ... princes, and great generals; of ancient ... their lives. Together ... founders. As also, the fabulous ... withal ... Morery ... his eighth ... in folio. To which are added ... of several illustrious ... nobility, and gentry, and most famous ... all professions, arts and ... present time. The second edition revis'd, corrected and enlarged to the year 1688; by Jer. Collier, A.M.
- LONDON; printed for Henry ... Fleetstreet; Thomas Newborough at the Golden-Ball in St. Paul's Church-Yard; the assigns of L. Meredith, at the Star in St. Paul's Church-Yard; and Elizabeth Harris, at the Harrow in Little-Britain
- Folio, 39 cm., 2 vols.; Vol. I [letters A-L], pp. [801 of which 794 are text]; Vol. II [letters M-Z], pp. [626 of which 624 are text]
- *Libraries* GU, **L** : CBPac, CtW, CtY, CU-A, DFo, DLC, DNLM, InU, IU, KMK, KU, LU, MBAt, MdBP, MeB, MHi, MWA, NcD, NcU, NIC, NN, OC, PP, PPiPT, PPL, PU, TU, TxU, Vi : CaBVaU
- *Notes*

1. Collier provides a summary of the changes made both in the French 7th and 8th eds., and of the additions he has provided.

- **1705, supplement (01)**
- *A supplement to the great ... dictionary*: being ... in folio. By Jer. Collier, A.M. Together with a continuation from the year 1688, to this time, by another hand
- LONDON; printed for Henry ... Fleetstreet; and Thomas Newborough at the Golden-Ball in St. Paul's Church-Yard
- Folio, 38 cm.; 2 pts. in 1 vol.; Pt. 1, pp. [596, of which 589 are text]; Pt. 2, pp. [79, of which 76 are text]
- *Libraries* **L** : CSmH, CSt, CU-S, DLC, InU, KU, LU, MdBP, NcU, OC, OrU, PU, TxU, ViU : CaBVaU
- *Notes*

1. Both parts are arranged alphabetically, the second consisting of 'the lives and actions [since 1688] of several great men' (Preface, Pt. 2). Pt. 1 contains information about places.
2. Catalogued by BL under Collier.

▸ **1721, supplement (02)**
▸ *An appendix to the three English volumes in folio of Morery's great ... geography*. This appendix being collected from authors of character, and chiefly from the French supplement to the said dictionary, printed in Holland in the year 1716. And wherein foreign lives and transactions are continued to the said period
▸ LONDON; printed for Geo. James; and are to be sold by R. Sare and F. Gyles in Holbourn; B. and S. Tooke at the Middle-Temple Gate; G. Strahan over-against the Royal Exchange; W. Taylor in Pater-Noster-Row; and J. Bowyer, and W. and J. Innys, in S. Paul's Church-Yard
▸ Folio, 38 cm.; pp. [398, of which 392 are text]
▸ ***Libraries*** CoP, E, LcP, **L**, LvU, MP, NcP, [IDT] : CSt, CtY, DFo, DLC, FU, IaU, KU, MdBP, MeB, OC, PU, TxU, ViU : CaBVaU
▸ ***Notes***
1. Contains some information about places. The 3 vols. referred to are the two of 1701 plus the supplement of 1705.
2. It is possible that some of the British regional-library holdings are the 2nd ed.; see (03) below.
3. BL catalogues under Collier.

▸ **{1721}, 2nd (to the Supplement) (03)**
▸ *A supplement to the great ... dictionary ...*
▸ LONDON; printed by W. Bowyer for C. Collier [etc.]
▸ 39 cm.; 341 l.
▸ ***Libraries*** : CLU, CtY, CU-A, DFo, DLC, IaU, ICN, KU, MeB, NcU, PU, TxU, WU
▸ ***Source*** NUC (NM 0771416)

MORGON, Charles Carroll **Y Am**
▸ **{1863} (01)**
▸ *J.H. Colton's American school geography*: comprising separate treatises on astronomical, physical, and civil geography ... Illustrated ... and accompanied by J.H. Colton's school atlas
▸ NEW YORK; Ivison, Phinney & co. Chicago; S.C. Griggs & co.
▸ 20 cm.; pp. vi, 588
▸ ***Libraries*** : DLC, MtBuM, NN
▸ ***Source*** NUC (NC 0569865). See note to entry (02), below.

▸ **1864 (02)**
▸ *J.H. Colton's* ... civil geography, with descriptions of the several grand divisions and countries of the globe. Illustrated with numerous engravings, and accompanied ...
▸ NEW YORK; Ivison, Phinney, Blakeman & Co. Chicago ...
▸ 12mo, 19.5 cm.; pp. [3], iv-vi, 588
▸ ***Libraries*** : IU, MH, **NN**, RPB, ViU
▸ ***Notes***
1. On the tp., quotes Arnold Guyot: 'Geography ought to be something different from mere description. It should not only describe, it should compare, it

should interpret, it should rise to the *how* and the *wherefore* of the phenomena which it describes.'

Morgan presents a succinct statement of his ambitions in the Preface: to be scientific; to avoid repetition; and to 'lay the foundation for solid attainments by inducing correct habits of thought and a spirit of philosophic inquiry, which, in geography, as elsewhere, must always have root in a thorough mastery of first principles.' As an example, he suggests that a scholar in geography should be able to infer from a knowledge of the rocks of a country 'a correct knowledge of many other of its leading physical characteristics' (pp. iii-iv).

The book was started by Prof. Alphonso J. Robinson, with Morgan as an assistant; R. retired because of poor health, and M. finished the job. M. also thanks Prof. Sanborn Tenney, Ambrose Eastman, and Richard S. Fisher (for statistics), and G. Woolworth Colton, 'from whom he has received many favours' (p. vi); acknowledges drawing on textbooks by Hughes (HUGHES 1852b?) and Woodbridge (WOODBRIDGE 1824?, 1844?), and the *Pronouncing gazetteer of the world* by Thomas and Baldwin.

The text is laid out in numbered paragraphs; questions on the text are placed at the foot of the page; they do not have set answers. Many of them can be answered only by consulting an atlas, e.g., the one issued with the work though bound separately.

The earth's astronomical situation, and also physical geography occupy pp. 1-118; 'civil' [i.e., systematic human] geography pp. 119-42, and 'local' [i.e., special] geography the remainder. North America is dealt with first, with the U.S. (pp. 192-324) receiving most attention; states are dealt with in regional groupings; there is no reference to the Civil War, even in the section on the southern states.

2. Discussed in Culler (1945, *passim*), where Morgan is treated as the senior author. See COLTON, J.W. (1865).

MORLEY, Charles **Y ? Am**

- **{1835) (01)**
- *The geographical key*: or, guide to an improved plan of studying geography
- HARTFORD; H. Benton
- 14 cm.; 32 p.
- ***Libraries*** : DLC

MORSE, Charles Walker **A n.c. Am**

- **{1840} (01)**
- *The diamond atlas* ... The eastern hemisphere
- NEW YORK; Samuel N. Gaston
- 239 p.
- ***Libraries*** : ICRL
- ***Notes***

1. For information about the content, see notes to (03), below.

- **{1856} (02)**
- *The diamond atlas*; with descriptions of all countries, exhibiting their actual

and comparative extent and their present political divisions, founded on the most recent discoveries and rectifications. The eastern hemisphere
- NEW YORK; S.N. Gaston
- *Libraries* : MiD

- **1857 (03)**
- *The diamond atlas*. With ... extent, and their present ... rectifications. By ... editor of 'Morse's General Atlas of the World.' The eastern hemisphere
- NEW YORK; Samuel N. Gaston, 115 and 117 Nassau, and, 15 Spruce Streets
- 18.5 cm.; pp. [vii], viii-xi, [2], 14-239
- *Libraries* : DLC, NN : **CaBVaU**
- *Notes*

1. On the cover: 'Morse and Gaston's Diamond Atlas. Oriental.' The text, which occupies the pages numbered in arabic numerals, dominates the maps and illustrations, so the work is far from being a conventional atlas. A volume dedicated to the western hemisphere and published earlier is implied in a statement in the Preface (pp. [ix]-xi); see note 1 to COLBY (1857a). A history of Classical geography and of exploration occupies pp. [13]-100. Historical information dominates that provided about the countries of Europe, except in the case of the United Kingdom, where a miscellany of information is provided about the larger cities and towns. The overall impression is of a discursive accumulation of widely available information decked out in a form better suited to please the eye than the mind (the pages are gilt-edged).
2. NUC (NM 0798812) records what appears to be a variant ed. in 2 vols.; see COLBY (1857a).

- **{1858} (04)**
- *The diamond atlas*, with descriptions ... divisions. The eastern hemisphere
- NEW YORK; S.N. Gaston
- Sm 4to; pp. (2), 242
- *Libraries* : MB, MH, NcA
- *Source* NUC (NM 0798814)

- **{1860} (05)**
- *The diamond atlas*, with descriptions ... countries: exhibiting ... rectifications
- NEW YORK; Samuel N. Gaston
- 19 cm; xi, 239 p.
- *Libraries* : CtY, MH
- *Source* NUC (NM 0798815)

MORSE, Charles W.
- *General atlas of the world* ... See COLBY (1856).

MORSE, Jedidiah **A Am**
- **1789a (01)**
- *The American geography*; or, a view of the present situation of the United States of America. Containing astronomical geography. Geographical defini-

tions. Discovery, and general description of America. Summary account of the discoveries and settlements of North America; general view of the United States; of their boundaries; lakes; bays and rivers; mountains; productions; population; government; agriculture; commerce; manufactures; history; concise account of the war, and of the important events which have succeeded. Biographical sketches of several illustrious heroes. General account of New England; of its boundaries; extent; divisions; mountains; rivers; natural history; productions; population; character; trade; history. Particular descriptions of the thirteen united states, and of Kentucky, the western territory and Vermont. - Of their extent; civil divisions; chief towns; climates, rivers; mountains; soils; productions; trade; manufactures; agriculture; population; character; constitutions; courts of justice; colleges; academies and schools; religion; islands; indians; literacy and human societies; springs; curiosities; histories. Illustrated with two sheet maps - one of the southern, the other of the northern states, neatly and elegantly engraved, and more correct than any that have hitherto been published. To which is added, concise abridgement of the geography of the British, Spanish, French and Dutch dominions in America, and the West Indies - of Europe, Asia, and Africa

- ELIZABETHTOWN; printed by Shepard Kollock, for the author
- 8vo, 19 cm.; pp. xii, 534, [3]
- *Libraries* **L** : CtY, DLC, FMU, GU-De, IaU, LU, MBAt, MeB, MH, MiU, MnU, MtBC, MU, MWA, NCH, NcU, NHi, NIC, NjR, NN, OCl, OClWHi, OFH, OSW, PBL, PHC, PNt, PPAmP, PPL, RPJCB, TxU, ViU : CaQMBN
- *Notes*

1. In writing this book, M. 'has aimed at utility rather than originality, and ... when he has met with publications suited to his purpose, he has made free use of them; and he thinks it proper here to observe that, to avoid unnecessary trouble, he has frequently used the words as well as the ideas of the writers, although the reader has not been particularly apprized of it' (p. vi; quoted by Wright [1959:153]).

   M.'s intention in a larger sense may be inferred from the fact that 438 pp. are devoted to the United States, 17 to all other part of America, 28 to Europe, 5 to Asia, and 3 to Africa.

- **1792, '2nd' (02)**
- *The American geography* ... America: containing astronomical geography. - Geographical definitions, discovery ... America and the United States:- of their boundaries; mountains; lakes, bays and rivers; natural history; productions ... manufactures and history. - A concise ... succeeded. With a particular description of Kentucky ... climates; soils; trade; character ... academies and schools; religion ... histories; etc. To which is added, an abridgment ... French, and Dutch ... West-Indies. - Of ...
- LONDON; printed for John Stockdale, Piccadilly
- 8vo, 21 cm.; pp. xvi, 536
- *Libraries* ExDel, **L**, LSe, LvU : CSmH, CtY, DeWI, DLC, FMU, IEG, MnU, MoU, MWA, NcD, NcU, NIC, NjP, NjNbS, NN, OClW, OFH, PHi, PPFr, PPL, PU, TxU, ViU, ViW : CaBVaU
- *Notes*

1. Though physically enlarged, spot-checking suggests no change in the text

until the section on Great Britain and Ireland; here footnotes provide a guide to supplementary sources of information (limited to those published by Stockdale). The only other changes noted were the inclusion, at the end of the appendix, of some information on political representation in Congress, and on the population of the United States obtained by the census of 1891.
2. Incorrectly identified as the 'second edition' on the title page, this is actually a reprinting in London of the 1789 edition (Brown 1941:214).

- **{1792}, 3rd (03)**
- *The American geography ...*
- DUBLIN; printed for J. Jones
- 21.5 cm.; pp. xvi, 536
- ***Libraries*** : CtY, DLC, KyU, MiU, OU
- ***Source*** NUC (NM 0799444)

- **{1793}, 2nd [authorized by Morse] (04)**
- *The American universal geography*, or, a view of the present state of all the empires, kingdoms, states, and republics in the known world, and of the United States in particular ... Illustrated with maps ...
- BOSTON; printed by Isaiah Thomas and Ebenezer T. Andrews. Sold at their bookstore, Faust's statue, No. 45, Newbury Street; by said Thomas in Worcester; by Berry, Rogers and Berry, in Newyork [sic]; by H. and P. Price, in Philadelphia; and by W.P. Young in Charleston
- 8vo, 22 cm., 2 vols.; Vol. I, pp. [?], 5-696; Vol. II, pp. 4, 552 [Brown 1941, 214]
- ***Libraries*** : CSt, CtY, CU-A, DLC, IU, KyLx, KyU, MB, MBAt, MeB, MHi, Mi, MiU, MWA, MWH, NBu, NcU, NjN, NN, OClWHi, PPL, PU, ViU
- ***Source*** NUC (NM 0799446)
- ***Notes***

1. In the Preface, which is reprinted in (08) below, M. states that this work 'may be considered as the second edition of the American Geography ... although it is so far renovated, and so much improved and enlarged that it was thought proper to give it a new title, corresponding to its more extensive design.' As the title implies, the book now includes reasonably full descriptions of all the countries in the world, not just of the U.S. In the Preface to Vol. II (which contains the descriptions of the countries in the 'eastern continent'), M. states that he took 'Chamber's quarto Dublin edition of Guthrie's Geographical Grammar' [i.e., (19) of GUTHRIE 1770] as his principal source.

   According to Wright (1959:157) the chapter on 'The rise and progress of geography' was drawn in large part from John Blair, *Fourteen maps of Ancient and Modern Geography ... to which is prefixed a Dissertation on the Rise and Progress of Geography* (London 1768, reprinted as *The History of the Rise and Progress of Geography* (London 1784). M. enlarged this chapter for the 6th ed. of 1812; see (14) below.

- **1794, revised ... (see below) (05)**
- *The American geography* ... astronomical geography; geographical definitions, discovery, and general description of their boundaries ... production;

populations ... succeeded: A particular ... territory, the territory south of the Ohio, and Vermont: of their extent ... histories; mines; minerals; military strength, etc. With a view of the British, Spanish, French, Portuguese, and Dutch dominions, on the continent, and in the West Indies, and of Europe, Asia, and Africa. ... A new edition, revised, corrected, and greatly enlarged by the author, and illustrated with twenty-five maps

- LONDON; printed for John Stockdale, Piccadilly
- 4to, 28 cm.; pp. [10], 715, [1]
- ***Libraries*** **L**, LDW, [IDT] : DLC, FTaSU, IHi, MWA, NjP, NN, OFH, OKentU, OrU, PHi, RPJCB
- *Notes*

1. Compared with the London ed. of 1792, the book is physically much larger. With the increase in size, however, came an increase in the size of the print, so that the book has expanded more than the text. Among the sections that are treated at greater length is the opening one; it deals with the solar system (where reference is made to Ptolemy's model of the universe), and with the related topic of the earth as a spherical body. The sections dealing with parts of America other than the United States have received, proportionately, more space than that dealing with the U.S. The treatment of Europe and Africa is the least changed, though even in these cases the number of pages has doubled.

- **{1794} (06)**
- *The American geography ...*
- LONDON; printed for J. Stockdale
- 28 cm.; pp. vi, [6]-715, [1]
- ***Libraries*** : CtY, DLC, MeB, MiU, NjP, NN, NRU, ViU
- ***Source*** NUC (NM 0799448), which notes that both this ed. and the preceding one have slight variations in the title, and that this one has only three maps
- *Notes*

1. This may be the ed. held by [IDT], not (05).

- **{1795}, new ... (see below) (07)**
- *A new and correct edition of the American geography*; or ... Kentucky, the western territory and Vermont ... To which is added an abridgment of the geography ...
- EDINBURGH; printed for R. Morison and son, Perth
- 8vo, 21.5 cm.; pp. xiv, [15]-531
- ***Libraries*** GU : AU, CtY, ICN, InU, IU, OClW, PHi, TKL, Vi
- ***Source*** ESTC and NUC (NM 0799550)
- *Notes*

1. 'This is a reprinting of the *American Geography*' [i.e., of (05)?] (Brown 1941:215).

- **1796, 3rd, corrected and considerably enlarged (08)**
- *The American universal geography* ... [as 1793] ... particular. In two parts. The first part treats of astronomical geography, and other useful preliminaries to the study of geography, in an enlarged and improved introduction - of the western, or American continent - of its discovery - its aboriginal inhabitants, and whence they came - its divisions - but more particularly of the United

States of America, generally and individually – of their situation, extent, civil divisions, rivers, lakes, climate, mountains, soil, produce, natural history, commerce, manufactures, population, character, curiosities, springs, mines and minerals, military strength, government, islands, history of the war, and the succeeding events. – And closes with a view of the British, Spanish, French, Portuguese, and other dominions, on the continent, and in the West Indies. The second part describes at large, and from the latest and best authorities, the present state, in respect to the above mentioned particulars, of the eastern continent – and its islands – as divided into Europe, Asia, and Africa – and subdivided into empires, kingdoms, and republics. To which are added, an improved catalogue of names of places, and their geographical situation, alphabetically arranged – an enlarged chronological table of remarkable events, from the creation to the present time – a list of ancient and modern learned and eminent men – and a table of all the monies of the world, reduced to the Federal currency. The whole comprehending a complete and improved system of modern geography. Calculated for Americans. Illustrated with twenty-eight maps and charts. By ... D.D. Minister of the congregation in Charlestown. Published according to Act of Congress. The introduction revised and amended by Samuel Webber, A.M. Hollins Prof. of Math. and Nat. Philos. in the University of Cambridge

- BOSTON; printed ... [as 1793] ... Worcester; by S. Campbell in New York; by M. Carey in Philadelphia; by Thomas, Andrews and Butler, in Baltimore; and by other booksellers in different parts of the United States
- 8vo, 22 cm., 2 vols.; Vol. I, pp. xiv, [2], 17-808; Vol. II, pp. iv, 692
- ***Libraries*** **L** : CSmH, CtY, DLC, ICA, InU, IU, KyLx, LU, MA, MB, MBAt, MeB, MH, MiU, MWA, MWiW, NbU, NcU, NIC, NRU, OC, OClWHi, OCU, OHi, PPL, RPB, RPJCB, TU, UU, ViU
- ***Notes***

1. On the tp. of Pt. 2 [i.e., Vol. II] of the BL copy is the statement: 'second edition of this volume.' The reason for the discrepancy in the numbering of the editions between Vols. I and II is a consequence of the fact that the 1st ed. consisted essentially of just the material that, from 1793, formed Vol. I; see (01) and (04) above.

- **{1796}, 2nd (09)**
- *The American universal geography ...*
- BOSTON; printed by Isaiah Thomas and Ebenezer T. Andrews
- 22 cm., 2 vols.
- ***Libraries*** : CPa, DSI, MWa, MWiW, OCl, PPL, ViU
- ***Source*** NUC (NM 0799451); not in Brown (1941)

- **{1801}, 3rd (10)**
- *The American universal geography ...*
- BOSTON; Isaiah Thomas ...
- 2 vols.
- ***Libraries*** n.r.
- ***Source*** Sahli (1941:382); not in Brown (1941)

- **{1802}, 4th, corr. and considerably enl. (11)**
- *The American universal geography*; or a view of the present state of all the

empires, kingdoms, states, and republics in the known world, and of the United States in particular. In two parts ... charts. The introduction revised and amended by Samuel Webber

- BOSTON; Isaiah Thomas ...
- 22.5 cm., 2 vols.
- ***Libraries*** : CSt,CtY, ICN, IU, MB, MeB, MH, MU, NN, NRU, OO, PHi, RPJCB, ScU, ViLxW
- ***Source*** NUC (NM 0799455)
- ***Notes***

1. 'Vol. I is 4th ed.; Vol. II is 3rd ed.' (Brown 1941:215); see note to (08), above.

- **{1805}, 5th (12)**
- *The American universal geography ...*
- BOSTON; printed by J.T. Buckingham, for Thomas & Andrews
- 23 cm., 2 vols.
- ***Libraries*** : CSt, CtY, DLC, GASC, InU, MiU, NcD, NcRS, NIC, NjNbS, NN, OClW, OKentU, OO, OrU, PBL, PBm, PHi, PU, ScU, T, TU, WaS
- ***Source*** NUC (NM 0799457)
- ***Notes***

1. 'Vol. I is 5th ed.; Vol. II is 4th ed.' (Brown 1941:216); see note to (08), above.

- **{1807} (13)**
- *The American universal geography*
- BOSTON
- ***Libraries*** : DLC
- ***Source*** NUC (NM 0799458); not in Brown (1941)

- **1812, 6th, arranged ... (see below) (14)**
- *The American universal geography* ... [as in 1802] ... all the kingdoms, states, and colonies in the known world. In two volumes. The first volume contains a copious introduction, adapted to the present improved state of astronomical science – a brief geography of the earth – a general description of America – an account of North-America, and its various divisions, particularly of the United States – a general account of the West-Indies, and of the four groups of islands into which they are naturally divided, and a minute account of the several islands – a general description of South-America, and a particular account of its various states and provinces – and a brief description of the remaining American islands. The second volume contains a geography of the eastern continent – a general description of Europe, and a minute account of its various kingdoms and states – a general description of Asia, its kingdoms, provinces, and islands – an account of the numerous islands arranged by modern geographers under the names of Austral(sic) Asia and Polynesia – a general description of Africa, and a particular account of its various states and islands. To which are added an abridgement of the last census of the United States – a chronological table of remarkable events from the creation to this time – an improved list ... men – and a copious index to the whole work. The whole comprehending a complete system of modern geography. Accompanied by a new and elegant general atlas of the world,

containing (in a separate quarto volume) sixty three maps, comprising, as far as they could be obtained, all the latest discoveries to the present time. Sixth edition. Arranged on a new plan, and improved in every part by a laborious examination of most of the late respectable Voyages and Travels, in Europe and Asia, by a free use of the information in the abridgement of Pinkerton's excellent Geography, and by the late admirable Statistical Tables of Hassel

- ▸ BOSTON; published by Thomas & Andrews, and sold, wholesale and retail, at their bookstore, no. 45, Newbury-Street - May, 1812. J.T. Buckingham, Printer, Winter-Street
- ▸ 8vo, 22 cm., 2 vols.; Vol. I, pp. viii, 8-872; Vol. II, pp. [2], 9-831, [1]
- ▸ ***Libraries* L** : AAP, CSmH, CtY, DeU, DLC, GA, MB, MdBP, MiU, NcD, NcU, NSyU, NWM, OKentU, PHi, PP, PPL, PSC, ViU, WU
- ▸ ***Notes***

1. The preface contains a summary of the publishing history. The only point additional to the information given above is that the publication of the 4th ed. began in 1801 [Vol. I?] rather than in 1802. This may explain the volume found by Sahli yet not recorded in Brown; see (10) above. The preface also identifies the principal sources drawn on the compilation of the work, and the sections in it that are new with this ed.
2. 'Vol. I is 6th ed.; Vol. II is 5th ed.' (Brown 1941:216); see note to (08), above.

- ▸ **{1819}, 7th (15)**
- ▸ *The American universal geography*; or, A ...
- ▸ BOSTON; Lincoln & Edmands
- ▸ 23 cm.; 2 vols.
- ▸ ***Libraries*** : CSt, CtY, DLC, DNLM, IaU, IdU, MB, MiU, MiU, NcU, Nh, NhNbS, OClW, OO, OU, PMA, PP, PPiPT, PU, ViU
- ▸ ***Source*** NUC (NM 0799462)
- ▸ ***Notes***

1. 'Vol. I is 7th ed.; Vol. II is 6th ed.' (Brown 1941:216).

- ▸ **{1819} (16)**
- ▸ *The American universal geography* or, a ... of S. America ... Australasia ... islands. The whole comprehending ... Accompanied by a general ... containing ... sixty-three maps; by Arrowsmith and Lewis
- ▸ CHARLESTON; Lincoln & Edmands, S.T. Armstrong, West, Richardson & Lord, Boston; G. Clarke, Charleston; S. Wood & Sons, Collins & Hannay, and J. Eastburn, & Co. New York; Webster and Skimers, Albany; Seward & Williams, Utica; M. Carey & Son, and W. Woodward, Philadelphia; Cushing and Jewett, and F. Lucas, Baltimore; and S.C. & J. Schenck, Savannah
- ▸ 8vo, 23 cm., 2 vols.; Vol. I, pp. iv, 9-898, [2]; Vol. II, pp. [2], 5-859
- ▸ ***Libraries* L** : DGU, NSyU, OC, VtU
- ▸ ***Notes***

1. Not in Brown (1941).
2. Brown (1941) is a would-be complete survey of the various special geographies written by Jedidiah Morse. Working before the publication of NUC, as he was, it is not surprising that he missed some of the editions. In addition to Brown (1941), Hauptman (1978:425-6), Wright (1959, *passim*), and

Warntz (1964, *passim*) all provide general discussions; Sahli (1941, *passim*) discusses (10) and (14).

MORSE, Jedidiah **A Am**

- **1814 (01)**
- *A compendious and complete system of modern geography*. Or a view of the present state of the world. Being a faithful abridgement of the American Universal Geography, (edition of 1812) with corrections and additions made from information since received. Illustrated by a representation of the solar system, and six maps of the principal divisions of the globe
- BOSTON; published by Thomas and Andrews, and sold at their bookstore, No. 45, Newbury-Street; by Eastburn, Kirk & Co., New York; M. Carey, Philadelphia; and the booksellers generally
- 8vo, 22 cm.; pp. viii, 9-670, [1]
- ***Libraries*** **L** : CSmH, CSt, DLC, ICU, MWA, NcD, NjP, NN, ViW
- ***Notes***

1. The allotment of space to the United States, to the rest of North America, and to the rest of the world is roughly that of the 6th (1812) ed. of MORSE (1789). It would be interesting to know why there was so little demand for this book that no later eds. were called for, when the much larger and presumably more expensive *Universal geography* continued in print until 1819.

MORSE, Jedidiah **Y Am**

- **{1795} (01)**
- *Elements of geography*, containing a concise and comprehensive view of that useful science ...
- BOSTON; printed by I. Thomas and E.T. Andrews
- 13 cm.; pp. 143
- ***Libraries*** : CtY, DLC, InU, MH
- ***Source*** NUC (NM 0799491). See (02) below.

- **{1796}, 2nd, cor. (02)**
- *Elements of geography ... science*, as divided into 1. astronomical 2. physical or natural 3. political geography. On a new plan. Adapted to the capacities of children and youth; and designed, from its cheapness, for a reading and classical book in common schools, and as useful winter evening's entertainment for young people in private families. Illustrated with a neat map of the United States, and a beautiful chart of the whole world
- BOSTON; printed ... Andrews. Sold by them at Faust's statue, and by I. Thomas, in Worcester; also in New York, by S. Campbell; in Philadelphia, by M. Carey; in Baltimore, by Thomas, Andrews and Butler; and by other booksellers, in different parts of the United States
- 24mo, 14 cm.; pp. xii, 13-143
- ***Libraries*** : CtY, DLC, InU, IU, MH, MiU, MWA, NSyU, RPJCB
- ***Source*** NUC (NM 0799493)

- **{1798}, 3rd, improved (03)**
- *Elements of geography ...*

- BOSTON; printed ... Worcester; in Albany, by Thomas, Andrews, and Penniman; in Baltimore ...
- 14 cm.; pp. xii, [13]-143, [1]
- ***Libraries*** : CtY, MH, N, OClW
- ***Source*** NUC (NM 0799495)
- ***Notes***

1. BL and NUC record further eds.: 1801, 4th, Boston; printed by I. Thomas and E.T. Andrews ..., xii, [13]-143, [1] pp. (: CSt, CtY, KyU, MH, PU, ViW); 1804, 5th, improved, Boston; printed for Thomas & Andrews, No. 45 Newbury-Street, xii, [13]-143, [1] pp. (L [lacks after p. 130] : MiU, NNC); and 1825, 6th, rev. and cor., New Haven; H. Howe, vi, [7]-162 pp. (: DLC, N, PU, ViU) ['Supposedly the 6th edition of this work, but with a different subtitle and content' (Brown 1941:217)].
2. Nietz (1961:209, 219) and Sahli (1941, *passim*) discuss (02).

MORSE, Jedidiah **Y Am**

- **{1784} (01)**
- *Geography made easy*. Being a short but comprehensive system of that very useful and agreeable science ...
- NEW HAVEN; Meigs, Bowen & Dana
- 12mo, 17 cm.; 2 p. l., [7]-214, [1] p
- ***Libraries*** : CtY, DLC, FTaSU, InU, MHi, MiU, MWA, NHi, NN, NNC, RPJCB, TxU
- ***Source*** NUC (NM 0799504)

- **1790, 2nd, abridged by the author (02)**
- *Geography made easy*: being an abridgement of the American Geography. Containing, astronomical geography - Discovery and general description of America - General view of the United States - Particular accounts of the thirteen United States of America, in regard to their boundaries, extent, rivers, lakes, mountains, productions, population, character, government, trade, manufactures, curiosities, history, etc. To which is added, a geographical account of the European settlements in America; and of Europe, Asia, and Africa. Illustrated with eight neat maps and cuts. Calculated particularly for the use and improvement of schools in the United States
- BOSTON; printed by Isaiah Thomas & Ebenezer T. Andrews
- 12mo, 18 cm.; pp. [1], viii, 9-322
- ***Libraries*** **L** : CtY, DeU, ICN, MH, MiU, MWA, NN, PPL, RPJCB : CaSU
- ***Source*** Discussed in Nietz (1961:213-15, 217, 220).
- ***Notes***

1. NUC records further eds.: 1791, 3rd, corr., Boston; printed by Samuel Hall (DLC, MiU, NN, NNC, NSyU, OU, PHi, RPJCB); and 1794, 4th, abridged, cor. and enl., Boston; printed by I. Thomas and E.T. Andrews, viii, [9]-432 pp. (CtY, DLC, IU, MB, MH, MiU, MWA, NN, NNC, PU, RPJCB).

- **1796, 5th (02)**
- *Geography made easy* ... containing astronomical geography; discovery of ... America; general ... States; particular ... the United ... America, and of

all the kingdoms, states and republics in the known world, in regard to their boundaries ... added, an improved chronological table of remarkable events, from the creation to the present time. Illustrated with maps of the countries described. Calculated ... schools and academies in the United States of America. Published according to act of Congress

- BOSTON; printed by I. Thomas and E.T. Andrews, Faust's Statue, No. 45, Newbury Street. Sold by said Thomas & Andrews, in Boston; I. Thomas, in Worcester; S. Campbell, New York; M. Carey, Philadelphia; and Thomas, Andrews & Butler, Baltimore
- 8vo, 18 cm.; pp. [1], viii, 9-432
- ***Libraries* L** : CtY, DLC, IU, MHi, MWA, T
- *Notes*

1. BL and NUC record further eds.: 1798, 6th, cor. by the author, Boston; printed by I. Thomas, viii, [9]-432 pp. (: DLC, MBAt, MH, MiU, NcD, NNC, RPJCB, ViU); 1800, 7th, Boston; printed by I. Thomas, viii, [9]-432 pp. (: CT, CtHT, CtY, FTaSU, FWpR, ICU, IU, M, MH, MiU, MMeT, MWA, NcU, NN, NNC, NNU, OO, RNHi, RPJCB, Vi, ViU : CaQQS) [NM 0799514; not in Brown (1941); 1802, 8th, Boston; printed by I. Thomas and E.T. Andrews, viii, [9]-432 pp. (: CtY, IHi, KyU, MB, MH, NcD, NN, NNC, OClWHi, PHi, PMA, PU) [NM 0799515; not in Brown (1941)]; 1804, 9th, Boston; printed by Joseph T. Buckingham for Thomas & Andrews, viii, [9]-432 pp. (: CoU, CtY, DLC, InU, MH, MiD, NcD, NhM, NN, NNC, NSyU, OClWHi, OrU, OU, PHi, ViU, WHi); 1806, 10th, Boston; printed by J.T. Buckingham for Thomas & Andrews, viii, [9]-432 pp. (: DGU, DLC, InU, MB, MeAu, MH, MiU, NcD, NcU, NcWsW, Nh, NNC, NSyU, OrU, PU, WHi); 1807, 11th, Boston; Thomas & Andrews; J.T. Buckingham, printer, viii, [9]-432 pp. (: AU, DLC, MB, Nh, NN, NNC, NSyU, OClWHi, ViU : CaOLU); 1809, Boston; Thomas & Andrews. J.T. Buckingham, printer, viii, [13]-364 pp. (: CtY, CU-S, MH, NIC, NSyU) [NM 0799519; not in Brown (1941)]; 1811, 14th and 2nd of this new abridgment, Boston (: DLC, KyU, MB, NNC, OU, ViU); 1812, 15th and 3rd, Boston; Thomas & Andrews, viii, [9]-360 pp. (: DLC, INU, MB, Nh, NNC, NRCR, NSyU, ViU); 1813, 16th and 4th, Boston; Thomas & Andrews, 360 pp. (: COU, CtY, DGU, IU, KyBgW, MB, N, NBuHi, NIC, NSyU, ViU : CaAEU); 1814, 17th and 5th, Boston; Thomas & Andrews, viii, 364 pp. (: CU, DLC, MB, Nh, NNC, OClWHi, OU, ViU); 1814, From the 16th Boston ed., Troy; Parker and Bliss, viii, [10]-360 pp. (: MH, NcD, NjNbS, NN, NSyU); 1816, 18th and 6th, Boston; Thomas & Andrews, viii, [9]-364 pp. (L : CSt, DLC, IHi, InU, MB, NcA-S, Nh, NIC, ODW, PMA : CaBVaU) [Sahli (1941, *passim*)]; 1816, 2nd Troy, from the 16th Boston, Troy; Parker and Bliss, viii, [9]-364 pp. (: CtY, DLC, MWH, NBuHi, NSyU, OkU : CaBViPA); and 1817 [Known from illustration in Nietz (1961: 199)].

- **1818, 19th and 7th (03)**
- *Geography made easy* ... America. Illustrated by maps of the world, and of North America. Also, by a separate atlas of nine select maps. - To which is added a list of questions, to relieve the labors of the instructor, and guide the inquiries of the pupil
- BOSTON; Thomas & Andrews. Sold at their bookstore, No. 45, Newbury-

Street, and by West and Richardson, No. 75, Cornhill; by S. Wood and Sons, New-York, M. Carey, Philadelphia; and by the principal booksellers in the United States. Ezra Lincoln, printer
- 12mo, 17 cm.; pp. viii, 9-364
- *Libraries* L : DLC, MWH, NcD, NNC, NSyU, OU, PPPrHi : CaAEU
- *Notes*

1. NUC records a further ed.: 1819, 20th, Utica and Boston, 364 pp. (CtY, DLC, INS, MH, MShM, MWA, NcU, NN, NNC, NSyU, NUt, OClWHi, PPAmP, PU).

- **{1820}, 22nd (04)**
- *Geography made easy ...*
- BOSTON; Richardson, & Lord, 75 Cornhill; J.H.A. Frost, printer, Congress Street
- 17.5 cm.; × p., 1 l., [13]-368 p.
- *Libraries* : DLC, IU, OClWHi
- *Source* NUC (NM 0799529)
- *Notes*

1. According to the Preface of MORSE, S.E. (1844 [i.e., *A system of geography, for the use of schools*], reprinted in the ed. of 1858), he, Sydney E. Morse, joined his father, Jedidiah, in preparing this ed. of *Geography made easy* for publication. See MORSE, J. and MORSE, S.E. (1822).
2. Not in Brown (1941).
3. Carpenter (1963:246-9) provides a general discussion; Nietz (1961, facsimile of tp. [196], 200, 206-7, 212-15, 217-20, 225, 227) discusses (01), (02), and (03); Sahli (1941, *passim*) discusses (01), and 1800, 7th, and 1814, Troy, and 1816, Boston, and (04), and 1819; Hauptman (1978:425, 430) draws on 1804, 9th.
4. Microform versions of (02), (04), and the eds. of 1800, 1807, 1813, and 1820 are available from CIHM.

MORSE, Jedidiah **Dy Am**

- **{1789}b (01)**
- *A new geographical dictionary*
- ELIZABETHTOWN
- *Libraries* : NjR
- *Source* NUC (NM 0799555); not in Brown (1941)

MORSE, Jedidiah, and Richard C. MORSE **Dy**

- **{1821}, 3rd, rev. and corrected (01)**
- *A new universal gazetteer, or geographical dictionary ...* Accompanied with an atlas. By ... and Richard C. Morse
- NEW HAVEN; S. Converse; [etc.]
- 24.5 cm.; pp. 832
- *Libraries* : CU, DeU, DLC, MnHi, NN, PHi, PPL, PSC, PU, Vi, WaSpG
- *Source* NUC (NM 0799564)
- *Notes*

1. It seems probable that this is a successor ed. to MORSE (1789b).

- **{1823}, 4th, rev. and cor. (02)**
- *A new universal gazetteer ...*
- NEW HAVEN; S. Converse
- 25 cm.; pp. 856
- ***Libraries*** : DGU, DLC, ICRL, KyU, MiU, MU, MWA, NcD, NcU, NNBG, NNUT, OU, PPL, ViU
- ***Source*** NUC (NM 0799565)
- ***Notes***

1. Discussed in Brown (1941, 1951), Rumble (1943), and Wright (1959).

MORSE, Jedidiah, and Sidney Edwards MORSE **Y Am**

- **{1822}, 23rd (01)**
- *A new system of geography, ancient and modern*, for the use of schools; accompanied with an atlas ... By ... and Sidney Edwards Morse. 23rd ed.
- BOSTON; Richardson & Lord
- 18 cm.; viii, 14-278, p. [Atlas separate, 100 p.]
- ***Libraries*** : CSmH, CtY, DLC, IU, MB, MiU, NNC, ViU
- ***Source*** NUC (NM 0799556)
- ***Notes***

1. According to the Preface of MORSE, S.E. (1844 [i.e., *A system of geography, for the use of schools*], reprinted in the ed. of 1858), he, Sydney E. Morse, joined his father, Jedidiah, in preparing the 22nd ed. of the *School geography* (Boston 1820) for publication. This must refer to *Geography made easy* [MORSE, J. 1784 (05)]. S.E.M. continues: 'Between that date [1820] and 1828 (two years after the death of Dr. Morse) five editions of the School Geography were published, the number of each edition varying from 10 to 20,000 copies.' Judging by the 25th ed. (1826, see below), it had been very extensively rewritten. Following NUC, it is treated here as a new work; but see Brown (1941:216).
2. BL and NUC record further eds.: 1824, 24th, Boston; Richardson & Lord (: DGU, MH, NcU, NjNbS, NNC, PHi); 1826, 25th, Boston; Richardson & Lord, J.H.A. Frost, printer, ix, [1] p., 1 l. [13]-342 pp. (L : CtY, DLC, MB, MiU, MtHi, NcD, NIC, NN, NRU, ViU); and 1828, 26th, Boston; Richardson & Lord, pp. xii, 13-300, 1-23 (L : CtHt, CtY, DLC, MB, MH, NNC, ViW, WU).
3. Elson (1964:398) cites the 24th ed. of 1824, and the 26th of 1828.

MORSE, Sidney Edwards **Y Am**

- **{1822} (01)**
- *A new system of modern geography*; or, a view of the present state of the world
- BOSTON; G. Clark
- 23 cm.; vi, 676 p.
- ***Libraries*** : DGU, DI-GS, DLC, IU, MH, MiU, NcU, NjR, NNC, OO, OU, TU, UU, ViU, ViW
- ***Source*** NUC (NM 0800185). Cited by Elson (1964:397).

MORSE, Sidney Edwards **Y Am**
- **{1844} (01)**
- *A system of geography, for the use of schools*
- NEW YORK; Harper & brothers
- *Libraries* : MB, MH, NWM
- *Source* NUC (NM 0800197)
- *Notes*

1. According to the Preface (seen in the ed. of 1858, below), publication of the edition of Jedidiah Morse's *Geography made easy*, which had been revised by his son, Sydney, was suspended in 1828, 'chiefly because ... [there was not] time ... for its proper revision.' The work to which this is a note is the product of that revision.
2. NUC records further eds. (all New York; Harper & Brothers): 1845, 72 pp. (CtY, MH, MiU, NN, NNC, OClW, OClWHi, OFH, TxU); 1846, (CtY, MH, OClW); 1847 (MH : CaQSherU); 1848 (ODW); 1849 (MH, NcD, PU : CaOONL); 1850 (CtY, DLC, OFH); 1851 (MH); 1852 (CtY, DHEW, MH); 1853 (CtY, MH, OClWHi); 1855 (MB, MH); c1857 [c1844] (OClWHi); 1858 (: **CaNSPH** [imperfect]) [c1844]; and 1864 (MH, NcA).
3. Hauptman (1978:430) refers to the ed. of 1845, and (1978:434) includes it in a general discussion; Culler (1945, *passim*) discusses the ed. of 1850; Elson (1964:401) cites the ed. of 1845.
4. Microform versions of the eds. of 1847 and 1849 are available from CIHM.

[MORTIMER, Favell Lee {born BEVAN}] **Y mSG**
- **{1849} (01)**
- *Near home; or, the countries of Europe described.* With anecdotes and numerous illustrations ...
- LONDON; Hatchard and co.
- 17 cm.; pp. xvi, 402
- *Libraries* n.r.
- *Source* St. John (1975, 1:188)
- *Notes*

1. Included because the author followed this book with companion vols., titled *Far off; or, Asia described* (1852), and *Far off; or, Africa and America described* (1854).

MORTON, Francis **A**
- **[1861] (01)**
- *A manual of geography*: being a description of the natural features, climate, and productions of the various regions of the earth
- LONDON; Robert Hardwicke, 192, Piccadilly
- 24mo, 15 cm.; pp. [2], iii-iv, [5]-192
- *Libraries* **L**
- *Notes*

1. Uses tiny print to pack masses of facts of the types identified in the title, into continuous prose, organized by country, or, in the case of the UK by county.
2. BL supplies the date.

- **{1862} (02)**
- *Geography*
- ? Hardwicke
- ***Source*** ECB, Index (1856-76:130)

MOTHER'S GEOGRAPHY ... **Y Am**

- **{1842} (01)**
- *The mother's geography*, a series of preparatory lessons adapted to the capacity of very young children ...
- NEW YORK; W.A. Le Blanc
- 16.5 cm.; 2 p. l., [9]-68 p.
- ***Libraries*** : DLC
- ***Source*** NUC (NM 0822772)
- ***Notes***

1. Discussed in Culler (1945, *passim*), and Brigham and Dodge (1933:19-20).

MOXON, Joseph **OG**

- **1659 (01)**
- *A tutor to astronomie and geographie*: or an easie and speedy way to know the use of both the globes, cœlestial and terrestrial. In six books. The first teaching the rudiments of astronomy and geography. The 2. shewing by the globes the solution of astronomical and geographical probl. The 3. [shewing ... solution of] problemes in navigation. The 4. [...] astrological problemes. The 5. [...] gnomonical problemes. The 6. [...] spherical triangles. More fully and amply than hath ever been set forth either by Gemma Frisius, Metius, Hues, Wright, Blaew, or any others that have taught the use of the globes: and that so plainly and methodically that the meanest capacity may at first reading apprehend it; and with a little practise grow expert in these divine sciences. Whereunto is added the antient poetical stories of the stars: shewing reasons why the several shapes and forms are pictured on the cœlestial globe. Collected from Dr. Hood. As also a discourse of the antiquity, progress and augmentation of astronomie
- LONDON; printed by Joseph Moxon: and sold at his shop on Corn-hill, at the signe of Atlas
- 4to, 19 cm.; pp. [15], 224, 40
- ***Libraries*** BmP, C, **L**, O, NP, PetC : ICU, INU, IU, MB, MH, MiU, N, NIC, NN, PPL, RPJCB
- ***Notes***

1. As the full title shows, this is not a special geography. In addition to the topics referred to in the tp., it provides definitions of basic geographical terms, plus an exposition of Ptolemaic cosmology, as modified by Tycho Brahe. In the Preface, M. refers to an English version of Willem Blaeu's *Institutio astronomica* that he, M., had published previously (BLAEU 1654), and which he subsequently republished (MOXON 1665) using the same title as the work to which this is a note.

   The 'discourse' is paginated separately.

- **1670, 2nd, corrected and enlarged (02)**
- *A tutor ... geographie.* Or ... astronomie and geographie. The 2 ... The 6. [...] trigonometrical problemes ... apprehend it, and with ... practice ... sciences. With an appendix shewing the use of the Ptolomaick sphere. The second edition, corrected and enlarged. Whereunto ... globes. As also ...
- LONDON; printed by Joseph Moxon: and sold at his shop in Russel Street, at the signe of the Atlas
- 4to, 20 cm.; pp. [8], 272, [8]
- ***Libraries*** BfQ, C, DC, EU, **L**, LLI, LSM, O, SwR : CLU, CtY, DLC, MH, OkU, RPJCB, WU
- ***Notes***

1. Still presents the Ptolemaic cosmology; this is surprising in that in the *Tutor to astronomy* (MOXON 1665) he had favoured the Copernican.

- **1674, 3rd, corrected and enlarged (03)**
- *A tutor to astronomy and geography.* Or ... The 1. teaching ... astronomy and geography. The 2. ... than hath yet been set forth, either ... methodically, that ... augmentation of astronomy
- LONDON; printed by Tho. Roycroft, for Joseph Moxon: and sold at his shop on Ludgate-Hill, at the sign of the Atlas
- 4to, 20 cm.; pp. [7], 271, [9]
- ***Libraries*** C, E, LeP, **L**, NoU, [IDT] : CtY, CU, DLC, FU, MH, MnU, NcD, NcU, RPJCB
- ***Notes***

1. As was common, the claim to be corrected and enlarged (or words to that effect) is carried through from one ed. to the next. The pagination, supported by spot-checking, suggests that the content of this ed. is essentially unchanged from 1674.
2. Facsimile reprint, 1968, New York; Burt Franklin.
3. Blackwell's Catalogue 976, *Antiquarian and Secondhand Books on the Sciences* records two versions of this ed. The first, which it identifies as (Wing M 3024) has pp. [vi +] 272 + [viii]; the second is described as (Variant of Wing M 3024) has pp. [viii +] 271 + [xi].

- **1686, 4th, corrected and enlarged (04)**
- *A tutor to astronomy and geography ...*
- LONDON; printed by S. Roycroft, for Joseph Moxon ... Ludgate Street ...
- 4to, 20 cm.; pp. [7], 271, [9]
- ***Libraries*** C, E, ES, LinC, **L**, LDW, LPO, MaPL, MRu, O : CoU, CSt, CtY, DLC, MH, MnU, NcD, NIC, NjP, PPL, RPJCB, ScU, ViU, WU
- ***Notes***

1. Pagination and spot-checking suggest no change.
2. Wing (1982) shows a copy held by GC, but fails to identify the library to which this code refers.

- **1698, 5th, corrected and enlarged (05)**
- *A tutor to astronomy and geography* ... Ptolemaick sphere
- LONDON; Printed for James Moxon, at the sign of the Atlas in Warwick Lane

- 4to, 20 cm.; pp. [7], 271, [9]
- *Libraries* E, EU, **L**, LPO, OChCh, [IDT] : CLU, CtY, PU
- *Notes*

1. BL catalogues as 1699, as does EU; see (07) below. Once again the work appears unchanged.

- **{1698}, 'Fifth' (06)**
- [*A tutor to astronomie and geographie* ...]
- LONDON? For W. Hawes
- 4to
- *Libraries* BmP, L, NcP, RgtP : CSmH, CLU, IU, MBAt, PPL
- *Source* Wing (1982, M3026A); did not find in BL or NUC

- **1699, 5th, corrected ... (see below) (07)**
- *A tutor to astronomy and geography* ... augmentation of astronomy. The fifth edition, corrected and enlarged by Phillip Lee
- LONDON (neither printer nor publisher appears on the tp.)
- 4to, 20 cm.; pp. [7], 271, [8]
- *Libraries* **L**, O : CLU, DeU, DSI, MeB, MWA, OrU, RPJCB
- *Notes*

1. Still unchanged from 1670.
2. Blackwell's Catalogue 958 *Antiquarian and Secondhand Books on the Sciences* states, 'printed for W. Hawes,' and identifies as (Wing M 3027).

MOXON, Joseph **OG tr**

- **1665 (01)**
- *A tutor to astronomy and geography*, or, the use of the Copernican spheres; in two books. The first being an explanation of the Copernican hypothesis and spheres. The second proving the phœnomena solved by the earths motion, as well as by its supposed stability; as appears by the application of these spheres to problemes astronomical, geographical, nautical, astrological, gnomonicaly, and trigonometrical
- LONDON; printed for Joseph Moxon, and sold at his shop on Ludgate-Hill neer Fleet-Bridge, at the signe of the Atlas
- 4to, 20 cm.; pp. [4], 184
- *Libraries* **L**, O, OB : CtY, DLC, MH, PPL
- *Notes*

1. Not a special geography. In contrast to the views he expressed in 1659, Moxon here favours the cosmology of Copernicus over that of Ptolemy. He presents the current arguments in favour of the former, and also deals carefully with the theological arguments put forward in favour of the latter by traditionalists.
2. According to BL the author of this work is Blaeu (BLAEU 1654).

MUIR, F. **?**

- **{1864} (01)**
- *A system of universal geography*
- EDINBURGH

- *Libraries* n.r.
- *Source* Allford (1964:247)

MÜNSTER, Sebastian **mSG tr**

- **1572 (01)**
- *A brief collection and compendious extract of strau[n]ge and memorable thinges*, gathered oute of the Cosmographye of Sebastian Munster. Where in is made a playne descrypsion of diuerse and straunge lawes rites, manners, and properties of sundry nacio[n]s, and a short reporte of straunge histories of diuerse men, and of the nature and properties of certayne fowles, fishes, beastes, monsters, and sundrie countries and places
- LONDON; imprinted at London in Fleet-Strete neare Sainct Dunstanes Church by Thomas Marthe
- 8vo, 14 cm.; pp. [26], 202 (numbered by leaves, some misnumbering, the last leaf being Fol. 102)
- *Libraries* **L** (lacks last 2 leaves)
- *Notes*

1. A miscellaneous collection of information, well summarized in the title, chosen because of their strangeness. Some countries have short sections devoted to them; in order of appearance they are: Ireland, England and Scotland, Sardinia, Iceland, Lapland, and Cathay. Among peoples the Turks receive most attention, being dealt with on Fol. 34-45 and 46-53. Though similar in size to the selection made from MÜNSTER (1553), see below, the items in this collection are generally different

- **1574 (02)**
- *A briefe collection* ... straunge ... out of .... Wherein ... plaine description of diuers ... lawes, rites, maners and properties of sondrye nations, ... report ... diuers ... certaine ... sondry countryes ...
- LONDON; imprinted at London in Fleetstreat by Thomas Marthe
- 8vo, 14 cm.; pp. [6], 202 (numbered by leaves; most of the errors of 1572 corrected, but some new ones introduced; last leaf again Fol. 102)
- *Libraries* **L** : DFo, MB
- *Notes*

1. The text has been reset, but apart from changes in spelling it seems to be identical to that of 1572.
2. Discussed in Wilcock (1975).

MÜNSTER, Sebastian **A tr**

- **1553 (01)**
- *A treatyse of the newe India*, with other new founde landes and ilandes, aswell eastwarde as westwarde, as they are knowen and found in these oure dayes, after the descripcion of Sebastian Munster in his boke of universall cosmographie; wherein the diligent reader may see the good successe and rewarde of noble and honeste enterpryses, by the which not only worldly riches are obtayned, but also God is glorified, and the Christian fayth enlarged. Translated out of Latin into Englishe. By Rycharde Eden
- LONDON; imprinted at London, in Lombard Strete, by Edward Sutton

- 15 cm.; pp. [204]
- ***Libraries*** **L** : MiU, NN, PHi, PPT, RPJCB
- ***Notes***

1. Of the total, 170 pages are text, 21 are preface, and 7 a dedication. As the title hints, only part of the original, that dealing with the new discoveries, is included.
2. A modern reprint can be found in Arber, Edward, ed., *The First Three English Books on America*, Birmingham, 1885.
3. Discussed in Beazley (1901) and Hodgen (1954).

MURBY, [Thomas?] **?**

- **{1871} (01)**
- *Geography, and atlas*
- ? Murby
- ***Libraries*** n.r.
- ***Source*** ECB, Index (1856-76:131)

MURBY, Thomas **Y SG?**

- **{[1874]} (01)**
- *Murby's scholars' home lesson book of geography*
- LONDON
- 8vo
- ***Libraries*** L

MURRAY, Hugh **Y**

- **1833, 4th, thoroughly revised ... (see below) (01)**
- *A catechism of geography*; comprising all the leading features of that important science, and including the most recent discoveries. With a vocabulary of geographical terms. Illustrated by eight engravings. Fourth edition, thoroughly revised and considerably enlarged
- EDINBURGH; published by Oliver & Boyd, Tweeddale Court; and Simpkin & Marshall, London
- 12mo, 15 cm.; pp. 108
- ***Libraries*** **L**
- ***Notes***

1. The first question is: 'What is geography?' And so the work proceeds through, 'How do geographers know the earth?' and, 'What are the boundaries of Sweden?' to, 'What is remarkable in the external appearance of Owhyhee [Hawaii]?' There is a section on the use of globes and geographical problems at the end.
2. BL records a further ed.: 1842, 7th, Edinburgh, 108 pp.

MURRAY, Hugh **? Am**

- **{1859 [1858]} (01)**
- *The encyclopedia of all nations*: comprising a complete physical, statistical, civil and political description of the world ... including the late discoveries

and travels by Dr. Kane, Dr. Barth and Dr. Livingston ... Edited by Elbridge Smith ...
- NEW YORK; Henry Bill
- 26 cm., 2 vols.
- ***Libraries*** : CtY, LU, MWA
- ***Source*** NUC (NM 0898332)
- ***Notes***

1. The similarity of both the title and the size of the work suggest that this may be essentially an ed. of MURRAY (1834). If it was that, it was not the first American ed., there having been at least 13 eds. of MURRAY (1834) published in Philadelphia between 1837 and 1853.
2. NUC records further eds.: 1860, New York, Henry Bill (KMK); and 1863, San Francisco, H. Bill for F. Dewing, 2 vols. (OkU).

MURRAY, Hugh **A**

- **1834 (01)**
- *An encyclopaedia of geography*: comprising a complete description of the earth, physical, statistical, civil, and political; exhibiting its relation to the heavenly bodies, its physical structure, the natural history of each country, and the industry, commerce, political institutions, and civil and social state of all nations. By ... assisted by the following gentlemen, in their respective departments of science; in astronomy and mathematical geography, William Wallace; in geology and the distribution of minerals, Robert Jameson; in botany and the distribution of plants, W.J. Hooker; in zoology and the distribution of animals, William Swainson. Illustrated by eighty-two maps, drawn by Sidney Hall, and upwards of a thousand other engravings on wood, by R. Branston, from drawings ... representing the most remarkable objects of nature and art in every region of the globe
- LONDON; Longman, Rees, Orme, Brown, Green, & Longman
- 8vo, 22 cm., 2 vols.; Vol. I, pp. xii, 850; Vol. 2, pp. 851-1567 (16 pp. supplement bound with Vol. 2)
- ***Libraries*** **L** : ICRL, KyU, NIC
- ***Notes***

1. Despite its title, which suggests an alphabetically ordered listing of information, this is a special geography, preceded by a history of geography (pp. 2-67), and a section on physical geography and the related fields of astronomy, geology, and the distribution of plants and animals, and also systematic human geography (pp. 68-279). The balance of Vol. 1 deals with Europe, and Vol. 2 with the rest of the world.

- **{1837}, rev., with additions, by Thomas G. Bradford (02)**
- *The encyclopaedia of geography* ... and about eleven hundred ... together with a new map of the United States
- PHILADELPHIA; Carey, Lea and Blanchard
- 26.5 cm., 3 vols.
- ***Libraries*** : DI, DLC, ICRL, KyU, MB, MWA, NBuU, NcD, NjNbS, NjR, OrU, ViU : CaBViP
- ***Source*** NUC (NM 0898337)

- ***Notes***

1. NUC records further eds. (Philadelphia, Lea and Blanchard, except as noted): 1838 (NWM, TU, ViU); 1839 (AAP, DLC, IdPI, NcRS, NjR); 1840, [2nd, 'thoroughly rev. and brought down to the present time' {CIHM}], 3 vols. (AAP, DLC, FTaSU, MB, MWA, N, OCl : CaNBFU); 1841, 3 vols. (CtY, ViU); 1842, 3 vols. (NcA); 1843, 3 vols. (AU, DN-Ob, MB, MH, ViU); 1844, 2nd, thoroughly revised and with a supplement bringing down the information to the present time, London; Longman, Brown, Green, and Longman, [iii]-xii, [2], xiii-xiv, 1578, 16 pp. (DLC); 1845, 3 vols. (CU-S, DLC, DN, DNLM, I, LU, NN : CaOKU); 1846, 3 vols. (DLC); 1847, 3 vols. (NcU, NN); 1848, 3 vols. (ICJ, NcCU); 1852, 3 vols. (DLC, OrPR : CaNBSaM); and 1853, 3 vols. (DLC, NcU); CIHM records 1857, 3 vols. (: : CaOKU).
2. Elson (1964:401) cites the ed. of 1843.
3. Microform versions of the eds. of 1840, 1845, 1852, and 1857 are available from CIHM.

MYERS, Thomas **A**

- **1812 (01)**
- *A compendious system of modern geography, historical, physical, political, and descriptive*: accompanied with many interesting notes; and a series of correct maps: being adapted to the use of the higher classes of pupils under both public and private tuition
- LONDON; printed or G. Wilkie and J. Robinson, Paternoster-Row
- 8vo, 22 cm.; pp. vii, 520
- ***Libraries*** **L**
- ***Notes***

1. A typical special geography of the time, with very little history; intermediate in size between PINKERTON (1802) and GUY (1810).

MYERS, Thomas **A**

- **1822 (01)**
- *A new and comprehensive system of modern geography, mathematical, physical, political, and commercial*; comprising a perspicuous delineation of the present state of the globe, with its inhabitants and productions; preceded by the history of the science; interspersed with statistical and synoptic tables; and accompanied with a series of coloured maps, a great variety of appropriate views, and numerous other engravings illustrative of the manners, customs, and costumes of the nations
- LONDON; Sherwood, Neely & Jones
- 4to, 26.5 cm., 2 vols.
- ***Libraries*** : **NN** (Vol. I only), NWM, PPL
- ***Notes***

1. Vol. I contains the Introduction and the account of Europe.

- **{1829} (02)**
- *A new and comprehensive system of modern geography ...*
- LONDON; Henry Fisher, son, & P. Jackson

- 28 cm., 2 vols.
- *Libraries* : LNT.

N

NANCREDE, Paul Joseph Guérard de NG

- **{1801} (01)**
- [*A complete system of universal geography*]
- *Notes*

1. 'Boston, Sept. 1st, 1801. Sir, having undertaken the publication of a complete system of universal geography of America, and of the world, we take the liberty to recommend our design to your notice and patronage' [signed: Joseph Nancrede and Bernard B. Macnulty] [Boston]. Broadside ([3] p., 27.5 cm. From NUC (NN 0013691), location (MH).

NARRATIVE ... NG

- **{1679} (01)**
- *Narrative and deduction of the several remarkable cases of Sir William Courten*, Sir Paul Pyndar, William Courten ... together with their surviving partners and adventurers to the East Indies, China and Japan, and divers other parts of Asia, Europe, Africa, and America: recollected out of the original writings and records
- LONDON
- Folio
- ***Libraries*** L
- ***Source*** BL and Cox (1935:74)

NEW AND COMPLETE ... A

- **1796 (01)**
- *A new and complete system of universal geography*: containing a full survey of the natural and civil state of the terraqueous globe: exhibiting all the latest and most authentic information concerning Europe, Asia, Africa, and America; the seas by which they are divided, and the isles scattered amid those seas; with an ample apparatus of tables, maps, etc. as also an accurate explanation of those principles of geography which depend upon the discoveries of astronomy; and a philosophical view of universal history; the last article written by Robert Heron
- EDINBURGH; printed for R. Morison and Son, Booksellers, Perth
- 4to, 21 cm., 2 vols.; each vol. contains 2 pts. bound separately; Vol. I, Pt. 1, pp. [1], xii, 154, 300; Pt. 2, [2], 301-888; Vol. II, Pt. 1, pp. [2], 528; Pt. 2, [2], 529-1067
- ***Libraries*** **OSg** : NBuG, NcU, ViU
- ***Notes***

1. The universal history is a blending of biblical and classical elements, ending with the fall of the Roman empire. Vol. I, Pt. 1, deals with northern, east-

ern, and central Europe; Pt. 2 with the British Isles and France; Vol. II, Pt. 1 with southern Europe and the United States, Pt. 2 with British and Spanish possessions in North America and on the mainland of South America (15 pp.), islands off the east coast of America (106 pp.), Asia (120 pp.), Africa (52 pp.), and the newly discovered islands in the Pacific (140 pp.). Vol. I is more or less a special geography, though for Scotland, England, and France it contains lengthy histories and unusually detailed current information (e.g., the French constitution of June 1795 is reproduced in full, occupying 20 pp.). In Vol. II extensive use is made of direct quotations; e.g., the description of Virginia, which occupies 39 pp., is taken almost entirely from Thomas Jefferson, *History of Virginia*, and the section on the islands in the Pacific comes entirely from the account given by Captain Cook of his voyage of 1776-7.

2. Bowen (1981:315) lists under Heron, as does NUC. Both BL and NUC (Vol. 242, p. 557) note that Robert Heron was a pseudonym used by John PINKERTON.

- **{1797} (02)**
- *A new and complete system of universal geography*, containing ... globe, exhibiting ... America, the seas ... and the islands in those seas ... to which is added a philosophical ... history
- LONDON; printed for D. Ogilvy and Son
- 22 cm.
- ***Libraries*** : MnHi, PPL
- ***Source*** NUC (NH 0313250)

NEW AND EASY ... **Y**

- **{1751} (01)**
- *A new and easy system of geography*, for the instruction of youth, and those of riper years, adapted to Mr Locke's and Monsieur Rollin's notions of initiating any person into the knowledge of the science ... Divided into lessons, by way of question and answer. With a curious map of the world, and a list of maps necessary for youth, etc. To which is added a new geography of Ireland
- DUBLIN; R. Buckley
- 13.5 cm.; pp. v, [3] 108, [2], ii, 43
- ***Libraries*** : NNC
- ***Source*** NUC (NN 0142650)

- **{1762} (02)**
- *A new and easy system of geography ...*
- ***Libraries*** : PPWa

NEW DESCRIPTION ... **Gt?**

- **{1689} (01)**
- *A new description of the world*, or, a compendious treatise of the empires, kingdoms, states, provinces, countries, islands, cities ... of Europe, Asia, Africa, and America, in their scituation, products, manufactures and com-

modities, geographical and historical; and an account of the natures of the people, their habits, customs, wars, religion, policies ... as also of the wonders and rarities of fishes, beasts, birds, rivers, mountains, plants ...
- LONDON
- 12mo
- ***Source*** Cox (1935:75); not in BL or NUC. Probably a ghost. See CLARK, S. (1689), and note that IU catalogues its copy under the title.

NEW GAZETTEER ... ?
- **{1793} (01)**
- *The new gazetteer; or, modern geographical index.* Containing a concise description of the empires, kingdoms, cities, towns ... in the known world, etc.
- EDINBURGH; David Ramsey
- 8vo
- ***Libraries*** L
- ***Notes***

1. BL catalogues under Gazetteer.

NEW GENERAL ATLAS ... **A**
- **1721 (01)**
- *A new general atlas*, containing a geographical and historical account of all the empires, kingdoms, and other dominions of the world. With the natural history and trade of each country. Taken from the best authors, particularly Cluverius, Brietius, Cellarius, Bleau, Baudrand, Hoffman, Moreri, the two Sansons, Luyts, the Atlas Historique, Sir John Chardin, Le Brun, Tournefort, etc. To which is prefix'd, an introduction to geography, rendering the principal parts of that science easy, and containing all that is necessary for the ready understanding of maps. Together with a copious alphabetical index. The maps, which are all engraven or revised by Mr. Senex, are laid down according to the observations communicated to the English Royal Society, the French Royal Academy of Sciences, and those made by the latest travellers; and the descriptions suited to the course of each map, which has not been observed in any other atlas
- LONDON; printed for Daniel Browne without Temple-Bar, Thomas Taylor over-against Serjeants-Inn in Fleet-Street, John Darby in Bartholomew-Close, John Senex in Salisbury-Court, William Taylor in Pater-Noster-Row, Joseph Smith in Exeter-Change, Andrew Johnston[,] engraver in Round-Court, William Bray next the Fountain-Tavern in the Strand, Edward Symon in Cornhill
- Folio, 54 cm.; pp. [8], xiii, [1], 272, 141-261, [10]
- ***Libraries*** **L** : AU, CSmH, CtY, DFo, DCL, ICN, ICU, InU, IU, LNT, MdBP, MH, MiD, MiU, NN, NNH, OCl, PPL, ViU : CaBViPA
- ***Notes***

1. The xiii pp. carry the coats of arms of some of the subscribers, starting with the Prince of Wales; the subscribers are also listed in the preliminary matter. Europe and its countries occupy pp. 11-272 (of which the British Isles occupy 185-272); Asia occupies 141-225 (second series), Africa 226-36, and Ameri-

ca. 237-61. Despite the subtitle, there is little history; most of the information is chorographical or, in the case of the British Isles, topographical. To verify the claim made at the end of the full title would call for comparisons with MERCATOR (1635) and ORTELIUS (1606).

This work probably deserves to be classified with the coffee-table books that proliferated towards the end of the eighteenth century - ADAMS, M. (1794), BALDWYN (1794), BANKES and others (1787), COOKE (1790), MIDDLETON, (1777-8), MILLAR (1782), PAYNE, J. (1791) - rather than with roughly contemporary works of equivalent bulk: ATLAS GEOGRAPHUS (1711-17) and *Compleat geographer*, i.e., (04) of THESAURUS GEOGRAPHICUS.

2. A microform version is available from CIHM.
3. See SENEX (1708-25).

NEW GEOGRAPHICAL DICTIONARY ... (1737) **Dy**

- **1737 (01)**
- *A new geographical dictionary*: containing a brief description of the countries, empires, kingdoms, provinces, cities, towns, mountains, rivers, lakes, gulfs, straits, isles, bayes, capes, etc. of the world. Translated from the French; with great improvements from the best modern books of voyages and travels, and from the most accurate maps and sea-charts
- LONDON; printed for D. Midwinter, at the Three Crowns in St. Paul's Church-Yard; and sold by E. Owen, in Amen-Corner
- 20.5 cm.; pp. [4], [714]
- *Libraries* **EU**
- *Notes*

1. An annotated gazetteer without maps. The particular areal subunit within which any given town or village is located is identified. In other words, locations are specified by their geographical context, not, as is done today, by their coordinates of latitude and longitude. Countries are described in terms of the political subunits they contain, except England; that country is said to be: 'the largest and best part of the isl. of Great Britain, so well known to its inhabitants, and so largely described by very many others, that it is needless to say anything of it here' (p. x).

- **{1745}, 2nd (02)**
- *A new geographical dictionary* ... description of all the countries ... world. With the situation and distances of the cities and towns, etc. from the most considerable places in each country; as also a more particular description of the course of the principal rivers ... Translated ... travels ... The 2d ed. To which is added, the latitude and longitude of the most considerable cities
- LONDON; printed for D.M. and sold by S. Birt [etc.]
- 20.5 cm.; pp. [734]
- *Libraries* : DLC
- *Source* NUC (NN 0147977)

NEW GEOGRAPHICAL DICTIONARY ...
See BARROW (1759-60).

NEW GEOGRAPHICAL DICTIONARY ... (1836) **Dy? Am**

- **{1836} (01)**
- *New geographical dictionary*. To which is added a geographical dictionary of the Bible
- BOSTON; Mussey
- 17.5 cm.; pp. (2), 304
- ***Libraries*** : MB, ODW, PU
- ***Source*** NUC (NN 0147979)

NEW HISTORICAL ... GEOGRAPHY ... **A**

- **1800 (01)**
- *A new historical and commercial system of geography*: containing a comprehensive history and description of the present state of all the kingdoms of the world: including the most recent discoveries of the latest voyagers and travellers. The whole forming a complete guide to geography, astronomy, the use of globes, maps, etc. To which are added, a new and copious geographical index, with the names and places alphabetically arranged; and a genealogical table of all the sovereigns of the world. The whole compiled from the best authorities and embellished with maps, and superb engravings
- MANCHESTER; printed by Sowler and Russell, 125, Deansgate
- Folio, 37 cm.; pp. xxiv, 693, [3]
- ***Libraries*** : : **Capc**
- ***Notes***

1. The usual wide range of topics is presented, but with little of the system or order provided by the contemporary PINKERTON. Like BANKES (1787?) and COOKE, G.A. (1790?), it provides much information on recent voyages of exploration in the Pacific, especially those of Cook.

- **{1803} (02)**
- *A new historical and commercial system of geography ...*
- MANCHESTER
- Folio
- ***Libraries*** n.r.
- ***Source*** Beeleigh Abbey Books, *A catalogue of voyages and travels*, BA/23, W. & G. Foyle Ltd, London, n.d., item 60
- ***Notes***

1. NUC records further eds. as follows: 1806, Manchester, S. Russell, 693 pp. (OU); 1808, Manchester (MWA); 1812, Manchester, printed and sold by Russell and Allen, [5], ix-xxiv, 693, [3] pp. (CLU, NIC : CaAEU); and 1815, Manchester; printed and sold by Russell and Allen, xxiv, 693, [3] pp. (CLU, DLC).

NEW INTRODUCTION TO GEOGRAPHY ... **Y**

- **1803, 5th edition ... (see below) (01)**
- *A new introduction to geography*; in a series of lessons: for youth. In which every division of the known world, its longitude, latitude, length, breadth, and capital, are exhibited at one view: with a concise description of the produce, manufactures, constitutions, laws, customs, manners, and religion of

the several countries. To which is prefixed, a correct map of the world. The fifth edition, corrected and enlarged
- ▸ LONDON; printed for S. Sael and Co. No. 192 Strand
- ▸ 15 cm.; pp. [5], vi-viii, 134, [2]
- ▸ ***Libraries*** **L**
- ▸ ***Notes***

1. Provides the facts of size and location, with emphasis on Europe (92 pp.), but the 'description' provides only scraps of the information promised in the title.
2. Catalogued by BL under Introduction. This might be the earliest ed. of BUTLER, John Olding (1854*) so far identified.
3. BL records further eds. as follows: 1813, 8th, enlarged; 1821, 10th, enlarged.

[NEWBERY, John] **Y**
- ▸ **1748 (01)**
- ▸ *Geography made familiar and easy to young gentlemen and ladies*. Being the sixth volume of the Circle of the Sciences, etc. Published by the King's authority
- ▸ LONDON; printed for J. Newbery, at the Bible and Sun, in St. Paul's Church-Yard
- ▸ 16[or 32]mo, 10 cm.; pp. [3], xv, 319
- ▸ ***Libraries*** **L** : CtY, ICU, MH, NNC
- ▸ ***Notes***

1. Catalogued by both BL and NUC under Newbery, John, and under Circle.
2. For the usefulness of geography, the author cites the philosopher John Locke 'in his excellent treatise on education ... the late M. Rollin ... [and] the Rev. Dr Watts' (p. ii). Following the opening 68 pp. on general geography, there is a chapter on 'geographical paradoxes,' with their solutions.
2. Allford (1964:240) attributes a book with this title to Newberry [sic], J. in 1746.
3. NUC records further eds. as follows: 1748 (ICU) [NC 0437539 notes: 'Second issue. Cf Welch, Charles. A bookseller of the last century ... John Newbery']; 1752, 3rd, Dublin, Wilson, 315 pp. (PU); 1769, 3rd, London; printed for Newbery and Carnan, 319 pp. (ICU); 1770, 4th, Dublin, H. Saunders [etc], 315, [5] pp. (CtY); 1776, 4th, London; printed for T. Carnan and F. Newbery, junior, 10 p. l., 319 (1) p., 61 (NN) [NN 0224337 notes: 'See, Toronto Public Library: Osborne Coll. p. 132-33']; and 1783, 5th, London; printed for T. Carnan (MH).

- ▸ **1793 (02)**
- ▸ *Geography made easy for children*, improved from the Circle of the Sciences, containing the new discoveries, etc.
- ▸ LONDON; printed for Darton & Harvey
- ▸ 14 cm.
- ▸ ***Libraries*** : CtY
- ▸ ***Source*** NUC (NN 0224331). Treated here as an ed. of NEWBERY (1748) on the basis of the second phrase in the title.

- **{1805}, 3rd (03)**
- *Geography made easy for children*; with a short and familiar account of the principal new discoveries. From ... sciences
- LONDON
- 12mo
- ***Libraries*** n.r.
- ***Source*** Watt (1824, 2:700)

NEWTON, John **mSG**
- **1679 (01)**
- *Cosmographia, or a view of the terrestrial and coelestial globes*, in a brief explanation of the principles of plain and solid geometry, applied to surveying and gauging of cask. The doctrine of the primum mobile. With an account of the Julian [sic] and Gregorian calendars, and the computation of the places of the sun, moon, and fixed stars from such decimal tables of their middle motion, as supposeth the whole circle to be divided into an hundred degrees or parts. To which is added an introduction unto geography
- LONDON; printed for Thomas Passinger, at the Three Bibles on London-Bridge
- 8vo, 18 cm.; pp. [15], 510, [16]
- ***Libraries*** **L** : CtY, CU, DLC, MB, MiU, NjP
- ***Notes***

1. The Introduction to Geography occupies pp. 411-58, and the special geography pp. 426-53. Information is extremely condensed, with the whole of America dealt with in less than two pages.

NICHOLS, F. **Gt**
- **(1811)**
- *An abridgement of a compend of geography ...*
- ***Notes***

1. A ghost; in Watt, (1824, 2). See note to (02) in NICHOLS (1811), below.

NICHOLS, Francis **Y Am**
- **{1809} (01)**
- *A compend of geography*, containing a concise description of the different countries of the earth. Also, a compend of astronomy; for the use of students
- PHILADELPHIA; F. Nichols
- 17.5 cm.; pp. xii, 190, 60
- ***Libraries*** : DLC, INU, MH, PU
- ***Source*** NUC (NN 0244926)

NICHOLS, Francis **Y Am**
- **{1811} (01)**
- [*An elementary treatise of geography ...*]
- ***Notes***

1. Known from the copyright statement in the next entry.

- **1813, new ... (see below) (02)**
- *An elementary treatise of geography*. Containing a concise description of the different countries of the world. Compiled from the best modern travels. Also, a compendious view of the solar system. For the use of schools. A new edition corrected and improved
- PHILADELPHIA; printed for Francis Nichols, William Fry, printer
- 14.5 cm.; pp. [2], 3-162
- *Libraries* : DeU, **NN**, NNC, OClWHi, PPWa
- *Notes*

1. One of the few authors not to puff his subject: 'Geography is so plain and simple that it may be taught to children of nine years of age, or to those of a more advanced age as a relaxation from severe studies. It is merely an object of memory, and therefore the acquisition of it depends chiefly on the care and application of the learner. It is one of the most easy and useful branches of literature that are taught at school; and may be learned before judgment is expanded, and fit to encounter subjects of reasoning and reflection' (p. 5).

   N. describes this book as an abridgement of NICHOLS (1809). He admits drawing on the work of AIKIN (1806) and PINKERTON (1802).

- **{1818}, cor. and improved (03)**
- *An elementary treatise of geography ...*
- PHILADELPHIA; printed for Francis Nichols, William Fry; printer
- 15 cm.; pp. 162
- *Libraries* : PSC
- *Notes*

1. Elson (1964:396) cites (02).

NICOLAY, Charles Grenfell, ed. **A**

- **1852, 1859 (01)**
- *A manual of geographical science, mathematical, physical, historical and descriptive*
- LONDON; J.W. Parker & son
- 8vo, 22 cm., 2 vols.; Vol. I (1852), pp. [5], vi-xvi, 445, [1]; Vol. II (1859), pp. [5], vi-xii, 615, [1]
- *Libraries* L, **[IDT]** : CtY, DLC, MnHi, NIC, NN, NWM, PU, T : **CaAEU**
- *Notes*

1. Vol. I contains: M. O'Brien, 'Mathematical geography' (pp. 1-142); J.R. Jackson, 'Chartography' [sic] (pp. 143-84); D.T. Ansted, 'Physical geography' (pp. 185-413); C.G. Nicolay, 'Theory of description and geographical terminology' (pp. 414-45). Vol. II contains: W.L. Bevan, 'Ancient geography' (pp. 1-144); C.G. Nicolay, 'Maritime discovery and modern geography' (pp. 145-548); an index (pp. 549-615).

   The objective is stated as: 'Hitherto [books such as this] intended to be used in education have been rather compendious works of reference, than introductions to the study of a science, and are often overloaded with details, while general principles are omitted. In the present work, an attempt has been made to avoid these evils, and so to classify, arrange, and systematize the information contained in it, that it may be immediately available both to the teacher and to the scholar; and by omission of all non-essential details, whether political, statistical, or topographical, to confine attention to the principal subject' (p. v).

Both the scholarly standards of the book, and the diverse areas of expertise that were perceived, in the nineteenth century, to be required if geography were to be carried on at a scholarly level, are shown by the qualifications of the men Nicolay recruited as his fellow workers. Nicolay himself was Librarian of King's College, London; O'Brien was Professor of Natural Philosophy and Astronomy, while Amsted was Professor of Geology at the same institution; Jackson had been lately the Secretary of the Royal Geographical Society. Only Bevan was not, visibly, a member of the academic establishment, being Vicar of Hay, Brecon.

While *The Manual* is both longer than Strabo's *Geography*, and also differs from its better-known predecessor in being more accurate in a factual sense, Nicolay comes no closer than the scholar of antiquity to finding in the 'facts' of geography a structural coherence that will enable the observer to understand why places have the character that they do.

NICOLAY, Charles Grenfell **OG**

- **{1849} (01)**
- *'On history and geography,'* in: Introductory lectures delivered at Queen's College
- LONDON; John W. Parker
- 8vo; pp. viii, 352
- ***Libraries*** L
- ***Source*** BL, catalogued under London, III, Queen's College, as well as under Nicolay

NILES, Sanford, ed. **Y Am**

- **{1885}b (01)**
- *Merrill's elementary geography.* Sanford Niles, editor
- ST. PAUL, MN; D.D. Merrill
- 24 cm.; pp. 88
- ***Libraries*** : DHEW

NILES, Sanford **Y Am**

- **{1887} (01)**
- *Niles's advanced geography.* Mathematical, physical, political
- ST. PAUL, MN; D.D. Merrill
- 31 cm.; 2 p. l., 134 p.
- ***Libraries*** : DHEW, DLC
- ***Source*** NUC (NN 0270479), which notes: 'On cover: Minnesota text book series.' NUC records another ed. in 1890, noting: 'Merrill's school book series.'

NILES, Sanford **Y Am**
- **{1885}a (01)**
- *Niles's elementary geography* including the geography, history and resources of Minnesota
- ST. PAUL, MN; D.D. Merrill
- 24 cm.; pp. 134
- ***Libraries*** : DLC
- ***Source*** NUC (NN 0270482), which notes: 'Minnesota text book series'
- ***Notes***

1. Discussed in Culler (1945, *passim*).

O

O'BRIEN, Matthew **?**
- **{1855?} (01)**
- *Manual of geographical science*
- Place/publisher n.r.
- ***Libraries*** n.r.
- ***Source*** Allibone (1854). This may have been a contribution made to NICOLAY (1852, 1859).

OF THE NEWE LANDES ... **mSG els**
- **{[1511 or 1521 or 1522]} (01)**
- *Of the newe landes* and of ye people founde by the messengers of the kynge of Portyngale named Emanuel. Of the X dyuers nacyons crystened. Of Pope John and his landes and of the costely keyes and wonders melodyes that in that lande is
- ANTWERP; John of Doesborowe
- ***Libraries*** L : CSmH, DLC, NN
- ***Notes***

1. The ed. of 1511 is listed by E.G.R. Taylor (1930) and Parks (1928, but not 1961). Parks added: 'This is not really of the new lands. It is mainly about Prester John and other medieval items' (Parks 1928:269). Taylor (1930:7), following Arber (1885), observes that this work is made up of three distinct tracts that were brought together by Doesborowe. They were looked on by Arber, who is the authority for 1511 being the date of publication, as being the first book in English to mention America. NUC (NO 0033671), while noting this information in summary form, also notes: '(Harrisse, Bibl. amer. vetus., no. 116, gives date of publication as 1521 or 1522).'
2. A modern facsimile ed., reduced in size, was published at Boston, 1922 (NO 0033673); (DLC, MiU, NN, RPJCB). The text is reproduced in Arber (1885) (L : AzTeS, CoCC, CoD, CtY, CU, CU, DLC, DN, FJ, I, KMK, MBBC, MdBP, MH, MiU, MsU, NjP, NN, OO, PPULC, UPB, UU, ViW, WaSp, WaU : CaBVa, CaBVaU, CaBViPA).

OGLE, M.J. **Y**

- **{[1870]} (01)**
- *First teachings about the earth*; its lands and waters; its countries and states, etc.
- LONDON; Simpkin
- 8vo
- ***Libraries*** L
- ***Source*** BL and ECB, Index (1856-76:130)
- ***Notes***

1. BL and NUC record further eds. as follows: 1870, 2nd, London and Dorking; Simpkin, Marshall (L : CtY); and [1875], 3rd, London (L).

OLNEY, J[essie] **Y Am**

- **{1847}a (01)**
- *An elementary geography*, adapted to Olney's outline maps, for common schools and general use
- NEW YORK; Pratt, Woodford & co.
- 15.5 cm.; pp. 93
- ***Libraries*** : DLC
- ***Notes***

1. CIHM records 1852, 4th (: : CaOLU).
2. Microform versions of both eds. are available from CIHM.

OLNEY, J[essie] **Y Am**

- **{[1849]} (01)**
- *Olney's quarto geography ...*
- NEW YORK; Pratt, Woodford & co.
- 30 cm.; pp. 69
- ***Libraries*** : DLC, OFH, MH
- ***Source*** NUC (NO 0081084)
- ***Notes***

1. NUC records further eds. (all New York; Pratt, Woodford): [1851?], 10th (MH); [1851?], 12th (MH, NNUT); [1856, c1849], 14th (MH); and [1856?, c1849], '15th' (MH).

OLNEY, Jessie **Y Am**

- **{1827} (01)**
- *Practical geography for the use of schools*, accompanied with problems for the use of the globes
- HARTFORD; D.F. Robinson & co.
- ***Libraries*** : MH

OLNEY, Jessie **Y Am**

- **{1847}b (01)**
- *A practical introduction to the study of geography ...*

- NEW YORK; Pratt, Woodford & co.
- 15 cm.; pp. vi, 108
- ***Libraries*** : DLC
- ***Notes***

1. According to Brigham and Dodge, the 'first edition dates from Hartford in 1828.' They also state, however, that the 55th ed. (1847) of the next work (i.e., *A practical system of modern geography*) 'was essentially the same volume' (Brigham and Dodge, 1933:18). Given the difference in the number of pages between the book to which this is a note and the *Practical system*, a problem is apparent.

OLNEY, Jessie **Y Am**

- **{1828} (01)**
- *A practical system of modern geography*; or, a view of the present state of the world, simplified and adapted to the capacity of youth. Accompanied by a new and improved atlas
- HARTFORD; D.F. Robinson
- 15 cm.; pp. 270
- ***Libraries*** : DLC, MB, OClWHi
- ***Source*** NUC (NO 0081036). See notes to (02), below.
- ***Notes***

1. NUC records further eds. (Hartford, CT, unless stated otherwise): [1829], New York, Robinson & Pratt, 288 pp. (FTaSU, MH); 1830, 4th, 268 pp. (DCU, DLC, KEmT, MH); [Sahli (1941:382) records, 1830, 6th]; 1831, 7th, 283 pp. (CoU, IU, MH, NNC, OCl, PU); 1831, 8th, 83 pp. (CtY, PU); 1832, 10th, 283 pp. (CtY, NNC); 1832, 11th, 283 pp. (CtY, DLC, MB, MH, NNC, ViU); 1833, 12th, 288 pp. (MH, OkU); and 1833, 13th, 288 pp. (MH, NjP, OClWHi).

- **1833, 14th (02)**
- *A practical system of modern geography* ... world. Simplified ... youth. Containing numerous tables, exhibiting the divisions, settlement, population, extent, lakes, canals, and the various institutions of the United States and Europe[;] the different forms of government, prevailing religions, the latitude and longitude of the principal places on the globe. Embellished with numerous engravings of manners, customs, etc. Accompanied ...
- HARTFORD; D.F. Robinson & co. Sold by all the principal booksellers in the U. States
- 14.5 cm.; pp. [5], vi, [7]-288
- ***Libraries*** : MH, NN
- ***Notes***

1. Proceeds by way of question and answer; assumes access to an atlas, as shown by the fact that the questions exceed the text in length. The book is justified on the grounds that while formerly geography had not been taught until the later years of schooling, it had, in recent years, been introduced with increasing frequency in primary classes; books suited to the limited capacities of children of that age were needed.
2. NUC records further eds. as follows (all New York, Robinson & Pratt, and, except as noted, 288 pp.): 1834, 15th, Hartford, D.F. Robinson (CtY, MH,

NcD); 1834, 16th, Hartford, Robinson & Pratt (DLC, ViU); 1835, 18th (CtHT, CtY, MH, N, NNC, PWcS, Vi); 1835, 19th, 284 pp. (MH, NNC); 1835, 20th, pp. n.r. (MH, NNC); 1836, 21st (OrU); 1836, 22nd (MH, NNC); 1836, 23rd (AU, DLC, MtHi, OrPR); 1837 [c1828], 24th, *A practical system of modern geography* ... youth. Containing numerous tables ... Embellished with numerous engravings of manners, customs, etc. Accompanied ... (NBuG); 1837, 25th (CtY, ICU, MB, Nh); 1838, 26th, pp. n.r. (MH, PHi); 1838, 27th (MiU, NBuG); 1838, 29th, pp. n.r. (MH); 1839, 30th, pp. n.r. (CtY, MH); 1839, 31st (PPStCh); 1839, 32nd (CtY, PU); 1839, 32nd (DAU); 1840, 33rd (MH, MiU, N, NN, NNC, OClW, ViU); 1841, 35th (CtY, DLC, LU, MH); 1842, 38th (DLC, OKentU, OrU); 1843, 40th, pp. n.r. (IU, MH); and 1844, 40th, pp. n.r. (MH).

- **{1844}, 41st (03)**
- *A practical system of modern geography* ...
- NEW YORK; Pratt, Woodford & Co.
- ***Libraries*** : MH
- ***Notes***

1. BL and NUC record further eds. as follows (all New York, Pratt, Woodford, and, until 1845, 294 pp.; 300 pp. thereafter, except as noted): 1844, 42nd (: DLC, MtU); 1844, 46th (: CtY, MB, MH, NcA); 1845, 47th (: NcU); 1845, 48th (: CU); 1845, 49th, pp. n.r. (L); 1846, *A practical system of modern geography* ... tables, exhibiting the divisions, settlement, population, extent, lakes, canals, and the various institutions of the United States and Europe, the different forms of government and prevailing religions ... (: CtY, MH); 1847, 51st, (: MeB); 1847, 55th (: DLC); [CIHM records 1848, 57th (: : CaQQS]; 1849, 60th, pp. n.r. (: MH); 1849, 62nd, 286 pp. (: OkU) {NO 0081077 notes: '[c1844]'}; 1850, 63rd, 296 pp. (: DLC, OrU); 1851, 63rd, 296 pp. (: ViU); 1851, 68th, 296 pp. (: ICRL); 1851, 69th, pp. n.r. (: CtY); 1852, 76th, pp. n.r. (: MH); 1860, 100th, Pratt, Oakley, 326 pp. (: DLC); and 1873, 105th (ECB, Index 1856-76:131).
2. Carpenter (1963:257-8) provides a general discussion; Hauptman (1978:430) quotes from (02), refers to it in general discussion (1978:433-4), and draws on it (1978:436); Sahli (1941:382), and Nietz (1961:209, 215, 217, 220, 223) probably refer to the 6th ed. of 1830; Sahli (1941, *passim*) discusses the 13th ed. of 1833, and the 35th ed. of 1841; Culler (1945, *passim*), and Nietz (1961:210, 211) probably discuss the 33rd ed. of 1840. Elson (1964:400) cites 11th, 1832; 18th, 1835; and 22nd, 1836.
3. A microform version of the 57th ed. of 1848 is available from CIHM.

O'NEILL, John **A Am**

- **{1808} (01)**
- *New and easy system of geography and popular astronomy*; or, an introduction to universal geography ...
- BALTIMORE; printed for the publisher, by G. Dobbin and Murphy
- 18.5 cm.; pp. xii, [13]-504
- ***Libraries*** : CsT, DLC, MdBE, MdHi, MdW, MWA, NSyU, PPL
- ***Source*** NUC (NO 0094542)

▸ *Notes*
1. 'Designed for mature readers and studious youth ... uses the questions-and-answer method, but many answers are a paragraph or even a page in length' (Brigham and Dodge 1933:13).

▸ **{1812}, 2nd, with ... (see below) (02)**
▸ *New and easy system of geography* ... geography ... The whole arranged in a catechetical form ... Illustrated by a map of the world, a map of North America and a plan of the solar system. 2nd ed., with considerable additions by Joseph James
▸ BALTIMORE; Fielding Lucas, jun. and Edward J. Coale. Fry and Kammerer, printers
▸ 18.5 cm.; iv pp., 1 l. [7]-450 p.
▸ ***Libraries*** : CtY, DeU, DLC, FTaSU, MdU, MH, NcD, OU, TxU, ViU
▸ ***Source*** NUC (NO 0094543)
▸ *Notes*
1. NUC records further eds. as follows (all Baltimore, F. Lucas): 1814, 3rd, 450 pp. (AU, CtY, MdU, MnU, NNC); 1816, 4th, 359 pp. (DGU, DLC, IU, MdW, MWA); 1819, 5th, 450 pp. (DGU, DLC, MdBS, MdW, NjR.) [Nietz (1961:219-20); Sahli (1941, *passim*)].
2. Elson (1964:396) cites ed. of 1814.

ORTELIUS, Abraham **A tr**
▸ **{1602?} (01)**
▸ *An epitome of Ortelivs his theatre of the world.* Wherein the principal regions of the earth are described in smalle mappes. With a brief declaration annexed to ech mappe. And donne in more exact manner, then the lyke declarations in Latin, French, or other languages. It is also amplyfied with new mappes wanting in the Latin editions
▸ LONDON; printed by Iohn Norton
▸ 13 × 18 cm.; 4 pp. l., 110 f., 3 l., 13 f., 2 l.
▸ ***Libraries*** C, L, OChch : CSmH, DLC, ICN, MH, MN, RPJCB
▸ ***Source*** NUC (NO 0139224)
▸ *Notes*
1. Pollard and Redgrave (1976) give the date as 1601?.

▸ **1603 (02)**
▸ *Abraham Ortelius his epitome of the Theatre of the worlde* nowe latlye, since the Latine[,] Italian, Spanishe, and Frenche editions, re:newed [sic] and augmented, the mappes all newe grauen according to geographicall measure. By Michael Cognet. Mathematician of Antwerpell[,] being more exactlye set forth, and emplefyed with larger descriptions, then any done heere to fore
▸ LONDON; printed for Ieames Shawe, and are to be solde at his shoppe nigh Ludgate
▸ Obl. 8vo, 12 × 16 cm.; 2 pts. (the additional material by Cognet being paginated separately), [17], 110 l. (i.e., 110 numbered maps); [2], 13 l. (i.e., numbered maps), [3]
▸ ***Libraries*** C[imp], CStJ, Eu, **L**, [IDT] : DFo, DLC, ICN, IU, MH, MiU, MnU, NN, RPJCB

- *Notes*

1. Each map is accompanied, on the facing page, by text describing the area mapped. There is also an introduction containing definitions of geographical terms, and the principles of drawing a 'table' (a map based on a sound projection).

- **1610 (03)**
- *An epitome of Ortelius his Theatre of the world*, wherein ... [as in 1602]
- LONDON; printed by Iohn Norton
- Obl. 8vo, 13 × 18 cm.; 2 pts., paginated separately; pp. [7], 110 l. (i.e., numbered maps); [6], 13 l. (i.e., numbered maps), [2]
- ***Libraries*** C, **L** : CSmH, DFo, DLC, ICN, MH, NN, RPJCB, TxU
- *Notes*

1. The date is supplied by BL. Judging by the pagination, this is another issue of the ed. of 1602. Coignet's introduction of 1603 is not present; in its place are representations of the celestial and terrestrial globes, with accompanying text. The layout of the book is as in 1603, but the maps are drawn differently (in both parts), and the accompanying text is usually shorter.

- **{1616} (04)**
- *An epitome of Ortelius his ...*
- LONDON; Norton
- Obl. 32mo
- ***Libraries*** : MH

ORTELIUS, Abraham **A tr**

- **1606 (01)**
- *Theatrvm orbis terrarvm. Abrahami Orteli Antverp. Geographi regii.* The theatre of the whole world; set forth by that excellent geographer Abraham Ortelius
- LONDON; printed for Iohn Norton, Printer to the Kings most excellent maiestie in Hebrew, Greeke, and Latine [on the final page the name of Iohn Bill is added to that of Norton as joint publisher]
- Folio, 50 cm.; pp. [12], followed by the body of text, which has 115 maps (according to BL), but these are pages with maps; there are 168 separate maps, of which 3 have inserts containing small portions of the main map at an enlarged scale. Also in the main body there are 134 pages of written text. There is then an addendum, entitled *Parergon, sive veteris geographiae...*, containing 46 maps (38 according to BL), including 1 with inserts, illustrating ancient history or classical mythology, 5 miscellaneous plates, and 4 maps of European countries, of which only one (England) is readily classified as historical, plus 81 pp. of text. Finally, there is a letter from Humfrey Lloyd to Ortelius concerning the island of Mona ("the ancient seat of the druids") of 7 pp.
- ***Libraries*** C, E, **L**, LVu, O : CSmH, DFo, DLC, ICN, MiU, MiU, PBL, RPJCB
- *Notes*

1. According to Pollard and Redgrave (1976) the maps were printed in Antwerp by the Officiana Plantiniana. The prefatory material includes a life of Ortelius

'written first in Latine by Francis Sweert ... and now translated into English by W.B.' Pollard and Redgrave tentatively identify the latter as W. Bedwell, and attribute the translation of the entire text to him. There is also a short preface by Ortelius, dated 1570; this includes comments on the order in which the material is presented.
2. The text that accompanies the maps includes a discussion of the etymology of the name of the region mapped, its relative situation as shown by the regions neighbouring it, and the sources on which Ortelius drew, plus a miscellany of other information. The Preface includes a justification for the order in which the maps are presented. Two rules apply. First, the whole is given before the parts, so the first map is that of the whole world. Second, at the level of individual countries, Ptolemy's method is used; this involves the unspecified assumption that the Atlantic ocean forms the western limit to the world; countries are then presented in order, from west to east (or so it is asserted). I have argued elsewhere (Sitwell 1980) that behind this procedure lies the assumption that world is pictured as an open book, to be read from left to right.
3. A facsimile ed. was published by TOT, 1968. They state that the original was the only ed. of Ortelius's *Theatrum Orbis Terrarum* with an English text, and was the first large world atlas to be published in England.
4. As a major figure in the history of cartography, and the successor to Mercator in the compilation of atlases, Ortelius is discussed in all standard histories of the subject (e.g., Bagrow 1964, Crone 1978).

OSBORNE, Thomas **mSG**
- **{1747} (01)**
- *An introductory discourse concerning geography*, navigation, government, commerce, religion, travel and a geographical description of Europe
- LONDON
- ***Libraries*** : MB
- ***Source*** NUC (NO 0147720), which notes: 'In A collection of voyages and travels. Vol. I, pp. i-lviii, 7. London: MDCCXLVII'
- ***Notes***

1. Other entries in NUC make it clear that Osborne was the compiler of the *Collection of Voyages and Travels*. The 1st ed. was published in 1745. The ed. of 1747 referred to in NO 0147720, was part of the 3rd ed. of the *Collection of Voyages and Travels* compiled by A. and J. Churchill.

OUISEAU, J. **Y**
- **{1791} (01)**
- *The practical geography, for the use of school*[s.] With an epitome of ancient geography and an introduction to the science of the globes
- LONDON; printed at the Logographic Press and sold by B. Law, etc.
- 21 cm.; pp. vi, 146
- ***Libraries*** : CtY

OUISEAU, J. **Y**

- **1794 (01)**
- *Practical geography*, with the description and use of the celestial and terrestrial globes
- LONDON; printed for C. Macrae, Orange Street, Leicester Square
- 12mo, 18 cm.; pp. ix, [3], 312
- ***Libraries*** **GP**, L
- ***Notes***

1. Refers in the Preface to 'an elementary Geography for very young people' published two years previously (presumably OUISEAU 1791). This book is intended for pupils of a 'more advanced age.' Pp. 1-158 form special geography of an arid kind. The modern reader will be astonished to read that 'care has been taken not to load the page with too many names' (p. vi). Pp. 159-233 are devoted to an alphabetical list of places, with their locations, plus some miscellaneous information. Pupils are expected to absorb all this information, learning it by heart. After that comes an eight-page list of Classical place names, and then the description and use of globes.
2. Peddie and Waddington (1914:227) record an ed. in 1804, and BL the 9th in 1828.

P

P., D.
See P[OWELL] D[avid].

P., R. **A**

- **1659 (01)**
- *A geographical description of the world*. Describing Europe, Asia, Africa, and America. With all its kingdoms, countries, and common-wealths. Their scituations, manners of the people, customs, fashions, religions, and governments. Together with many notable historicall discourses therein contained
- LONDON; printed by John Streater
- Folio, 29 cm.; pp. [4], 154, [5]
- ***Libraries*** C, EU, **L**, O, [IDT] : DLC, MeB, MChB, MWA, NcU, NIC, NjP, NN, NNH, ViU : CaBViP
- ***Notes***

1. Bound at the back of Denis Petau (Petavius, Dionisus), *The History of the world*, edited by R.P. The *Geography*, which is not found in the French counterpart, is not reminiscent of other English works of the period. The treatment of countries and continents is relatively well balanced, though some bias is shown towards those featured in the Bible. The small islands of Europe receive more attention than usual.

PAGE, Robert **Gt**

- **{1671}**

- *Cosmographie, or, a description of the whole world*, represented (by a more exact and certain discovery) in the excellencies of its scituation, commodities, inhabitants, and history: of their particular and distinct governments, religions, arms, and degrees of honour used among them. Enlarged with many and rare additions. Very delightful to read in so small a volume
- LONDON; H.B. for John Overton
- 12mo
- *Notes*

1. Neither in BL nor NUC. Listed as item 676 in Maggs (1941), where it is stated that 'Pages 112-116 relate to America, giving firstly, short accounts of the various British colonies, and then the King of Spain's dominion and Brazil.' This is a ghost, someone at some stage having taken an F for a P; see (05) in FAGE (1658).

PARISH, Elijah **Y Am**

- **{[1807]} (01)**
- *A compendious system of universal geography*, designed for the use of schools ... Compiled from the latest European and American [travellers, voyagers and] geographers
- NEWBURYPORT, MA; Thomas & Whipple
- 12mo, 17.5 cm.; 3 pp. l., [5]-213 p.
- ***Libraries*** L : DLC, IU, NSyU, OClW, OClWHi, TU
- ***Source*** BL and NUC (NP 0095253)
- ***Notes***

1. NUC records further eds. as follows (all Newburyport, MA): [1808], 2nd (MH, NNC, NNU-W, PU); [1809], 2nd, 212 pp. (IEdS); and [1810?], 2nd, 212 pp. (CtY, NRU).
2. Brigham and Dodge (1933:12-13) discuss (01).

PARISH, Elijah **Y Am**

- **{1810} (01)**
- *A new system of modern geography*; or a general description of all the considerable countries in the world. Compiled from the latest European and American geographies, voyages and travels. Designed for schools and academies ... Ornamented with maps
- NEWBURYPORT [MA]; Thomas and Whipple, no. 2, State-street. Sold, wholesale and retail, at their bookstore, and by all the principal booksellers in the New-England states. Greenough & Stebbins, printers
- 12mo, 18.5 cm.; pp. 370
- ***Libraries*** L : CSt, CtY, DGU, DLC, InU, MiU, OCH, OCl
- ***Source*** BL and NUC (NP 0095271)
- ***Notes***

1. NUC records further eds. (both Newburyport): 1812, 2nd, 366 pp. (CSmH, CSt, CtY, DLC, DSI, ICJ, MB, NSyU); and 1814, 3rd, 366 pp. (CSt, CTY, ICN, MB, MeB, MH, N, NjR, NN, NSyU, OO, OrHi, ViU).
2. Nietz (1961:207-9, 212-16, 219-20, 228), and Sahli (1941, *passim*) discuss (01); Sahli (1941, *passim*) discusses the 2nd ed. of 1812. Dryer (1924:118)

discusses the ed. of 1812. Though he omits *modern* from the title, and implies that it was first published in 1812, Carpenter's general discussion (1963:251-3) presumably applies to this book.

PARKER, Richard Green **Y Am**

- **{1842} (01)**
- *Parker's geographical questions* ... prepared particularly for Worcester's Atlas, etc.
- BOSTON
- 12mo
- ***Libraries*** L

- **{1847} (02)**
- *Parker's geographical questions*. Questions in geography, adapted for the use of Morse's ... or any other respectable collection of maps ... To which is added, a concise description of the terrestrial globe
- NEW YORK; Harper & brothers
- 19 cm.; pp. 60
- ***Libraries*** : DLC
- ***Notes***

1. Allibone (1854) records an ed. in 1848, and BL and NUC record further eds. as follows (all New York; Harper & brothers): 1855, 114 pp. (L : DLC, MH, MiU, NBuG, NIC); 1855, 68 pp. (: MH, ViU); and 1863, 114 pp. (: NBuG).
2. Culler (1945, *passim*) discusses (02).

PARLEY, Peter [pseudonymous] **Y**

- **1838 (01)**
- *A grammar of modern geography*. By Peter Parley, author of Tales about Europe, Asia, etc. etc. With maps and engravings
- LONDON; printed for Thomas Tegg and Son, Cheapside; Tegg and co. Dublin; Griffin and co. Glasgow; and J. and S.A. Tegg, Sydney, and Hobart Town
- 16mo, 14 cm.; pp. xvi, 304
- ***Libraries*** **L** : ICU, IU, MH
- ***Notes***

1. 'Often attributed to Samuel Griswold Goodrich but listed as spurious in his "Recollections of a Life Time". He lists a similar title among those written by George Mogridge' (NUC [NP 0107364]). However, NUC does not attribute this title to Mogridge.

   A comparison with GOODRICH (1835), which is the work referred to in the second sentence of the title, and which was issued by the same publisher in London, is interesting in that the tone in each of the two books is very different. GOODRICH (1835) is filled with vulgar prejudice, racism, and ethnocentrism; this work is largely free of those vices.

- **1854, 4th, revised and corrected (02)**
- *A grammar of modern geography* ...

- LONDON; William Tegg and Co., 85, Queen Street, Cheapside
- 16mo, 14 cm.; pp. xii, 351
- ***Libraries*** **L**
- ***Notes***

1. The changes since 1838 take the form of the addition of new 'lessons'; spot-checking suggests that existing material is left unaltered (e.g., Ireland, where there is no mention of the famine).

PASCHOUD, [M]. **A**

- **1722, 1724 (01)**
- *Historico-political geography*: or, a description of the names, limits, capitals, divisions, descriptions of particular provinces, situation, extent, air, soil, commodities, rarities, rivers, chief towns, inhabitants, manners, language, populousness, dominions, pretentions, government; kings or princes genealogy, titles, revenues, residence; states or courts of justice, laws, nobility, order of knighthood, clergy, arch-bishopricks, bishopricks, universities, religion, advantages, defects, interests of the several countries in the world. Collected from the best authors. By the Reverend Mr. ... chaplain to the Right Honble William-Anne Earl of Albemarle
- LONDON; printed by J. Read; and sold ... (see below)
- 8vo, 20 cm., 2 vols.; Vol. I (1722, sold at Mr. Peter du Noyer's in the Strand, and at Mr. Abel Rocayrol's in St. Martin's Lane], pp. [2], xvi, 375, [8]; Vol. II (1724, sold by W. Taylor at the Ship in Pater-Noster-row, Peter du Noyer in the Strand, Abel Rocayrol in Green-Street near Leicester-Fields; and at the author's boarding school in Little Chelsea), pp. [2], x, 395, [5]
- ***Libraries*** **L** : MnU
- ***Notes***

1. Vol. I is devoted to Europe but does not cover the whole continent; the first 107 pages of Vol. II are also devoted to it. Makes some use of question and answer.

- **{1726} (02)**
- *Historico-political geography* ... or, a particular description of the several countries in the world, in their situation ... soil, divisions, provinces, rivers, commodities, rarities, capital cities, chief towns, inhabitants, manners, language, populousness, etc; the genealogy, pretensions, government, titles, revenues, residence, etc. of their kings and princes ...
- LONDON; F. Clay
- 20 cm., 2 vols.
- ***Libraries*** : DLC.
- ***Source*** NUC (NP 0124923). This entry provides the author's initial.

- **1729, 2nd, with additions (03)**
- *Historico-political geography* ... world; in ... etc. The ... princes. Their respective states, courts of justice, laws, nobility, orders of knighthood, clergy, archbishopricks, bishopricks, universities, and religion
- LONDON; printed for William France, at the Meuse-Gate, near Charing-Cross
- 8vo, 19 cm.; pp. [2], xvi, 395, [5]

- ▸ ***Libraries*** **L** : CLU
- ▸ ***Notes***

1. This appears to be Vol. II of the ed. of 1724. Vol. I was apparently abandoned. In addition to that omission, however, the countries are dealt with in a different order, making quick comparisons difficult.

PAYNE, Isaac **Y**

- ▸ **{1806} (01)**
- ▸ *Introduction to geography*
- ▸ Place/publisher n.r.
- ▸ ***Libraries*** n.r.
- ▸ ***Source*** Peddie and Waddington (1914:226)

- ▸ **1809, 2nd, considerably enl. and improved (02)**
- ▸ *An introduction to geography*; intended chiefly for the use of schools; with a large collection of geographical questions and exercises; an outline of the solar system; and a selection of the most useful problems on the globes
- ▸ LONDON; printed for Darton and Harvey, Grace-Church-Street
- ▸ 17.5 cm.; pp. [iii], iv-xii, [1], 2-348
- ▸ ***Libraries*** : PHC, PU : **CaBVaU**
- ▸ ***Notes***

1. In the Preface the author plays down the need for memorization, and justifies his book on the grounds that the information is laid out more clearly and in a more consistent format than in other books of the same type. He acknowledges drawing on 'the works of Pinkerton [PINKERTON 1802], Cruttwell [CRUTTWELL 1798], Guthrie [GUTHRIE 1770], Aikin [AIKIN 1806], Walker [WALKER, A. 1808?], and Adams [probably ADAMS, M. 1794]' (p. vi). In the *Advertisement to the Second Edition*, it is noted that articles are now devoted to 'population, face of the country, natural curiosities, religion, and history' (p. vii). By inference the following articles were present in the first ed.: 'soil & productions,' 'manufactures & commerce,' 'government & laws.' Although articles with these headings are provided for countries in every continent, the amount of information provided, and the number of countries for which it is provided, varies from continent to continent; there is a strong bias in favour of Europe.
2. 'Commended by Lon. Month. Rev., Aug 1811' (Allibone 1854).

PAYNE, John **OG**

- ▸ **1796 (01)**
- ▸ *Geographical extracts, forming a general view of earth and nature*. In four parts: Part I. Curious particulars respecting the globe – various phaenomena of nature – winds waters – the electric fluid. Part II. Natural productions of the earth – mines, minerals, and fossils – vegetables. Part III. Animal productions – reptiles – fishes – insects – birds and fowl – quadrupeds. Part IV. Peculiarities in the human species. Illustrated with maps
- ▸ LONDON; printed for G.G. and J. Robinson, Paternoster-Row
- ▸ 8vo, 21 cm.; pp. [1], xiv, 530
- ▸ ***Libraries*** **L** : DLC, MeB, PMA, PPL

▸ ***Notes***
1. As the full title shows, a systematic not a special geography.

PAYNE, John **A Am**
▸ **{1798-1800} (01)**
▸ *A new and complete system of universal geography*; describing Asia, Africa, Europe and America; with their subdivisions of republics, states, empires, and kingdoms: the extent, boundaries and remarkable appearances of each country; cities, towns, and curiosities of nature and art, also giving a general account of the fossil and vegetable productions of the earth. The history of man, in all climates, regions, and conditions: customs, manners, laws, governments, and religions: the state of arts, sciences, commerce, manufactures, and knowledge. Sketches of the ancient and modern history of each nation and people, to the present time. To which is added, a view of astronomy, as connected with geography; of the planetary system to which the earth belongs: and of the universe in general ... Being a large and comprehensive abridgement of universal geography. With additions, corrections and improvements from the latest and best authors. By James Hardie ...
▸ NEW YORK; printed for, and sold by John Low, book-seller, at the Shakespeares Head, no. 332 Water-street
▸ 22.5 cm., 4 vols.
▸ ***Libraries*** : CSt, CtHT, CtY, DLC, GU-De, MeB, MiU, MiU, MNBedf, MWA, NCanHi, NIC, NjR, OOxM, PHi, PP, PPAmP, PPCCH, WAU
▸ ***Source*** NUC (NP 0161881), which notes: 'Vols 1 and 2 revised by James Hardie. Vol. 3: 1800. Vol. 4 relates to America and contains Supplement: Description of the Genesee country.' If it were not for the change in title, this would be treated as (02) of PAYNE, John (1791).

PAYNE, John **A**
▸ **1791 (01)**
▸ *Universal geography formed into a new and entire system*; describing Asia, Africa, Europe, and America; with their subdivisions of empires, kingdoms, states, and republics: the extent, boundaries, and remarkable appearances of each country; cities, towns, and curiosities, of nature and art. Also giving a general account of the fossil and vegetable productions of the earth, and of every species of animal; the history of man, in all climates, regions, and conditions; customs, manners, laws, governments, and religions: the state of arts, sciences, commerce, manufactures and knowledge. Sketches of the ancient and modern history of each nation and people, to which is added, a short view of astronomy, as connected with geography; of the planetary system to which the earth belongs; and of the universe in general. With a set of maps, drawn from the best materials, every one of which is neatly coloured; and a great variety of copper-plates
▸ LONDON; printed for the author, and sold by J. Johnson, No. 72, St. Paul's Church Yard, and C. Stalker, Stationer's Court, Ludgate Street
▸ Folio, 37 cm., 2 vols.; Vol. I, pp. [1], iv, 624; Vol. II, pp. [1], 766, [38]
▸ ***Libraries*** L, SanU : CLSU

▸ ***Notes***
1. In the preface P. states that he began work on this book in 1772, and that it had occupied him fully for the preceding ten years. He acknowledges that his book follows roughly the same plan as that of Collyer, published in 1765. I have been unable to find any record of this, or any other book, for which Collyer was the principal author; but see FENNING, COLLIER, and others (1764-5). P. treats the continents in the following order: Asia, Africa, Europe, America. Included with Asia and Africa in Vol. I are two appendices (pp. 521-624) devoted to recent English voyages of discovery, and a description of the newly discovered islands in the Pacific. Europe (pp. 3-624) and America occupy Vol. II. There are two indices, one geographical and one historical and miscellaneous, which seems to indicate that the author expected the book to be used as a work of reference.
   For contemporary works of a comparable scope and scale, see ADAMS, M. (1794); BALDWYN (1794); BANKES and others (1787); COOKE, G.A. (1790). A comparison with *A new and complete system of universal geography*, 2 vols., Edinburgh, 1796 (See NEW AND COMPLETE ... 1796) might also be interesting.
2. ESTC and NUC record further eds. as follows: 1792, London, 2 vols. (: NmU) [NP 0161887 records what appears to be a version of this work that appears to be made up from one vol. from each ed. (WaU)]; 1793-4, Dublin, 3 vols. (LonU : DLC, OO, OrP). Dryer (1924:117) records 'a large and comprehensive abridgment' ed., 1798-9, New York, 4 vols., 'with additions, corrections and improvements by James Hardie' (see PAYNE, J. 1798-9, above).

▸ **{1809} (02)**
▸ *Universal geography* ... [as 1791] ... kingdomes ... earth; and .. .climates and conditions ... people. A short view of astronomy. As connected ... general. With a set ... copper-plates; descriptive of the most remarkable curiosities in the world ...
▸ DUBLIN; printed by T.M. Bates, (no. 89) Coombe
▸ 29.5 cm., 3 vols.
▸ ***Libraries*** : PU
▸ ***Notes***
1. Dryer (1924:117) notes that the names of 1,500 subscribers to the American edition of 1798-9 are listed.

PEARCE, S.E. **Y SG?**
▸ **{1873} (01)**
▸ *Geography, child's earliest glimpse of*
▸ ? Relfe
▸ ***Libraries*** n.r.
▸ ***Source*** ECB, Index (1856-76:130)

PELHAM, Cavendish ?
▸ **{1808} (01)**
▸ *The voyages and travels of Capt. Cook,* Mungo Park, La Perouse and others;

... with a ... geographical description of the world. Embellished with ... engravings and maps. (The world or the present state of the universe ... Vol. II)
▸ LIVERPOOL, London
▸ 4to, 2 vols.
▸ ***Libraries*** L
▸ ***Source*** BL, which notes: 'Vol. I has an additional engraved titlepage which reads: "the world or the present state of the universe, etc. published ... 1814;" and also a frontispiece with the date 1820. Vol. I was printed at Liverpool.'
▸ ***Notes***
1. According to NUC (NP 0187766), which lists the contents, this is only a collection of voyages, and contains no special geography as such.
2. Downes (1971:384) mentions a work by Pelham in the text, but does not include it in his bibliography.

▸ **{1810} (02)**
▸ *The world, or present state of the universe*
▸ LONDON
▸ ***Libraries*** n.r.
▸ ***Source*** Allford (1964:242)

PELHAM, Cavendish **A n.c.**
▸ **1819 (01)**
▸ [engraved tp. begins] *The world or the present state of the universe.* [tp. proper begins] The voyages and travels of Capt. Cook, Mungo Park, La Perouse, and others ... With a geographical description of the world. Embellished with numerous engravings and maps. In two volumes
▸ LONDON; printed at the Caxton Press by Henry Fisher, (Printer to His Majesty) Sold at 87, Bartholomew-Close
▸ 4to, 27 cm., 2 vols.; Vol. I, pp. [6], 764; Vol. II, pp. [2], 5-870, iv, iv [sic]
▸ ***Libraries*** L : CtY, CU-B, DLC, LU, NcU, NN, TU : CaBVaU, CaBViPA
▸ ***Notes***
1. Special geography occupies pp. 649-852 and 868-70 of Vol. II. There are no tables, lists, or statistics. There is almost nothing on commerce, nor on any economic activity other than farming. The topics presented are those that could be expected to interest a landed gentleman with a taste for travel and an interest in nature as well as in the people of the countries he visits. There are occasional footnotes: some provide historical information, or even a chronology, but in other cases it is difficult to see what distinguishes them from the text. Africa is ignored. Pp. 853-68 are devoted to summaries of voyages of discovery not previously mentioned. No explanation is offered for the inclusion, after the last of this supplementary set of voyages, of descriptions of Greenland and the Grecian archipelago.

PELTON, Cale **mSG Am**
▸ **{1851} (01)**
▸ *Key to Pelton's hemispheres*, designed as an introduction to the key to his

complete series of outline maps ... arranged in verse for musical recitations. To which is added a brief description of the present state of the world ...
- PHILADELPHIA; Sower & Barnes
- 22 cm.; 176 p.
- ***Libraries*** : DLC, ICJ, MH, WHi

PELTON, Cale **mSG Am**
- **{1848} (01)**
- *Key to Pelton's new and improved series of outline maps*, containing all the important geographical names in the known world; to which is added a brief description of the present state of the world ... By ... A.M.
- PHILADELPHIA; J.H. Jones, printer
- 22 cm.; 208 p.
- ***Libraries*** : DLC, KyHi, OO
- ***Notes***

1. NUC records further eds. (both Philadelphia): 1849 (MH) and 1850 (MH).

- **{1853} (02)**
- *Key to Pelton's new and improved series of outline maps* ... world ... arranged in verse for musical recitations; to which is added ...
- PHILADELPHIA; Sower & Barnes
- 22 cm.; 192 p.
- ***Libraries*** : DLC, MH, ViU
- ***Notes***

1. NUC records further eds. (all Philadelphia, and 192, 48 pp. unless otherwise stated): 1854 (DLC); 1855 (CtY, MH, OClStM, OClWHi); 1856 (MiU, OC, PP, PPL); and 1873, pp. n.r. (PPWA).

PEMBLE, William **Y OG**
- **1630 (01)**
- *A briefe introdvction to geography* containing a description of the grovnds, and generall part thereof, very necessary for young students in that science. Written by that learned man, Mr ... Master of Arts, of Magdalen Hall in Oxford
- OXFORD; printed by Iohn Lichfield printer to the famous vniversity for Edward Forrest
- 4to, 18 cm.; pp. [4], 46 (last page misnumbered 64)
- ***Libraries*** **L**, LU, O : CtY, DFo, NIC
- ***Notes***

1. A short guide to general, as opposed to special, geography. Written before 1618, according to Baker (1928:271). As was normal in England at that time, it follows the Ptolemaic cosmology, but admits that the earth might rotate on its axis while remaining at the centre of the universe.

- **{1635} (02)**
- *A briefe introduction to geography* ... thereof. Very necessary ...
- LONDON; printed by M. Flesher, for Edward Forrest, and are sold by R. Royston

- Folio, 28 cm.; pp. 28
- ***Libraries*** C : CtY, MH, MnU, NN, NNUT
- ***Source*** NUC (NP 0195360)

- **1658 (03)**
- *A briefe introduction to geography* ... the grounds ...
- LONDON; printed by William Hall, for John Adams, Edw: Forrest, and John Forrest
- Folio, 28 cm.; pp. 28
- ***Libraries*** **L** : MH, NN

- **{1675}, fifst [sic] ed. (04)**
- *A brief introduction to geography* ...
- OXFORD; printed by Leonard Lichfield, for Edward Forrest
- 4to, 20 cm.; pp. [4], 41
- ***Libraries*** C, **L**, O, OMa : CLU, PBL

- **{1685} (05)**
- *A briefe introduction to geography* ...
- OXFORD; printed by Leonard Lichfield. Printer to the University for Anthony Stephens
- 4to, 19 cm.; pp. [4], 47
- ***Libraries*** **L**, LPO, O : CLU, KU, MnU
- ***Notes***

1. Discussed in Baker (1935:131), Gilbert (1972:53-54), and Bowen (1981:76).

PENNANT, Thomas **A n.c.**

- **1798-1800 (01)**
- *Outlines of the globe*
- LONDON
- 4to, 30 cm., 4 vols.; Vol. I (1798), pp. [6], v, [11], 263, [10]; Vol. II (1800), pp. [14], 374, [13]; Vol. III (1798), pp. xi, [11], 284, [13]; Vol. IV (1798), pp. [8], 317, [22].
- ***Libraries*** **L**, LRag, SanU : NN, WU
- ***Notes***

1. The subtitles of the vols. are as follows:
   Vol. I. The view of Hindoostan: western Hindoostan. Printed by Henry Hughes.
   Vol. II. The view of Hindoostan: eastern Hindoostan. Printed by ...
   Vol. III. The view of India extra Gangem, China, and Japan. Printed by Luke Hanford, Great Turnstile, Lincoln's-Inn Fields; and sold by John White, Horace's Head, Fleet Street.
   Vol. IV. The view of the Malayan isles, New Holland, and the spicy islands. Printed ...
2. In the Advertisement to Vol. I, P. states that it, along with Vol. II, are composed from the 14th and 15th of his *Outlines of the Globe*. He died shortly after their publication. Vols. III and IV were prepared for publication by his son, David. No more were published. The work as a whole contains a wide range of information, including considerable history. Places are dealt with

more or less contiguously, with little obvious effort to provide information about them in a systematic way.

Pennant's interest in botany shows from time to time, notably in Vol. IV, which contains an 80-page 'Flora Indica.'

PENNELL, Alice Hart **Y Am**

- **{1831} (01)**
- *A key to the questions in Adam's geography*; together with an account of the principal countries, kingdoms, states, cities and towns; with a description of the most remarkable mountains, rivers, etc., etc., of the world
- BALTIMORE; J.J. Harrod
- 18 cm.; pp. 63
- ***Libraries*** : DLC, MH, PHi, PP
- ***Source*** NUC (NP 0202714). Cited by Elson (1964:399).

PERCIVAL, James G. **Gt**

- **{1829}**
- *A geographical view of the world*
- NEW YORK; D.M. Jewett
- ***Libraries*** n.r.
- ***Source*** Elson (1964:399)
- ***Notes***

1. Information provided by Elson in parentheses shows that this book is a subsequent ed. of PHILLIPS (1826).

PERIER, DU
See BELLEGARDE (1708).

PERKS, William **Y**

- **1792 (01)**
- *The youth's general introduction to Guthrie's geography*. Being a complete pocket atlas; containing a description of the several empires, kingdoms, and states, in the world, their extent, bearings, air, soil, produce, commerce, strength, government, religion, etc. Accompanied with twenty-seven maps on a new plan, shewing the situation of above 3,000 cities and towns, the particulars of which are mentioned in the work, with their corresponding ancient and French names, etc. To which are prefixed, elementary chronology and astronomy: with an account of various phenomena in the atmosphere, viz. thunder, lightening, hail, rain, snow, wind, etc. The most useful geometrical figures defined and constructed: and the use of maps. Illustrated by a variety of examples. The whole adapted to the capacity of youth. By ... Pimlico
- LONDON; printed for the author: and sold by Messrs. Robinson, Paternosterrow[sic]; and Scatcherd and Whitaker, Ave-Maria-Lane
- 8vo, 18 cm.; pp. vi, iii-vi, 7-415, [1]
- ***Libraries*** L

▸ *Notes*
1. Follows the plan of the grammars of the eighteenth century that emphasized the British Isles. Much of the text is little more than a list of places. Whether it is, as its title implies, a shortened version of GUTHRIE (1770), or whether it merely capitalizes on his reputation, has not been established.

▸ **1793, 2nd, enlarged and improved (02)**
▸ *The youth's general introduction to Guthrie's geography*. Containing ... states in ... world, accompanied with twenty-eight maps on a new plan, comprising a complete pocket atlas; to which are prefixed: elementary chronology and astronomy: the use of the globes and maps. Illustrated ...
▸ LONDON; printed for the author and G.G. and J. Robinson, Pater-Noster-Row
▸ 4to, 22 cm.; pp. viii, 351, [1]
▸ ***Libraries*** **L** : CtY, KyU, NjP, PPL
▸ ***Notes***
1. Although reference to them is omitted from the tp., the meteorological phenomena are still dealt with. The book contains the new constitution of the Republic of France, proclaimed on 10 August 1793 (pp. 186-92), and also a list of the new French *departements*.

PERSPECTIVE GLASS ... **?**
▸ **{1676} (01)**
▸ *A perspective glass*, by which you may see the situation of all the countreys and islands in the world. Together with a description of each part distinct; as likewise all the kingdoms and islands
▸ LONDON
▸ ***Libraries*** n.r.
▸ ***Source*** Cox (1938:344)

PETAU, Denis [PETAVIUS, Dionisius]
See P., R. (1659).

PEYTON, C.G. **Y Am**
▸ **{1833} (01)**
▸ *Geography made easy and interesting ...*
▸ BALTIMORE; J.N. Lewis. Philadelphia; T.L. Bonsal
▸ 16.5 cm.; pp. 94
▸ ***Libraries*** : DLC
▸ ***Notes***
1. Discussed in Brigham and Dodge (1933:15).

PHILLIPS, C. **Y SG?**
▸ **{1854a} (01)**
▸ *Geography for children*
▸ [LONDON]; Simpkin

- 18mo
- ***Libraries*** n.r.
- ***Source*** BCB, Index (1837-57:110)

PHILLIPS, Charlotte **Y SG?**
- **{1854b} (01)**
- *Questions in geography* for the use of schools and evening classes
- LONDON; Simpkin, Marshall & co.
- 12mo; 44 pp.
- ***Libraries*** n.r.
- ***Source*** (misplaced)

PHILLIPS, Charlotte **Y SG?**
- **{1854} (01)**
- *Questions in geography*, with answers, etc.
- LONDON; Simpkin, Marshall & co.
- 16mo
- ***Libraries*** L

PHILLIPS, Richard (Goldsmith, Rev. J.) **Y Am**
- **{1804} (01)**
- *An easy grammar of geography*, for the use of schools; with maps
- PHILADELPHIA; printed for Benjamin Johnson
- 15 cm.; pp. 108, 38
- ***Libraries*** : MB, NNC, PHi, PU
- ***Source*** NUC (NP 0328085). See note to (01) of PHILLIPS (1803), below, for a possible ed. in 1803; see also notes to (02) and (04) below.

- **{1805}, new (02)**
- *An easy grammar of geography*. Intended as a companion and introduction to the Geography on a Popular Plan for Schools and Young Persons ... With maps. A new edition
- LONDON; Richard Phillips
- 12mo; pp. vi, 144
- ***Libraries*** L
- ***Notes***

1. This book is taken to be a successor ed. to (01) in this entry, despite the change in the full title, because subentry (04), below, which has the same title as (01), also has 144 pp.

- **{1806} (03)**
- *An easy grammar of geography*, intended as a companion ...
- LONDON; printed for Richard Phillips
- 12mo, 15 cm.; pp. vi, 144
- ***Libraries*** L (imperfect) : NNC
- ***Source*** BL and NUC (NP 0328092)

▸ ***Notes***
1. Discussed in Nietz (1961:206).

▸ **{1807} (04)**
▸ *An easy grammar of geography*, for the use of schools ...
▸ PHILADELPHIA; printed for J. Johnson
▸ 15 cm.; pp. 144
▸ ***Libraries*** : MB, MH, Mi, NNC, PV
▸ ***Source*** NUC (NP 0328086)
▸ ***Notes***
1. Elson (1964:395) records an ed., 1807, Boston, William Norman; NUC records further eds.: 1810, Philadelphia, 144 pp. (ICN, NN, ICN); 1811, 31st, London, 144 pp. (DLC, MB, MH); 1811, 2nd, Boston, 124 pp. (MHi); 1812, Philadelphia, 203 pp. (MH, NjR, PPL) [NP 0328093 and NSP 0014281, which supplies the size]; 1813, 3rd, An easy grammar of geography for the use of schools. With maps. 3d improved ed., carefully rev., with considerable additions, and a map of the United States. Boston, 119 pp. (MH, MWH, N, PU); 1813, 45th, London (MH); 1815, 47th, London (MH); 1816, New ... improved by a citizen of Philadelphia. Philadelphia, 179 pp. (NcU, ViFreM, ViU); 1817, New, [Philadelphia]? (ViU); 1818, New, Philadelphia, 195 pp. (DLC, NcD, ViU, ViW); and 1820, Philadelphia, 192 pp. (CtY, OC).
2. Elson (1964:394) cites (01); Brigham and Dodge (1933:8) discuss the ed. of 1818, Philadelphia.

PHILLIPS, Richard (Goldsmith, Rev. J.) **Y SG? Am**
▸ **{1807}, 4th (01)**
▸ *Elements of geography*, principally compiled with a view to teach children, at an early age, the geography of the United States. 4th ed.
▸ PHILADELPHIA; Kimber & Conrad
▸ 15 cm.; pp. 36
▸ ***Libraries*** : NN, NSyU, PP
▸ ***Source*** NUC (NP 0328109)

PHILLIPS, Richard (Goldsmith, Rev. J.) **Gt**
▸ **{1822}* (01)**
▸ *A general view of the earth*
▸ NEW HAVEN
▸ 2 vols.
▸ ***Libraries*** : RP
▸ ***Source*** NUC (NP 0328153), which notes: 'Reprint.' Might be the 1822 ed. of the next entry. RP report that they cannot locate this book in their collection.

PHILLIPS, Richard (Goldsmith, Rev. J.) **Y Am**
▸ **1810 (01)**
▸ *A general view of the manners, customs and curiosities of nations*; including

a geographical description of the earth. The whole illustrated by fifty-four maps, and other engravings. In two volumes

- PHILADELPHIA; Johnson & Warner; for sale at their stores in Philadelphia, Richmond, (V.) and Lexington (K.) J. Bouvier, printer. Philadelphia
- 18 cm., 2 vols.; Vol. I, pp. [3], iv-xliv, [45]-292; Vol. II, pp. [3], 4-314, [2]
- ***Libraries*** : KyHi, **MH**, **MWA**, MWH, NcAS, PHi, ViLxW, ViU
- ***Notes***

1. Intended as an aid to instruction: 'No practice can be more absurd ... than to attempt to communicate a knowledge of the divisions of the earth, and the relative positions of countries, by mere verbal descriptions. A hundred volumes in folio could not so clearly illustrate these objects as a single map! ... *To become acquainted with maps should, therefore, be the primary purpose of every student of geography*' (pp. iii-iv, emphasis in the original). 'The obvious, the easiest, and the only successful mode of communicating a knowledge of maps, is by making a pupil copy or draw them' (p. iv). The Preface ends (p. vi) with a list of seventy authoritative works from which the information supplied in the book has been drawn.
2. NUC records further eds. as follows: 1813, Philadelphia; Johnson and Warner, 2 vols. (OCl, PU); and 1817, Philadelphia, B. Warner, 2 vols. (NcD, NcU, ScU).

- **{1818} (02)**
- *A general view of the manners ...*
- PHILADELPHIA; B. Warner
- 18.5 cm., 2 vols.; Vol. I, pp. xlii [43]-252; Vol. II, pp. n.r.
- ***Libraries*** L : NBu, PP
- ***Source*** NUC (NP 0328157), according to which the American libraries hold only Vol. I, and BL, which lists '2 vol.,' and adds: 'This is a spurious edition, altered and rearranged, of the same author's 'Geography illustrated on a popular plan.' The reference is to PHILLIPS (1805), see below.
- ***Notes***

1. NUC records further eds. as follows: 1822, New Haven, John Babcock, 2 vols. (CtY, DHU, MeB, PP, PU); and 1825, Revised by the senior publisher, New Haven, J. Babcock ... Charleston ..., 2 vols. (CtY).

PHILLIPS, Richard (Goldsmith, Rev. J.) **Y Am**

- **1826, 1st American ... (see below) (01)**
- *A geographical view of the world*, embracing the manners, customs, and pursuits, of every nation; founded on the best authorities. 1st American edition, revised, corrected, and improved, by James G. Percival, M.D. Illustrated by eight copperplate views
- NEW-YORK; E. Hopkins and W. Reed
- 18.5 cm.; pp. [3], 4-406, [2], 46, [2]
- ***Libraries*** : DLC, DGU, MB, **MWA**, NjR, OCl, ODW, OrHi, PlatS, PP
- ***Notes***

1. Contains neither preface, introduction, nor advertisement. A quick comparison of this work with PHILLIPS (1810), i.e., *A general view of the manners, customs and curiosities of nations*, suggests that this work could easily be

taken as a paraphrase of the other. The style in this book tends to be more prosaic and matter-of-fact than in the other.

Belongs to the tradition of special geography in that an effort is made to provide some information about every country in the world; moreover, the description of each country begins with the specification of its location and boundaries. On the other hand, the information about countries is limited largely, though not exclusively, to the topics identified in the full title, namely, manners, customs, and pursuits. The last includes the ways and means of making a living. To those are often added some notes on the appearance and dress of the inhabitants, and, less often, some information on the climate. Very occasionally dramatic features of the landscape, such as geysers and volcanoes are described. On the whole the book seems intended for the vicarious tourist rather than the classroom.

Much of the text takes the form of quotations from the writings of recent travellers, though the sources are identified in so sketchy a fashion that it would be hard to identify the originals. A propensity to generalize about the behaviour of social classes and national character is highly visible, e.g., 'The common people [of Sweden] are orderly and industrious, sober, loyal, and religious; yet when intoxicated, furious and ungovernable' (p. 28); and, 'visit a Russian, of whatever rank, at his country seat, and you will find him lounging about, uncombed, unwashed, unshaven, halfnaked, eating raw turnips, and drinking *quass*' (p. 35).

The 46 pp. separately numbered at the end carry an appendix prepared by J.G. Percival, entitled 'Varieties of the human race.' The categorization of human groups into races is based on criteria of 'physical conformation' and language. The Caucasian race is dealt with first (pp. 1-25); that section ends with the following eulogy: 'It includes the most civilized nations, and indeed all, that have made any great progress, or have showed any high inventive power. It is not only the most enterprising and intelligent, but the most elegant of all the races, excelling them in complexion, features, and form. The civilization of the other races, after gaining a certain point, has continued stationary ... Wherever they have come in contact with Caucasians, the latter have prevailed, except in the short triumphs of the Mongols, under Genghis and Timur.'

This section (once allowance has been made for its ethnocentrism), when taken together with the general range of topics dealt with in the remainder of the text, might induce some anthropologists to claim that the book belongs to their tradition.

- **1826, 2nd American, by ... (see below) (02)**
- *A geographical view of the world* ... 2nd American edition, by James G. Percival
- NEW YORK; William Reed
- 12mo, 18.5 cm.; pp. [3], 4-406, [2], 46
- ***Libraries*** : MB, MWA
- ***Notes***

1. Spot-checking suggests that the text is the same as in (01).
2. NUC records another ed. in: 1827, 2nd, Boston, N.Y., [sic] printed for Reed and Jewett, 406 pp. (MiU).

- **1827, 3rd, rev. by J.G. Percival (03)**
- *A geographical view of the world*, embracing the manners, customs, and pursuits, of every nation; founded on the best authorities. Revised, corrected, and improved, by James G. Percival, M.D. Illustrated by engravings
- BOSTON; printed for Reed and Jewett, New York
- 12mo, 18 cm.; pp. [3], 4-406, [2], 46
- ***Libraries*** : MSaE, **MWA**
- ***Notes***

1. The prefatory p. [3] now carries an advertisement as well as the copyright: 'Two editions consisting of 8000 copies, of this work, having been sold in the United States, in the space of eighteen months, and a third being called for, it has been again revised, and a large number of wood engravings introduced.' Spot-checking reveals no changes in the text.
2. NUC records further eds. (Boston except as noted): 1827, 7th, 406 pp. (KyU); 1827, 8th, 406, [2], 46 pp. (MB, ViU); 1827, 9th, 406, [2], 46 pp. (MiU); 1828, 30th (MH); 1829, 2nd American, New York, 406, 48 pp. (DLC, NNC); 1831, 6th American, Hartford, 406 pp. (DeU, InU); 1833, 2nd American, Hartford, 406, [2], 46 pp. (FU, KyU); 1835, 2nd American, Hartford, 406, [2], 46 pp. (LU, WU); 1836, 2nd, New York, 406, [2], 46 pp. (NcU); and 1838, 3rd American, New York, 406, [2]., 46 pp. (CU, DLC, PPPD).

- **1841, 6th American, rev. (04)**
- *A geographical view of the world ...*
- NEW YORK; Robinson, Pratt & co.
- 19.5 cm.; pp. [3], 4-406, [2], 46
- ***Libraries*** : MB, ViU, WU : **CaAEU**
- ***Notes***

1. The CaAEU ed. has 15 unnumbered pages with illustrations preceding the tp.
2. NUC records further eds. as follows (all New York): 1843, 7th American, 406 pp. (MiU); 1845, 8th American, 406, [2], 46 pp. (CtY, ICN, ODW); 1849, 8th American, 406, 46 pp. (NIC, OCl); and 1851, 10th American, 406, 1 l., 46 pp. (NN).
3. Elson (1964:399) cites the ed. of 1829.

PHILLIPS, Richard (Goldsmith, Rev. J.) **Y**

- **{1802} (01)**
- *Geography for the use of schools*. By Rev. J. Goldsmith [Sir R. Phillips]
- [LONDON]; Phillips
- ***Libraries*** n.r.
- ***Source*** Peddie and Waddington (1914:227)

- **{1803} (02)**
- *Geography for the use of schools* and the young persons in general
- LONDON; printed for R. Phillips, by T. Gillett
- 17 cm.; pp. xlii, 560 [i.e., 580]
- ***Libraries*** : NN

PHILLIPS, Richard (Goldsmith, Rev. J.) **Y**

- **{1805} (01)**
- *Geography, illustrated on a popular plan*[: for the use of schools and young persons, with thirty-five engravings]
- LONDON; [Longman, Hurst, Rees, Orme, Brown, and Green]
- ***Libraries*** n.r.
- ***Source*** Allford (1964:242)
- ***Notes***

1. The words in [ ] are derived from CIHM. The relationship of this title to the next, i.e., *Geography, on a popular plan*, has not been established.
2. BL and NUC record further eds. as follows: 1815, London (L); 1818, 8th, London, (L); and 1824, New, London, 740 [4] pp. (: CoU, DLC : CaOLU).
3. An ed. of 1826 is known from Peter Barrie, Sutton Coldfield, UK, *A Catalogue of Maps and Atlases*, Catalogue M. 7, item 832.

- **{1829}, revised ... (see below) (02)**
- *Geography, illustrated on a popular plan*; or, a geographical view of the world, embracing the manners, customs, and pursuits of very nation; founded on the best authorities ... Revised, corrected, and improved, by James G. Percival
- BOSTON; I.R. Buts
- 19.5 cm.; pp. 406, [2]
- ***Libraries*** : CtY, MiU, NcD, OClW, OWorP
- ***Source*** NUC (NP 0328183)
- ***Notes***

1. A microform version of the ed. of 1824 is available from CIHM.

PHILLIPS, Richard (Goldsmith, Rev. J.) **Y**

- **{1802} (01)**
- *Geography, on a popular plan*
- LONDON?
- ***Libraries*** n.r.
- ***Source*** St. John (1975, 1:188-89), which notes: 'First published in 1802'
- ***Notes***

1. NUC records an ed.: 1805, new, London, 645 pp. (IU).

- **{1806}, new (02)**
- *Geography, on a popular plan*, illustrated with sixty copper-plates, for the use of schools, and young persons: being a sequel to the Grammar of Geography
- LONDON; printed for Richard Phillips, and sold by Tabart & co.; Wilson and Spence, York; H. Mozley, Gainsborough; and Martin Keene, Dublin
- 17.5 cm.; pp. xii, 645
- ***Libraries*** : : CaOTP
- ***Source*** St. John (1975, 1:188-89)
- ***Notes***

1. St. John (1975, 1:189) notes: 'The "Advertisement to the second edition", dated march, 1805, reads: "The author, having in some measure newly arranged his work, and having revised with great care every part of it,

may, in connection with his smaller volume entitled, *A grammar of geography*, venture to recommend it as the most complete and pleasing system of geographical instruction ever presented to the world".'
2. James Thin, Edinburgh, Catalogues 332 and 348, *Antiquarian and general second-hand* [n.d., ca. 1970, item 385, and ca. 1978, p. 60] records: 1808, 5th.

PHILLIPS, Richard (Goldsmith, Rev. J.) **Y**
- **{1803} (01)**
- *A grammar of general geography for the use of schools* and young persons with maps and engravings
- LONDON; [Longman, Rees, Brown, and Green]
- 14.5 cm.
- ***Libraries*** n.r.
- ***Source*** NUC (NP 0328197), which notes: 'First ed., 1803'
- ***Notes***
1. The words in [ ] are derived from CIHM.
2. For information on the content of the book, see (02), below.
3. BL and NUC record further eds. (all London, Longman): 1819, 189 pp. (L); 1822, 188 pp. (: NBuG); 1823, 188 pp. (: NR); {CIHM records [1824?] (: : CaOONL)}; {St. John (1975, 1:189) records [1825?], 192 pp. (: : CaOTP)}; [1829], 191 pp. (L); and [1835?], 212 pp. (L : : CaAEU); [1837 reported by E.J. Tulley, Antiquarian and Second-Hand Bookseller, 8, Gloucester Road, Brighton, *Miscellany*, no. 2, 1980, p. 7]; [1839?] (: DLC).
4. Allford (1964:242) records An *easy* grammar of general geography. Can he have confused *A grammar of general geography* (i.e., the work to which this is a note) with *An easy grammar of geography* (i.e., PHILLIPS 1804)?

- **1840? new, corrected and modernized ... (02)**
- *A grammar of general geography ...*
- LONDON; Longman, Orme, Brown, Green, and Longman
- 14 cm.; pp. [7], iv, 212, 16
- ***Libraries*** : : CaAEU
- ***Notes***
1. N.d., but the dates of the eds. of the books that are listed in the separately numbered 16 pp. at the end (which are devoted to advertisements], point to publication in the period 1839-41.

   Special geography occupies pp. 13-103; 106-24 are devoted to the systematic exposition of topics in both physical and human geography. Subsequent sections are devoted to: the use of globes; of maps; questions and exercises; heights of mountains; a gazetteer; and a vocabulary of the names of places. The last is a guide to the pronunciation of place names, with a supplementary phrase identifying the nature of the place; e.g., 'Al-tai; a great range of mountains in Siberia, very rich in mines.'
2. BL, NUC, and St.John (1975) record further eds. (all London): [1841?], 242 [i.e., 246) pp. (: NNC); [1844?], Rev. and corr. by Hugh Murray, 230 pp. (: CLU); [1846?], Rev. and corr. by Hugh Murray, 230 pp. (: : CaOTP); [1851], Rev. (L); 1852, Rev., 335 pp. (: TU, ViU); and [1855?], Rev., 335 pp. (: OU). BCB, Index (1837-57:110) records an ed. in 1855. BL and NUC

record further eds. (both London): 1860, Corrected and improved by W.S. Kenny (L); and 1861, New, corrected and revised, by W. Webster (L). ECB, Index (1856-76:130) records an ed. in 1865, [revised?] by Webster.

- **{1868}, revised ... (see below) (03)**
- *Kenny's Goldsmith's grammar of geography* ... Revised and brought down to the present time by F. Young, etc.
- LONDON, Guildford [printed]
- 12mo
- ***Libraries*** L

- **{1868}, new ... (see below) (04)**
- *A grammar of geography* ... With maps and illustrations. A new edition ... corrected to the end of 1867 [by W. Webster]
- LONDON, Newton [printed]
- 12mo
- ***Libraries*** L
- ***Notes***

1. ECB, Index (1856-76:130) records an ed. in 1872, [London], Longman.
2. NUC (NP 0328197) records, without a date, 'A grammar of general geography for the use of schools and young persons with maps and engravings. A new edition, corrected and modernized by Rev. J. Goldsmith ..., London' (CtY, IaSlB).
3. Microform versions of the eds. of [1824?], [1835?], and [1839?] are available from CIHM.

PICKET, Albert, and John W. Picket **Y Am**

- **{1816} (01)**
- *Geographical grammar*. Combining the interrogative mode of instruction, with concise definitions, the use of maps, and the terrestrial globe
- NEW YORK; printed and published by Smith and Forman, at the Franklin Juvenile Bookstore, 190 and 195 Greenwich-Street
- 21 cm.; pp. 80
- ***Libraries*** : N, NNC, PHi
- ***Source*** NUC (NP 0347191), which identifies Albert Picket as the sole author. See note to next entry for information about the contents.

- **1817, 2nd, with alterations and improvements (02)**
- *Geographical grammar* ...
- NEW YORK; Daniel D. Smith, at ... no. 190 Greenwich ...
- 22.5 cm.; pp. 72
- ***Libraries*** : DLC, NN, ViU
- ***Notes***

1. Pp. 3-20 contain questions. The answers to most of them could be found by looking at maps. Such maps, however, would have to be far more detailed than those bound at the end of the text. Pp. 20-41 present highly condensed special geography in a semi-tabular form. Pp. 42-65 deal with the terrestrial

globe (i.e., a globe representing the earth), problems related to its use, and the information that can be obtained from it. The modern and Classical names for places are listed on pp. 66-72.
2. Elson (1964:396) cites (01).

PICQUOT, A. **Y**
- **{1812} (01)**
- *Elements of universal geography, ancient and modern ...*
- LONDON; printed for the author
- 19 cm.; pp. vii, 312
- ***Libraries*** : ICN

- **{1817}, 2nd, enlarged, etc. (02)**
- *Elements of universal geography, ancient and modern*: containing a description ... of the several countries, states, etc., to which are added historical, classical and mythological notes ...
- LONDON
- 12mo
- ***Libraries*** L
- ***Notes***

1. BL records 4th, enlarged, 1826, London.

PICTET and ST. QUENTIN, Dominiqe de **Y tr**
- **{1791} (01)**
- *Complete system of geography*, adapted to M. l'Abbé Gaultier's method of teaching that science. The maps and every article necessary for the game published and sold by Mr Dudley Evans
- READING; Smart & Cowslade
- 2 vols.?
- ***Libraries*** : L (incomplete, Vol. II only)
- ***Notes***

1. According to BL, the book contains parallel English and French texts. British Museum (1881-1900) catalogues this title under ADAMS, Dudley.
2. See GAULTIER (1792).

PIERSON, David Harrison **Y Am**
- **{1854} (01)**
- *A system of questions in geography*, adapted to any modern atlas ...
- NEW YORK; Kiggins and Kellogg
- 18 cm.; iv, p., 1 l., [5]-189 p.
- ***Libraries*** : DLC, MB, MH, OO
- ***Source*** NUC (NP 0357904)
- ***Notes***

1. BL and NUC record further eds. (New York, Kiggins and Kellogg except as

noted): 1854, 2nd (L : MH); 1855, 5th (: MH, N) [According to Culler (1945:310), the 5th ed. was published in 1853]; 1865, 8th (: MH); and 1874, Rev., New York, Mason, 191 pp. (: DLC, MH).

PILLANS, James **Y mSG**
- **{1847} (01)**
- *Outlines of geography, principally ancient*; with interesting observations on the system of the world, and of the best manner of teaching geography
- EDINBURGH
- 12mo
- ***Libraries*** L

PINKERTON, John **Y**
- **{1820} (01)**
- *An introduction to Mr. Pinkerton's abridgement of his modern geography*, for the use of schools. By John Williams
- LONDON, Halifax [printed]; printed for Longman, Hurst, Rees, Orme, and Brown, Paternoster-Row; and Cadell and Davies, Strand
- ***Libraries*** **L**
- ***Notes***

1. An abridgement of an abridgement. The words are those of Pinkerton; the choice is that of Williams.

PINKERTON, John **A**
- **1802 (01)**
- *Modern geography*. A description of the empires, kingdoms, states, and colonies; with the oceans, seas, and isles; in all parts of the world: including the most recent discoveries, and political alterations. Digested on a new plan. The astronomical introduction by the Rev. S. Vince, A.M., F.R.S. and Plumian Professor of Astronomy, and Experimental Philosophy, in the University of Cambridge. With numerous maps, drawn under the direction, and with the latest improvements, of Arrowsmith, and engraved by Lowry. To the whole are added, a catalogue of the best maps, and books of travels and voyages, in all languages: and an ample index. In two volumes
- LONDON; printed by A. Strahan, Printers Street; for T. Cadell, jun., and W. Davies, Strand; and T.N. Longman and O. Rees, Paternoster-Row
- 4to, 28 cm., 2 vols.; Vol. I, pp. [3], xiv, [2], cvii, [1], 666; Vol. II, pp. [1], viii, 835, [1]
- ***Libraries*** L, OSg, **[IDT]** : CtY, CU, MeB, MH, MnHi, NIC, NNC, OClWHi
- ***Notes***

1. A more scholarly work than most special geographies, with footnotes in which P. cites more than 130 authorities for his material. He also put more thought into the organization of his book than anyone since Pat GORDON (1693); e.g., the sequence in which countries are described is related to their power; the most powerful are described first, the least powerful last (see BEAUMONT).

Vol. I is devoted to Europe, with the British Isles receiving most attention; there is much less historical information provided than in such contemporary books as the later eds. of GUTHRIE (1770).

- **{1804} (02)**
- *Modern geography* ... Vince ... The article America, corrected and considerably enlarged, by Dr. Barton, of Philadelphia. With numerous ... Arrowsmith, and engraved by the first American artists. To the whole ...
- PHILADELPHIA; J. Conrad, & co. [etc.]
- 23.5 cm., 2 vols.
- ***Libraries*** : CoU, CtY, DI-GS, DLC, DGU, KyU, MdBJ, MdBP, MWA, NcD, NjNbS, OClW, PP, PPA, PPF, PU, TU, ViU
- ***Source*** NUC (NP 0375491)

- **1807, new, greatly enlarged (03)**
- *Modern geography* ... [as in 1802] ... maps, revised by the author. To the whole ... index. A new edition, greatly enlarged. In three volumes
- LONDON; printed for T. Cadell and W. Davies, Strand; and Longman, Hurst, Rees, and Orme, Paternoster-Row
- 4to, 27 cm., 3 vols.; Vol. I (subtitled Europe), pp. [3], iv-li, [2], ii-cxxxi, [2], 2-739; Vol. II (subtitled Asia), pp. [5], iv, [3], ii-iii, [2], 2-820; Vol. III (subtitled America and Africa), pp. [7], 2-1006
- ***Libraries*** L : : CaAEU, CaOONL
- ***Notes***

1. The length of the text has been increased by 40 per cent. Most of the additional space is devoted to Asia, including the islands of the south Pacific, and America. In addition, each volume now ends with a section devoted to the fauna of the continents. Unlike (01), this ed. contains lengthy sections composed entirely of direct quotations.

- **{1811}, 3rd, cor., in two vols. (04)**
- *Modern geography* ... astronomical introduction by M. La Croix ... tr. by John Pond ... With numerous ... author, and engraved by Mr. Lowry. To the whole ...
- LONDON; printed for T. Cadell and W. Davies, [etc.]
- 28 cm., 2 vols.
- ***Libraries*** : CSt, CtY, DLC, DNLM, IaU, ICN, IU, MH, MiU, T, TU : CaBVaU, CaBViPA, CaOOU
- ***Source*** NUC (NP 0375493)
- ***Notes***

1. Judging by the 'Advertisement to the Third Edition' included in the ed. of 1817 (05), below, the reduction to two volumes from three was achieved by the use of smaller type, along with some relatively modest deletions. Space was even found for a little additional material.

- **1817, new ... (see below) (05)**
- *Modern geography* ... [as in 1803] ... introduction by M. La Croix, member of the Institute of France; translated by John Pond esq. Astronomer-Royal. With numerous maps, revised by the author, and engraved by Mr. Lowry.

To which ... languages; and an ample index. A new edition, with additions and corrections to the year 1817
- LONDON; printed for T. Cadell and W. Davies, Strand; and Longman, Hurst, Rees, Orme, and Brown, Paternoster-Row
- 28 cm., 2 vols.; Vol. I, pp. [3], vii, [2], vi-xxxi, [1], lxxxi, 778, [1]; Vol. II, pp. vii, [1], 850, [1]
- ***Libraries*** **L**: NNC
- ***Notes***

1. Compared with the 1st ed., which in terms of number of pp. is close in size, there has been a considerable reallocation of space devoted to different countries: England has shrunk from 128 pp. to 83; Scotland from 65 to 54; Europe taken as a whole from 646 to 466; Asia from 520 to 384. America, on the other hand, now has 494 pp. compared with the 189 of the 1st ed. Africa remains relatively neglected, though the number of pp. devoted to it has increased from 68 to 81.

- **{1864} (06)**
- *Modern geography ...*
- PHILADELPHIA
- 2 vols.
- ***Libraries*** : ODW
- ***Notes***

1. Discussed in Bowen (1981), Sitwell (1972), Wilcock (1974), Wise (1970:48-50), and Wright (1959, *passim*).
2. A microform version of (02) is available from Lost Cause Process of Louisville, KY; CIHM supplies versions of (03) and (04).
3. See NEW AND COMPLETE ... (1796).

PINKERTON, John **Y**

- **1803, abridged ed. (01)**
- *Modern geography* ... Cambridge. Carefully abridged from the larger work, in two volumes quarto. With maps, drawn under the direction, and with the latest improvements, of Arrowsmith. To the whole is added, a catalogue ... languages
- LONDON; printed by ... Cadell and ...
- 8vo, 21 cm.; pp. [1], xx, lxvii, [1], 61.
- ***Libraries*** **L** : CtY, PMA
- ***Notes***

1. The preface contains the guidelines used in producing a cheaper version of the original work 'for general use as a school book.' 'The geographical discussions interspersed throughout the work, as being the least necessary to the young students in this science, have been curtailed with less reserve [than the author's style in general]; much also of the technical and least interesting parts of botany and mineralogy has been omitted ... But the most sedulous care has been uniformly exerted to entrench as little as possible on the topics that more peculiarly belong to geographical science' (p. vii).

- **1806, 2nd abridged, revised by the author (02)**
- *Modern geography* ... [as in 1803] ... languages, and an ample index

- LONDON; printed for T. Cadell ... Strand; and Longman, Hurst, Rees, and Orme, Paternoster-Row
- 8vo, 22 cm.; pp. [1], xx, lxvii, [1], 676
- ***Libraries*** **L**
- ***Notes***

1. Apart from the addition of an index (pp. 643-76), only slight changes have been made to the text since 1803.

- **{1811}, 3rd (abridged), rev. and enl. by the author (03)**
- *Modern geography ...*
- LONDON; printed for T. Cadell and W. Davies [etc.]
- 22 cm.; pp. xx, cxix, 834
- ***Libraries*** : ViU

PINKERTON, John **Y Am**

- **{1805} (01)**
- *Pinkerton's geography, epitomized for the use of schools*. By David Doyle. Wherein the arrangement of the original is faithfully preserved. In the description of the United States, each state is more systematically described, than in any work of the kind heretofore published. This compilation is published at the request, and under the inspection of many respectable teachers
- PHILADELPHIA; printed for Samuel F. Bradford, No. 4, S. Third Street
- 18.5 cm.; pp. [v], vi, [1], 8-348
- ***Libraries*** : CtY, DLC, InU, MH, OClWHi : **CaBVaU**
- ***Notes***

1. D. states in the Preface (p. vi) that he chose to abridge Pinkerton's *Geography* because of its excellent reputation.
2. See NEW AND COMPLETE ... (1796).

[PINNOCK, William?] **Y**

- **{1821} (01)**
- *A catechism of geography*
- Place/publisher n.r.
- ***Libraries*** n.r.
- ***Source*** Allford (1964:243). The identification of Pinnock as the author is no more than a surmise; it is not made by Allford.
- ***Notes***

1. At least two different wordings are associated with the head title *A catechism of geography*. See (02), which follows immediately, and PINNOCK (1831a), below. Then there is PINNOCK (1831b), below. Their mutual relations are not clear.

- **{1823}, 29th, cor. ... (see below) (02)**
- ... *A catechism of geography*; being an easy introduction to the knowledge of the world, and its inhabitants. For the use of children ... 29th ed., cor. and improved by many valuable additions
- LONDON; printed for G. and W.B. Whittaker
- 14 cm.; pp. 70

- ▸ *Libraries* : MB, OClWHi
- ▸ *Source* NUC (NP 0376986), which shows the book as still published anonymously
- ▸ *Notes*

1. BL and NUC record further eds. in which the author is identified: 1825, 33rd, London, 70 pp. (: NNC); 1827, 40th, London (L).

[PINNOCK, William] **Y**

- ▸ **{1831a}, 47th ... see below (01)**
- ▸ *A catechism of geography*; being an easy introduction to the knowledge of the world, and its inhabitants; the whole of which may be committed memory at an early age. Forty-seventh edition, corrected, and improved by many valuable additions
- ▸ LONDON; printed for Whittaker, Treacher & co.
- ▸ 13.5 cm.; pp. 72
- ▸ *Libraries* : : CaOTP
- ▸ *Source* St. John (1975, 2:811), which asserts that the 40th ed. was published in 1827

- ▸ **{1833} (02)**
- ▸ *A catechism of geography* ... knowledge of the wold [sic] ... age
- ▸ MONTREAL; H.H. Cunningham
- ▸ Pp. [3], 4-72
- ▸ *Libraries* : : CAQQS
- ▸ *Source* CIHM
- ▸ *Notes*

1. CIHM records 1839, 1st Cdn, Montreal, 72 pp., 'published and sold by W. Grieg' (: : CaOOA); BoC 7700 records 1843, 7th Montreal; Armour and Ramsay, 72 pp. (CaOTP); 1846, 8th Montreal; R. & A. Miller (CaOTP).

- ▸ **{1847}, 9th (03)**
- ▸ *Catechism of geography* ...
- ▸ MONTREAL; Campbell Bryson
- ▸ *Libraries* : : CaBVaU
- ▸ *Notes*

1. The attribution to Pinnock is speculative; the book itself is anonymous.
2. CaBVaU also hold a copy of *Pinnock's catechism*, but carrying the name (as editor?) of Chas. P. Watson on the tp.; it is catalogued as rev., 1851, Montreal, R. & A. Miller. CaBVaU reports that these two 'titles are basically the same, [although] the chapter numbers differ slightly as they go along, leading to differing chapter numbers from Chapter 5 and on.'
3. NUC records 1854, 10th, Montreal, Ramsay, 69 pp. (CaOTP, CaQQS).
4. See PINNOCK (1831b), below, for further problems with *Pinnock's catechism*.
5. Microform versions of (01) and of the eds. of 1839 and 1854 are available from CIHM.

PINNOCK, William **Y**

- **{1828} (01)**
- [*A comprehensive*?] *Geography and history, grammar of modern*
- Place/publisher n.r.
- ***Source*** Peddie and Waddington (1914:227)
- ***Notes***

1. This is presumed to be the 1st ed. of (02). For a possible American ed., see PINNOCK, ed. E. Williams (1835), below.

- **[1830?], 2nd (02)**
- *A comprehensive grammar of modern geography and history*. For the use of schools and for private tuition. With maps, views, costumes, etc. By ... author of Pinnock's Catechisms, History of England, Rome, Greece, etc.
- LONDON; printed for Poole & Edwards. (Successors to Scatcherd & Letterman) 12, Ave Maria Lane
- 15 cm.; pp. [5], ii-iv, [1], 2-464
- ***Libraries*** : : **CaBVaU**
- ***Notes***

1. Intended as a school textbook at the secondary level. The use of tiny print makes it possible for the book to contain much more information than the small number of pages suggests is likely. In pp. [1]-66, terms are defined and the astronomical relations of geography are presented. Europe occupies pp. 67-324, although some of that space is devoted to European colonies in other continents. No reason is given for this, but it is in keeping with the overall approach, which is discursive rather than orderly.
2. NUC records an ed. in 1834, London; Holdsworth & Ball (MH).

PINNOCK, Wm. **?**

- **{1827} (01)**
- *Geography and general history*
- Place/publisher n.r.
- ***Libraries*** n.r.
- ***Source*** Peddie and Waddington (1914:227)

PINNOCK, W. **?**

- **{[c1820]} (01)**
- *Geography and history, modern*
- Place/publisher n.r.
- ***Libraries*** n.r.
- ***Source*** Peddie and Waddington (1914:227)
- ***Notes***

1. Treated as a separate title from the previous because Peddie and Waddington record both this title, in 1820, and a *Grammar of modern geography and history* in 1828, as well as *Geography and general history*, 1827, which precedes this entry.

PINNOCK, W. ?

- **{[c1830]} (01)**
- *Geography made easy*
- Place/publisher n.r.
- ***Libraries*** n.r.
- ***Source*** Peddie and Waddington (1914:227)
- ***Notes***

1. BL records further eds.: 1844, 15th, Geography made easy ... continued to 1839; 1847, Geography made easy ... to 1845. BCB, Index (1837-57:110) records an ed. in 1851.

PINNOCK ?

- **{1854} (01)**
- *Geography, Pinnock's*
- ? Allman
- ***Libraries*** n.r.
- ***Source*** BCB, Index (1837-57:109)

PINNOCK, William **A Am**

- **{1853}, enlarged ... (see below) (01)**
- *Panorama of the old world and the new*. Comprising a view of the present state of the nations of the world, their manners, customs, and peculiarities, and their political, moral, social, and industrial condition. Interspersed with historical sketches and anecdotes. Enlarged, revised, and embellished with several hundred engravings, from designs of Croome, Devereux, and other distinguished artists
- BOSTON; L.P. Crown
- 24 cm.; vii p. [sic]
- ***Libraries*** : NNC
- ***Source*** NUC (NP 0377056), which notes: 'Dummy of excerpts from the complete work, with publisher's advertisement'

- **1854 (02)**
- *Panorama of the old world and the new ...*
- BOSTON; L.P. Crown & Co. 61 Cornhill. Philadelphia; J.W. Bradley, 48 North Fourth St.
- 22 cm.; pp. [3?], iv-vii, 8-616
- ***Libraries*** : **MB**, NNC, RPB
- ***Notes***

1. The tp. of the MB copy is missing, being replaced by a manuscript version. Intended for family reading.
2. NUC records an ed. in 1859, Philadelphia, 616 pp. (UU).

PINNOCK, William **Y Can**

- **{1831b}, revised and enlarged (01)**
- *Pinnock's catechism of geography*; being an easy introduction to the knowl-

edge of the world and its inhabitants, the whole of which may be committed to memory at an early age
- MONTREAL; R. & A. Miller
- 18cm.; pp. 72, [1]
- ***Libraries*** : : CaQQS
- ***Source*** CIHM
- ***Notes***

1. The relationship of this book to PINNOCK (1821) and PINNOCK (1831a) above, both of which are entitled *A catechism of geography*, is unclear. There are also problems in the relationship of this work to (02) and (03) of PINNOCK (1831a), also published in Montreal, but by different firms.

- **{1853}, revised ... (see below) (02)**
- *Pinnock's catechism of geography*; revised and adapted for use in this country. For the use of the schools of the Christian Brothers
- NEW YORK; D. & J. Sadlier & Co.
- 17 cm.; pp. [2], 124, 14
- ***Libraries*** : IU, PLatS, PPiU : CaBVaU, CaQQS
- ***Source*** NUC (NP 0376991), which notes: 'copy 2: 1855'
- ***Notes***

1. CaBVaU report that the text of this ed. and the next (i.e., 1870) are similar to each other, but that they differ to a somewhat greater degree from the eds. of 1847 and 1851 listed under PINNOCK (1833), above.
2. CaBVaU report an ed. in 1870, New York; D. & J. Sadlier.
3. It seems likely that there is only one author called Pinnock who wrote books related to special geography, though he is identified variously as W., Wm., William, and even without an initial.
4. Culler discusses the ed. of 1853 (1945, *passim*). Nietz (1961:213) is probably referring to this work.
5. Microform versions of the eds. of 1831 and 1853 are available from CIHM.
6. See MANNING (1864).

PINNOCK, William; ed., Edwin WILLIAMS **Y Am**

- **1835 (01)**
- *A comprehensive system of modern geography and history*. Revised and enlarged from the London edition of Pinnock's Modern Geography, and adapted to the use of academies and schools in the United States. By Edwin Williams. Author of the New Universal Gazetteer, New-York Annual Register, etc.
- NEW YORK; Bliss, Wadsworth & co. No. 111 Fulton-Street
- 12mo, 17.5 cm.; pp. [3], iv-xii, [13]-502, [2]
- ***Libraries*** L : DLN, **MB**, N
- ***Notes***

1. Includes the Preface of the 'London edition' on which it is based (See PINNOCK 1828, above). In contrast to the author's 'more juvenile publications,' this one is intended for 'more advanced pupils, with the view of meeting the progressive knowledge of the age' (p. v).

   The text is organized by numbered paragraphs, with questions provided at the end of sections.

2. NUC records an ed. in 1836, New York and Boston, 502 pp. (MB, RPB). BCB, Index (1837-57:109) records an ed. in 1845, [London]; Simpkin.

PITT, Moses **A n.c.**

- **1680-3 (01)**
- *The English atlas ...*
- OXFORD; printed at the Theater, for Moses Pitt, at the Angel in St. Pauls-Church-Yard
- Folio, 54.5 cm., 5 vols.; Vol. I (1680), pp. 9, 52, 28, 72, [62], [1]; Vol. II (1681), pp. [4], 151, [81]; Vol. III (1683), pp. [4], 157-312, [74]; Vol. IV (1682), pp. [4], 244, [42]. For Vol. V, see note 4, below.
- ***Libraries*** L, OJ : CLU, CSmH, **CtY**, MdAN, **MH**, MiD, NBu, NN, PBL RPJCB
- *Notes*

1. Each volume has its own tp. That for Vol. I in the CtY and MH copies reads: 'The English Atlas. Volume I. Containing a description of the places next the North-Pole, as also of Muscovy, Poland, Sweden, Denmark, and their several dependencies. With a general introduction to geography, and a large index, containing the longitudes and latitudes of all the particular places, thereby directing the reader to find them readily in the severall maps.'
2. NUC (NP 0392162) records a slightly different wording: 'The English atlas, volume first; containing a description ... North Pole ... to find them readily in the maps. Published according to the directions of William Lord Bishop of St Asaph, Sir Christopher Wren, Dr Isaac Vossius, Dr Po. Pell, Dr Tho. Gale, and Mr Robert Hooke.'
3. The CtY copy differs from that at MH in that the tp. is followed by 6 pp. containing the 'Proposal for printing the English Atlas,' which is signed by the men who are listed in the (NP 0392162) tp.
4. According to BL, of the projected 11 volumes only 4 were published in full, together with the text of Vol. V. Only L is known to have that incomplete Vol. V.

   The titles of the other vols are as follows:

   Vol. II. The English Atlas: Containing the description of part of the empire of Germany. viz. The Upper and Lower Saxony: the Dukedom of Mecklenburg, Bremen, Magdeburg, etc. The Marquisates of Brandenburg, and Misnia, with the territories adjoining. The Palatinate of the Rhine: and the Kingdom of Bohemia. By William Nicolson, M.A. Fellow of Queen's College, Oxon.

   Vol. III. The English Atlas. Containing ... of the remaining part of the empire of Germany. Viz. Schwaben, the Palatinate of Bavaria, Arch-Duke-dom of Austria, Kingdom of Hungary, Principality of Transylvania, the Circle of Westphalia; with the neighbouring provinces. By Will. Nicolosn, M.A. Archdeacon of the Diocese of Carlisle, and Fellow of Queen's College, Oxon.

   Vol. IV. The English Atlas. Containing the description of the Seventeen Provinces of the Low Countries, or the Netherlands. By Richard Peers, M.A. and a Superior Beadle in the University of Oxon.
5. The text that accompanies the maps and describes the countries is reminiscent of HEYLYN (1652), although there seems to be less history.

6. Discussed in Taylor, E.G.R. (1940).

PLAYFAIR, James **A**

- **1808-14 (01)**
- *A system of geography, ancient and modern*; containing 1. The history of geography from its origin to its latest improvements. - Physical geography. - A review of theories of the earth. 2. Ancient and modern lineal measures reduced to the English standard. The extent and population of the globe. - A survey of the ocean, etc. - Longitudes and latitudes of places alphabetically arranged. 3. A review of all the empires, kingdoms, and provinces in Europe, Asia, Africa, and America; ascertaining their boundaries, extent, subdivisions, and dependencies; tracing chains of mountains, rivers, bays, promontories, etc.; specifying the climate and soil of every country; its products, population, and manners of its inhabitants; giving an account of its manufactures, commerce, literature, religion, government, revenue, etc. - its ancient and modern history; together with the situation, magnitude, antiquities of every city, remarkable town and edifice; including recent discoveries, political alterations, etc. 4. A complete atlas, ancient and modern. In folio, accurately constructed, and engraved by the most eminent artists
- EDINBURGH; printed for Peter Hill, Edinburgh; and Vernor Hood, and Sharpe, London. Alex Smellie, printer
- 4to, 28 cm., 6 vols.; Vol. 1, 1808; pp. 22, [2], cccxx, 502 [contains the history and the systematic aspects of geography (to p. cccxx), plus descriptions of Iberia and Gaul, and also their modern counterparts]; Vol. 2, 1809; pp. viii, 807 [contains the description of the United Provinces, England, Wales, and Scotland]; Vol. 3, 1810; pp. vii, [1], 693 [contains Ireland, Scandinavia, Italy, Switzerland, and Russia]; Vol. 4, 1812; pp. iv-viii, 724 [contains Germany, Hungary, Poland, Prussia, and European Turkey]; Vol. 5, 1813; pp. vii, [3], 827 [contains the countries of Asia other than Arabia]; Vol. 6, 1814; pp. viii, [2], 867 [contains Arabia, Africa, and North and South America, including under the last the islands of the Pacific]
- ***Libraries*** **L** : DLC, MH, PPAmP, PPL
- ***Notes***

1. The work as a whole contains more history and less physical geography than PINKERTON (1802); it is less scholarly in form (no sources are cited), and, at least when compared with the later eds. of Pinkerton, more Eurocentric.

PLINY (the younger) **mSG tr**

- **1566 (01)**
- *A summarie of the antiquities, and wonders of the worlde*, abstracted out of the sixtene first bookes of the excellente historiographer Plinie, wherein may be seene the wonderfull workes of God in his creatures, translated oute of the French into Englishe by I.A.
- LONDON; imprinted by Henry Denham, for Thomas Hacket, and are to be solde at his shop in Lumbert Streate
- 13 cm.; pp. [128]
- ***Libraries*** **L** : CSmH

▸ *Notes*
1. 118 pp. of text. The book includes the dedication to the Cardinal Bishop of Orleans by Blaysee de Changy, who made the French translation of the Latin original. The 1st book is dealt with in a line; the 2nd, which deals approximately with present-day physical geography, is reduced to 8-9 pp.; the 3rd, 4th, and 5th are special geography (dealing with Asia, Africa, and Europe, respectively) and receive 7-8 pp. each; the 7th book (18-19 pp.) deals with human biology and character; the 8th book (23 pp.) deals with land mammals, the 9th book (7-8 pp.) deals with marine life; the 10th book (20 pp.) with birds; the 11th book (17 pp.) with insects and the like; and the 12th to 16th books (16 pp.) with many varieties of trees.

▸ **1585 (02)**
▸ *The secrets and wonders of the world*. A book right rare and straunge, containing many excellent properties, giuen to man, beastes, foules, fishes and serpents, trees, and plants. Abstracted out of that excellent naturall historiographer Plinie. Translated out of French into English
▸ LONDON; printed for Thomas Hacket, and are to be solde at his shop in Lumbard Streete, under the Popes head
▸ 4to, 19 cm.; pp. [62] of which 58 are text
▸ ***Libraries*** L : CSmH
▸ *Notes*
1. The French dedication of the 1st ed. is gone. The body of the text, though reset, appears to be unchanged, spelling apart, except for the conclusion; that appears to have been shortened by two paragraphs.

▸ **1587 (03)**
▸ *The secrets and wonders of the world* ... ryght ... contayning ... fishes, and ...
▸ LONDON; printed for T. Hacket ...
▸ 4to, 18 cm.; pp. [62]
▸ ***Libraries*** L, GU : DFo, DLC
▸ *Notes*
1. Though reset and with some alteration in spelling, the text seems unchanged.

PORTER, Thomas **A**
▸ **1659 (01)**
▸ *A compendious view, or cosmographical, and geographical description of the whole world*. With more plain general rules, touching the use of the globe, than have yet been published. Wherein is shewed the situation of the several countries and islands: their particular governments, manners, commodities, and religions. Also a chronology of the most eminent persons, and things that have been since the creation, to this present: wherein you have a brief of the gospel, or a plain, and easie table, directing readily where to find the several things, that were taught, spoke, done and suffered, by Jesus Christ, through out the Gospel. The which is not onely pleasant, and delightful, but very useful, and profitable; for all. But chiefly for those who want, either time, to read, or money to buy, many books

- [LONDON]; are to be sold by Robert Walton, at the Globe and Compass, in S. Paul's Churchyard, on the North-side
- 8vo, 16.5 cm.; pp. [6], 138, [2]
- ***Libraries*** L : **CtY** (imperfect, lacks covers)
- ***Notes***

1. Special geography for the poor. Descriptions of all the countries of the world occupy pp. 29-92.

POVOLERI, Giovanni **Dy**

- **1775 (01)**
- *New geographical tables*. Exhibiting at one view all the empires, kingdoms, states, republics, provinces, titles, position, situation, extent, climates, boundaries, sub-divisions, square miles, cities, chief towns, coronation places[,] villages, latitude, longitude from Lond., bearing from Lond., forts, ports, oceans, seas, gulphs, bays, streights, islands[,] isthmuses, capes, promontories, rivers, lakes, mineral waters, mountains, numb. of inhabit.[,] languages, religions, form of governmt.[,] patriarchates, archbishopricks, bishopricks, universities, academies, ord. of knighth.[,] commodities, productions, curiosities, etc. in Europe, Asia, Africa, and America: including an authentic list of all the counties, cities, and boroughs, of Great Britain and Ireland which return members (and what number) to their respective Houses of Parliament. To which are prefixed, I. Directions for the use of globes, geographical definitions, climates, horary circle, and grand divisions of the world, on a new plan. II. Chronological tables of the sovereigns of England, Germany, France, Spain, Russia, Denmark, and Sweden, from the year 768 to 1775. With a list of uninhabited islands, ruins of celebrated places, countries and places known by different names, and ancient names of remarkable seas and rivers, in the appendix.
- LONDON; printed for, and sold by, the author, No. 12, Noel-Street, Soho: sold also by T. Cadell, and P. Elmsly, in the Strand
- 12mo, 15.5 cm.; pp. [12], 3, [2], 4-40, [442]
- ***Libraries*** **L** : DLC, IU
- ***Notes***

1. 'My principal design ... is to present the young ladies I have the honor to instruct with a short pleasing method how to improve in geography' (p. 5). If, as he soon implies he did, P. expected his young ladies to learn the facts presented in the tables that constitute the entire body of his book, I pity them. As a single example, here is one line of facts; they are set out in eight columns, headed *cities, towns, etc; countries, states, etc.; quar.; rivers, seas, etc.; sovereigns; relig.; long.; lat.*: 'Minori Bp [i.e., bishoprick] Principato; Italy; Eur.; Salerno G. Mediter.; K. of Naples; Cathol; 14E. 41N.'

P[OWELL], D[avid] **OG**

- **{1573} (01)**
- *Certaine brief and necessarie rules of geographie*, seruing for the vnderstanding of the chartes and mappes.
- LONDON; imprinted by Henry Binneman
- 8vo, 15 cm.; pp. [13]

- ▸ ***Libraries*** L, O
- ▸ ***Notes***

1. BL catalogues as P., D. E.G.R. Taylor identifies the author's name, and says of this work that 'it deals in [an] elementary fashion with the circles, parallels and meridians, and other conventions of the map' (Taylor 1939:34).
2. Facsimile ed., 1976 (Johnson/TOT).

POWER, A. Bath
See note to ALLISON, M.A. (1828).

PRESENT STATE ... **Gt**

- ▸ **{1739}**
- ▸ *[The] present state of all nations*. Describing their respective situations, persons, habits, buildings, manners, laws and customs, religion and policy, arts and sciences, trades, manufactures and husbandry, plants, animals and minerals
- ▸ DUBLIN; printed by and for George Grierson, Printer to the King's most excellent majesty, at the King's-Arms and Two Bibles in Essex-street
- ▸ 4to, 5 vols.; Vol. V, 22 cm.; pp. x, 558
- ▸ ***Source*** Cox (1938:351) and letter from IEN
- ▸ ***Notes***

1. IEN has the 5th vol. and supplied the following information.
   The title given above is followed by: 'Vol. Vth. and last. Containing the present state of Africa and America. Wherein are described, the present state of Africa and America, and of the following kingdoms and nations in particular (viz.) Ethiopia, Zanguebar, Caffraria and the Hottentot nations; Congo, Angelo, Guinea, Nigritia or Negroland, Zaara, Biledulgerid, Morrocco, Algiers, Tunis, Tripoli, and the African islands, with an abstract of the ancient history of Africa. A desertation on the first peopling of America. The discovery thereof by Columbus, the present state of Mexico or new-Spain, with the history of the conquest by Cortez, the ancient history of Mexico, with an account of the French settlement at Florida; shewing the advantage of an alliance with Spain; a description of Florida, Terra-Firma and the Province of Darien. A description of the Spanish islands of Cuba, Hispaniola and Porto-Rico. A summary of the state of Spanish America. A description of Brazil or the dominions of Portugal in America. A state of the British plantations, particularly Virginia, Maryland, New-England, Nova Scotia and Acadie, New-Britain, New York, and New-Jersey, Pensylvania, Carolina and Georgia, Jamaica, Barbadoes, the Caribbee islands, the Bahama islands, Bermudas, Newfoundland, and French-America. Illustrated with curious copper-plates of the habits and animals; with maps of the several countries described in this volume, accurately drawn, according to the geographical part of this work, by Herman Moll.'
   The Introduction (which occupies pp. i-x) is devoted to the controversy over the question of whether or not Africa had been circumnavigated by the Phoenecians. The wording of the opening two paragraphs suggests that this is not the first edition of this work.
   Accepting the information provided by IEN, the date, the title, and the

authorship become an issue. One candidate is SALMON (1725-38). The evidence in favour of this opinion consists of three points:

(1) The title of SALMON (1725-1738), which was published in 33 vols., 8vo, starts, 'Modern history: or .... ,' and then continues, 'the present state of all nations. Describing their respective situations, persons, habits, buildings, manners, laws and customs, religion and policy, arts and sciences, trades, manufactures and husbandry, plants, animals and minerals.' In other words, apart from the opening three words, the titles of the two works are identical.

(2) A small 4to ed. of SALMON (1725-38) was published anonymously in Dublin, over the years 1724-30. The NUC entry (NS 0063591) records 4 vols., but then adds a note that reads: '*Contents*. Africa, V. 5 - America, V. 5 - Asia, V. 1 - Europe, V. 2, 4.' It seems that it is because Vol. 3 is missing that the cataloguers at MdBP classified this as a 4-vol. work.

So both works are anonymous, published in Dublin, and made up of 5 vols.

(3) In the Introduction to Vol. 30 of the ed. of 1725-38, Salmon rebuts criticisms made of his history of Africa. He adds that he intends to publish a supplement to 'bring down the history of every country to the year 1737; and to add to it such further discoveries and improvements as have been made by other travellers since I entered on this work' (p. 31). In the Introduction to the anonymous *Present State*, of 1739, the author writes: 'I shall take an opportunity of considering some objections that have been made to the modern history of Africa' (p. i); and: 'I intend, as soon as America is finish'd, to take a survey of the whole work ... and bring down the history of every country to the year 1737; and to add to it such further discoveries and improvements as have been made by other travellers since I enter'd on this work' (p. x).

PROCLUS **OG tr**

- **1550 (01)**
- *The descripcion of the sphere or frame of the worlde*, right worthy to be red and studyed on of all noble wyttes. Spetyally of all those that be desyrous to attayne any perfect knowledge in cosmographie ... Englysshed by me Wyllyam Salysbury
- [LONDON]; imprinted by me Robert Wyer: dwellynge at the sygne of St. John Euangelyst in St. Martyns parysshe besyde Charynge Crosse [at the back of the book]
- 12mo, 13 cm.; pp. 48
- ***Libraries*** CMp, **L** : DNLM
- ***Notes***

1. As the title suggests, this work is concerned with the lines that geographers ascribe to the surface of the earth on the basis of astronomical observations. It was translated from Latin.
2. 'Proclus was an Athenian scientist of the fifth century A.D., his book a standard medieval text' (Park 1928:270). CUNINGHAM (1559), for example, used it as such in the first book of his *Cosmographical glass*.

PTOLEMY **mSG tr**

- **{[1530? 1532? 1535?]} (01)**
- *Here begynneth the compost of Ptholomeus*, prince of astronomye: translated oute of Frenche into Englyshe for them that wolde have knowledge of the compost
- [LONDON]; imprynted by me Robert Wyer
- 8vo, 14 cm.; pp. [136]
- ***Libraries*** L : CSmH, DFo (imp)
- ***Notes***

1. British Museum (1959-66) gives the date as 1535?; NUC (NP 0630176) as 1532?; Pollard and Redgrave (1976) as 1530?
2. *Compost* here has the meaning of compute or calculate. This work is primarily concerned with astrology, and so with astronomy, and hence with those aspects of geography that have to do with the division of the earth's surface by lines derived from astronomical observations.

- **{[1540?]} (02)**
- *The compost, of Ptholomeus*
- [LONDON]; R. Wyer
- 8vo
- ***Libraries*** : CSmH
- ***Source*** Pollard and Redgrave (1976)

- **{[1550?]} (03)**
- *The compost, of Ptholomeus*
- [LONDON]; R. Wyer
- 8vo
- ***Libraries*** L
- ***Source*** Pollard and Redgrave (1976)

- **{[1562?]} (04)**
- *The compost of Ptholomeus* prince of astronomye: uery necessarye, vtile, and profytable, for all suche, as desyre the knowledge of the science of astronomye
- [LONDON]; imprinted by Thomas Colwell
- 8vo, 15 cm.; 4 p. l., 95[i.e., 96] f.
- ***Libraries*** : MH, NN
- ***Source*** NUC (NP 0630184)

- **{[1638?]}, anr ed. corrected (05)**
- *The compost, of Ptholomeus*
- ? M. P[arsons] f. H. Gosson, sold by E. Wright
- 8vo
- ***Libraries*** L
- ***Source*** Pollard and Redgrave (1976)
- ***Notes***

1. Every history of geography that deals with the sixteenth century or the Middle Ages, as well, of course, as the Classical period, makes some reference to Ptolemy. An introduction to his cartography can be found in Thompson (1948), together with an indication of the research that had been

done on Ptolemy's *Geography* to that time (n. 1, p. 230). A seemingly definitive assessment of Ptolemy's contribution to the history of cartography is given by Dilke (1987); however, those who are curious about the capacity of deeply entrenched ideas to withstand criticism may be struck by the nature of Dilke's response to an extremely vigorous assault on Ptolemy, albeit in his role as an astronomer rather than as a geographer, made by Newton (1977).

PUBLIC SCHOOL ... **Y Can**

- **1887 (01)**
- *The public school geography.* Authorized for use in the public schools, high schools and collegiate institutes of Ontario, by the Department of Education
- TORONTO; Canada Publishing Co.
- 25 cm.; 164 pp.
- ***Libraries*** : : **CaAEU**, CaBVaU
- ***Notes***

1. Content is strongly biassed in favour of North America, and of Canada in particular.
2. A microform version is available from CIHM, which also records an ed. of [1900?].

PUFFENDORF, Samuel **NG tr**

- **{1695} (01)**
- *An introduction to the history of the principal kingdoms and states of Europe.* [By ... Counsellour of State to the late King of Sweden]. Made English from the original[, High-Dutch]
- LONDON; M. Gilliflower
- 19 cm.; pp. 538
- ***Libraries*** : CSt, CLU, DLC, MH, NcU
- ***Source*** NUC (NP 0641195)
- ***Notes***

1. Included, despite its limitation to Europe, because of the reference made to it in CRULL (1705). The words in the title set in [ ] are taken from the 3rd ed. (**L**).
2. Both BL and NUC spell his name PUFENDORF, but the spelling given above is that recorded on the tp. of the 3rd ed., 1705, held by L.
3. BL and NUC note later eds.: 1697, 2nd (: CLU, CtY, OCl, PPiU, PSt, PU : CaBVaU); 1699, 3rd (L : CLU); 1700, 4th (: CLU, DFo, ICN, IdU, MnU, NIC); 1702, 5th (: NcD); 1706, 6th, (: CLU, CU-S, FU, ICN, IU, KU, PBm, TxDaM, TxU, ViW, WaU); 1711, 7th (: CU, MA, NcD); 1719, 8th (: DLC, MWA, OClJC, OU, PPL, RPB : CabVaU); and 1728, 9th (: ICN, MnU, PPL, PU).
4. NUC also notes an English translation based on revised versions of Pufendorf's original, the revisions having been made by B. de la Martinière. Copies of the ed. of 1748 are located at (CLU, DLC, KyLx, MiU, OO, PU); copies of the ed. of 1764 are held at (LU, MH, MiU, MoU, NIC, NjP, NN, PPL, PU); a copy of the 1774 ed. is held at MeB. Lastly, NUC notes an ed. of Puffendorf ... Introduction ... principal ... states of Europe, 1782; a copy is held by (PPL).

5. Francis Edwards, Ltd reports, in Catalogue No. 975, *Voyages and Travels* (London 1973) as having for sale: Puffendorf (S), *An introduction to the history of the kingdoms and states of Asia, Africa, and America both ancient and modern*, 1705. It is likely that this is a ghost; see CRULL (1705).

PUTZ, Wilhelm **? tr**

▸ **{1850} (01)**
▸ *Handbook of modern geography and history ...* Translated from the German by R.B. Paul
▸ LONDON
▸ 12mo
▸ ***Libraries*** L

▸ **{1851}, 1st American, rev. and cor. from the London ed. (02)**
▸ *Manual of modern geography ...* by the Rev. R.B. Paul
▸ NEW YORK; D. Appleton & company; [also] Philadelphia, G.S. Appleton
▸ 19 cm.; pp. xi, 336
▸ ***Libraries*** : DLC, ICarbS, MH, ODW, OkU, PP
▸ ***Source*** NUC (NP 0640915)
▸ ***Notes***
1. NUC records further eds. (both New York, D. Appleton): 1858, 2nd American ... from the London ed., 336 pp. (OClWHi, ViU); 1870, 336 pp. (NN).

R

R., E. **Y**

▸ **1790 (01)**
▸ *Geography and history*. Selected by a lady, for the use of her own children
▸ LONDON; printed for B. Law, No. 13, Ave-Marie Lane, Ludgate Street
▸ 12mo, 18 cm.; pp. x, 366, [2]
▸ ***Libraries*** L
▸ ***Notes***
1. The bulk of the text is made up of lists of countries, towns, rivers, and the like, though in the form of text rather than tables. There is a strong bias in favour of Europe in general and the British Isles in particular. The section on the globes and geographical problems comes at the back rather than, as is usual, at the front (see note to entry (02) of KEITH, T. 1787). History is reduced to a chronology of events, plus lists of eminent people, all acknowledged as being taken from Guthrie (presumably GUTHRIE 1770). To judge from the preface, the book was intended for children as young as five years old.

▸ **1794, 2nd (02)**
▸ *Geography and history ...*
▸ SOUTHAMPTON, [for London]; printed and sold by T. Baker. For B. Law, Ave-Maria-Lane, Ludgate Street

- 12mo, 18 cm.; pp. viii, [6], 387, [1]
- ***Libraries*** **L**
- ***Notes***

1. A dedication has been added. Spot-checking revealed no changes in the text. The tables have been extended to note events since the 1st ed.

- **1797, 3rd (03)**
- *Geography and history ...*
- LONDON; printed for B. Law, No. 5, Stationers-Court; and C. Law, No. 13, Ave-Marie-Lane, Ludgate Street
- 12mo, 18 cm.; pp. viii, [6], 392
- ***Libraries*** **L**
- ***Notes***

1. The section on France has been adjusted to take note of the newly created *departements*; the old provincial names are, however, still presented. The tables have been extended to note events since the previous ed.
2. NUC records a further ed.: 1801, 4th, London, 362 pp. (CtY); Peddie and Waddington (1914:227) record: 1811, 8th, London (library n.r.); then NUC records further eds. (both London): 1818, 11th (MH) [NG 0127315 catalogues under geography]; 1828, 15th, 388 pp. (CU).

- **1834, 17th (04)**
- *Geography and history ...*
- LONDON; printed for Longman, Rees and Co.; Baldwin & Craddock; J.G. & F. Rivington; J. Booker; J.M. Richardson; J. Duncan; Hamilton & Co.; Whittaker & Co.; Sherwood & Co.; Simpkin & Marshall; S. Hodgson; J. Souter; Darton & Harvey; Holdsworth & Ball; Houlston & Son; and S. Poole
- 12mo, 18 cm.; pp. xii, 379, [10]
- ***Libraries*** **L**
- ***Notes***

1. At the end of the preface, which remains as it was in the 1st ed., there is a note stating that in the 16th ed. 'the historical notices have been considerably enlarged, and brief sketches of recent revolutions added, in order to supply youthful students with a correct outline of the present political state of the world' (p. vii). It is signed W.C.T. A further note to the 17th ed. along the same lines is signed by M.; presumably this is Samuel Maunder (see (05) below). The organization and form of the book are essentially unchanged. A section on the solar system and one on ancient geography have been added.

- **1859, 22nd (05)**
- *Geography and history ...* Revised and augmented by Samuel Maunder ...
- LONDON; Longman and Co.; Hamilton and Co.; Simpkin and Co.; Whittaker and Co.; F. and J. Rivington; Houlston and Wright; C.H. Law; J.S. Hodgson; Tegg and Co.; Hall and Co.; Piper and Co.; and Relfe Brothers
- 12mo, 18 cm.; pp. viii, 399, [1]
- ***Libraries*** **L**
- ***Notes***

1. No obvious changes since 1834.

RALEIGH, Walter **NG**

- **1636 (01)**
- *Tvbvs historicvs: an historicall perspective; discovering all the empires and kingdomes of the world*, as they flourisht respectively under the foure imperiall monarchies. Faithfully composed out of the most approved authors, and exactly digested according to the supputation of the best chronologers. (With a catalogue of the kings and emperours of the chiefe nations of the world)
- LONDON; printed by Thomas Harper, for Benjamin Fisher
- 19 cm.; pp. [26]
- ***Libraries*** C, **L**, O : CLU, CSmH, CtY, DFo, IU, NcU, NN
- ***Notes***

1. Not geography. It contains a supposed chronology of all kingdoms, and a list of the kings 'of the chiefe nations of the world.' Published posthumously.

RAND, E.H. **?**

- **{1872} (01)**
- *Geography, physical and political*
- ? Lockwood
- ***Libraries*** n.r.
- ***Source*** ECB, Index (1856-76:131)

RANDALL, Joseph **A**

- **1744 (01)**
- *A system of geography*; or, a dissertation on the creation and various phænomena of the terraqueous globe: as it consists of subterraneous caverns, subterraneous waters, mountains, vallies, plains, and rocks. With an hypothesis concerning their causes. A description of all the empires, kingdoms etc. of the world. Exhibiting their boundaries, situation, division, subdivision, square miles, antient geography, chief towns in each division, distance & bearing from the capital, climate, government, remarkable laws, policy, trade, revenues, forces, curiosities, persons of the inhabitants, character, religion, customs, ceremonies. With extracts of antient and modern history, and of some of the most celebrated voyages and travels, interspersed throughout the whole. To which is prefixed, an introduction to those parts of the mathematics, necessary to a thorough knowledge of the subject of geography; viz. algebra, geometry, plain trigonometry, the use of the globes, projection of the sphere, spherical trigonometry, geometrical and physical astronomy, great variety of geographical and astronomical problems, the construction of maps; digested into definitions, problems, and theorems, and fully demonstrated
- LONDON; printed for Joseph Lord, bookseller, in Wakefield; and sold by him at his shop in Barnsley, and at Pontefract; and J. Rivington, at the Bible and Crown in St. Paul's Church-yard, London
- 19 cm.; pp. [14], lx, 676, [1]
- ***Libraries*** **L** : **MH**, N, NBuG, NcU, NjP
- ***Notes***

1. An independent voice from the north of England. Pp. 170-88 contain a 'system of geography' in the tradition of Burnett (Taylor, E.G.R. 1948). The

balance of the book is devoted to special geography. Though explicitly intended for the 'young beginner' (p. [5]), the mathematics presented in the book, which include simultaneous equations and the extraction of cubes, makes it plain that it is students at university level whom the author has in mind. His enthusiasm for Newtonian science, together with a conviction that the earth was created, was typical of the time, but his commitment to the former is deeper than that of most of his contemporaries among special geographers. He is also, and this time knowingly, typical of his time in the matter of using other people's words: 'Several of the heads, mentioned on the title-page, being subjects that do not admit of new relations, I reckon myself no plagiary to grant, that I have made several extracts from other volumes: it is what those authors themselves have done ... when they found [them] succinctly worded by a credible and able pen' (p. [10]). His independence shows most clearly in his political and religious views, for they are radical in terms of the English society of his day. Pp. 587-600 (within the description of the British empire in North America) is devoted to a presentation of the Quakers' own view of themselves.

[RASTELL, John ?] **NG**

- **{[1520?]} (01)**
- *A new interlude and a mery, of the nature of the iiij elements*, declarynge many proper poyntes of phylosophy naturall, and of divers straunge landys
- 8vo
- ***Libraries*** L
- ***Notes***

1. A play rather than special geography. 'In [it] Experyence discourses with Studyous Desire on strange lands and marvels' (Cox 1935:69). NUC (NN 0074091) records a modern ed., edited by James O. Halliwell, London, 1848. It is the source of the full title given above.

READWIN, T.A. **Gt**

- **{1842}**
- *Geography*
- Place/publisher n.r.
- 18mo
- ***Libraries*** n.r.
- ***Source*** Allibone (1854)
- ***Notes***

1. Thomas Alison Readwin is recorded in both BL and NUC as the author of a number of books on geology. It is probable that this is a ghost, the key word of the title having been misspelled.

REDWAY, Jaques Wardlaw **Y SG? Am**

- **{[c1887]} (01)**
- ... *Butler's complete geography*
- PHILADELPHIA; E.H. Butler and company
- 32 cm.; pp. 141, 17

- ***Libraries*** : CtW, CtY, MA, OClWHi, PP, PPD, PPF
- ***Source*** NUC (NR 0110117), which notes: 'Butler's geographical series' and 'Lettered on cover; New England edition'
- ***Notes***

1. NUC records further eds. (all Philadelphia; E.H. Butler): 1887, Tennessee. ed., 141, 12 pp. (T) [NR 0110119 notes: 'Geography of Tennessee, by J. Berrien Lindsley: 12 p. at end']; [c1887] 1892?, Illinois ed.,141, 10 pp. (DHEW, ICU) [NR 0110118 notes 'Geography of Illinois (10 p. at end) has special t.-p. dated c1892']; and [1895, c1887], New England ed. (MH).
2. Hauptman (1978:432) includes a reference to (01) in a general discussion; Elson (1964:405) cites (01).

REID, Alexander **Y SG?**

- **{1849} (01)**
- *A first book of geography*: being an abridgment of Dr. Reid's Rudiments of Modern Geography: with an outline of a geography of Palestine
- EDINBURGH
- 16mo
- ***Libraries*** L
- ***Notes***

1. BL records further eds. (all Edinburgh): 1853, 4th; 1855, 5th; 1855, 6th; 1857, 7th; 1858, 8th; 1859, 9th; 1861, 10th; 1861, 11th; 1863, 13th; 1864, 14th; 1865, 15th; 1866, 16th; 1866, 17th; 1871, 22nd; 1876, 26th; and [1882], 27th. BL also records a 30th ed. in 1890, and two eds. translated into the 'Nagari character for the use of the lower English classes in Indian schools.'

REID, Alexander **Y SG?**

- **{1837} (01)**
- *Rudiments of modern geography*; with an appendix containing an outline of ancient geography, [problems on the use of globes, and directions for the construction of maps. For the use of schools]
- EDINBURGH; Oliver & B[oyd]
- 16mo, 16 cm.; pp. 108
- ***Libraries*** L : MB
- ***Source*** BL and NUC (NR 0138496)
- ***Notes***

1. The words in the title set in [ ] are taken from the NUC entry cited in (02) below.
2. BL records further eds. (all Edinburgh): 1842, 4th; 1850, 7th, 1854, 12th; 1854, 13th; 1855, 14th; 1855, 15th; 1856, 16th; 1857, 17th; 1857, 18th; 1858, 19th; 1859, 20th; 1860, 21st (L : MH) [in NUC]; 1862, 23rd; 1863, 24th; 1864, 26th; 1866, 27th; 1866, 28th; 1870, 33rd; 1871, 34th; and 1872, 36th.

- **{1873}, 37th, revised (02)**
- *Rudiments of modern geography*; with an appendix, containing an outline of sacred geography ... schools. 37th ed., rev.

- EDINBURGH; Oliver and Boyd
- 12mo, 16 cm.; pp. 180
- ***Libraries*** L : NIC
- ***Source*** BL and NUC (NR 0138498)
- ***Notes***

1. Since the first ed., *ancient* geography has been either replaced by, or relabelled as, *sacred* geography.
2. BL records further eds. (all Edinburgh): 1877, 39th; 1877, 40th; 1879, 41st, 180 pp.; 1881, 43rd, 180 pp.; 1882, 44th, 180 pp.; 1885, 47th, 180 pp. BL records other eds. in 1888, 1889, 1891, 1891, 1892, 1893, 1897, and 1899 (56th).

REID, H[ugo] **Y Can**

- **1856 (01)**
- *Elements of geography*; adapted for use in British North America, containing the geography of the leading countries of the world, with British North America fully developed and the outlines of physical and astronomical geography. By ... Professor of Language and Logic, and Principal of the Day Schools, Dalhousie College, Halifax
- MONTREAL; B. Dawson. Halifax, N.S., A. & W. Mackinlay. St. John, N.B., J. Bowes & Sons
- 15 cm.; [v]-vii, [8]-152 p.
- ***Libraries*** : : CaNSWA, CaOTP
- ***Source*** BoC 5726
- ***Notes***

1. Amtmann (1971) records: 1858, 2nd, Halifax, N.S., Bowes.
2. A microform version is available from CIHM.

REID, H[ugo] **Y SG?**

- **{1849} (01)**
- *Geography, first book of*
- ? Grant & G.
- 18mo
- ***Libraries*** n.r.
- ***Source*** BCB, Index (1837-57:110)
- ***Notes***

1. ECB, Index (1856-76:130) records a further ed.: 1865.

REID, Hugo **Y SG?**

- **{1852} (01)**
- *A system of modern geography* ... with exercises of examination. To which are added treatises on astronomy and physical geography
- EDINBURGH
- 8vo
- ***Libraries*** L
- ***Notes***

1. BL records further eds. (both Edinburgh): 1853, 2nd; 1857, 3rd.

RHIND, William **Y SG?**
- **{1858} (01)**
- *Class-book of elementary geography*
- EDINBURGH; Simpkin
- 8vo
- ***Libraries*** L
- ***Source*** BL and ECB, Index (1856-76:130)

RING, L. **Y**
- **{1831} (01)**
- *A grammar of modern geography*, on a new and easy plan, illustrated by an accompanying atlas, for the use of schools and private tuition
- LONDON; Simpkin & Marshall, Stationers' Court
- 12mo, 15 cm.; pp. iv, 5-189
- ***Libraries*** L
- ***Notes***

1. Very brief descriptions of the countries of the world, with questions for a teacher to ask at the end of major sections; followed by sections on the use of globes and astronomical terms; followed by a gazetteer (modern sense); followed by a table of place names, each being identified as being a town, river, etc., and where located (e.g., Spain).

RITTER, Karl **OG tr Am**
- **1865 (01)**
- *Comparative geography* by ... late Professor of Geography in the University of Berlin. Translated ...
- PHILADELPHIA; J.B. Lippincott & co.
- 19cm; pp. [3], iv-xxx, 31-220
- ***Libraries*** : **CtY**, DLC, ICarbS, ICJ, ICRL, KEmT, MB (missing), MiU, Nh, NN, NRU, OClW, OCU, PPAN, PPFr, PPL, ViU, WU
- ***Notes***

1. The Preface, by Gage, the translator, provides a guide to Ritter's published work, and Gage's evaluation of their appeal to a general, educated, American audience. This book is an example of the comparative method that Ritter used in his attempt to make of geography something that was more than a mere description of places, given the unsuitability of his material from the point of view of science, narrowly defined. The tone is strongly theistic (e.g., 'The earth was made to be the home of mind, soul, character' [p. xvi]). It could also be characterized as high-minded nature mysticism, with an occasional tendency towards numerology, e.g., the relating of the length of coast of an island to its area, presumably seeing in this ratio a clue to the destiny of the people who live there. Not special geography.
2. BL and NUC also record four other eds. of this work, distinguished from it by their publishers and place of publication. NUC also records eds. in 1874, 1881, and 1890?.
3. Discussed in Hauptman (1978:427).

RIVETT, John **OG**

- **1794, 2nd, enlarged (01)**
- *An introduction to geography, astronomy, and the use of globes* to which are added, a chronological table of remarkable events, discoveries and inventions from the creation to the year 1794; and a large collection of questions, designed for the use of young persons. Second edition, enlarged
- NORWICH
- 8vo, 21 cm.; pp. [4], v-xvi, 163
- ***Libraries*** **L**
- ***Notes***

1. A variation on the theme of special geography occupies pp. 11-58. Pp. 11-44 are devoted to 'the grand divisions of land'; here information is provided at the level of continents rather than of countries, except for England and Wales. Features of terrain are the focus of interest. Pp. 45-58 are devoted to 'the grand divisions of water.' There is some speculation about the origins of various land forms.

ROBBINS, Royal **mSG Am**

- **1830 (01)**
- *The world displayed, in its history and geography*; embracing a history of the world, from the creation to the present day. With general views of the politics, religion, military and naval affairs, arts, literature, manners, customs, and society, of ancient as well as modern nations. To which is added, an outline of modern geography. Two volumes in one
- NEW YORK; W.W. Reed & co.
- 18.5 cm., 2 vols. in 1; pp. [5], vi, [7]-228; [5], vi, [7]-408, [60]
- ***Libraries*** : CSmH, CtY, **MH**, MWA, NcD, OClW, TU, ViU
- ***Source*** NUC (NR 0314396)
- ***Notes***

1. As the tp. implies, more space is given to history than geography. Vol. II, pp. 397-408, contains a highly condensed special geography, consisting chiefly of an enumeration of the division of the lands, both natural and political, plus snippets of information about places and people.
2. NUC records further eds. (all New York, W.W. Reed until 1832, H. Savage thereafter, 2 vols. in 1): 1831 (NN, OO); 1832 (MiU, MoU, OCl, OU, TxU, Vi) {NR 0314399 notes: '"A system of modern geography": v. 2, p. [397]-408'}; 1833 (DLC, ICU, IU, MiU, NIC, OCl, OrP, OrU, OU, ViU); and 1834 (ICRL, MiAlbC, OFH, OO); 1836 (TNJ).

- **1837 (02)**
- *The world displayed ...*
- NEW YORK; published by H. Savage
- 8vo, 18.5 cm., 2 vols. in 1; pp. [5], vi, [7]-228; [5], vi, [7]-431, [1], [30] l.
- ***Libraries*** : **MWA**
- ***Notes***

1. The final 30 leaves are illustrations, which are blank on the reverse.
2. NUC records an ed. in 1838, New York, H. Savage, 2 vols. in 1 (KyLoU).

ROBERTS, George ?
- **{1834}, new ... (see below) (01)**
- *The elements of modern geography and general history*, on a plan entirely new ... New edition ... improved, etc.
- LONDON
- 12mo
- ***Libraries*** L
- ***Source*** BL, which identifies the author as a schoolmaster
- ***Notes***

1. The relationship of this title to the next two is not known.

ROBERTS, G. ?
- **{1843} (01)**
- *Geography and history*
- [LONDON]; Bohn
- 12mo
- ***Source*** BCB, Index (1837-57:109)

ROBERTS, George ?
- **{1843}, 3rd (01)**
- *Pinnock's elements of history and geography improved*
- Place/publisher n.r.
- 12mo
- ***Libraries*** n.r.
- ***Source*** Allibone (1854), who identifies the author as 'formerly mayor of Lyme Regis.' May be the same work as the previous entry.

ROBERTSON, Charles **Y OG**
- **{1812} (01)**
- *A geographical exercise book*
- Place/publisher n.r.
- ***Libraries*** n.r.
- ***Source*** Peddie and Waddington (1914:226). See note to (02), below.

- **[1820] (02)**
- *A geographical exercise book* ... [the cover, which is also tp., is damaged, so the title is incomplete]
- LONDON; for Lackington, Hughes, Harding, Mavor, and others
- Oblong 4to, 18 × 22 cm.; pp. [46]
- ***Libraries*** L
- ***Notes***

1. Not special geography. Contains geographical problems; e.g., [how to] find the sun's declination for any given day; for a particular hour at a given place, to find what hour it is at any other place; to find those places in the torrid zone to which the sun is vertical on any given day. The method to be used is given, and then exercises in its use are presented. Space is provided for students using the book to enter their answers.

ROUGEAT, A. **?**

- **{1846} (01)**
- *Geography, amusing*
- ? Ackermann
- ***Libraries*** n.r.
- ***Source*** BCB, Index (1837-57:109)

ROUND WORLD ... **Y OG**

- **1883 (01)**
- *The round world*: a reading book of geography for Standard II
- LONDON; Marcus Ward & Co., 67, 68, Chandos Street and at Belfast and New York
- 8vo, 17 cm.; pp. 122, [4]
- ***Libraries*** L
- ***Notes***

1. As would be expected from a book intended for the Standard II curriculum, a systematic rather than a special geography.

ROWBOTHAM, John

- *The geography of the globe* ... See (02) of BUTLER, John O. (1826).

ROWSON, Susanna **Y Am**

- **{1806} (01)**
- *An abridgment of universal geography, together with sketches of history.* Designed for the use of schools and academies in the United States
- BOSTON; printed for John West ... David Carlisle, printer
- 12mo, 17.5 cm.; pp. iv, [13]-302
- ***Libraries*** L : DeU, DLC, IU, LU, MB, MH, MiU, NjP, OU, ViU : CaQQS
- ***Source*** BL and NUC (NR 0476997)
- ***Notes***

1. A microform version is available from CIHM.

ROWSON, [Susanna] **? Am?**

- **n.d. (01)**
- *System of geography*
- Place/publisher n.r.
- ***Libraries*** n.r.
- ***Source*** Allibone (1854)

ROWSON, Susanna **Y SG? Am**

- **{1818} (01)**
- *Youth's first steps in geography*, being a series of exercises making the tour of the habitable globe. For the use of schools
- BOSTON; Wells and Lilly
- 18.5 cm.; 2 p. l., [13]-178 p.

- ***Libraries*** : CtY, MH, OU
- ***Source*** NUC (NR 0477203)

RUDD, John Churchill **Y Am**

- **{1816} (01)**
- *A compendium of geography*, containing, besides the matter usual in such works, a short system of sacred geography, intended to aid the young in acquiring a knowledge of the places mentioned in the Holy Scriptures. To which is added, an introduction to the study of astronomy. Designed for the use of schools. By the Rev. ...
- ELIZABETHTOWN, [NJ]; printed by J. and E. Sanderson, for Mervin Hale
- 18 cm.; vi, [2], [13]-192 p.
- ***Libraries*** : CtY, NjP, NNC
- ***Source*** NUC (NR 0495610)
- ***Notes***

1. NUC records further eds.: 1819, 2nd, 288 pp. (CtY, InU, NjP); 1826, 3rd, Auburn, 220 pp. (CtY, MB, NNU, OClWHi).
2. Elson (1964:396) cites (01).

RUPP, J. Daniel **Y SG?**

- **{1836} (01)**
- *Geographical catechism*
- Place/publisher n.r.
- ***Libraries*** n.r.
- ***Source*** Allibone (1854)
- ***Notes***

1. NUC cross references to Rupp, Israel Daniel, but records no special geographies under that author.

S

S., M.
See SYMSON (1702, 1704).

SADLER, T. **Y**

- **{1804} (01)**
- *Geographical lessons for children*
- Place/publisher n.r.
- ***Libraries*** n.r.
- ***Source*** Peddie and Waddington (1914:226)

SALMON, Thomas **Dy**

- **1746 (01)**

▸ *The modern gazetteer: or, a short view of the several nations of the world.* Absolutely necessary for rendering the public news, and other historical occurrences, intelligible and entertaining. Containing, I. An introduction to geography; with directions for the use of the terrestrial globe. II. The situation and extent of all the empires, kingdoms, states, provinces, and chief towns, in Europe, Asia, Africa, and America: also, a description of the most considerable seas, lakes, rivers and mountains, all rang'd in alphabetical order. III. The produce, manufactures, trade, constitution, forces, revenues, and religion, of the several countries. IV. The genealogies and families of the emperors, kings, and princes, now reigning
▸ LONDON; printed for S. and E. Ballard, in Little-Britain; J. and P. Knapton, in Ludgate-Street; D. Browne, without Temple-Bar ; C. Hitch, in Pater-Noster-Row; and A. Millar, in the Strand
▸ 12mo, 17 cm.; pp. [466]
▸ ***Libraries*** **L**, LNC : CLU, ICN
▸ ***Notes***
1. There is a 25-p. introduction to geography, of which 15 pp. are devoted to a 'general description of Old England our native country.' The rest of the work is, as the title states, a geographical dictionary devoted to the description of places, rather than a gazetteer in the modern sense, though many locations in terms of longitude and latitude are given. Not as large as SALMON (1749).
2. BL and NUC record further eds. (all London, S. and E. Ballard until 1759, unpaged): 1747, 2nd (: DLC, ViU); 1756, 3rd, [501] pp. (L : DLC, MWA, PHi); 1757, 4th, [489] pp. (: CtY, DLC, ICRL, MiU, ViU); 1758, 5th, [499] pp. (: CtY, ICU, KU, MH, MnU, NCH); 1759, 6th, [504] pp. (L : MH, NIC, OCl); 1762, 7th, E. Ballard, [506] pp. (: DLC, ICRL, KyHi, MChB, MeB, OClWHi, RPJCB); and 1769, 8th, E. Ballard, [508] pp. (: DLC, NcD, PPL).

▸ **{1773}, 9th, carefully corrected ... (see below) (02)**
▸ *The modern gazetteer ... world ...* 9th ... corrected, with considerable improvements. By Mr Potter. And a new set of maps, viz. the world, Europe, Asia, Africa, North America, South America, and Germany
▸ LONDON; printed for E. Ballard, Bowyer and Nichols, J. Beecroft, W. Strahan, J. Hinton, J. Rivington, W. Johnston, T. Longman, G. Keith, Hawkes, Clarke and Collins, S. Crowder, B. Law, T. Lowndes, T. Caslon, E. and C. Dilly, T. Becket, H. Baldwin, T. Cadel, G. Robinson, W. Domville, H.S. Woodfall, and R. Baldwin
▸ 18 cm.; pp. [528]
▸ ***Libraries*** LSM : MnU : **CaBViPA**
▸ ***Notes***
1. ESTC for British holdings. The Introduction is now 35 pp., of which England occupies the last 25.
2. BL and NUC record further eds. (all unpaged): 1777, 10th, with considerable improvements and alterations. By a gentleman in ... Edinburgh, (: DSI); 1782, 10th, London; E. Ballard (L, LvU : CtY, KyU, MH, ViU). ESTC for extra-L holdings.

SALMON, Thomas **A**
- **1725-38 (01)**
- *Modern history: or, the present state of all nations*. Describing their respective situations, persons, habits, buildings, manners, laws and customs, religion and policy, arts and sciences, trades, manufactures and husbandry, plants, animals and minerals
- LONDON
- 8vo, 20 cm., 31 vols.
- ***Libraries*** GU, **L** : CLU, ICN, MiD, NN, PMA, PPL
- ***Notes***

1. Vols. I to X contain special geography.
2. The earliest complete set housed at the British Library is made up of the volumes listed below. The subtitle of Vol. I is given, together with bibliographic information; thereafter, unless stated otherwise in a note, the contents are summarized.
3. Vol. I is 3rd ed.; the remainder of the set seem to be 1st ed.

Vol. I, 1725, 3rd ed. In which the empire of China; the kingdoms of Japan, Tonquin, Cochin-China and Siam; the Ladrone and Philippine islands, and that of Macassar are comprehended. Illustrated with cuts, and maps, accurately drawn, according to the geographical part of this work, by Herman Moll. Printed for James Crokatt at the Golden Key, near the Inner-Temple-Gate in Fleet-Street. Pp. [24], 464, [10].

There is a second tp., which reads: 'Modern history ... minerals. No. 1. For the month of June, 1724. The second edition. Printed for James Crokatt at the Golden Key, near the Inner-Temple Gate in Fleet-Street; and sold by J. Graves in St. James's-Street, C. King in Westminster-Hall, C. Rivington in St. Paul's Church-Yard, J. Brotherton at the Bible, and J. Clark under the Royal Exchange in Cornhill.'

Vol. II [1725]. Describes Amboyna, Banda, and Molucca islands, Borneo, Java, Sumatra; starts India. Pp. [6], 469, [13].

Vol. III [1726]. Describes India, Pegu, Arragan, Ceylone, Persian Empire. Pp. [8], 469, [15].

Vol. IV [1726]. Describes Persia Arabia, Asiatick Tartary, Turkish empire in Asia, Palestine. Pp. [8], 516, [12].

Vol. V [1717 (sic)]. Describes Egypt, Crim and Little Tartary, Romania, Greece, remainder of Turkey in Europe. Pp. [8], 465, [12].

Vol. VI [1727]. Describes Russian empire, Sweden, Denmark, Norway, and Greenland. Pp. [8], 461, [18]. (The last 35 pp. are devoted to what the author admits is a digression on deep-sea fishing and related topics.)

Vol. VII [1728]. Describes Poland, Bohemia, Silesia, Moravia, Hungary, Transilvania, Sclavonia, Servia, Croatia, German empire in general, and the circles of Austria, Bavaria, Franconia, and Upper Saxony. Pp. [8], 468, [12].

Vol. VIII [1733]. Describes Upper Saxony (northern part), Lower Saxony, Upper and Lower Rhine, Westphalia, United Netherlands. Printed for T. Wotton, at the Three Daggers and Queens Head, over against St. Dunstan's Church, J. Shuckburgh, next the Inner Temple Gate in Fleetstreet, and T. Osborn Jun. in Grays-Inn. Pp. [6], 431, [10].

Vol. IX [1729]. Describes the United Provinces, Austrian and French Netherlands, Switzerland, part of Italy (i.e., Savoy, Piedmont, Genoa, Milan, Mantua, Monteferrat, Modena, Parma, Tuscany). Printed for Tho. Wotton, at ... Head, against St. ... Shuckburgh, next the Rainbow Coffee-House, at the Inner-Temple-Gate; both in Fleet-Street; and T. Osborne. Pp. [8], 460, [20].

Vol. X [1729]. Describes the remainder of Italy (i.e., Republic of Venice, Pope's dominions, kingdom of Naples, Italian islands, islands in the Adriatick, Ionian, and Tuscan seas). Pp. [6], 418, [7]. (Beginning with Vol. X, the history of countries is given, as well as their chorography.)

Vol. XI [1730]. France. Pp. [9], 462, [18]. (Chorography occupies pp. 1-320; the remainder is history.)

Vol. XII [1730]. The history of France to 1730. Pp. [8], 495, [13].

Vol. XIII [1731]. Describes Spain and Portugal. Pp. [7], 550, [10].

Vol. XIV [1731]. Pp. [4], 466, [14].

The tp. is revised to read: 'Modern ... minerals. Vol. XIV. Treats of the islands of Britain, and more at large of the south part of Great Britain call'd England, describing the face of the country, the mountains, forests, seas, rivers, fisheries, soil, vegetables, and the several species of animals it produces. 2. The persons and habits of the natives. 3. Their genius and temper, virtues and vices. 4. Their diet, rural sports and other diversions. 5. Their husbandry and gardening: and, 6thly and lastly, treats of the several antient divisions of England, and the modern division of it into circuits and counties, and herein more particularly of the county of Middlesex and city of London, shewing the antient as well as present state of that metropolis. Illustrated ...

Vol. XV [1732]. Describes London and Westminster [including their histories]. Pp. [4], 468, [12].

Vols. XVI to XXVI are devoted to the history of England:

Vol. XVI [1732]; pp. [7], 492, [14]
Vol. XVII [1732]; pp. [4], 488, [12]
Vol. XVIII [1733]; pp. [4], 456, [24]
Vol. XIX [1733]; pp. [8], 442, [32]
Vol. XX [1734]; pp. [8], 453, [19]
Vol. XXI [1734]; pp. [9], 436, [28]
Vol. XXII [1734]; pp. [8], 443, [28]
Vol. XXIII [1734]; pp. [8], 446, [26]
Vol. XXIV [1734]; pp. [4], 430, [22]
Vol. XXV [1734]; pp. [3], 476
Vol. XXVI [1735]; pp. [8], 468, [30]

Vol. XXVII [1735]. Describes Africa in general; Ethiopia, Zanguebar, Caffraria, Hottentot nations, Congo, Angola, Guinea, Nigritia, Zaara, Biledulgerid, Moroco, Algiers, Tunis, Tripoli, African islands. Pp. [4], 477, [7].

Vol. XXVIII [1736]. Describes Mexico and Florida. Printed for the author: and sold by J. Roberts, in Warwick Lane; and the booksellers in town and country. Pp. [4], 460, [14].

Vol. XXIX [1737]. Describes Peru and Darien. Pp. [4], 457, [13].

Vol. XXX [1737]. Pp. [2], 468, [12].

The tp. is revised to read: 'Modern ... minerals. To which is added, an abstract of the antient history of every nation. Describes: Chili, La Plata, Amazons, Spanish islands, Brazil, British plantations' (Virginia occupies 155 of 157 pp.).

Vol. XXXI [1738]. Describes Virginia con't., remainder of North

America, Caribbean islands (except Spanish), other islands. Pp. [4], 587, [7]. (In this book alone there is no subtitle.)

4. In the Introduction to Vol. I, S. claims to provide the results of a comparative analysis of the conflicting accounts provided by different travellers. He criticizes the ATLAS GEOGRAPHUS (1711-17) for being a compilation of unresolved conflicts, Moll [sic] (presumably the *Compleat geographer*, i.e., (03) and (04) of THESAURUS GEOGRAPHICUS, for being too brief, and the collections of travels by Churchill and Harris for containing too much fabulous material, and also too much trivia.

   In the Preface to Vol. XII, S. notes that his book has been popular, stating: 'I was obliged to reprint my first essays within three months, and now after six years the sale continues so brisk.'

   It seems safe to conclude that the first edition of Vol. I was published in 1724, perhaps in installments. The bound Vol. I, which forms part of the BL holding, may be a third printing (hence the two tps.), but it seems likely that the first eds. of all the later vols. are the ones given here, with the 1717 of Vol. V being a misprint of 1727.

   In the Preface to Vol. XII, S. reports a change in the organization of the contents. His original plan had to been to write the geography first, in 15 vols.; this was to be followed by the modern history of all the countries, starting with that of England in 6 or 7 vols. Starting with this vol. (i.e., XII, France), he is presenting the histories of countries along with their geographies.

   In the Introduction to Vol. XXX, S. rebuts criticisms of his history of Africa. He adds that he intends to publish a supplement to 'bring down the history of every country to the year 1737; and to add to it such further discoveries and improvements as have been made by other travellers since I entered on this work' (p. 31).

5. NUC (NS 0063593) concurs with BL in recording Vol. I as being of the 3rd ed.
6. NUC has 19 entries. Judging by BL, the principal one is NS 0063593, though it records only 32 vols. (The libraries holding copies are those listed above.)

- **{1724-30} [anon] (02)**
- *Modern history; or, the present state of all nations* ... Illustr. with several copper-plates ... with maps of the several countries accurately drawn by Herman Moll
- DUBLIN
- Sm. 4to, 4 vols. (or 5; see note below)
- ***Libraries*** : MdBP
- ***Source*** NUC (NS 0063591), which notes: '*Contents*. Africa, V. 5 – America, V. 5 – Asia, V. 1 – Europe, V. 2, 4'
- ***Notes***

1. The Preface to this ed. is reprinted in the folio ed. of 1739; see note to (03), which follows.
2. For a possible subsequent ed. of this anonymous, 4to version of the work, see PRESENT STATE ... (1739).

- **1739 (03)**
- *Modern history: or, the present state of all nations* ... minerals. Being the most complete and correct system of geography and modern history extant in any language. Illustrated with cuts and maps accurately drawn according to the geographical part of this work, by Herman Moll
- LONDON; printed by Messrs Bettesworth and Hitch in Pater-noster-Row; J. Clarke under the Royal Exchange in Cornhill; S. Birt in Ave Maria Lane; Tho. Wotton over against St. Dunstan's Church, and J. Shuckburgh next the Inner Temple Gate, both in Fleetstreet; and T. Osborne in Gray's Inn
- 25 cm., 3 vols.; Vol. I, pp. [3], iv-xvi, [2], 877, [15]; Vol. II, pp. [3], iv-viii, [1], 900, [23]; Vol. III, pp. [2], iii-xxvi, 784, [23]
- ***Libraries*** BrP : **CtY**, DN, ICN, InU, IU, MiU, NIC
- ***Notes***

1. Vol. I (pp. v-vi) contains 'The Preface to the Quarto Edition.' This is essentially a review of the work that Salmon has done in preparing *The present state*. In the course of this review, there is mention of 'the state of England, which is a distinct work comprehended in thirteen volumes octavo. The two first of which contain a particular description of this kingdom, and its modern inhabitants, with an examination of the characters given of the English by foreigners.' This is clearly a reference to Vols. XIV to XXVI of (01) above. At first sight, it would seem to mean, in turn, that the quarto ed. must have been published between 1735 and 1739. However, a quarto ed. was being published in Dublin - (02) above - in parallel with the octavo first ed. It seems reasonable to interpret Salmon's statement that the section devoted to England constituted a 'distinct work' as a statement of intention, made before 1731, and that, in the final event, the section on England was incorporated in *The present state* as published in octavo. It also seems likely that the Dublin ed. does not include the material presented in the 13 vols. devoted to the British Isles. When that material is excluded, it seems possible that the material presented in the remaining 20 vols. octavo could have been presented in 5 vols. quarto. That supposition is supported by the fact that this, folio ed. contains descriptions of the countries of every continent with the notable exception of the British Isles.

- **1744-6, 3rd (04)**
- *Modern history: or, the present state of all nations* ... [as 1st ed.] ... Moll. The 3rd ed. With considerable additions and improvements, interspersed in the body of the work: also the history and revolutions of each country, brought down to the present time. In three volumes
- LONDON; printed for T. Longman in Pater-noster-Row. T. Osborn, in Grays-Inn. J. Shuckburgh, in Fleet-street. C. Hitch, in Pater-noster-Row. S. Austen, in Newgate-street. And J. Rivington, in St. Paul's Church-Yard
- 37 cm., 3 vols.; Vol. I, pp. [10], v-xii, 777, [12]; Vol. II, pp. [7], 832; Vol. III, pp. [3], ii, 628, [10]
- ***Libraries*** LeU, L, LNC, MP, SanU, SwnU : **CtY**, CU, DLC, FU, GHi, IEG, MWA, NcD, NjP, NjR, OU, Txu, ViU

- **{1739-55}, [Vol. 1, 1755] (05)**
- *Modern history* ...

- DUBLIN; W. Williamson
- 23 cm., 5 vols. in 6
- ***Libraries*** : ICU
- ***Source*** NUC (NS 0063611), which notes: 'Vols. 2-5 have imprint: Dublin, Printed by and for G. Grierson, 1739'

- **{n.d.} (06)**
- *Modern history* ... [as (03)]
- LONDON; printed for Messrs Bettesworth and Hitch, etc.
- 25.5 cm., 3 vols.
- ***Libraries*** : Vi
- ***Source*** NUC (NS 0063590), which notes: '"The present state of Virginia": p. 425-503'

SALMON, Thomas **A**

- **1749 (01)**
- *A new geographical and historical grammar*: wherein the geographical part is truly modern; and the present state of the several kingdoms of the world is so interspersed, as to render the study of geography both entertaining and instructive. Containing, I. A description of the figure and motion of the earth. II. Geographical definitions and problems, being a necessary introduction to this study. III. A general division of the globe into land and water. IV. The situation and extent of the several countries contained in each quarter of the world; their cities, chief towns, history, present state, respective forms of government, forces, revenues, taxes, revolutions, and memorable events. Together with an account of the air, soil, produce, traffic, curiosities, arms, religion, language, universities, bishopricks, manners, customs, habits, and coins, in use in the several kingdoms and states described. Illustrated with a set of twenty-two new maps of the several countries, drawn by the direction of Mr. Salmon, and ingraved by Mr. Jefferys, geographer to his Royal Highness the Prince of Wales
- LONDON; printed for William Johnston, at the Golden-Ball in St. Paul's Church-Yard
- 8vo, 21 cm.; pp. [16], 24, 550, [16]
- ***Libraries*** **L** : CU, IU, MH, ViU : CaBVaU
- ***Notes***

1. A typical special geography, except that the historical section on England includes a chronology that is unusually detailed, with an entry for most years from the accession of Henry VIII in 1509 to the 1630s, and five to ten entries per year thereafter.

   Carries a royal licence for the sole right to print, publish, and vend for fourteen years.

- **1751, 2nd, with very great additions and improvements (02)**
- *A new geographical and historical grammar* ... twenty-three ...
- LONDON; printed for ...
- 8vo, 20 cm.; pp. 608, [16]
- ***Libraries*** LcP, **L** : MH, NcD, NN

- ***Notes***

1. Despite the increase in the number of pages, the book is little altered; only the description of America has received significant additions.

- **{1752}, 3rd, with additions (03)**
- *A new geographical and historical grammar* ... twenty-two ...
- DUBLIN; P. Wilson, J. Exshaw, J. Edsall, R. James, S. Price and M. Williamson
- 20 cm.; pp. 520
- ***Libraries*** : CtY

- **{1754}, 3rd, with very great additions and improvements (04)**
- *A new geographical and historical grammar* ... twenty-three ...
- LONDON; printed for W. Johnston
- 8vo; 640 p., 8 l.
- ***Libraries*** HuC : CtY, ICN, MWA, NN, PP
- ***Source*** NUC (NS 0063626)

- **{1756}, 4th, with ... (05)**
- *A new geographical and historical grammar* ...
- LONDON; W. Johnston
- 640, [16] p.
- ***Libraries*** : CtY, ICU, PHi
- ***Source*** NUC (NS 0063627)

- **1757, 5th, with ... (06)**
- *A new geographical and historical grammar* ...
- LONDON; printed for ...
- 8vo, 20 cm.; pp. 640, [16]
- ***Libraries*** HuC, **L** : CtY, IaU, IU, NGlc
- ***Notes***

1. Judging by the pagination, the description of the world outside Europe is unchanged since 1751.

- **1758, 6th, with ... (07)**
- *A new geographical and historical grammar* ...
- LONDON; printed for William Johnston, in Ludgate-Street and P. Davey and B. Law, in Ave-Mary Lane
- 8vo, 20 cm.; pp. 740, [16]
- ***Libraries*** **L** : CoU, DLC, KU : CaQMNB
- ***Notes***

1. Spot-checking and the pagination suggest that, apart from the chronology of England being brought up to date, no changes have been made since 1754.

- **{1760}, 7th, with ... (08)**
- *A new geographical and historical grammar* ...
- LONDON; W. Johnston [etc.]
- 21 cm.; pp. xiv, 15-640, [16]

- ***Libraries*** : DLC, MH, NcU
- ***Source*** NUC (NS 0063631)

- **{1762}, 8th, with ... (09)**
- *A new geographical and historical grammar ...*
- LONDON; printed .... Ludgate-Street. H. Woodful, J. Hinton, R. Baldwin, W. Strahan, J. Richardson, B. Law, and S. Crowder and Co.
- 21 cm.; pp. 640, [16]
- ***Libraries*** : FU, NNH, PPL, ViU
- ***Source*** NUC (NS 0063635)

- **1764, 9th, with improvements to these present times (10)**
- *A new geographical and historical grammar ...*
- LONDON; printed for W. Johnston in Ludgate-Street; H. Woodfall, J. Hinton, R. Baldwin, W. Strahan, J. Richardson, B. Law, S. Crowder, J. and T. Pote, and W. Nicholl
- 8vo, 21 cm.; pp. xiv, 15-605 [213 misnumbered], [18]
- ***Libraries*** L, LUBd : DLC, OrU
- ***Notes***

1. More type per page probably accounts for most of the reduction in the number of pages. Some additions have also been made, including the proclamation of George III, following the peace treaty of 1763, that provided for the administration of the territories in North America newly acquired by Great Britain.

- **{1766}, 10th, with ... (11)**
- *A new geographical and historical grammar ...*
- LONDON; W. Johnston [etc.]
- 21 cm.; pp. xiv, 15-605, [15]
- ***Libraries*** LeU : DLC, Ia, IU, NNC, OrU, PMA, ScU
- ***Source*** NUC (NS 0063636 and 0063637)

- **{1766}, 12th, with ... (12)**
- *A new geographical and historical grammar ...*
- DUBLIN; P. Wilson [etc.]
- 21 cm.; pp. 646
- ***Libraries*** LeP : Vi
- ***Notes***

1. The Dublin publisher may have decided that this should be counted as the 12th ed. because he knew that, in addition to the ten published in London, there had also been one published in Dublin prior to this one; see (03), above.

- **1767, new ed., with large ... (see below) (13)**
- *A new geographical and historical grammar ...* illustrated with a new set of maps of the countries described, and other copper plates, ten whereof were not in any former edition. A new edition, with large additions, which bring the history down to the present time
- EDINBURGH; printed by Sands, Murray, and Cochran, for James Meuros, bookseller in Kilmarnock

- 8vo, 21 cm.; pp. [1], 7-601, [15]
- ***Libraries*** **L**, SanU : CU, DLC, RPJCB, ViU
- ***Notes***

1. Compared with 1764, the changes are minor, being noted only in the Netherlands, Germany, Poland, and Florida. An advertisement states that the 'intended additions respecting Scotland' were omitted for lack of space.

- **1769, 11th, with very great additions ... (14)**
- *A new geographical and historical grammar* ... illustrated with a set ... present times
- LONDON; printed ... Ludgate Street; W. Strahan; J. Hinton; R. Baldwin; L. Hawes, W. Clerke, and R. Collins; S. Crowder; T. Caslon; T. Longman; B. Law; T. Lowndes; J. and T. Pote; Z. Stuart; W. Nicoll; G. Robinson and J. Roberts; T. Cadell, and S. Bladon
- 8vo, 21 cm.; pp. xiv, 15-615, [15]
- ***Libraries*** **L**, ShP : CtY, CU, DLC, MWiW, NBu, NN, NPV, OCU, PHi, TxU
- ***Notes***

1. This ed. has the additional material referred to in the title of the Edinburgh ed. of 1767, plus further additions.

- **{1771}, 14th, with ... (see below) (15)**
- *A new geographical and historical grammar* ... countries described. The 14th ed., with very large additions and improvements brought down to the present time
- EDINBURGH; Wilson and Darling
- 22 cm.; pp. xii [2] 7-603 [13]
- ***Libraries*** : NcD

- **1772, 12th, with great amendments and improvements (16)**
- *A new geographical and historical grammar*; containing the true astronomical and geographical knowledge of the terraqueous globe: and also the modern state of the several kingdoms of the world; under these four heads: I. The astronomy of the solar system, and particularly of the earth. II. Universal geography, shewing the diverse circumstances relating to the earth, water, and atmosphere. III. Geographical elements exemplified in definitions, problems, theorems, and paradoxes. IV. Particular geography, concerning the natural and political parts of Europe, Asia, Africa, and America; wherein the climate, productions, and people; the customs, policy, manufactures, and traffic; the religion, strength, history and other particulars, of all nations, are treated of, in a manner tending to render the study of geography entertaining and instructive. Illustrated with twenty-five maps and plates elegantly executed. The twelfth edition, with great amendments and improvements, by Mr. Robertson
- LONDON; printed for C. Bathurst; W. Strahan; J. and F. Rivington; W. Johnston; J. Hinton; L. Hawes, W. Clarke, and R. Collins ; T. Davies; S. Crowder; T. Longman; T. Caslon; B. Law; T. Lowndes; J. and T. Pote; Z. Stuart; W. Nicoll; G. Robinson; T. Cadell; and R. Baldwin
- 8vo, 21 cm.; pp. xv, [1], 696
- ***Libraries*** **L** : CLU, CtY, MiU, MnU, OGaK, RPJCB, ViU : CaAEU

▸ *Notes*

1. The short introductory section on general geography has been entirely rewritten, and increased in length from 23 to 72 pp. Some additions to, and reorganization of, material in the chronological material have been made in the case of some countries.

▸ **{1780}, new, with large additions (17)**
▸ *A new geographical and historical grammar* ... and the present state of the several kingdoms of the world ... Illustrated with a new set of maps ... and other copper-plates
▸ EDINBURGH; Murray and Cochran
▸ 22 cm.; 1 p. l., [v]-xii, [2], 7-629, [15] p.
▸ *Libraries* : DLC, NcD
▸ *Source* NUC (NS 0063644)

▸ **1785, 13th (18)**
▸ *Salmon's geographical and astronomical grammar*, including the antient and present state of the world; and containing 1. The Newtonian system of the planets. 2. A particular view of the earth. 3. Geographical elements, exemplified in definitions, problems, theorems, and paradoxes. 4. The grand divisions of the globe. 5. The extent of empires, kingdoms, states, provinces, and colonies; with an account of their climates, animals, birds, metals, minerals, rivers, bays, and natural curiosities. 6. Origin and history of nations, forms of government, religion, laws, revenues, commerce, and taxes. 7. Their language, genius, revenues, customs, and public buildings. 8. An account of the new discoveries in the South Seas. 9. A geographical table, in which is given the longitude, latitude, and bearings, of the principal places in the world. 10. The coins of the various nations, and their value in English money. 11. A chronological table of remarkable events. 12. A list of the men of learning and genius. The thirteenth edition; with considerable corrections and additions, in which the history of the various countries in every quarter of the globe is continued to the year 1785, including a full account of the new discoveries, and illustrated with maps, and other plates, elegantly executed
▸ LONDON; printed for C. Bathurst ... Rivington, S. Crowder, T. Longman, B. Law, J. and T. Pote, C. Dilly, G.G.J. and J. Robinson, T. Cadell, R. Baldwin, J. Nichols, J. Sewell, W. Goldsmith, W. Nicoll, J. Murray, J. Bew, W. Lowndes, Scatcherd and Whitaker, and W. Stuart
▸ 8vo, 21 cm.; pp. vi, 770
▸ *Libraries* L : MChB, MiD, MH, MWA, N : CaAEU, CaOTP
▸ *Notes*

1. The change to the opening half-dozen words of the title, followed by the further changes in the body of the title, all suggest that this is a new work. However, the fact that the book is identified as the 13th ed. on the tp. argues against that conjecture. An examination of the contents confirms the opinion that this is an ed. of Salmon's *New geographical and historical grammar*. The opening section on astronomy and physical geography is very little, if at all, altered from 1772. There are greater changes in the special geography, but in general they can be considered as revisions rather than as completely new text. The section on Europe has been reduced from 347 to 284 pp., Asia from 96 to 90; Africa is unchanged at 50 pp., and America increased from

115 to 145 pp. In addition, 75 pp. are devoted to the new discoveries in the Pacific.
2. Discussed in Warntz (1964:59, 128-9). Wright (1959:150-1) discusses (01).
3. Microform versions of (01) and (07) are available from CIHM.

SALMON, Thomas **A**

- **1752-3 (01)**
- *The universal traveller: or, a compleat description of the several nations of the world.* Shewing I. The situation, boundaries, and face of the respective countries. II. Number of provinces and chief towns in each. III. The genius, temper, and habits of the several people. IV. Their religion, government, and forces by sea and land. V. Their traffick, produce of their soil, animals, and minerals. VI. An abstract of the history of each nation. Brought down to the present time. And illustrated with a great variety of maps and cuts
- LONDON; printed for Richard Baldwin, at the Rose in Pater-Noster-Row
- Folio, 36 cm., 2 vols.; Vol. I (1752), pp. [11], 3-692; Vol. II (1753), pp. [10], 3-773, [4]
- ***Libraries*** **L**, NoP : CtY, DLC, IaU, PSt
- ***Notes***

1. Vol. I deals with Asia and eastern, central, and northern Europe. Some history is usually, though not invariably, given. Vol. II deals with western and southern Europe, except for the British Isles, which are entirely omitted, and also Africa and America; varying amounts of history are presented. Of all Salmon's works this has the least history. The contrast with the *Grammar* (SALMON 1749) is marked; so too is the subsequent record of publication for the two works.

- **{1753-5} (02)**
- *The universal traveller* ... several foreign nations ... II. The number ...
- LONDON
- ***Libraries*** GU
- ***Source*** Letter from GU

- **{1755} (03)**
- *The universal traveller* ... IV. The religion ... V. The trafick ...
- LONDON; printed for Richard Baldwin
- Folio, 2 vols.
- ***Libraries*** : MH
- ***Source*** NUC, and Beeleigh Abbey Books, *A Catalogue of Voyages and Travels*, Catalogue BA/23, item 67, n.d. [1975]

SALMON, Thomas; ed., J. TYTLER **A**

- **{1777} (01)**
- *The new universal geographical grammar* ... and history of all the different kingdoms of the world ... Illus. with a new and correct set of maps of the countries described. The whole being an improvement and continuation of Mr. Salmon's Grammar. Brought down to the present time by J. Tytler
- EDINBURGH; W. Darling

- 21 cm.
- ***Libraries*** : CtY (imperfect, all after 696 wanting)
- ***Source*** NUC (NS 0063648)

- **1778, new (02)**
- *A new universal geographical grammar*: wherein the situation and extent of the several countries are laid down according to the most exact geographical observations, and the history of all the different kingdoms of the world, is interspersed in such a manner, as to render the study of geography both useful and entertaining. Under these three heads: I. A compendious study of astronomy. II. The geographical definitions, problems, and general divisions of the earth, necessary to be understood as an introduction to this grammar. III. A particular description of the countries contained in each quarter of the world; their cities, chief towns, respective forms of government, forces, revenues, taxes, and history. Together with an account of the air, soil, produce, traffic, curiosities, arms, religion, language, universities, bishopricks, manners, customs, habits, and coins, in use in the several kingdoms and states treated of. Illustrated with a new and correct set of maps of the countries described. The whole being an improvement and continuation of Mr. Salmon's Grammar. Brought down to the present time by J. Tytler
- EDINBURGH; printed and sold by W. Darling:– J. Milliken, Whitehaven; and most other booksellers in Great Britain
- 8vo, 22 cm.; pp. xii, 704
- ***Libraries*** BrP, **L**
- ***Notes***

1. Compared with the 1772 ed. of SALMON (1749), the astronomical section is shorter and much of the physical geography is omitted. In their place 'a short abridgement of ancient [i.e., biblical and classical] history' has been added. Spot-checking suggests that the special geography has been much revised. It is because of the latter changes that Tytler's edition is treated as an independent work in this guide.

- **{1781}, 2nd, with large additions (03)**
- *A new universal geographical grammar ...*
- EDINBURGH; J. Spottiswood
- 22 cm.; pp. xii, 13-770
- ***Libraries*** : ViLxW
- ***Source*** NUC (NS 0063649)

- **1782 (04)**
- *The new universal geographical grammar ...* treated of. And a chronological table of remarkable events from the creation to the present time
- EDINBURGH; printed for J. Spottiswood, bookseller
- 8vo, 22 cm.; pp. [1], xii, 13-770
- ***Libraries*** L : DLC
- ***Notes***

1. There are slight changes in the section on astronomy and physical geography; spot-checking reveals some changes in the special geography since 1778. The tp. identifies it as 'the second edition, with large additions.' When compared with the 1785 ed. of SALMON (1749), numerous passages are found to be

identical, but there are also sizable differences. Overall, the two books are closely equivalent in size, and countries are dealt with in almost the same order; those in Europe are dealt with in considerably greater detail in this work than in the one by Salmon; outside Europe the balance is reversed. See TYTLER (1788).

SARGEANT, Anne Marie **Y SG?**
- **{[1850]} (01)**
- ... *Papa and mamma's easy lessons in geography*; or, the elements of geography in a new and attractive form
- LONDON; Thomas Dean and son
- 12mo, 16.5 cm.; pp. vi, 90
- ***Libraries*** : : CaOTP
- ***Source*** St. John (1975, 2:812), which notes: 'At head of title: A Companion to Miss Corner's "Play Grammar"'
- ***Notes***

1. The date is confirmed by BCB, Index (1837-57:110).

SCHOLASTIC GEOGRAPHY **?**
- **{1875} (01)**
- *Scholastic geography*, with maps
- ? Stewart
- ***Libraries*** n.r.
- ***Source*** ECB, Index (1856-76, index to 1875)

SCOTT, Joseph **Y Am**
- **{1807} (01)**
- *Elements of geography, for the use of schools ...*
- PHILADELPHIA; printed by Kimber, Conrad & co.
- 17.5 cm.; 1 p. l., [ii]-iii, 165 p.
- ***Libraries*** : DGU, DLC
- ***Source*** NUC (NS 0344116)
- ***Notes***

1. Discussed in Nietz (1961:208, 211), and Sahli (1941, *passim*).

SCOTT, Joseph **Dy Am**
- **{1799-1800} (01)**
- *The new and universal gazetteer*; or, modern geographical dictionary ... To which is added, a new and easy introduction to geography and astronomy; with a nomenclature, explaining the essential terms in each science. Illustrated with twenty-five maps, an armillary sphere, and several diagrams
- PHILADELPHIA; printed by Patterson & Cochran, no. 108, Race-street
- 22 cm., 4 vols.
- ***Libraries*** : DLC, DNLM, IaU, KyU, MB, MWA, MWHi, N, NcD, NHi, NjR, NWM, O, OAk, OCl, OHi, PHi, PPF, PPL, PU, VtU
- ***Source*** NUC (NS 0344125)

SCOTT, William ?

- **{1804}, 2nd (01)**
- *Compendium of geography: containing* its general principles and an account of all the countries of the earth. 2nd ed.
- Place/publisher n.r.
- ***Libraries*** n.r.
- ***Source*** Allibone (1854), and James Thin, Catalogue 331, *General Second-hand and Antiquarian*, item 2069

SCOTT, William **Y**

- **{[1812]}, 5th, improved ... (see below) (01)**
- *A new compendium of geography*: containing its general principles, and an account of all the countries of the earth, [their divisions, towns, rivers, lakes, mountains, bays, straits, capes, islands, etc., with an appendix consisting of tables of latitude and longitude, population of countries, towns, etc.: intended chiefly for the use of schools] ... 5th ed., impr. by an entire set of new maps, and other important particulars
- EDINBURGH; printed by John Moir
- 18 cm.; pp. 270
- ***Libraries*** : ViU
- ***Source*** NUC (NS 0352161), which supplies the date
- ***Notes***

1. The words in [ ] are derived from CIHM, which records 1816, 6th (: : CAQQS).
2. Allibone (1854) also records Scot[sic], William, *System of Geography*, London, 12mo, n.d.
3. A microform version of the ed. of 1816 is available from CIHM.

SCOTTISH SCHOOL-BOOK ASSOCIATION **Y**

- **[1842] (01)**
- *A complete system of modern geography for the use of schools*, with copious exercises on each section; also, an outline of sacred geography, and numerous problems on the globes. Drawn up for the Scottish School-Book Association. Being No. XII of their new series of school-books
- EDINBURGH; William Whyte, & Co. booksellers to the Queen Dowager, 13, George Street: Longman, Brown & Co., London; W. Grapel, Liverpool; and W. Carey, Jun., & Co., Dublin
- 8vo, 18 cm.; pp. [3], iv, 332
- ***Libraries*** L
- ***Notes***

1. It is a special geography, but of the most arid kind, being little more than a list of facts, rather than normal prose. There is a short astronomical section at the beginning, and one on the use of globes at the end.
2. BL supplies the date.

SCRIBONIUS, Gulielmus Adolphus **OG**

- **1621 (01)**

▸ *Naturall philosophy: or a description of the world*, namely, of angels, of man, of the hauens, of the ayre, of the earth, of the water: and of the creatures in the whole world
▸ LONDON; printed by I.D. for Iohn Bellamie, and are to be sould at south entrance of the Royall Exchange
▸ 4to, 18 cm.; pp. [4], 64
▸ ***Libraries*** **L**, Lwe, O, OL : CSmH, CtY, DFo, NjP, PP
▸ ***Notes***
1. Not a special geography. It contains a fair amount of what would now be classified as physical geography. There was a later ed. in 1631.

SEALLY, John **Dy**
▸ **{[1781]} (01)**
▸ *A complete geographical dictionary*; or, universal gazetteer of ancient and modern geography, containing a full ... description of the known world in Europe, Asia, Africa and America ... The geographical parts by ... [and] the astronomical parts from the papers of Israel Lyons
▸ LONDON; J. Fielding
▸ 27 cm., 2 vols.
▸ ***Libraries*** : CtY
▸ ***Source*** NUC (NS 0365123)

▸ **1787 (02)**
▸ *A complete geographical dictionary* ... gazetteer; of ... geography: containing a full, particular, and accurate description of the known world; in Europe, Asia, Africa, and America: comprising a complete system of geography, illustrated with correct maps and beautiful views of the principal cities, etc. and chronological tables of the sovereigns of Europe. The geographical parts by ... member of the Roman Academy; author of the Histoire chronologique, sacré et profane; Elements of Geography and Astronomy, etc. etc. Interspersed with extracts from the private manuscripts of one of the officers who accompanied Captain Cook in his voyage to the southern hemisphere. By the King's royal licence and authority
▸ LONDON; printed for Scatcherd and Whitaker, Ave-mary Lane
▸ 4to, 25.5 cm., 2 vols.; Vol. I, pp. [4], iv, and collates A-Z, Aa-Zz, Aa-Zzz, 4A-4Z ... 12A (verso); Vol. II, a-Z ... 10A-10O, xxiv
▸ ***Libraries*** **L**
▸ ***Notes***
1. A geographical dictionary. The entries vary in length from two lines for a small town to four or more pages for a country or a continent. The longest entry noted (7 pp.) was for Senegal; this unusually great length was apparently inspired by the recent publication, in Paris, of a book written by a Mr Adanson, who had recently returned from a stay of five years in that country.

SELLER, John **At**
▸ **[1679] (01)**
▸ *Atlas minimus; or, a book of geography* shewing all the empires, monarchies,

kingdoms, regions, dominions, principalities and countries, in the whole world
- LONDON; And are to be sold at his [i.e., Seller's] house at the Hermitage in Wapping, and in Pope's Head Alley in Cornhill, Lon.
- 16mo, 12 cm.; [54] l.
- ***Libraries*** **L**, O : CLU, CSmH, CtY, DFo, DLC, IU, NBu, ViU, WU
- ***Notes***

1. An atlas rather than a special geography. The pages facing the maps carry some textual material, but it is chiefly tables of some of the places to be found on the maps opposite.
2. See ATLAS MINIMUS (1758).

SELLER, John **At?**

- **{[1676?]} (01)**
- *Atlas terrestris*: or a book of mapps, of all the empires, monarchies, kingdoms, regions, dominions, principalities, and countreys in the whole world accommodated with a brief description, of the nature and quality of each particular countrey
- [LONDON]; and are to be sold at his shopps, in Wapping at the Hermitage: and in Exchange-Alley near the Royall-Exchange
- Folio, 52 cm.; 1 p. l., 28 pp. of text [some numbered, some not], in sets of 4 pp. between maps
- ***Libraries*** : RPJCB
- ***Source*** NUC (NS 0404133), which supplies the date, and notes: 'The 4-page texts inserted between the maps are titled respectively: A description of the world, and of the moon, and of the three systems of Ptolemy, Tycho, and Copernicus; A geographical description of the earth, and first of Europe; A relation of the French conquests in the Netherlands, in the years 1672, and 1673; A description of the Seventeen Provinces ...; A geographical description of Asia; A geographical description of Africa; A geographical description of America.'

- **{[1700?]} [anon] (02)**
- *Atlas terrestris ...*
- LONDON; J. Seller
- Obl. 24mo, 2 vols.
- ***Libraries*** : DLC
- ***Source*** NUC (NS 0414136), which notes: 'Accompanying text: A new system of geography. 40 pp. This was announced in the Term catalogue, February, 1685.' Also: 'Map no. 2 V. 1, dated 1700; no. 21, by Seller, June.'
- ***Notes***

1. NUC also records three other versions of this work, all dated about 1700 (CtY; CtY, IU; MiU). In one of those at CtY, the text amounts to little more than captions to the maps.

SELLER, John **mSG**

- **[1685] (01)**
- *A new systeme: of geography, designed in a most plain and easy method*, for

the better understanding of that science. Accommodated with new mapps, of all the countreys, regions, empires, monarchies, kingdoms, principalities, dukedoms, marquesates, dominions, estates, republiques, soveraignties, governments, seignories, provinces, and countreys in the whole world. With geographical tables, explaining the divisions in each mapp

- LONDON; Are to be sold at his [i.e., Seller's] shop on the west-side of the Royal Exchange
- 8vo, 15 cm.; pp. [4], 112 [numbered, but 30 followed by 33]
- *Libraries* L, O : CtY, DLC, MiU : **CaAEU**
- *Notes*

1. BL supplies the date, presumably on the basis of the announcement in the Term catalogue; see note to (02) of SELLER (1676?).

   The descriptions of the countries are brief, and are unusual in that they do not vary much in length from one country to another. It resembles ORTELIUS (1602?) in being an atlas with text rather than a lengthy text with some maps. However, in this work the maps and text do not face one another but are separated, with the text occupying the front of the book, and the maps the back.
2. NUC also records what may be another version of this issue (NS 0404176; PPiU). In this entry there is no colon after system in the title.

   ESTC and NUC record further eds. (all London): 1690, 112 [i.e., 110] pp. (LDW : CtY); 1690, J. Seller, 112 pp. (: CtY, DLC, PPiU); 1694?, Wapping [i.e., London], 112 pp. (C : CtY, CU, DLC, MH, N); 1703, 3rd , J. Seller, C. Price, and J. Senex, 40 pp. (: CSmH, ICN); and 1769, Seller (: PPiU).

SENEX, John **mSG**

- **1717 (01)**
- *An introduction to geography in two parts*. Part I. Of the definition of geography; of the figure of the earth; definitions; of the latitude of places, or the elevation of the Pole, and of the division of the earth into five zones; of climates; of the four seasons of the year; of the divisions of the earth according to the shadows of bodies; of longitude; of the dimensions or greatness of the earth; of the motions of the earth, and its situation in the system of the world; of mountains; of mines, woods and desarts; of the properties of the ocean, and its parts; of lakes and marshes; of rivers; of mineral waters; of the change of the watery surface into that of dry land, and the contrary. Part II. Of the natural division of the earth, made by the investing ocean; of the division of the ocean by the earth. Of the several religions profess'd in the several parts of the earth. A short account of the several considerable nations in Europe, and the interest they have in each other. A curious description of Asia, Africa, and America; in which is contain'd all that is in any ways remarkable, either with respect to its inhabitants, its antiquities, or its natural history. N.B. These sheets being design'd to be pasted under the maps of the world and quarters publish'd by J. Senex, are now made up into books, for the use of those who have bought the maps without them
- LONDON; sold by J. Senex, at the Globe in Salisbury-Court near Fleetstreet
- 39.5 cm.; Part I, pp. [18]; Part II, pp. [17]
- *Libraries* [IDT]

- ***Notes***

1. The IDT copy is made up of pages mounted individually. Part I is entitled 'An introduction to general geography.' Part II is entitled 'A general description of the world.'
2. See SENEX (1718), below, for what is presumably the complete work.

SENEX, John **At**

- **{1708-25} (01)**
- *Modern geography: or all the known countries of the world.* Laid down from the latest observations and discoveries, communicated to the Royal Society of London and Academy of Sciences at Paris. To which is added a geography of the antient world ...
- LONDON; for T. Bowles
- 69 × 29 cm.; 34 folding maps
- ***Libraries*** : ICN, MWiW
- ***Source*** NUC (NS 0414508), which notes: 'Title-page wanting; title supplied from Library of Congress. A list of geographical atlases. No. 550.'
- ***Notes***

1. MWiW confirms that this work is an atlas, not a special geography. ICN reports that the dates of publication mark the earliest and latest dates of publication of the maps that form the atlas.

SENEX, John

- **1721**
- *A new general atlas* ... See NEW GENERAL ATLAS.

SENEX, John **mSG**

- **{1718} (01)**
- *A treatise of the description and use of both globes*, to which is annexed a geographical description of our earth
- LONDON; printed for J. Senex and W. Taylor
- Pp. 114
- ***Libraries*** : CU
- ***Source*** NUC (NS 0414525)

SERVICE, John Paterson **Dy**

- **1787 (01)**
- *Recreation for youth: a useful and entertaining epitome of geography and biography.* The first part comprising a general view of the several empires, kingdoms, republics, states, remarkable islands, mountains, seas, rivers, and lakes; with their situation, extent, capitals, population, produce, arts, religion, and commerce: including the discoveries of Capt. Cook, and others. The second part including the lives of the most eminent men who have flourished in Great-Britain, and its dependencies.
- LONDON; printed for G. Kearsley, No. 46, Fleet-Street
- 12mo, 14.5 cm.; pp. [3], iv-xi, 396

- *Libraries* **L**
- *Notes*

1. The Preface is at odds with the title, in that it suggests that the intended audience were adults concerned with self-improvement, though malapropisms make the interpretation uncertain. Pp. 1-72 contain a geographical dictionary; pp. 73-88 a 'treatise on natural geography,' much of which is taken up with lists of features of physical geography. The remainder of the book is a dictionary of biography.

SHAND, Alexander **Y**

- **1833 (01)**
- *Outlines of modern and ancient geography*: with an introduction to astronomy by the use of globes, and rules for ascertaining the places of the principal fixed stars, to which are added tables of ancient measures, adapted to burgh and parochial schools
- EDINBURGH; A. Macredie
- 18 cm.; pp. vi, 126
- *Libraries* : : **CaQQS**
- *Notes*

1. Special geography of a condensed type occupies pp. 3-96.
2. A microform version is available from CIHM.

SHARMAN, John **A**

- **1794, 2nd ... (see below) (01)**
- *An introduction to astronomy, geography, and the use of the globes*. The second edition, considerably enlarged and improved. By ... teacher of geography, etc.
- DUBLIN; printed by John Cherrurier, No. 128, Capel-Street
- 12mo, 17 cm.; pp. [3], 4-348
- *Libraries* **L**
- *Notes*

1. The introduction to astronomy occupies pp. [3]-46; the special geography, pp. 72-276, and the use of globes, pp. 282-336. Proceeds by way of question and answer. Relatively readable, with the text outweighing tables and quasi-lists. Provides information with respect to location and boundaries, selected features of the physical environment, chiefly rivers and those related to other bodies of water, plus a miscellany of features of human geography.
2. BL records two other eds. (both Dublin): 1801; 1818, 7th.

SHARPE, Gregory **mSG tr**

- **1758, 2nd, corrected and enlarged (01)**
- *An introduction to universal history*. Translated from the Latin of Baron Holberg. The second ... with notes historical, chronological, and critical ... To which is prefixed a short system of geography, with maps, etc.
- LONDON; printed for A. Millar in the Strand, J. Ward in Cornhill, and A. Linde in Catherine Street
- 21 cm.; pp. [7], vii-xxi, 4, [8], 341

- *Libraries* **L** : NbU, CtY, ICU, IEN, MH, PPL, PU
- *Notes*

1. 'The short system of geography prefixed is a translation from Holberg, in which the same freedom is used with him, as he took with Jacoboeus, who was the author of his geographical compendium' (p. 2, 1st series). The geography, which is special and strongly Eurocentric, occupies pp. 1-42.

SHAW, Benjamin F. **Y Am**

- **{1854}, 4th (01)**
- *Primary geography on the basis of the object method of instruction* ed. 4 [sic]
- PHILA[DELPHIA]; Lippincott
- Pp. 56
- *Libraries* : OO
- *Source* NUC (NS 0477588), which notes: '(Lippincott's geographical ser.)'

- **{1862} (02)**
- *Primary geography* ..
- PHILADELPHIA; J.B. Lippincott & co.
- 25.5 cm.; pp. 55 [1]
- *Libraries* : DLC
- *Source* NUC (NS 0477589), which notes: 'Written by Shaw but published under Allen's name'
- *Notes*

1. NUC records 3rd ed., 1863 (CU). See Allen, F. (1862).

SHAW, Benjamin F., and Fordyce A. ALLEN **Y Am**

- **{1864} (01)**
- ... *A comprehensive geography* combining mathematical, physical, and political geography with important historical facts ...
- PHILADELPHIA; J.B. Lippincott & co.
- 34 cm.; pp. 114
- *Libraries* : DLC
- *Source* NUC (NS 0477585), which notes: 'L.C. copy replaced by microform'
- *Notes*

1. ECB, Index (1856-76:130) records an ed. in 1865, Philadelphia.

- **{1866}, 2nd (02)**
- *A comprehensive geography* ... historical facts designed to promote the normal growth of the intellect ...
- PHILADELPHIA; J.B. Lippincott
- 34 cm.; pp. 114
- *Libraries* : TNJ
- *Notes*

1. TNJ report that the Preface states that this book is intended as a school book to act as a sequel to 'the Primary geography' (presumably SHAW 1854*, above). The table of contents shows that the book is divided into seven parts: natural history, ancient history, medieval history, mathematical geography,

physical geography (occupying the first 42 pp.), and physical and political geography. The facts presented in the last section are organized in terms of continents.

[SHAW, Benjamin F.] and Fordyce A. ALLEN **Y Am**

- **{1862} (01)**
- *An oral geography for junior and primary schools* on Pestalozzian principles, containing only the pictorial maps and natural history engravings of the author's primary geography ...
- PHILADELPHIA; J.B. Lippincott & co.
- 24.5 cm.; pp. 4, [7]
- ***Libraries*** : DLC
- ***Source*** NUC (NS 0477587), which notes: '(On cover: Lippincott's geographical series)'

SHERWOOD, M.M. **?**

- **{1818} (01)**
- *Introduction to geography*
- Place/publisher n.r.
- ***Libraries*** n.r.
- ***Source*** Peddie and Waddington (1914:226)

SHOBERL, Frederic, and W.H. PAYNE, eds. **mSG**

- **1821-5 (01)**
- *The world in miniature*
- LONDON; printed for R. Ackerman, 101, Strand, and to be had of all booksellers
- 12mo, 15 cm., 37 vols.
- ***Libraries*** **L** : CSmH, CtY, CU-S, DLC, MnU, NBU, NjP, NN, NNC, OClWHi, OrP, PHi, PPF, Vi, ViU, WaS, WU : CaBVaU
- ***Notes***

1. In general this work is less concerned with political and economic topics, and more with sociological ones, than most special geographies. Popular rather than scholarly in tone. See FRANSHAM (1740) for a similar title, and possibly a source. Although both BL and NUC list Shoberl as the sole editor, W.H. Payne edited the volumes on England, Scotland, and Ireland.

   The basic entry in NUC is NS 0514083, which at first sight seems to list both the libraries holding complete sets of the work and also the subsections of the complete work. The libraries in question are those listed above. The entry also states the year of publication and number of volumes devoted to each of the following regions: Africa, Asiatic islands and New Holland, Austria, Hindoostan, Illyria and Dalmatia, Japan, Netherlands, Persia, Russia, South Sea islands, Spain and Portugal, Switzerland, Tibet and India beyond the Ganges, and Turkey.

   A comparison of this list with that of the countries given below shows that it does not include the volumes on China, or on England, Scotland, and Ireland. Most of the volumes devoted to each of the countries listed in NS

0514083 are also listed separately under their respective titles (e.g., Africa is NS 0514002).

In the list that follows, the subtitles of the individual volumes are given, as well as other information derived from all the sources available in NUC, not just from NS 0514083.

Africa, containing a description of the manner and customs, with some historical particulars of the Moors of the Zahara, and of the negro nations between the rivers Senegal and Gambia: illustrated with two maps, and forty-five coloured engravings. Printed for R. Ackermann ... 4 vols. [1821 according to NUC]. Vol. I, pp. [5], iv-xii, [6], 180 (includes the table of contents for all 4 vols. on Africa); Vol. II, pp. [2], 170; Vol. III, pp. [2], 168; Vol. IV, pp. [2], 184.

The Asiatic islands and New Holland: being a description of the manners, customs, character, and state of society of the various tribes by which they are inhabited: illustrated by twenty-six coloured engravings. In two volumes. Printed for R.A. Ackermann, Repository of Arts, Strand, and to be had of all booksellers. 2 vols. [1824 according to NUC]; Vol. I, pp. [5], iv-vii, [3], 291; Vol. II, pp. [2], 289.

Austria, containing a description ... character and costumes of the people of that empire. Illustrated by thirty-two coloured engravings. In two volumes. Printed ... 2 vols. [1823 according to NUC]; Vol. I, pp. [3], iv-xii, 161, [1]; Vol. II, pp. [2], 200.

China, containing illustrations of the manners ... empire. Accompanied by thirty coloured engravings. In two volumes. Printed ... 2 vols. [1823 according to NUC]; Vol. I, pp. [3], iv-xiv, 208; Vol. II, pp. [2], 257.

England, Scotland, and Ireland, edited by W.H. Payne. Containing a description of the character, manners, customs, dress, diversions, and other peculiarities of the inhabitants of Great Britain. Illustrated with eighty-four coloured engravings. Printed ... 4 vols., 1827; Vol. I, pp. [3], iv-viii, [6], 288; Vol. II, pp. [3], 288; Vol. III, pp. [2], 288; Vol. IV, pp. [2], 252.

Hindoostan, containing a description of the religion, manners, customs, trades, arts, sciences, literature, diversions, etc. of the Hindoos. Illustrated with upwards of one hundred coloured engravings. In six volumes. Printed ... 6 vols. [1822 according to NUC]; Vol. I, pp. [5], vi-xxxix, [1], 187, [1]; Vol. II, pp. [2], 273; Vol. III, pp. [2], 324; Vol. IV, pp. [2], 216; Vol. V, pp. [2], 234; Vol. VI, pp. [2], 240.

Illyria and Dalmatia; containing a description of the manners, customs, habits, dress, and other peculiarities characteristic of their inhabitants, and those of adjacent countries; illustrated with thirty-two coloured engravings. Printed ... 101, Strand, and to be had ... 2 vols. [1821 according to NUC. An advertisement (pp. [vi]-xviii) identifies these as the first volumes of the set to be published]; Vol. I, pp. [7], viii-xviii, 146; Vol. II, pp. [2], 168.

Japan, containing illustrations of the character, manners, customs, religion, dress, amusements, commerce, agriculture, etc. of the people of that empire. With twenty coloured engravings. Printed ... Repository of Arts, Strand ... 1 vol. [1821, another ed., 1823 according to NUC]; pp. [3], iv-xii, [1], 286.

The Netherlands, containing a description of the character, manners, habits, and costumes of the inhabitants of the late seven United Provinces, Flanders and Brabant. Illustrated with eighteen coloured engravings. Printed ... 1 vol. [1823 according to NUC]; pp. [3], iv-viii, [2], 241.

Persia, containing a brief description of the country; and an account of its government, laws, and religion, and of the character, manners, customs, arts, amusements, etc. of its inhabitants. In three volumes. Printed ... 3 vols. [1822 according to NUC]; Vol. I, pp. [3], iv-xviii, 240, [1]; Vol. II, pp. [2], 236, [1]; Vol. III, pp. [3], 233.

Russia, being a description of character, manners, customs, dress, diversions, and other peculiarities of the different nations inhabiting the Russian empire. In four volumes. Illustrated with seventy-two coloured engravings. Printed ... 4 vols. [1822-3 according to NUC]; Vol. I, pp. [3], vii, [1], 181, [2]; Vol. II, pp. [2], 294, [1]; Vol. III, pp. [2], 204, [1]; Vol. IV, pp. [3], x-xvii, [1], 267.

South sea islands: being a description of the manners, customs, character, religion, and state of society among the various tribes scattered over the great ocean, called the Pacific, or the South Sea: illustrated with twenty-six coloured engravings. In two volumes. Printed ... 2 vols. [1824 according to NUC]; Vol. I, pp. [3], iv-xvi, 320; Vol. II, pp. [2], 325.

Spain and Portugal, containing a description of the character, manners, customs, dress, diversions, and other peculiarities of the inhabitants of those countries. In two volumes; illustrated with twenty-seven coloured engravings. Printed ... 2 vols. [1825 according to NUC]; Vol. I, pp. [3], iv-x, 303; Vol. II, pp. [3], 281, [1].

Switzerland; containing .... customs, diversions, dress, etc. of the people of that country in general, and of the inhabitants of the twenty-two cantons in particular. Illustrated by eighteen coloured engravings. Printed ... 1 vol., n.d.; pp. [3], iv-vi, [2], 287, [1].

Tibet and India beyond the Ganges; containing a description of the character ... customs, dress, religion, amusements, etc. of the nations inhabiting those countries: illustrated with twelve coloured engravings. Printed ... 1 vol. [1821? and 1824 according to NUC]; pp. [1], xi, [1], 352.

Turkey, being a description of the manners, customs, dresses, and other peculiarities characteristic of the inhabitants of the Turkish empire; to which is prefixed a sketch of the history of the Turks: translated from the French of A.L. Castellan, author of Letters on the Morea and Constantinople, and illustrated with seventy-three engravings containing upwards of one hundred and fifty costumes. In six volumes. Printed for R. Ackermann, 101, Strand. 6 vols. [1821 according to NUC]; Vol. I, pp. [3], ii-xv, [1], 211; Vol. II, pp. [2], 237; Vol. III, pp. [2], 264; Vol. IV, pp. [2], 300; Vol. V, pp. [2], 234; Vol. VI, pp. [2], 244.

2. NUC also lists an ed. of 1827 [NS 0514084 (PPL)]; see also NS 0514005, Austria, 1827 (MB).

   In addition, NUC notes that for Persia there were eds. in 1828 [J. Grigg, Philadelphia; NS 0514055 (CLSU, CU, DGU, DLC, ICJ, MB, MH, NBu, NcD, NN, OrP, OrPR, OU, ViU, VtMiM, WaU)]; 1834 [J. Grigg, Philadelphia; NS 0514056 (PSC)]; and 1845 [Grigg & Elliot, Philadelphia; NS 0514058 (MB, PP, PPF)]. Finally, NUC lists an 1829 ed. for Turkey (H. Cowperthwait, Philadelphia, no NUC number).

SHORT COMPENDIUM ... **? tr**

▸ **{1791} (01)**

- *A short compendium of ancient and modern historical geography*. Tr. from the French ... by Mr. de Lanségüe
- LONDON; printed at the Logographic press and sold by T. Cadell [etc.]
- 21 cm.; 1 p. l., [v]-viii, 478 p.
- ***Libraries*** : DLC

SINCLAIR, Archibald **? Am**

- **{1843} (01)**
- *A system of modern geography* ... according to a new arrangement, etc.
- PARSONSTOWN
- 12mo
- ***Libraries*** L
- ***Notes***

1. BCB, Index (1837-57:110) records an ed. in 1844.

SKETCH OF GEOGRAPHY **mSG**

- **1775 (01)**
- *A small sketch of geography*
- DUBLIN; printed by R. Marchbank, for James Porter, (No. 14.) Skinner Row
- 16.5 cm.; pp. [5], 6-43, [5]
- ***Libraries*** **L**
- ***Notes***

1. A skeletal special geography; essentially reduced to tables; most list countries or provinces with their cities and remarkable features. Ireland is placed first.

SLATER, Eliza **Y SG?**

- **{1840} (01)**
- *Lessons in geography, ancient and modern*, with notes
- LONDON; Suttaby
- 12mo
- ***Libraries*** L
- ***Source*** BL and BCB, Index (1837-57:110), which identifies her as 'Mrs J. Slater'
- ***Notes***

1. BL records an ed. in 1854.

SLOMAN, W.H. **?**

- **{1835} (01)**
- *Geography*
- Place/publisher n.r.
- ***Libraries*** n.r.
- ***Source*** Peddie and Waddington (1914:226)
- ***Notes***

1. This might be Sloman, W.H., *Five hundred questions on geography*, London, 12mo [1835]; in BL.

SMILEY, Thomas Tucker **Y Am**

- **{1823-4} (01)**
- *An easy introduction to the study of geography, on an improved plan ...*
- PHILADELPHIA; printed by Clark & Raser
- 15 cm.; pp. 243
- ***Libraries*** : DLC, PHi
- ***Source*** NUC (NS 0620011)
- ***Notes***

1. NUC records further eds. (all Philadelphia): 1824, 2nd, 243 pp. (ViU); 1825, 3rd, 250 pp. (CtHT, PPeSchw, TxSaO); 1827, 5th, 250? pp. (ViU); 1828, 6th, 252 pp. (NNC, PPL); 1830, 7th, 256 pp. (DLC, NcU); 1832, 14th, 256 pp. (N, NcU, PPAN); 1832, 16th, 256 pp. (CtY, PU); 1832, 17th (PPL); 1834, 22nd, 252 pp. (DLC); 1835, 26th, 252 pp. (NcU); and 1837, 35th, 252 pp. (MiU).
2. Sahli (1941, *passim*) and Brigham and Dodge (1933:14) discuss (01); Sahli (1941, *passim*) discusses the 22nd ed. of 1834; Elson (1964:399) cites the 6th and 7th eds.

SMILEY, Thomas Tucker **A Am**

- **{1838} (01)**
- *The encyclopaedia of geography ...*
- HARTFORD; Belknap & Hammersley
- 17 cm.; pp. 248
- ***Libraries*** : DeU, DLC, MtHi
- ***Source*** NUC (NS 0620024)
- ***Notes***

1. For information on this book, see notes to (02), below.

- **1839 (02)**
- *The encyclopaedia of geography*, comprising a description of the earth, exhibiting its relation to the heavenly bodies, its physical structure, the natural history of each country, and the industry, commerce, political institutions, and civil and social state of all nations. Illustrated by numerous engravings, representing the most remarkable objects of nature and art in every region of the globe on the plan of Murray's Encyclopaedia of geography. Adapted to the use of schools and families
- PHILADELPHIA; Hogan & Thompson. Sold by booksellers generally throughout the United States
- 17 cm.; pp. [3], 4-264
- ***Libraries* L** : KyU, MH, NN, OC, PHi
- ***Notes***

1. What's in a name? This is a conventional special geography, with pp. 7-45 devoted to systematic aspects of the subject. Questions related to the text are included at the foot of pages and at the end of chapters.
2. NUC (NS 0620026 and 0620029) add 'Cincinnati; Ely and Strong.'
3. See MURRAY (1834) for the model on which this work is based.
4. BL and NUC record further eds.: 1839, Hartford, Belknap & Hammersley; Philadelphia, Grigg & Elliot, 264 pp. (L : MH, MnU : CaBVaU); and 1839, Hartford, G. Robins, 264 pp. (: CtY, MH).

5. Brigham and Dodge (1933:14) discuss (01).

SMITH, Charles **Y Am**
- **{1795} (01)**
- *Universal geography made easy*; or, a new geographical pocket companion: comprehending a description of the habitable world, [with maps]
- NEW YORK; printed by Wayland & Davis, for the author & L. Wayland
- 13.5 cm.: 192 p.
- ***Libraries*** : DLC, ICN, MWA, NN, PU, RPJCB
- ***Source*** NUC (NS 0625111)
- ***Notes***

1. The words in [ ] are taken from the 2nd ed., as recorded in NUC.
2. NUC records 2nd ed., 1800, New York, C. Smith, 192 pp. (DSI, NNC).
3. Elson (1964:394) cites (02).

SMITH, John **Y SG? Am**
- **{1816} (01)**
- *A new compend of geography*: treating principally of America. With an introduction, explaining the astronomical part of geography. In question and answer. Designed for the use of schools. Compiled from the latest authorities
- COOPERSTOWN, [NY]; printed by H. and E. Phinney, and sold by L. and B. Todd, Hartwick
- 17 cm.; pp. 216
- ***Libraries*** : CSmH, MH, N, NBC, NBuG, NcD, OKentU, PPL, Wy-Ar
- ***Source*** NUC (NS 0640500)
- ***Notes***

1. Discussed in Nietz (1961:196, 216-17, 219) and Sahli (1941, *passim*).

SMITH, John **A**
- **1810-11 (01)**
- *A system of modern geography*; or the natural and political history of the present state of the world. With numerous engravings
- LONDON; printed for Sherwood, Neely, and Jones, Paternoster-Row; by Gillet and Son, Crown-court, Fleet-street
- 4to, 27 cm., 2 vols.; Vol. I, pp. [3], iv-viii, lvi, 914; Vol. II, pp. [3], iv-xi, [1], 1046, [10]
- ***Libraries*** L
- ***Notes***

1. Makes the common claim to have avoided 'the ancient relations of faithless travellers' (p. i, 2nd series), and the 'idle tales, gross errors, and misrepresentations' (p. ii, 2nd series) that are alleged to be present in other, unnamed 'general' geographies. Cites sources in footnotes. The whole of Vol. I is devoted to the countries of northern Europe, of which the UK, excluding Ireland, occupies pp. 390-914 (England alone, pp. 390-730). The rest of Europe, including Ireland, occupies Vol. II, pp. 1-759.

   Sources are rarely cited once Europe has been left behind. Pp. 609-10 contain an assessment of the German character that is prescient in reaching the conclusion that, if Germany were ever to be united, it would become the predominant country of Europe.

SMITH, R[ichard] M[cAllister] **Y Am**
- **{1849} (01)**
- *The child's first book in geography*, designed as an introduction to R.M. Smith's New common school geography ...
- PHILADELPHIA; Grigg, Elliot & co.
- 15.5 cm.; pp. 160
- ***Libraries*** : DLC
- ***Notes***

1. Discussed in Culler (1945, *passim*).

SMITH, R[ichard] M[cAllister] **Y Am**
- **{1848} (01)**
- *Modern geography, for the use of schools, academies, etc.* on a new plan, [by which the acquisition of geographical knowledge is greatly facilitated, illustrated with maps and numerous engravings]
- PHILADELPHIA; Grigg, Elliot & co.
- 30.5 cm.; pp. 80
- ***Libraries*** : DLC
- ***Source*** NUC (NS 0648868)
- ***Notes***

1. The words in [ ] are derived from CIHM, which records 1849, Philadelphia (: : CaOLU).
2. Discussed in Culler (1945, *passim*).
3. A microform version of the ed. of 1849 is available from CIHM.

SMITH, R[ichard] M[cAllister] **Y Am**
- **{1846} (01)**
- *A new system of modern geography, for the use of schools, academies, etc.*, particularly designed for the south and west ...
- PHILADELPHIA; Grigg & Elliot
- 19 cm.; pp. 168
- ***Libraries*** : PBL, ViU
- ***Source*** NUC (NS 0648869)

SMITH, Roswell Chamberlain **Gt?**
- **{1852} (01)**
- *A concise and practical system of geography for common schools, academies, and families*. Designed as a sequel to the 'First Book.' Illustrated with thirty-two steel maps and numerous engravings
- NEW YORK; Burgess
- Pp. vi, [7]-80
- ***Libraries*** : OO

▸ *Notes*
1. Culler records *A concise and practical system of geography*, Philadelphia: J.B. Lippincott, 3rd ed., 1857. As noted below, this is the subtitle of Smith's *Quarto ... geography* (i.e., SMITH, Roswell 1846b, below). It seems likely that both the book recorded by Culler and the book to which this is a note are eds. of the latter work, especially as NUC occasionally records Lippincott as an associate publisher along with Burgess of New York. Even so, Culler's '3rd ed.' remains to be explained.

SMITH, Roswell Chamberlain **Gt**
▸ **{1836}**
▸ *Geography on the productive system*
▸ PHILADELPHIA; W. Marshall and Co.
▸ ***Libraries*** n.r.
▸ ***Source*** Elson (1964:400)
▸ *Notes*
1. See the body of the title of SMITH, Roswell C. (1835), below.

SMITH, Roswell Chamberlain **Gt**
▸ **{1851}, 13th**
▸ *An introductory geography*
▸ NEW YORK; Cady and Burgess
▸ ***Libraries*** n.r.
▸ ***Source*** Elson (1964:402)
▸ *Notes*
1. See the subtitle of SMITH, Roswell C. (1846a), below.

SMITH, Roswell Chamberlain **Y Am**
▸ **{1862} (01)**
▸ *New geography*, containing map questions interspersed with such facts as an observing tourist would notice ... for the use of common schools in the United States and Canada
▸ PHILADELPHIA; J.B. Lippincott
▸ 35 cm.; pp. 92
▸ ***Libraries*** : NN
▸ ***Source*** NUC (NS 0650148)

SMITH, Roswell Chamberlain **Y Am**
▸ **{1846}a, 3rd (01)**
▸ *Smith's first book in geography.* An introductory geography, designed for children ... 3rd ed. ...
▸ NEW YORK; Paine & Burgess
▸ 16 cm.; pp. viii, [9]-176
▸ ***Libraries*** : NRU (imperfect)
▸ *Notes*
1. What happened to the first and second eds.? An explanation of their seeming

non-existence may be found in the unusual practice Smith adopted in his *Geography* of 1835 (see below); there he combined in one book material for students at beginning, intermediate, and advanced levels. Perhaps he gave up doing that in 1846, and replaced the one book with three. It was in 1846 that publication ceased in Hartford and began in New York. See SMITH, Roswell C. (1846b, i.e., *Smith's quarto, or second book in geography*), below; like the book to which this is a note, it was published by Paine & Burgess.
2. NUC records further eds. (all New York, 176 pp., except as noted): 1846, 3rd (MH); 1846, 4th, Portland, pp. n.r. (MH); 1847, 6th, 175 pp. (OClWHi, PHi); 1848, 7th, 175 pp. (MB, MH); and 1849, 9th, pp. n.r. (MH, NIC). See notes to (02) and (03) below.

- **{1850}, 10th (02)**
- *Smith's first book in geography ...*
- NEW YORK; Cady & Burgess
- 17 cm.; pp. viii, [9]-176
- ***Libraries*** DLC
- ***Notes***

1. Culler (1945:310) lists this as *Introductory Geography*, and states that it is the 2nd ed.
2. NUC records further eds. (all New York, Cady & Burgess, 176 pp): 1851, 12th (MH); 1851, 13th (DLC, NN, NNC); and 1852, 14th [c1846] (MeB, MH).

- **1852, 16th, revised (03)**
- *Smith's first book in geography.* An introductory geography, designed for children. Illustrated with one hundred and twenty-six engravings, and twenty maps. Sixteenth edition, revised
- NEW YORK; published by Cady and Burgess
- 17 cm.; pp. [i-iv], v-vii, [viii], 176
- ***Libraries*** : : CaAEU
- ***Notes***

1. The preface, pp. [iv]-vii, is devoted largely to praise of the idea that the purpose of education is to get children to understand the facts that they are obliged to learn by heart.
   The text takes the form of question and answer. Superficially, the objective is to provide students with a basic geographical vocabulary (consisting chiefly of elements in the physical environment, plus toponyms). In addition, cultural attitudes are explicitly inculcated; e.g., the 6th question is: 'What do the works of creation plainly show?' Answer: 'The wisdom and goodness of the Creator' (p. 9). Ethnocentric bias is strongly marked throughout, especially with respect to religion, which receives more attention than is common in the genre; e.g., the 'Deluge' is asserted to have begun on 7 December 4,200 years before 1852.
   The illustrations are not referred to in the text, and their relationship to it is not always clear.
2. NUC records further eds. (all New York, D. Burgess, 180 pp., except as noted): 1853, 18th, pp. n.r. (MH); 1853, 19th, pp. n.r. (MH, NjP); 1853, 20th [recorded in Elson (1964:402)]; [c1854], 29th [sic], Boston, 180 pp. (CtY, MH); 1854, 23rd (MB); 1855, 25th (DLC, OClW); 1856, 27th (OU);

1856, 29th (MB); 1856, 29th, Ivison & Phinney (MH, NNC); 1859, 29th, Philadelphia, Lippincott (MU); 1860, 29th, Philadelphia, Lippincott, 178 pp. (MB, MH, OU); and [1860?, c1854], 29th, Philadelphia, Lippincott, pp. n.r. (MH).
3. Discussed by Culler (1945, *passim*) and Nietz (1961:202, 211); Hauptman (1978:431, 434-5) refers to the 13th ed. in general discussion, and Elson (1964:402) cites the eds. of 1851 and 1853.

SMITH, Roswell Chamberlain **Y Am**
- **{1835} (01)**
- *Smith's geography*. Geography on the productive system ...
- PHILADELPHIA; W. Marshall & co. Hartford, D. Burgess & co. [etc.]
- 16 cm.; 2 p. l., 315 p.
- *Libraries* : DLC, ICJ, MB
- *Source* NUC (NS 0650098)
- *Notes*

1. For information on the contents and organization of the book, see note 1 to (02), below.

- **1836, revised and improved (02)**
- *Smith's geography*. Geography on the productive system; for schools, academies, and families, revised and improved. Accompanied by a large and valuable atlas
- PHILADELPHIA; W. Marshall & co. Hartford, D. Burgess & co. Sold by booksellers throughout the United States
- 16.5 cm.; pp. [3], 6-272
- *Libraries* : DLC, NN, NNC, OClWHi
- *Notes*

1. Divided into three parts, which are intended for beginners, for older pupils, and for those considerably advanced in geography. It thus contains within one work material that other authors present in a series of the three books. S. advocates rote learning, and makes no claim to inculcate morality. See note to SMITH, Roswell C. (1846a, i.e., *Smith's first book in geography*), above, and SMITH, Roswell C. (1846b, i.e., *Smith's quarto, or second book in geography*), below.
2. Resembles MITCHELL, S.A. (1840c) in that he includes as part of his section on Asia references to Biblical events (pp. 220-1). There is, however, a marked contrast in tone. S. avoids the condemnation of the Jews in which M. indulges.
3. BL and NUC record further eds. (Hartford, 274 pp., except as noted): 1837, Philadelphia (: CtY, MH, NNC); 1838 (: CtY, DLC, IU, MH, MtHi, OkU); 1838, Philadelphia (: CtY); 1839 (: DeU, MH : CaBVaU); and 1840, 312 pp. (L : DLC, MB, NNC, OU : CaBVaU). Sahli (1941:383) records an ed. in 1840, New York. NUC records further eds. as follows (Hartford, 312 pp., until 1845, except as noted; thereafter, New York): 1841 [c1840] (MB, MH, OClW); 1842 (CtY, MH); 1843, pp. n.r. (CtY, MH, NIC); 1844 (CtY, IU,

MH, OClW) [According to Culler (1945:310) this is the 2nd ed.]; 1845 (DLC, OU); 1846 (N); 1846 [c1840], Portland (MeB); 1847 (NBuU); 1847, pp. n.r. (MH); 1848 (MH, NC); 1850 (DLC MH); and 1851 (NNC).

- **{1854}, latest ... (see below) (03)**
- *Geography on the productive system* for schools, academies, and families. Latest rev. and improved ed. containing the addition of ancient geography
- NEW YORK; D. Burgess
- 18 cm.; pp. 356
- ***Libraries*** : DLC, MiU, Nh
- ***Source*** NUC (NS 0650117)
- ***Notes***

1. NUC records a further ed.: 1855, New York, 356 pp. (MH, NNC). Culler (1945:310) records a 3rd ed., 1857, Philadelphia [Culler, library n.r.]. NUC records further eds. (Philadelphia except as noted): 1860, New York, pp. n.r. (MH); 1860, 356 pp. (CtY, NNC); and 1867, pp. n.r. (MH).
2. Brigham and Dodge (1933:20), and Nietz (1961:229) discuss (01); Nietz (1961:217, 221) are probably references to the ed. of 1839; Sahli (1941, *passim*) discusses the New York ed. of 1840; Culler (1945, *passim*) discusses the eds. of 1844 and 1857; Hauptman (1978:434) quotes from the ed. of 1851; Elson (1964:400-3) cites (02) and the eds. of 1837, 1840, 1848, 1851, 1855, and 1860.

SMITH, Roswell Chamberlain **Y Am**

- **{1846}b** (01)
- *Smith's quarto, or second book in geography.* A concise and practical system of geography ... designed as a sequel to the 'First Book' ...
- NEW YORK; Paine & Burgess
- 30.5 cm.; pp. vi, [7]-72
- ***Libraries*** : DLC
- ***Notes***

1. NUC records further eds. (New York, Cady & Burgess [until 1852, then D, Burgess], pp. n.r., except as noted): 1847, Paine & Burgess, 4th (MH); 1849, 9th, 75 pp. (MH, NNC); 1850, 12th (MH); 1850, 13th, 76 pp. (DLC, MiU); 1851, 14th, 76 pp. (MB, MH); 1851, 15th (MH); 1852(?), Rev. (MH); 1852, Burgess, 22nd, 80 pp. (IU, MH, OO); 1853, D. Burgess, 26th (MH); 1853, 28th, 84 pp. (ICRL); 1853? 41st [sic], 88 pp. (ICU); 1854, 29th (MH); 1854, 32nd (MH); 1855, Rev., 84 pp. (DLC, OKentU); 1855, 34th (MH); 1855, 37th (MH); 1856, 38th (MH); 1856, 39th, 84 pp. (NNC); and 1858, 41st, 88 pp. (NNC).
2. Probably discussed by Culler (1945, *passim*); see note to SMITH, Roswell C. (1852), above.

SMITH, Thomas **A**

- **{1806} (01)**
- *Elements of geography* ... By the Rev. ...
- LONDON; printed for J. Wallis, sen.; J. Wallis, jun.; and J. Harris
- 12.5 cm.; pp. 152

- *Libraries* : : CaOTP
- *Source* St. John (1975, 1:189), which notes: '(Scientific Library)'

- **{1815} (02)**
- *Elements of geography* ... Enlarged and improved by Donald M'Donald
- NEW-YORK; printed by Largin & Thompson, for D. Macdonald & J. Gillet
- 11.5 cm.; 4 p. l., 152 p.
- *Libraries* : CtY
- *Source* NUC (NS 0652867), which notes: '(Added t.-p.: The Scientific Library ... Vol. II)'
- *Notes*

1. Given the pagination, the assertion of the tp. that this ed. is enlarged rouses scepticism.
2. NUC records a further ed., 1818, New-York, 142 pp. (NNC).
3. On the basis of the copy seen at BL, this is taken to be Vol. 2 of Thomas Smith, *The scientific library; or, repository of useful and polite literature: comprising astronomy, geography, mythology, ancient history, modern history, and chronology*, London, 6 vols.
4. It is a small-scale special geography; countries are generally allotted 4-5 pp. each.
5. Elson (1964:397) cites the ed. of 1818.

SMITH, Thomas **Y**

- **{1802}a** (01)
- *The universal atlas, and introduction to modern geography*: in which are described, the most celebrated empires, states and kingdoms of the world. With a general view of astronomy; the solar system; the fixed stars and constellations; definition of geography; figure and motions of the earth; vicissitudes of the seasons, etc; a description of the terrestrial and celestial globes; with geographical problems; eastern and western hemispheres, etc. Also the method of adverting the time of day in distant nations, is clearly elucidated on a new geographical clock. The whole illustrated with thirty-one maps and plates, accurately delineated by an eminent geographer. Engraved by John Cooke. The introduction, and geographical descriptions by the Rev. ...
- LONDON; printed for J. Harris, successor to E. Newbery, corner of St Paul's Church-Yard, and J. Cooke, by H. Bryer, Bridewell-Hospital, Bridge-Street
- 24 cm.; pp. [9], viii-xxiv, 87
- *Libraries* : **CtY**, InU
- *Notes*

1. Intended for the instruction of children, and in particular for girls.

SMITH, Thomas **A**

- **{[1802?]}b** (01)
- *The wonders of nature and art; or, a concise account of* whatever is most curious and remarkable in the whole world ...
- [LONDON; J. Harris]

- V [i.e., vols.]
- *Libraries* : DP

- **{1803-4} (02)**
- *The wonders* ... world; whether relating to its animal, vegetable, and mineral productions, or to the manufactures, buildings and inventions of its inhabitants, compiled from historical and geographical works of established celebrity, and illustrated with the discoveries of modern travellers
- LONDON; printed for J. Walker
- 8vo, 14 cm., 12 vols.
- *Libraries* L : CU, MH, NcD, **NN**, NNU-W, OkU, PHi
- *Notes*

1. The subtitles and pagination of the separate vols. follow.

   Vol. 1. Great Britain and Ireland; France, including Lorraine and Alsace, etc.; Spain and Portugal. Pp. [3], iv-x, [2], 275.

   Vol. 2. Italy; Germany, including Hungary and Bohemia; Switzerland, and the canton of Grisons; the United Provinces; Denmark, Norway, and the adjacent isles. Pp. [4], 304.

   Vol. 3. Greenland; Sweden, and Lapland; Poland; Muscovy, or the Russian empire; Turkey in Europe. Pp. [4], 263.

   Vol. 4. Turkey in Asia; Persia; Asiatic Tartary. Pp. [4], 282.

   Vol. 5. India; Indian islands. Pp. [4], 283.

   Vol. 6. China; Tonquin, Tibet, Cochin-China, and the island of Formosa. Pp. [4], 282.

   Vol. 7. Egypt; Barbary; Negroland and Guinea. Pp. [4], 284.

   Vol. 8 (1804). Nubia, or Sennar, and Abyssinia; Congo, including the kingdoms of Angola, Benguela, and Loango; Caffraria and the Cape of Good Hope; African islands. Pp. [3], iv, 276.

   Vol. 9 (1804). South America; North America. Pp. [3], iv, 284.

   Vol. 10 (1804). North America (cont); West India islands; New discoveries; Otaheiteau. Pp. [3], iv, 284.

   Vol. 11 (1804). New Discoveries [i.e., Otaheitean islands, Sandwich islands; New South Wales]. Pp. [3], iv, 252, [8].

   Vol. 12 (1804). New Discoveries: Pelew islands; Nootka sound; New Zealand. Pp. [3], iv, 252.
2. Tends to avoid references to location, especially in the early vols., and thus is not, according to the usage of the time, truly geographical. However, the content of the text is much like that found in PINKERTON (1802), though more attention is paid to the description of buildings than was the case in most special geographies. Where information of a geological type is given, it comes first. Although not mentioned on the tp., information on customs, manners, government, and religion is provided, especially outside Europe. References to sources are made in the text, which, overall, takes the form of a discursive ramble through a wide variety of topics and places.

- **{1806-7}, revised ... (see below) (03)**
- *The wonders* ... *world*; compiled ... travellers ... Revised, corrected, and improved by James Mease
- PHILADELPHIA; printed for Robert Carr, for Birch & Small: sold also by M. Carey, Kimber, Conrad & co. and Jacob Johnson

- 15.5 cm., 14 vols.
- ***Libraries*** : CSt, CtY, DeGE, InU, IU, KyLx, MH, NBu, PHi, PP, PPL, PSC, PV, RPB, ViU, ViW : CaBViP
- ***Source*** NUC (NS 0652904)

- **{1838}, 8th (04)**
- *The wonders of nature and art*, comprising upwards of three hundred of the most remarkable curiosities and phenomena in the known world, with an appendix of interesting experiments, in different arts and sciences ... by J. Taylor
- LONDON; J. Chidley
- Pp. vi, [2], 568
- ***Libraries*** : OCl
- ***Source*** NUC (NS 0652905)
- ***Notes***

1. Because this ed. is only one volume in length, it must be an abridged ed. Judging by the title, as recorded in NUC, it seems likely that it is no longer a special geography.
2. NUC records later, single-volume eds. in: 1839, 9th, London, J. Chidley, 568 pp. (OCl); 1839, Halifax (PBL); 1839, London, 436 pp. (PP, RPB).

SMOLLETT, Tobias George **A**

- **{1764} (01)**
- *The present state of all nations*, containing a geographical, natural, commercial, and political history of all the countries in the known world
- LONDON
- 8vo, 8 vols.
- ***Libraries*** n.r.
- ***Source*** DNB (1937-8, 18:590)

- **1768-9 (02)**
- *The present state of all nations ...*
- LONDON; printed for R. Baldwin, No. 47, Paternoster-row; W. Johnston, No. 16, Ludgate-street; S. Crowder, No. 12, and Robinson and Roberts, No. 25, Paternoster-row
- 8vo, 22 cm., 8 vols.
- ***Libraries*** E, L : CLU, CtY, CU, NjP, NNC, RPB : **CaAEU**, CaBVaU, CaNSHD, CaOTU
- ***Notes***

1. The pagination and countries described in the separate vols follow.

   Vol. I, 1768. Deals with northern Europe, emphasizing islands in the Atlantic, including those off the coast of Scotland. Pp. [4], iv-viii, 510.

   Vol. II, 1768. Scotland and England. Pp. [4], 478.

   Vol. III, 1769. England (cont.) and Ireland. Pp. [3], 4-480.

   Vol. IV, 1769. Parts of east central Europe, including the Turkish Balkans; Germany. Pp. [3], 4-479.

   Vol. V, 1769. Remainder of Germany, remainder of the Balkans, Switzerland, United Netherlands. Pp. [3], 4-479.

Vol. VI, 1769. Austrian Netherlands, France, west and central Mediterranean Europe. Pp. [3], 4-488.

Vol. VII, 1769. Asia, except India, Japan, and nearby islands. Pp. [3], 494.

Vol. VIII, 1769. The remainder of Asia, Africa, and America. Pp. [3], 4-524, [2].

2. The preface contains a justification of the work and an outline of its contents; much of it is a list of what other contemporary authors put on the tp. Also: 'We begin with the polar regions, from whence we advance towards the equator, describing each country in its turn; a method which indeed order and regularity seem naturally to suggest' (p. vii). As the summary of the contents of the vols. provided above shows, this principle is followed in only an intermittent and haphazard fashion. It is also said that sources will be cited, but this is done only in Vol. I. A likely place to start a search for the sources used would be BÜSCHING (1762), supplemented for the world outside Europe by either the German original or its French translation. Shows some bias towards islands, and a lesser one towards the artistic side of culture. Is consistent in providing relatively little history.

- **{1768} (03)**
- *The present state of all nations ...*
- LONDON; printed for R. Baldwin, W. Johnston, and Robinson and Roberts
- 22 cm., 2 vols. in 1
- ***Libraries*** : PU
- ***Source*** NUC (NS 0660918)
- ***Notes***

1. According to PU, this is not an ed. separate from (02), above, but rather seems to be vols. 1 and 2 of that ed. bound together.

SOLINUS, Julius Gaius **mSG tr**

- **{1587} (01)**
- *The excellent and pleasant worke of Iulius Solinus Polyhistor*. Contayning the noble actions of humaine creatures, the secretes and prouidence of nature, the description of countries, the maners of the people: with many meruailous things and strange antiquities, seruing for the benefitt and recreation of all sorts of persons. Translated out of Latin into English, by Arthur Golding, Gent.
- LONDON; printed by I. Charlewoode for Thomas Hacket
- 4to, 17 cm.; pp. [228]
- ***Libraries*** C, GU, L, O : CSmH, CtY, ICN, MiU, NNC
- ***Notes***

1. Starts with the history of the city of Rome; moves on to the character/quality of human beings – dealt with by giving instances, usually of extreme examples of a type. Then describes the countries of the world, giving miscellaneous information, much of it suggesting gossip in character. Regarded by modern critics as a work of inferior quality, though representative of the time. It was popular in the Middle Ages, and continued to be so well into the sixteenth century, as shown by this translation. 5 pp. are devoted to a life of Solinus.

2. Facsimile ed. 1955, Gainesville, FL; Scholars' Facsimiles & Reprints, with an introduction by George Kish.

- **{1590} (02)**
- *The excellent ... Polyhistor*
- LONDON
- 4to
- ***Libraries*** : MWA
- ***Source*** NUC (NS 0711636)
- ***Notes***

1. In 1590 an ed. of Solinus, dated 1587, was issued, bound with the 2nd ed. of MELA (1585). The only difference between the two issues of Solinus is the heading to the first chapter.
2. MWA reports that it does not have the book recorded in NS 0711636; nor does it have a record of having disposed of it.

SPAFFORD, Horatio Gates **Y Am**

- **{1809} (01)**
- *General geography, and rudiments of useful knowledge.* In nine sections ... Illustrated with an elegant improved plate of the solar system ... a map of the world ... of the United States ... and several engravings on wood. Digested on a new plan, and designed for the use of schools
- HUDSON; printed by Croswell & Friary
- 12mo, 18 cm.; pp. xii, 381
- ***Libraries*** L : CtY, DLC, MHi, MnHi, N, NjR, NSyU, OClWHi, PMA, PPC, RPJCB
- ***Source*** BL and NUC (NS 0774694)

SPEED, John **A**

- **1627 (01)**
- *A prospect of the most famovs parts of the world.* Viz Asia, Africa, Europe, America. With these kingdomes therein contained, Grecia, Roman empire, Germanie, Bohemia, France, Belgia, Spaine, Italie, Hungarie, Denmarke, Poland, Persia, Turkish empire, Kingdo:[sic] of China, Tartaria, Sommer islands, ciuill warrs, in England, Wales, and Ireland. You shall finde placed in the beginning of the second booke marked with these ** and (5) Together with all the prouinces, counties and shires, contained in that large theater of Great Brittaines empire
- LONDON; printed by Iohn Dawson for George Humble, and are to be sold at his shop in Popes-Head Pallace
- Folio, 43 cm., 2 pts.; pp. [2], 48 of text, plus 22 double-page maps; then the 2nd book, with its own tp. [see below], pp. [16], 146 of text (incl. subordinate tps. and gazetteers), plus 66 double-page maps, [10]
- ***Libraries*** C, CheR, HeP, HlU, LeP, **L**, NP, SwnU : DLC
- ***Notes***

1. The second tp. reads: 'The theatre of the empire of Great Britaine: presenting

an exact geography of the kingdomes of England, Scotland, Ireland, and the iles adioyning: with the shires, hundreds, cities and shire-townes, within ye kingdome of England, divided and described by Iohn Speed. Imprinted at London. Anno cum privilegio 1627. Are to be sold by George Humble at the Whit horse in Popes-head Alley.'

The subordinate tps., referred to above, introduce the sections on Wales, Scotland, and Ireland, but the pagination is continuous.

The first two pages of the text provide a 'generall description of the world'; they present a pre-Copernican view of the universe.

2. A note in NUC (NS 0804054) reads: 'Maps comprising pts 2-5 were originally published with this title [i.e., Theatre of Great Britaine]. 1st ed. appeared in 1611, 2nd ed., 1614; 3rd ed., in 1627 with the title, "A prospect of the most famous parts of the world".'
3. Facsimile ed., 1966, Amsterdam; TOT.

- **1631 (02)**
- *A prospect of the most famovs parts of the world ...*
- LONDON; printed ...
- Folio, 42 cm., 2 pts.; pp. [3], 48 (plus maps); [16], 146 (plus maps), [10]
- ***Libraries*** C, Gp, **L**, Luu, O : CSmH, DFo, DLC, MB, MH, MiU, NNC
- ***Notes***

1. The second tp. is dated 1627, but those to the sections on Wales, Scotland, and Ireland are all dated 1631. This is the issue that BL dates 1631-7. It appears to be identical to that of 1627.

- **1646 (03)**
- *A prospect of the most famous parts of the world ...* Kingdome of China ... civill warres ... provinces ... counties ...
- LONDON; printed by John Legatt, for William Humble, and are to be sold at his shop in Popes-head Palace
- Folio, 41 cm., 2 pts.; pp. [3], 48 (plus maps); [16], 146 (plus maps), [10]
- ***Libraries*** L : ICN
- ***Notes***

1. The second tp. is dated 1627, but those to the sections on Wales, Scotland, and Ireland, are all dated 1646. The work seems unchanged since 1627.

- **1662 (04)**
- *A prospect of the most famous parts of the world ...*
- LONDON; printed by M. and S. Simmons, and are to be sold by R. Rea the elder, and R. Rea, the younger
- Folio, 44 cm., 2 pts.; pp. [3], 48, [10] (incl. tp. of Pt. 2, bound between pp. 44 and 45); 146 (plus maps), [10]
- ***Libraries*** **L**, NP : CtY, MB, MiU
- ***Notes***

1. The second tp. is dated 1650 [? or 1652? or 1654?], but those of the sections on Wales, Scotland, and Ireland are all dated 1662. The text still continues apparently unchanged.

- **1676 (05)**
- *The theatre of the empire of Great-Britain*, presenting ... kingdom ... adjoyn-

ing: as also the shires ... townes within the kingdom of England and the principality of Wales; with a chronology of the civil-wars in England, Wales and Ireland. Together with a prospect of the most famous parts of the world, viz. Grecia, Roman-empire, Germany ... Spain, Italy, Hungary, Denmark ... Turkish empire, kingdom of China ... Summer islands. In this new edition are added; in the theatre of Great-Britain, the principal roads, and their branches leading to the cities and chief towns in England and Wales; with their computed distances. In a new and accurate method. The market towns wanting in the former impressions. A continuation of all the battels fought in England, Scotland, Wales and Ireland; with all the sea-fights. The arms of all the dukes and earls, whose titles of honour were wanting in each particular county, to the last creation. The description of his Majesty's dominions abroad; with a map fairly engraven to each description, viz. New-England, New-York, Carolina, Florida, Virginia, Mary-land, Jamaica, Barbados. In the prospect of the world. The empire of the Great Mogul, with the rest of the East Indies, Palestine, or the Holy-land, the empire of Russia

- ▸ LONDON; printed for Thomas Basset at the George in Fleet-Street, and Richard Chiswel at the Rose and Crown in St. Paul's Church-Yard
- ▸ Folio, 45 cm., 2 pts.; [19], 146 (plus maps); [2], 56 (plus maps), [11]
- ▸ ***Libraries*** C, EU, LcP, LcU, LeP, **L**, LvU, NP, O, RchP, StkP : CSmH, CtY, DLC, ICN, InU, MB, MBAt, MH, MiD, MWiW, N, NBu, NhD, Nhi, NjP, NN, NNC, NNH, PPiU, RPJCB, TxU, Vi, ViU
- ▸ ***Notes***

1. The change in title from the 1st ed., with the *Theatre* coming before the *Prospect*, has led cataloguers in some libraries to conclude that this, 1676, ed., is an entirely different work from that of 1627, rather than merely a revision. The 'principal roads' are shown on five double-page schematic diagrams that are intermediate in form between maps and lists. On each page, London is shown at the bottom centre; from it, lists of towns radiate, indicating the places through which travellers will pass on their way to the towns shown at the end of the each of the lists.

   The description of the universe given at the beginning of the *Prospect* still follows the Ptolemaic model.
2. NUC (NS 0804097 to 0804109) detail variations in bibliographic information, but they all seem to refer to the same work.

SPEED, John **A**

- ▸ **1646 (01)**
- ▸ *A prospect of the most famovs parts of the world.* Viz. Asia, Affrica, Europe, America. With these kingdomes therein contained, Graecia, Rowane[sic] empire, Germany, Bohemia, France, Belgia, Spain, Italy, Hungary, Denmark, Poland, Persia, Turkish empire, kingdome of China, Tartaria
- ▸ LONDON; printed by M.F. for William Humble, and are to be sold at his shop in Popes-head palace
- ▸ Obl. 8vo, 15 × 10 cm.; the 2nd pt. of a single work (the 1st pt. lacks the tp. but is the *Theatre of the Empire of Great Britain*, pp. [2], 206 (incl. 19 single-page maps)
- ▸ ***Libraries*** **L** (lacks the tp) : CtY, CU-I, DLC, MiU, N, NHi, NjP, NNH, TxU

▸ *Notes*
1. A scaled-down version of SPEED (1627), with the order of the parts reversed.

▸ **1668 (02)**
▸ *A prospect of the most famous parts of the world* ... Africa ... kingdoms ... Romans ...
▸ LONDON; printed for Roger Rea, at the Gilded Cross in Winchester-Street near Gresham-Colledge
▸ Obl. 8vo, 15 × 11 cm.; the 2nd pt. of ...; pp. [2], 206 (incl. 19 single-page maps)
▸ ***Libraries*** C, L : CLU, CtY, MB, MH, NIC, OU, RPJCB

▸ **1675 (03)**
▸ *A prospect of the most famous parts of the world* ... Roman ...
▸ LONDON; printed by W.G
▸ Obl. 8vo, 16 × 11 cm.; the 2nd pt of ... pp. [2], 276 (incl. 26 maps)
▸ ***Libraries*** **L** : CtY, DLC, MiU, NBuG
▸ *Notes*
1. Most of the additional material has to do with North America, but there is also more on India and Russia.
2. As a map maker, Speed is discussed in the standard histories of cartography (e.g., Bagrow 1964; Crone 1978).

SPEEDWELL, Robert **Y mSG?**
▸ **[n.d.]**
▸ *The costumes and customs of the world*, being a sketch of the manners and habits of the different nations of the globe. Designed for the instruction and amusement of children
▸ BALTIMORE; published by J. Moore, printer
▸ 14 cm; [5], 6-64 p.
▸ ***Libraries*** : DLC

SPENCE, Lancelot M. Dalrymple **Y**
▸ **1867 (01)**
▸ *The civil service geography*: being a manual of geography, general and political, arranged especially for examination candidates and the higher forms of schools. By the late ... Revised thoroughly by Thomas Gray, one of the assistant secretaries to the Board of Trade. With woodcuts, six maps, and a general index
▸ LONDON; Lockwood and co., 7 Stationers' Hall Court
▸ 8vo, 16.5 cm.; pp. [5], vi-viii, 136, 8, 8
▸ ***Libraries*** **L**
▸ *Notes*
1. In the Preface, G. reports that a draft of the book was ready by 1862, but Spence died then, leaving it in manuscript. The Introduction provides S.'s objective, namely, that the student preparing for the Civil Service examin-

ation should 'get mapped out in his mind the relative situation of places, and ... acquire a knowledge of the general, political, and physical characteristics of each country' (p. [v]). Contains tables and listlike text, with an index of some 3,000 place names.

2. BL and NUC record further eds. (all London): 1869 (L); 1871 (L); 1873, 4th (L); 1875, 5th (L) [ECB, Index (1874-80:67) notes: 'by T. Gray']; 1881, 7th (L); 1882, 8th (L); 1884, 9th (L); and also 1890 (L : CU). BL records 11th ed., 1903.

STACY, J. **?**

- **{1818} (01)**
- *A new book of geography on an improved plan* intended ... for children, etc.
- LONDON
- 16mo
- ***Libraries*** L

STAFFORDE, Robert **A**

- **1607 (01)**
- *A geographicall and anthologicall description of all the empires and kingdomes*, both of continent and islands in this terrestriall globe. Relating their scituations, manners, customs, prouinces, and gouernements
- LONDON; printed by T.C. for Simon Waterson, dwelling at the signe of the Crowne in Paules Church-yard
- 4to, 19 cm.; pp. [8], 67, [5]
- ***Libraries*** **L**, Lu, O, OT : DLC, CSmH, NjP, ViU
- ***Notes***

1. The author is not identified on the tp.; however, he signed the dedication. The apology for the slightness of the content, contained in the Preface, is justified.
2. Pollard and Redgrave (1976) identify T.C. as T. Creed, and note that the 1st ed. was entered in the Stationers' Register on 8 March 1608.

- **1618 (02)**
- *A geographicall and anthologicall description* ...
- LONDON; printed by N. Okes for S. W[aterson]
- 4to, 19 cm.; pp. [8], 67, [5]
- ***Libraries*** **L**, O : NN, RPJCB
- ***Notes***

1. Seems unchanged from 1607.

- **1618 (03)**
- *A geographicall and anthologicall description* ...
- LONDON; N.O. for Iohn Parker
- 4to, 20 cm.; pp. [8], 67, [5]
- ***Libraries*** E, **L** : CSmH, DFo, DLC, NjP, RPJCB : CaBVaU

▸ *Notes*
1. BL notes: 'to this copy a new titlepage is prefixed with this imprint, and the imprint of the other titlepage is torn away'

▸ **1634 (04)**
▸ *A geographicall and anthologicall description ...*
▸ LONDON; printed by N. Okes for S. Waterson, dwelling ...
▸ 4to, 21 cm.; pp. [6], 55, [3]
▸ ***Libraries*** C, L, **O**, OChch, WorC : CSmH, DFo, IU, PPL, RPJCB
▸ *Notes*
1. Despite the reduction in the number of pages, the text seems unchanged. A handwritten note on the tp. of the Bodleian copy states: 'This booke was for ye most part made by Jo. Prideux Rector of Exon. Coll. and since B[ishop] of Worcester but afterwards published by his scholler Rob. Stafford under his own name.'
2. Discussed in Baker (1928:261), and Bowen (1981:76).

STANFORD, Edward **A tr**
▸ **1878-85 (01)**
▸ *Standford's compendium of geography and travel* ... based on Hellwald's 'Die Erde and ihre Völker.' Translated by A.H. Keane
▸ LONDON; E. Stanford
▸ 8vo, 6 vols.
▸ ***Libraries*** **BfQ**, L
▸ *Notes*
1. BL records the dates as 1882-5, and also records an incomplete set of 1878, 1879, and the 2nd ed. of 1904. Judging by a statement in the preface to the 1879 ed. of the vol. *Australasia*, each volume had its own editor who supplemented the work provided by Keane. Probably because Hellwald's book was used only as a starting point, NUC does not look on this as a single work, but catalogues each volume separately.
   The short titles, authors, and the date of first publication for each of the six vols. are as follows:
   Africa; Johnston, Alexander Keith; 1878
   Asia; Keane, A.H., and Temple, R.; 1882
   Central and South America; Bates, Henry Wallace; 1878
   North America; Hayden, Ferdinand V.; 1883
   Australasia; Wallace, Alfred Wallace; 1879
   Europe; Rudler, F.W.; 1885
2. Sets of all six vols., though in no case entirely of the earliest ed., can be found at (: DLC, NN, OCl, WaS).

STAUNTON, T.H. **Y SG?**
▸ **{1860} (01)**
▸ *The family and school geography*
▸ LONDON; Bentleys

- 8vo
- ***Libraries*** L
- ***Source*** BL and ECB, Index (1856-76:130)

STAUNTON, T.H. **?**
- **{1861} (01)**
- *Geography*
- [LONDON]; Bentleys
- ***Libraries*** n.r.
- ***Source*** ECB, Index (1856-76:130)

STAUNTON, T.H. **Y SG?**
- **{1864} (01)**
- *The school and college geography*
- LONDON
- Pp. 387
- ***Libraries*** : PBL

STEERWELL, J.
See LITTLE TRAVELLER.

STEIL, Benjamin **?**
- **{1844} (01)**
- *Steill's pictorial geography*
- LONDON; Phelps
- 16mo; pp. 140
- ***Libraries*** : MWA
- ***Source*** NUC (NS 0893178) and BCB, Index (1837-57:110)

STEINWEHR, Adolph Wilhelm August Friedrich, von **Y Am**
- **{[1870]} (01)**
- *... A school geography, embracing a mathematical, physical, and political description of the earth*
- CINCINNATI; Wilson, Hinkle & co. Philadelphia; Claxton, Remsen & Haffelfinger; [etc.]
- 32 cm.; pp. 126
- ***Libraries*** : DLC, OO, PPL
- ***Source*** NUC (NS 0903092), which notes: '(The eclectic series of geographies [no. 3])'
- ***Notes***

1. NUC records further eds. (all Cincinnati): [1877], Indiana ed., 126, [2], 10 pp. (DHEW, DLC); [c1880], 126, [2], 12 pp. (OU) [NS 0903094 notes: '"Geography of Ohio, a supplement to the Eclectic series of geographies" (c1877) 12 p. (at end)']; [c1881], 126 pp. (Di-GS, DLC, InStme, MH) {NS

0903095 notes: '(eclectic series of geographies [no. 3])'}. NUC also records an ed. in [1898].

STEINWEHR, A.[W.A.F.], von and BRINTON, D.G. **Y Am**

- **{[1870]a} (01)**
- ... *An intermediate geography*, with lessons in map drawing
- CINCINNATI; Wilson, Hinkle & co. Philadelphia, Claxton, Remsen & Haffelfinger; [etc.]
- 32.5 cm.; pp. 92
- ***Libraries*** : DLC, OClWHi, OO, PPL
- ***Source*** NUC (NS 0903077), which notes: '(The eclectic series of geographies [no. 2])'
- ***Notes***

1. NUC records further eds. (all Cincinnati): [1877], Indiana ed., 90, [4], 10 pp. (CtY, CU, DHEW, DLC, DN, MH, PPL); {Elson (1964:404) records an ed. in 1878}; [1879], 90, [2] pp. (DLC); [1880], 96 pp. (MH) {NS 0903080 notes: (1) '(The electric [sic] series of geographies, 2)'; (2) 'Cover: Massachusetts ed.'}; [1881], 96 pp. (DI-GS, DLC, PBa). NUC also records an ed. in [1898].
2. Culler (1945, *passim*) discussed the ed. of 1881; Elson (1964:404) cites the ed. of 1878.

STEINWEHR, A.[W.A.F.], von and BRINTON, D.G. **Y Am**

- **{[1870]b} (01)**
- ... *Primary geography*
- CINCINNATI, NEW YORK; Van Antwerp, Bragg & co.
- 23.5 cm.; pp. 86
- ***Libraries*** : CtY, DHEW, DI-GS, DLC, InStme, MnHi, MoU, OC, OClWHi, ODW, PPL, ViU
- ***Source*** NUC (NS 0903086), which notes: '(Eclectic series of Geographies [no. 1])'
- ***Notes***

1. NUC records further eds. (all Cincinnati): [1870], 86, [6] pp. (DLC) {NS 0903087 notes: '"Primary geography of Missouri": [6] p. at end'}; [1870], 84 pp. (DLC, NN, OClWHi) {NS 0903088 notes: '(Eclectic series of Geographies [no. 1])'}; 1877? 86 pp. (OC) {NS 0903089 notes: (1) '[between 1877 and 1890]'; (2) 'At head of title: Eclectic series of geographies'}. NUC also records eds. in [c1898] and [c1912].
2. Culler (1945, *passim*) discusses (01).

STERNE, G.M. **Y SG?**

- **{1850} (01)**
- *A physical and political school geography*, etc.
- LONDON
- 8vo
- ***Libraries*** L

▸ ***Notes***
1. BL records further eds.: 1851, 2nd; 1853, 3rd.

STEVEN, William **Y SG?**
▸ **{1841a} (01)**
▸ *Basis of geography*
▸ Place/publisher n.r.
▸ ***Libraries*** n.r.
▸ ***Source*** Allibone (1854)

▸ **{1852}, 4th (02)**
▸ *Basis for geography ...*
▸ LONDON; Millington & Co.
▸ 12mo
▸ ***Libraries*** L
▸ ***Notes***
1. BL notes that this work is part of Dr. Bedford's Elementary school series; there is a cross-reference to Bedford, F.W.

STEVEN, William **?**
▸ **{1841b} (01)**
▸ *Geography, progressive*
▸ Place/publisher n.r.
▸ ***Libraries*** n.r.
▸ ***Source*** Peddie and Waddington (1914:226)

STEWART, Alexander **Y mSG**
▸ **{1828} (01)**
▸ *A compendium of modern geography* [with remarks on the physical peculiarities, productions, commerce, and government of the various countries; questions for examination at the end of each division; and descriptive tables in which are given the pronunciation and a concise account of all the places that occur in the work]
▸ EDINBURGH; [Oliver & Boyd]
▸ 12mo; pp. [7], 4-296
▸ ***Libraries*** L : : CaQQS
▸ ***Notes***
1. The words in the title set in [ ] are derived from CIHM.
2. NUC records a further ed.: 1830, 2nd, Edinburgh, 299 pp. (KMK).

▸ **{1833}, 4th, thoroughly rev. and extended (02)**
▸ *A compendium of modern geography* ... pronunciation, and a concise account of every place of importance throughout the globe. Illustrated by ten maps, and an engraving, showing the heights of the principal mountains in the world ...
▸ EDINBURGH; Oliver & Boyd
▸ 15 cm.; pp. 321

- ***Libraries*** L : CLSU
- ***Source*** BL and NUC (NS 0933594)

- **1835, 5th, carefully ... (as below) (03)**
- *A compendium of modern geography*: with remarks ... throughout the world. Illustrated by ... mountains on the globe. By the Rev. ... author of 'The history of Scotland,' etc. Fifth edition, carefully revised and enlarged
- EDINBURGH; published by Oliver & Boyd, Tweedale Court; and Simpkin, Marshall, & Co., London
- 15 cm.; pp. 8, [3], 4-323, [1]
- ***Libraries*** : **CtY**, IU
- ***Notes***

1. The title suggests what is nowhere stated explicitly, namely, that the book is intended for use in schools. It straddles the boundary between special geography and geographical dictionaries. It proceeds country by country; the description of each begins with the location and boundaries, followed by a table of subdivisions, and then 1-3 pp. of descriptive text. Then comes a set of questions about the country, followed by an alphabetical list of places in the country deemed to be of interest, with information provided about each.

   The opening 8 pp. carry extracts from published reviews, chiefly though not exclusively of this book, published in contemporary magazines and newspapers.
2. BL and NUC record further eds. (all Edinburgh): 1839, 6th (L : : CaNBSaM); 1843 7th, 324, 36 pp. (L : MH (imperfect, tp. missing) : **CaNSPH**) [The final set of 36 pp. carry a list of educational works published by Oliver & Boyd, along with extracts from reviews of those books]; 1846, 8th (L); 1850, 9th (L); 1854, 11th (L); 1854, 12th (L); 1855, 13th (L : : CaOOU); 1856, 14th (L); 1856, 15th (L); 1857, 16th (L); and 1861, 18th, 478 pp. (L : PU).

- **{1862}, 19th, revised (04)**
- *A compendium of modern geography*, political, physical, and mathematical; with a chapter on the ancient geography of Palestine, outlines of astronomy and geology, a glossary of geographical names, descriptive and pronouncing tables, questions for examination, etc. Containing 11 maps, of which 5 are newly drawn and engraved by W. and A.K. Johnson
- EDINBURGH; Oliver & Boyd
- 8vo, 18cm; pp. 478
- ***Libraries*** L : IU
- ***Source*** BL and NUC (NS 0933598)
- ***Notes***

1. BL and NUC record further eds. (all Edinburgh): 1864, 20th (L); 1867, 21st (L); 1868, 21st, 472 pp. (: IU); 1869, 22nd (L); 1871, 24th (L); 1874, 27th (L); 1877, 29th (L); 1879, 30th (L); 1880, 31st, 474 pp. (L); 1882, 32nd, 474 pp. (L); 1884, 33rd, 525 pp. (L); and 1887, 34th, 525 pp. (L). BL also records an ed. in 1889.
2. Microform versions of the eds. of (01), 1833, 1839, and 1855 are available from CIHM.

STEWART, Alexander **Y SG?**
- **{1839} (01)**
- *The first book of modern geography*, with numerous exercises
- EDINBURGH
- 12mo
- ***Libraries*** L

STEWART, K[ensey] J[ohn] **Y Am**
- **1864 (01)**
- *A geography for beginners*. By the Rev. ... Illustrated with maps and engravings
- RICHMOND, VA; J.W. Randolph
- 18 cm.; pp. [v], vi-viii, [1], 2-223
- ***Libraries*** : CtY, DLC, DNW, GEU, MBAt, MdBP, MH, MiU, MWA, NcD, NcU, NIC, NjP, NN, OC, OCL, OClWHi, ODW, OU, PHi, PP, PPL, PU, ViU : **CaBVaU**
- ***Notes***

1. The world as seen by the elite of the Confederate states. 'The first decided outbreak of the Revolution of 1776 occurred in Charleston ... The first collision of the War of Independence of the Southern States occurred at Charleston ... and was occasioned by the President, elected by the citizens of the Northern States, attempting to seize ... the forts in Charleston Harbour, and turn their guns upon the city they were designed to protect' (p. 41). 'Every effort ... was made by the Northern government to capture the capital and break up ... the Confederacy. But by the constant, evident and acknowledged by the God of Battles [sic] ... these efforts have all failed ...' (p. 43).

   'Education consists not merely in the acquisition of knowledge, but also, and chiefly, in learning to observe accurately and systematically ... Inasmuch as all valuable attainments *must be the result of labour* [emphasis in the original], and therefore [sic] the scholar must master certain things in order to obtain a knowledge of geography ... In acknowledging his obligations to preceding authors, [the present author] would state that it has not been his object to depart either from the facts or the words of the best geographers, except where greater conciseness might be gained' (pp. [iii]-iv).

   Although the God of the Bible is given credit for the events in human history until the division of languages at Babel, yet credit is also given to 'the diverse climates in which men live, and their different modes of life' for the fact that 'nations have been kept distinct; and men have become separated into races, of different colour, as well as government' (p. 31). Marx as well as Darwin had fertile ground in which to sow.

   There are sections devoted to the fauna and flora of the Confederate States; they reveal an awareness of the interdependence and interaction of the creatures of the earth that foreshadows ecology.
2. NUC records a further ed: 1868, New York, E.J. Hale, 226 pp. (ViLxW).

STOUT, Charles B.
See note to HART (1824).

STOUT, Charles B. and W.W. SMITH **Y SG? Am**
- **{1866} (01)**
- *The young geographer - a book for beginners*
- [NEW YORK?]; Ivison, Phinney, Blakeman & Co.
- 4to; 128 pp.
- ***Libraries*** n.r.
- ***Source*** Dryer (1924:121); not in NUC

STOWE, Harriet Elizabeth (Beecher) **Y**
- **1855 (01)**
- *A new geography for children.* Revised by an English lady, by direction of the author. With numerous illustrations and maps
- LONDON; Sampson Low, Son, & Co., 47, Ludgate Hill. English and American booksellers and publishers
- 8vo, 17 cm.; pp. [5], vi-xv, [2], 2-240
- ***Libraries*** L : CtY, NN, WU
- ***Notes***

1. The Preface opens with: 'This little book having been prepared for the use of children in America, there were many details of that country unnecessary for very young children in England to be much acquainted with ... The editor, in compiling the lessons on the British Isles, has endeavoured to follow ... the original plan of the work.'

   The text is divided into lessons (i.e., chapters); at the end of each there are questions on their content. Pp. 1-25 deal with measurement, maps, and geographical terms (e.g., island, lake); the tone is patronizing and sentimental. The section on the British Isles follows (pp. 27-96), England being dealt with on a county-by-county basis. The account of the U.S. (pp. 128-37) is dominated by a discussion of slavery; this takes the form of a reasoned polemic against it. The sections on Asia and Oceania contain a number of denigrating references to the native peoples. Pp. 178-240 deal systematically with topics in human and physical geography. Lesson XXXI, 'Religion' (pp. 205-18), makes the key distinction that between those who follow the Bible and those that do not. The former are divided into Catholics, Orthodox, and Protestants, the last being those who read the Bible and think for themselves. Two points are made about pagan countries. First, 'in all those countries where the Bible is unknown there is no such thing as *liberty*' (p. 210). The point is reinforced with grisly stories about the cruelty of despots. Second, 'in all countries where the Bible is unknown, females are despised and cruelly treated.' Once more, the point is illustrated with examples (pp. 212-16). Lesson XXXII, 'Government' (pp. 219-25) emphasizes the rights of individuals and the freedom of speech that prevail in Great Britain and the United States.

- **1855 (02)**
- *A new geography ...*
- LONDON; Sampson Low ...
- 18 cm.; pp. [5], vi-xv, [1], 237, [3]
- ***Libraries*** [IDT] : NNC

▸ ***Notes***
1. The p. following the tp. carries the statement: London: Bradbury and Evans, printers, Whitefriars.

STRUDWICK, E.P. **Y SG?**
▸ **{1837} (01)**
▸ *Geographical questions adapted to Keith's Geography and Butler's Atlas*
▸ LONDON
▸ 12mo
▸ ***Libraries*** L

SULLIVAN, R[obert?] **?**
▸ **{1844} (01)**
▸ *Geography and history*
▸ [LONDON]; Longman
▸ 18mo
▸ ***Libraries*** n.r.
▸ ***Source*** BCB, Index (1837-57:109)

▸ **{1881}, 4th [of the same work?] (02)**
▸ *An introduction to geography and history, ancient and modern*
▸ DUBLIN; Sullivan
▸ 12mo; pp. 3, 209
▸ ***Libraries*** : MB
▸ ***Notes***
1. BL also records ed. in 1893 (41st), 1916, and 1920; NUC records an ed. in 1919.

SULLIVAN, Robert **Y mSG**
▸ **{1840} (01)**
▸ *Geography generalized*
▸ DUBLIN
▸ ***Libraries*** n.r.
▸ ***Source*** Allford (1964:245). See note to (03).

▸ **{1848}, 9th, rev. ... (see below) (02)**
▸ *Geography generalized*; or, an introduction to the study of geography ... and an introduction to astronomy ... Ninth edition, revised and corrected
▸ DUBLIN; William Curry, Jun., & Co
▸ 12mo; pp. 288
▸ ***Libraries*** L
▸ ***Notes***
1. BL and NUC record further eds. (all Dublin): 1851, 15th, 288 pp. (: DLC); 1853, 17th (L : MH); 1854, 18th (L); and 1855, 20th, 320 pp. (: PV).

▸ **1858, 24th, revised and corrected (03)**
▸ *Geography generalized* ... geography on the principles of classification and

comparison, with maps and illustrations; and an introduction to astronomy. By ... LL.D., T.C.D.

- DUBLIN; Marcus and John Sullivan, 27, Marlborough-Street. Longman, Brown, Green, and Longmans, London; Fraser and Co., Edinburgh
- 15 cm.; pp. [5], v, [3], 8-320
- ***Libraries*** : : **CaBVaU**
- ***Notes***

1. Mathematical and physical geography, pp. 7-118; the method of teaching geography is presented on pp. 119-26 (credit given to Mr. Woodbridge 'the eminent American geographer' and Pestalozzi [p. 126]); introduction to astronomy, pp. 127-54; introduction to geography, pp. 155-272 (in the table of contents these pp. are identified as *Political Geography*; it is highly condensed special geography with a focus on toponyms); index, pp. 273-82; guide to pronunciation of names, pp. 283-90; introduction to geology, pp. 291-316.
2. BL and NUC record further eds. (both Dublin): 1859, 25th, 320 pp. (: MiU); and 1860, 26th (L : PBL).

- **1862, 29th (04)**
- *Geography generalized* ... and introductions to astronomy, history, and geology
- DUBLIN; Marcus and John Sullivan
- ***Libraries*** : : **CaBVaU**
- ***Notes***

1. BL and NUC record further eds. (all Dublin): 1863, 31st, 384 pp. (: TxDaM); 1866, 34th, 384 pp. (: ICRL, NN, NNU-W); 1868, 38th, 384 pp. (L : MH); 1873, 46th, 384 pp. (: NjP); 1874, 48th (L); 1875, 51st, 435 pp. (: CLSU); 1884, 67th, 448 pp. (: CU); 1887, 71st, 448 pp. (L). BL also records eds. in 1888, 74th; 1893, 78th; 1902; and 1905.

SULLIVAN, Robert **Y mSG**

- **{1846}, 6th (01)**
- *Introduction to geography.* Ancient, modern, and sacred; with an outline of ancient history
- MONTREAL; Armour & Ramsay
- ***Libraries*** : : CaBVaU
- ***Notes***

1. Judging by information provided in BL and NUC (NS 1050207), this work was first published in Dublin; see the information with respect to place of publication in (02).

- **1848, 7th (02)**
- *An introduction to geography* ...
- DUBLIN; Published by direction of the Commissioners of National Education, and reprinted by express permission, at Montreal, by Armour & Ramsay
- 18 cm.; 2 p. l., 140
- ***Libraries*** : : **CaNSWA**

▸ *Notes*
1. According to the Preface, this book is an introduction to SULLIVAN (1840), above. The table of contents shows that, as in the case of the more advanced book, the space devoted to special geography is relatively limited. Pp. 136-40 contain a Supplement added by the Canadian publishers, who thought 'the information respecting the British American provinces too meagre' (note at the foot of p. 136).
2. CIHM records 1848, 7th, Montreal; Armour & Ramsay (: : CaOLU).
3. BL and NUC record further eds. (all Dublin): 1848, 7th, 140 pp. (: : CaVBaU, CaNSWA); 1849, 11th, 143 pp. (L); 1853, 23rd (L).

▸ **1854, 23rd (02)**
▸ *An introduction to geography*, ancient ...
▸ NEW YORK; A. Cunningham
▸ Pp. [7], 12-180
▸ *Libraries* : : CaBVaU, CaOOA
▸ *Notes*
1. The special geography occupies pp. 33-124.
2. BL and NUC record further eds., all Dublin; 1855, 28th, 180 pp. (: MoU, PBL); 1861, 52nd (L); and 1869, 92nd (L).
3. Andrews states that this work was 'in its ninth edition by 1869,' and further that it 'was viewed by its author as an introduction to his *Geography Generalized* [see above]' (1986:229).
4. Microform versions of the 7th and 23rd eds. are available from CIHM.

[SULLIVAN, Robert] **Y SG?**
▸ **{1878} (01)**
▸ *Outlines of geography for junior classes*
▸ 12mo; pp. 84
▸ *Libraries* L
▸ *Notes*
1. BL records eds. in 1890 (32nd), 1916, and 1920.

SUMNER, Arthur
See note in sources to WARREN (1857), WARREN (1860b), and WARREN (1874).

SWAN, John **OG**
▸ **1635 (01)**
▸ *Speculum mundi*. Or a glasse representing the face of the world; shewing both that it did begin and must also end; the manner how, and time when, being largely examined. Where unto is joyned an hexameron, or a serious discourse of the causes, continuance, and qualities of things in nature; occasioned as matter pertinent to the work done in the six dayes of the worlds creation

- CAMBRIDGE; Printed by the printers to the Universities of Cambridge [frontispiece has: printed by T. Buck and R. Daniel]
- 4to, 18 cm.; pp. [16], 504, [26]
- ***Libraries*** C, Dt, Ercp, **L**, O : CSmH, CSt, CU-S, DFo, DLC, FU, IEN, IU, KyU, MH, MiU, MWiW, NjP, NN, NNUT, OClW, TxU
- ***Notes***

1. Not a special geography; rather, a sympathetic commentary on the account of the creation of the world, as recorded in Genesis. In other words, it deals with phenomena and processes now regarded as belonging to the field of physical geography, while accepting Genesis as though it provided a naturalistic description of the creation of the world. Supplements this source with information derived from authors of the Classical period. Uses the cosmology of Tycho Brahe as its model of the universe.

   The flavour of the period can be glimpsed in Swan's discussion of some points made by Mercator. The latter had argued that the world had been created in the summer time. Swan notes that Mercator's authority had been 'the Priests of Egypt, who, observing the river *Nilus* to overflow about the Summer Solstice, adored it as a God, esteeming the time of its inundation for an infallible beginning of divine actions in things created; and thereupon, for the beginning likewise of the yeare at the time of the worlds creation. But if this were the only cause, we may not unfitly say, that it was folly and superstition which first set this opinion abroach; and therefore he is worthy of blame who will go about to maintain it' (p. 29).
2. Later eds. in 1643, 1665, and 1670.

SWINTON, William **? Am**

- **{1875}a** (01)
- *A complete course in geography: physical, industrial, and political.* With a special geography for each state
- NEW YORK and Chicago; Ivison, Blakeman, Taylor, and company
- 31 × 25 cm.; pp. iv, [2], 134, [2]
- ***Libraries*** : DHEW, DLC, DI-GS, MiU, OO, OrU
- ***Notes***

1. NUC records further eds. (New York, 141 pp., except as noted): 1875, [New] (DHEW, DLC, DI-GS, MiU, OO, OrU); 1877, 134 pp. (ICJ, LU); [Elson (1964:404) records an ed. in 1878, pp. n.r.]; 1879 (MiU, NN); and [188–?] (IU). NUC also records eds. in [1890, c1875]; and 1903.
2. Elson (1964:404) cites the ed. of 1878.
3. Discussed in Brigham and Dodge (1933:24-25).

SWINTON, William **Y Am**

- **{[1875]}b** (01)
- *Elementary course in geography*: designed for elementary and intermediate grades, and as a complete shorter course
- NEW YORK and Chicago; Ivison, Blakeman, Taylor, and company
- 27.5 cm.; 2 p. l., 128 p.
- ***Libraries*** : DHEW, DLC, ICRL, MH, NN, OFH, OkU, OrU, PPT

- ▸ ***Source*** NUC (NS 1104926)
- ▸ ***Notes***

1. NUC records further eds.: [1877? c1875], New York and Chicago (MH); and 1903.
2. Brigham and Dodge (1933:24-25) and Culler (1945, *passim*) discuss; Hauptman (1978:429) discusses (01) and quotes from it (1978:433 , 435); Elson (1964:404) cites (01).

SWINTON, William **Y Am**

- ▸ **{1880} (01)**
- ▸ *A grammar-school geography, physical, political, and commercial*
- ▸ NEW YORK and Chicago; Ivison, Blakeman, Taylor, and company
- ▸ 32 cm.; pp. iv, [2], 118
- ▸ ***Libraries*** : DHEW, DLC, MB, NWM
- ▸ ***Source*** NUC (NS 1104959). See note to (02).
- ▸ ***Notes***

1. NUC records a further ed.: [c1880], New York and Chicago, Ivison, Blakeman, Taylor, 116, 32 pp. (MH, NN) [NS 1104963 notes: '"Middle states. Special geography of New York, New Jersey, Pennsylvania, Maryland, Delaware, and the District of Columbia. Designed to accompany Swinton's Grammar-school geography", 32 p.']; Culler (1945:310-311), records further eds. (all 1880, New York, and Ivison, Blakeman, Taylor, unless stated otherwise): New Jersey and Pennsylvania ed., American Book co., library n.r.; New England ed., library n.r.; Middle States ed., library n.r. NUC records further eds. (all 1881, New York and Chicago, Ivison, Blakeman, Taylor, unless stated otherwise): East-Central State ed., 118, 33 pp. (DHEW, DLC, PBa); Middle states ed. 118, 31 pp. (DLC); New England ed., [New York only], 118, 33 pp. (DHEW, DLC); New York state ed., 118, 9 pp. (DHEW, DLC); and West central states ed., 118, 38 pp. (DHEW, DLC).

- ▸ **1883, c1880 (02)**
- ▸ *A grammar-school geography: physical, political, and commercial.* By ... gold-medalist for geography, Paris Exhibition, 1878, and author of Swinton's Geographical Series, World-Book Series, Outlines of the World's History, etc.
- ▸ NEW YORK and Chicago; Ivison, Blakeman, Taylor, and company
- ▸ 30 cm.; pp. [iii], iv, [3], 2-118, 37
- ▸ ***Libraries*** : : **CaBVaU**
- ▸ ***Notes***

1. Intended to be the sequel, at a higher level, to Swinton's *Introductory geography* (see below, 1882); similar in level to the *Complete course in geography* (SWINTON 1875a, above), but differs in that special eds. of this work were issued containing 'the detailed geography of individual states or sections' (see above). Contains both physical and political geography, the two being blended so as to 'present themselves to the pupil's mind, not as isolated phenomena, but as a connected whole' (p iii). The text consists of numbered paragraphs. The information tends to be utilitarian, commercial, and administrative in nature, resembling that found in present-day works such as the *Statesman's Yearbook.*

The supplementary 37 pp. are devoted to: 'Southern States. Special geography of Virginia, West Virginia, North Carolina, South Carolina, Georgia, Alabama, Florida, Mississippi, Lousiana, Arkansas, Texas, Kentucky, and Tennessee.'

2. NUC records a further ed.: [1887?, c1880], New York and Chicago (MH). NUC also records eds. in [1888]; [1890?, c1880]; [1896?]; [189–?]; and 1908.
3. Culler (1945, *passim*) discusses the eds. of 1880; Elson (1964:404) cites (01).

SWINTON, William **Y Am**

- **{c1882} (01)**
- *Introductory geography in readings and recitations*
- NEW YORK and Chicago; Ivison, Blakeman, Taylor, and company
- 23.5 cm.; 1 p. l., 116 p.
- ***Libraries*** : DHEW, DLC, MB, MoU, NN, PBa, PHi
- ***Source*** NUC (NS 1104977)
- ***Notes***

1. According to NUC (NS 1104978), which refers to an ed. of [1890?, c1882], this work 'Forms introduction to the author's Grammar school geography.'
2. Discussed in Culler (1945, *passim*).

SWINTON, William **Y Am**

- **{1878} (01)**
- *Primary geography*
- NEW YORK; Ivison, Blakeman, Taylor, and Co.
- ***Libraries*** n.r.
- ***Source*** Culler (1945:310)

- **{[c1879]} (02)**
- *Primary geography*: introductory to Swinton's 'Grammar school geography'
- NEW YORK, Cincinnati; American book compay
- 23.5 cm.; pp. vi, 90
- ***Libraries*** : NBuG
- ***Notes***

1. NUC records further eds. (both New York and Chicago): 1879, 90 pp. (DLC, NN); and 1880 (MH).
2. Culler (1945, *passim*) discusses (01).

[SYMSON, Matthias] **Y n.c.**

- **1702 (01)**
- *Geography compendiz'd; or, the world survey'd*: being a system of geography, describing all the empires, kingdoms, and dominions of the earth. Volume 1, containing an introduction to geography, and a description of Europe. Collected from the most approv'd writers on that subject for the use of the Marquess of Douglass, by M. S.
- EDINBURGH; sold by Mr Henry Knox and John Vallange

- 8vo, 14.5 cm.; pp. [4], xxiv, 320
- ***Libraries*** **L**
- ***Notes***

1. The author's name appears at the end of the Dedication. Judging by a reference in the Preface, Symson was inspired by Eacherd. The first special geography published in Scotland? Dedicated to a young nobleman, presumably for his education.
2. Alex. M. Frizzell (bookseller) of Castlelaw, West Linton, Peebleshire, Catalogue #214, *Scotland in the 18th Century*, describes Vol. I as 'all published' (p. 19).

[SYMSON, Matthias] **A**

- **1704 (01)**
- *Encheiridion geographicum*. Or, a manual of geography. Being a description of all the empires, kingdoms, and dominions of the earth. Shewing their situation, bounds, dimensions, commodities, ancient and modern governments, divisions, chief cities, religions, languages, chief rivers, Arch-bishopricks, bishopricks, and universities. Collected from the best books on that subject
- EDINBURGH
- 8vo, 15 cm.; pp. [72]
- ***Libraries*** **L**
- ***Notes***

1. Little more than a pamphlet; no preface, index, or indication of printer or bookseller. The pages vary considerably in size.

SYNOPSIS ... (1785) **Y SG? Am**

- **{1785} (01)**
- *A synopsis of geography*, with the use of the terrestrial globe, intended for the benefit of youth, especially that of the students in the public grammar school in Wilmington
- WILMINGTON, [DE?]; printed by James Adams
- 8vo; pp. 58
- ***Libraries*** : DeWI, MWA
- ***Source*** NUC (NS 1120296)

SYSTEM OF GEOGRAPHY ... (1701)
See (02) of THESAURUS GEOGRAPHICUS (1695).

SYSTEM OF GEOGRAPHY ... (1805) **A**

- **1805 (01)**
- *A system of geography*; or, a descriptive, historical, and philosophical view of the several quarters of the world, and of the various empires, kingdoms, and republics which they contain: particularly, detailing those alterations which have been introduced by the recent revolutions. With an appendix,

containing those astronomical remarks which are necessary in forming a true notion of the nature of the earth. In four volumes. By a literary society
- GLASGOW; printed by Niven, Napier and Khull, Trongate, for W. and D. Browlie, booksellers
- 22 cm., 4 vols.; Vol. I, pp. [3], ii-xiv, 657; Vol. II, pp. [4], iv-viii, 704; Vol. III, pp. [4), iv-vi, 639; Vol. IV, pp. [4], iv-vi, 688
- ***Libraries*** EU : DLC
- ***Notes***

1. Vol. I is devoted to America, Vol. II to Africa and Asia, Vol. III to most of Europe, and Vol. IV to the British Isles and France. Contains more commercial, botanical, and zoological information than PINKERTON (1802), but lacks his statements of method and objective. Contains some extensive bodies of direct quotation. Might be a forerunner to GLASGOW GEOGRAPHY (1822).

- **{1812} (02)**
- *System of geography*
- GLASGOW
- ***Libraries*** n.r.
- ***Source*** Allford (1964:243), who states that the author is J. Bell. See BELL (1829-32).

SYSTEM OF GEOGRAPHY ... (1834) **Y**
- **1834 (01)**
- *A system of geography, compiled for the use of schools*
- DUBLIN; printed by Richard Davis Webb, 27, Great Brunswick-Street
- 12mo, 17 cm.; pp. [4], iv, 364
- ***Libraries*** L: NNC (imperfect)
- ***Notes***

1. Special geography, much of it in small type, occupies almost the whole book. No hint as to the author or publisher.

T

TAYLOR, E.W., *Mrs* **?**
- **{1868} (01)**
- *Geographical questions*
- ? Hamilton
- ***Libraries*** n.r.
- ***Source*** ECB, Index (1856-76:130)

- **{[1870]} (02)**
- *One thousand geographical questions*, at home and abroad ... with notes, etc.
- LONDON

- 16mo
- ***Libraries*** L

[TAYLOR, Emily] **?**
- **{1809} (01)**
- *The ball I live on; or, sketches of the earth*
- LONDON; John Green
- 12mo; pp. iv, 104
- ***Libraries*** L
- ***Notes***

1. Allibone (1854) records an ed. in 1841. BL and NUC record an ed. in [1846], 104 pp. (L : TxU).

TAYLOR, Isaac **Y mSG**
- **{1821} (01)**
- *Scenes all the world over*, for the amusement and instruction of little tarry-at-home travellers. By the Rev. ...
- LONDON; printed for Harris and son
- 16 cm., 2 vols.
- ***Libraries*** : : CaOTP
- ***Source*** St. John (1975, 2:812)
- ***Notes***

1. According to St. John (1975, 1:190-1), Taylor began by writing separate vols. entitled *Scenes in Africa, Scenes in America, Scenes in Asia*, and *Scenes in Europe*. In her Vol. II, p. 812 she reports that the 2-vol. set to which this is a note was made up of Vol. I, *Scenes in Europe*, 5th, and *Scenes in Asia*, 2nd; and Vol. II, *Scenes in Africa*, 2nd, and *Scenes in America*.

   According to the same sources, there were subsequent eds. of all the volumes separately, as well as *Scenes in England*.

TAYLOR, Isaac **Y mSG**
- **{1829} (01)**
- *Scenes in foreign lands*. From the portfolio and journal of a traveller in various parts of Europe, Asia, Africa, and America. Upon a plan arranged by the late Rev. Isaac Taylor
- LONDON; Grant and Griffith
- 18 cm.; pp. xviii, 328
- ***Libraries*** : DLC
- ***Notes***

1. NUC records further eds. (both London): 1841, J. Harris (MH); [1843?], Grant and Griffith, 328 pp. (: : CaOTP).

TAYLOR, Jefferys **Y NG**
- **{1848} (01)**
- *Glance at the globe, and at the world around us*

- LONDON; Houlston and Stoneman
- 17 cm.; pp. xv, 247
- ***Libraries*** L : : CaOTP
- ***Source*** St. John (1975) and letter from CaOTP. The book belongs to the tradition of natural philosophy rather than to geography.

TAYLOR, Jefferys **OG**

- **{1832} (01)**
- *A new description of the earth, considered chiefly as a residence for man*
- LONDON; printed for Harvey and Darton
- 17 cm.; pp. [6], 164
- ***Libraries*** : **NN** : CaOTP
- ***Notes***

1. The work of a pious mind. Primarily a systematic physical geography; however, there are sections on 'thought' and the Bible.

TAYLOR, John **NG**

- **1637 (01)**
- *The carriers cosmographie*. Or a briefe relation, of the innes, ordinaries, hostelries, and other lodgings in, and neere London, where the carriers, waggons, foote-posts, and higglers, doe usually come, from any parts, townes, shires and countries, of the kingdomes of England, principality of Wales, as also from the kingdomes of Scotland and Ireland. With nomination of what daies of the weeke they doe come to London, and what daies they return ... As also, where the ships, hoighs, barkes, t[i]ltboats, barges and wherries, do usually attend to carry passengers, and goods to the coast townes of England, Scotland, Ireland, or the Netherlands; and where the barges and boats are ordinarily to bee had that goe up the river of Thames westward from London
- LONDON; printed by A[nne] G[riffin]
- 4to; pp. 24
- ***Libraries*** L : MB, MdBP, MH, MnU, NIC, OU
- ***Source*** BL and NUC (NT 0066285). Only the short title suggests that this might be a special geography.

[TEACHER, A] (i.e., a teacher) **Y Am**

- **{1836} (01)**
- *The village school geography*. Embellished with numerous engravings and ten [neatly engraved] maps
- HARTFORD; White, Dwier & Co.
- 16mo
- ***Libraries*** L
- ***Source*** Sahli (1941:383) and BL
- ***Notes***

1. The words in the title set in [ ] are taken from the NUC entry for the ed. of 1840 (Vol. DLXXXV, p. 228).

2. Culler (1945:307) and NUC (Vol. DLXXXV, p. 228) record a further ed.: 1840, 4th, Hartford, library n.r.
3. Nietz (1961:210, 219), Sahli (1941, *passim*), and Brigham and Dodge (1933: 17) all discuss (01); Culler (1945, *passim*) discusses the ed. of 1840; Elson (1964:401) cites the ed. of 1837.

TEGG, Thomas **Y**

- **{1858} (01)**
- *Tegg's first book of geography for children*
- LONDON
- 18mo
- ***Libraries*** L
- ***Source*** BL and ECB, Index (1856-76:130)

TEMPLEMAN, Thomas **mSG**

- **[1729] (01)**
- *A new survey of the globe: or, an accurate mensuration of all the empires,* kingdoms, countries, states, principal provinces, counties, & islands in the world. The area is given in square miles, by which the extent, magnitude, and true proportion, that one country bears to another, are exactly known. The distant and separate territories, of every prince, and state, are collected together, and so regularly plac'd, that at one view, their whole dominions may be seen. The chief city, or town, of every kingdom, province, and island, the longitude, latitude, & nearest distance from London in British miles. The Protestant kingdoms and states, distinguish'd from those of the Roman Catholicks; a comparison between the greatness and extent of their several dominions; the difference ballanc'd and demonstrated. Also the antient Persian and Roman empires, compar'd with the present Russian, Turkish and other great empires: and what proportion the known and habitable earth bears to ye seas and unknown parts. A collection of all the noted seaports in the world, shewing in what country they are, & to whom subject, with their longitude, latitude, and distance from the port of London, by sea; also the settlements & factories, belonging to the English, Dutch, French, Portuguese, Spaniards, etc. in the East and West-Indies, Africa, and other parts. With notes explanatory & political, wherein the number of people in all ye principal countries and cities of Europe are severally calculated from the number of houses or bills of mortality
- LONDON; engraved by J. Cole in Great Kirby Street. Hatton Garden, [may refer only to the tp.] London
- 4to, 26 × 36 cm.; pp. [4], ix, [1], 35 l. of plates [i.e., tables, 1 per leaf]
- ***Libraries*** L : **CtY**, CU, DLC, MH, NIC, RPJCB
- ***Notes***

1. Primarily a statistical compilation, but a few lines of text accompany each table. The pagination of the CtY copy is: [4], ix, [5], and then the tables.
2. According to Wright (1959:156), this is the source of information of this type included by GUTHRIE (1770), and MORSE (1789a), who copied from Guthrie.

- **[1765?] (02)**
- *A new survey of the globe ...*
- LONDON; printed for John Bowles at No. 13 in Cornhill
- Obl. 4to, 24 × 35 cm.; pp. [4], ix, [1], 35 l. of plates
- ***Libraries*** **L** : PPULC
- ***Notes***

1. BL dates this ed. as 1730?. However, on plate 27, which is the only plate where the text that accompanies it has been changed, there is a reference to the 'Peace of 1763 ... soon after [which] ye governments of Canada, Quebec, and East and West Florida were established.'
2. NUC notes an ed. of 1720 (NT 0089988). PPL reports that this is a ghost.

THAYER, William A. **Y Am**

- **{1815} (01)**
- *Compendium of geography being a concise description of the various parts of the world*; adapted to the capacities of children and youth
- PITTSFIELD, MA; Allen
- Pp. 82
- ***Libraries*** : NN, PU
- ***Source*** NUC (NT 0135286)

- **{1817} (02)**
- *A compendium of geography ...*
- PORTLAND; Mussey and Whitman
- 18 cm.; pp. 86
- ***Libraries*** : DLC, MB, MWA
- ***Notes***

1. Discussed in Brigham and Dodge (1933:13).

THESAURUS GEOGRAPHICUS **A**

- **1695 (01)**
- *Thesaurus geographicus*. A new body of geography: or, a compleat description of the earth: containing I. By way of introduction, the general doctrine of geography. Being an account of the situation, figure and bigness of the earth in respect of the rest of the world, its division into land, water and air, with several remarks upon the nature and properties of each. Also the division of the surface of the land and water. Together with the doctrine of the sphere, the use of globes and maps, etc. II. A description of all the known countries of the earth: an account of their situation, bounds and extent, climate, soil and production, chief rivers, mountains and seas: together with the general history and succession of the princes and the religion, manners and customs of the people. Also analytical tables; whereby is shewn at a view, the division of every kingdom or state into provinces and counties, with their divisions into diocesses, bailywicks, etc. and the chief towns situated in each. III. The principal cities and most considerable towns in the world, particularly and exactly describ'd: shewing the magnitude, principal buildings, antiquity, state, condition, etc. of each place; as also the situation, with its distance and bearing from other towns, for the easier finding it in the map.

IV. Maps of every country of Europe, and general ones of Asia, Africa and America, fairly engraven on copper, according to the best and latest extant. And also particular draughts of the chief fortified towns of Europe. Collected with great care from the most approv'd geographers and modern travellers and discoveries, by several hands. With an alphabetical table of all the towns names

- LONDON; printed for Abel Swall and Tim. Child, at the sign of the Unicorn at the west-end of St. Paul's-Church
- Folio, 32 cm.; pp. [8], 506, [12]
- ***Libraries*** EU, HIU, **L**, NP, O, [IDT] : CLU, CtY, DFo, DLC, IaU, IEN, LU, NmU, NcU, NjN, OCl, OKentU, PPL, RPJCB : CaNSWA
- ***Notes***

1. The Preface contains a definition of geography as it was then understood; it is essentially that of Varen. Europe is given much more space than the other continents, but no country within Europe is particularly favoured. Both the Ptolemaic and Copernican accounts of the universe are given, with a slight preference being shown for the latter. In the Preface it is stated that the maps and plans are chiefly derived from continental sources (Sanson, De Wit, and Vischer being mentioned in the case of the maps, and du Fer in the case of the town views).
2. There was also a 4to ed., according to Beeleigh Abbey Books [W. & G. Foyle Ltd., London], *A Catalogue of Voyages and Travels*, Catalogue BA/23, n.d., item 71.

- **1701 (02)**
- *A system of geography: or, a new & accurate description of the earth in all its empires, kingdoms and states*. Illustrated with history and topography, and maps of every country, fairly engraven on copper, according to the latest discoveries and corrections, by Herman Moll. To which are added alphabetical index's of the names, ancient and as well as modern, of all the places mention'd in the work. And a general index of remarkable things
- LONDON; printed for Timothy Childe at the White Hart at the west-end of St. Paul's Church-yard
- Folio, 31 cm., 2 pts., pp. [31], 26, 444; [2], 230, [28]
- ***Libraries*** **L**, LUBd : DLC, FTaSu, InU, IU, MBrZ, MdAN, MiU, NjP, RPJCB, Vi
- ***Notes***

1. The tp. of Pt. 2 reads: 'A system ... earth, in all ... states. Part the second, containing the description of Asia, Africa, and America. Written in Latin by Joan Luyts Professor in Acad. Ultraj. English'd with large additional accounts of the East-Indies, and the English plantations in America. Illustrated with maps, fairly engraven on copper, according to the modern discoveries and corrections, by Herman Moll. Printed for Tim. Childe, at the White-Hart in St. Paul's Church-Yard.'

   In the Preface the editors state that this work is based on Cluverius 'as improv'd by Buno, Hekelius, and Reiskius,' Sanson, Luyts, and the English Atlas. The first ed. of Cluverius, *Introductionis in Universam Geographiam*, was published in 1624.

   Pt. 1 contains an 'Introduction to geography' by Robert Falconer.

2. The reason for looking on this work as being the second ed. of the *Thesaurus geographicus* is provided in the notes to (03), below.

- **1709, 3rd (03)**
- *The compleat geographer*: or, the chorography and topography of all the known parts of the earth. To which is premis'd an introduction to geography, and a natural history of the earth and the elements. Containing a true and perfect account of I. The situation, bounds and extent, climate, soil, productions, history, trade, manufactures: the religion, manners and customs of the people; with the revolutions, conquests and other changes of all the countries on the earth. II. The several provinces that every kingdom or state is divided into. III. The principal cities and most considerable towns in the world, the magnitude, principal buildings, antiquity, present state, trade, history, etc. As also the situation, with the distance and bearing from other towns: together with all necessary pieces of natural history. The whole containing the substance of at least an hundred and fifty books of modern travels, faithfully abstracted and digested into local order; whereby the present state of the most remote countries is truly shewn, and the obsolete and fabulous accounts of former writers wholly expung'd. To which are added maps of every country, fairly engraven on copper, according to the latest surveys, and newest discoveries, most engrav'd by Herman Moll. The third edition. Wherein the descriptions of Asia, Africa and America are compos'd anew from the relations of travellers of the best repute, especially such as have appear'd within thirty or forty years last past
- LONDON; printed for Awnsham and John Churchill, at the Black Swan in Pater-Noster-Row. And Timothy Childe, at the White-Hart, at the west-end of St. Paul's Church-Yard
- Folio, 32 cm.; Pt 1; pp. [26], lv, [5], 482; Pt 2; pp. [2], 341, [1], xxii
- *Libraries* **L**, LcP : **CtY**, ICN, MBAt, PBa : CaBViPA
- *Notes*

1. The tp. to the second part reads: 'Thesaurus geographicus: or the compleat geographer. Part the Second. Being the chorography, topography, and history of Asia, Africa, and America. Faithfully extracted from the best modern travellers and most esteemed historians: and illustrated with maps, fairly engraved on copper, according to the most modern discoveries and corrections by Herman Moll. The third edition very much enlarg'd.'

   So this work identifies itself as the third ed. However, no earlier eds. with exactly this title are known. It thus seems likely that the first two eds. had somewhat different titles. The evidence in favour of both this opinion, and also for identifying *The compleat geographer* as their successor is provided in the following extract from the 'Advertisement' that follows the tp.: 'A blame may be laid upon us for sending out the last edition, without the improvements that are now made. To that we answer, that these improvements were then intended, and would have been made, if an unforeseen incident had not precipitated that publication; for when the maps were all engrav'd, and part of the book printed, a sudden and unexpected notice was given of an intended new edition of Dr. Heylin's Cosmography, wherein it was suppos'd the new editor would have inserted the observations of modern travellers, and renew'd that learned author's descriptions. For this reason the design that was then on foot, namely, to abstract all the modern travels into Asia, Africa and America, whereby to make the descriptions of those parts of the world as compleat as that of Europe, was by necessity shorten'd. But now that editor has shewn himself, and let us see that our first design is perfectly unperform'd by him, we have resum'd it.'

An ed. of HEYLIN (1652) 'improv'd with an historical continuation to the present time by Edmund Bohun' was published in 1703. Two years earlier a folio 'description of the earth,' of comparable bulk, with maps by Moll, and printed by for Timothy Childe at the White Hart at the west end of St. Paul's churchyard, appeared in the book shops (see (02), above). Six years earlier still, a folio 'description of the earth,' only a little smaller in size and printed for Tim. Childe (along with Abel Swall), was published (see (01), above.

An Advertisement (pp. [5-6]) justifies special geography in the following words: 'As the knowledge of foreign countries is a science that no man of either learning or business can excusably be without, so there is no certain way of obtaining it, but by consulting the travellers that have been upon the spot. But the number of travellers is so great, and their writings so voluminous, that the study of them is tedious; and considering the many unnecessary things contain'd in such writings, the reading 'em is even irksome. Wherefore are abstracted ... to the purpose of geography, and their accounts of places dispos'd in right method, cannot but be acceptable to the publick.'

The authors 'abstracted in this work' are listed on pp. [6-7].

2. Catalogued by BL under Geographer.

▸ **1723, 1722, 4th (04)**
▸ *The compleat geographer* ... by Herman Moll ... Wherein the descriptions ... forty years last past
▸ LONDON; printed for J. Knapton, R. Knaplock, J. Wyat, J. and B. Sprint, J. Darby, D. Midwinter, E. Bell, A. Bettesworth, W. Taylor, W. and J. Innys, R. Robinson, J. Osborn, F. Fayram, J. Pemberton, J. Hooke, C. Rivington, F. Clay, E. Symon, J. Batley, J. Nix, and T. Combes
▸ Folio, 35 cm.; Pt 1, pp. [22], li, [5], 402; Pt 2, pp. [2], 288, xx
▸ *Libraries* L, O : CSt, **CtY**, DFo, DLC, IaU, IU, MH, MiD, MiU, MnHi, MnU, NN, NNC, OCl, OClW, OkU, PPL, RPJCB, Vi
▸ *Notes*

1. The tp. to the second part reads: 'Thesaurus ... travellers, and ... The fourth edition, much amended. 1722.'

The 'advertisement concerning this new edition' that follows the tp. is the same as that of 1709, though it is here squeezed onto one page (i.e., p. [5]).

An index of the places in Europe follows the table of contents in the opening series of unnumbered pages, as it did in the ed. of 1709. The two are not identical; a check of the contents of the letter K revealed a number of changes in spelling between the two eds. A search of the text of the two eds., in those cases where there were differences between the contents of the two indexes suggests that the ed. of 1723 was proofread more carefully than that of 1709.

The reduction in the number of pages achieved between the eds. of 1709 and 1723 seems to have been achieved chiefly by an increase in the total length of a line of text, which is set in two columns, from 15 to 16 cm. to 18 to 18.5 cm. This change in format makes a quick comparison of the text

of the two eds. impossible. Spot-checking, however, suggests that the changes were few and mostly minor.
2. For works of comparable size, see: ATLAS GEOGRAPHUS (1711-17), COMPLETE SYSTEM (1744), MOLL (1739).
3. A microform version of (03) is available from CIHM.

THOMAS, J. and BALDWIN, T., eds. **Dy Am**
- **{[1855]} (01)**
- *A complete pronouncing gazetteer, or geographical dictionary, of the world*
- PHILADELPHIA?
- ***Libraries*** n.r.
- ***Notes***
1. Known from the copyright statement in (02).

- **1861 (02)**
- *A complete pronouncing gazetteer, or geographical dictionary, of the world.* Containing a notice and the pronunciation of the names of nearly one hundred thousand places. With the most recent and authentic information respecting the countries, islands, rivers, mountains, cities, towns, etc. in every portion of the globe. Including the latest and most reliable statistics of population, commerce, etc. Also, a complete etymological vocabulary of geographical names, and many other valuable features, to be found in no other gazetteer in the English language. Edited by ... assisted by several other gentlemen. Revised edition, with an appendix containing nearly ten thousand new notices and new tables of population according to the latest census returns of the United States and foreign countries
- PHILADELPHIA; J.B. Lippincott & Co.
- 24.5 cm.; pp. [iii], iv-xviii, [1], 10-2182
- ***Libraries*** : : **CaBVaU**
- ***Notes***
1. NUC (NSC 0099631) catalogues under Complete ... NUC also records a further ed.: 1873, Philadelphia (: : CaBVaU).

THOMSON, James **Y**
- **1827 (01)**
- *An introduction to modern geography*: with an appendix, containing an outline of astronomy, and the use of the globes
- BELFAST; printed and published by Sims and M'Intyre, Donegall-Street
- 12mo, 18 cm.; pp. [6], 270
- ***Libraries*** L
- ***Notes***
1. The text, which is often not much more than a list of places, is supplemented by footnotes; these are sometimes more copious than the text itself.

- **1831, 3rd (02)**
- *An introduction to modern geography ...*
- BELFAST; Simms and M'Intyre
- Pp. [vi], 272, [4]

- *Libraries* : : CaQQS
- *Notes*

1. The special geography occupies pp. 9-213.
2. BL and NUC record further eds.: 1850, London (: NjNbS); and 1857, 27th, 288 pp. (L).
3. A microform version of the ed. of 1831 is available from CIHM.

[THORIE, John] **Dy**

- **1599 (01)**
- *The theatre of the earth*. Containing very short and compendious descriptions of all countries, gathered out of the cheefest cosmographers, both ancient and moderne, and disposed in alphabeticall order. For the benefit of all such as delight to be acquainted with the knowledge of strange countries, and the scituation thereof, and especially for trauellers, to whom the portabilitie of this small volume will not be little commodious. What is performed in this booke is more at large set downe to the readers view in the next leafe
- LONDON; printed by Adam Islip
- 4to, 19 cm.; pp. more than 113 [last entry Thinissa] in the BL copy
- *Libraries* **L**, MaPL
- *Notes*

1. As the full title states, this is a dictionary, not special geography; the first in English. According to the preface the information given is very much of the type identified in the long titles that became fashionable late in the seventeenth century. There is a strong bias in favour of place names recorded in Classical literature: England appears as Anglia, France as Francia, but to get the information provided about the latter, the reader must look up Gallia (there is no cross reference). The author is identified in Pollard and Redgrave (1976).

- **1601, anr ed. (02)**
- *The theatre of the earth* ... alphabetical order
- LONDON; [Printed by Adam Islip] for William Iones, and are to be sold at his shop near Holbourne conduit at the signe of the Gun
- 4to, 20 cm.; pp. A-Z^4, Aa-Ff^4
- *Libraries* : DFo
- *Notes*

1. In this ed. the dedication is signed by Giovanne Thorisi, hence the attribution of authorship [NUC (NT 0136600)].

TOMLINS, F[rederick] G[uest], comp. **A**

- **{1835} (01)**
- *A complete system of geography, ancient and modern*; comprising a full description of the world, physical, political and historical; founded on the works of Pinkerton, Guthrie, Goodrich, Adams, etc. corrected by reference to Malte-Brun, Balbi, Hassel, and other continental and American authors: including the most recent discoveries, and the latest territorial arrangements. Compiled and arranged by ...
- LONDON; Mayhew, Isaac, and Co.

- 24 cm.; pp. [iii]-xi, 1067, [1]
- ***Libraries*** : DLC
- ***Notes***

1. Further eds. are known: 1841, Royal 8vo [William Smith (Booksellers) Ltd., 35-39 London Street, Reading, UK, *Catalogue No. 205 (1980)*, item 635]; 1842, Halifax [UK], 1067 pp. (: DLC) [recorded in NUC]; 1844, Halifax [UK], Royal 8vo, 1080 pp. [Howes Bookshop Ltd, 3 Trinity Street, Hastings, Sussex, UK; *Catalogue 191*, item 1824].

- **1845 (02)**
- *A complete system of geography* ... arranged by ... editor of 'The History of the United States,' 'Ancient Universal History,' etc.
- HALIFAX; printed and published by William Milner, Cheapside
- 24 cm.; pp. [5], viii-xi, [1], 1067, [5]
- ***Libraries*** **L : : CaBVaU**
- ***Notes***

1. 'The chief ... advantage to be derived from ... geography being the acquirement of such a knowledge of the various regions of the globe as will elucidate history and politics, the following compendium is principally devoted to that object; and, consequently, all purely scientific matter, both geological and astronomical, has been avoided ... As the peculiar object of geographical works is to describe the natural and political circumstances, the inhabitants, the animals, the productions, the state, progress, and prospects of the nations of the earth, the editor of the present has has [sic] sought to furnish, from the latest accounts, the fullest particulars of those interesting subjects' (p. v). 'To the friends of education the present Geography is confidently recommended, as a means of cheaply disseminating knowledge, which will enlarge, and, consequently, refine the mind; whilst it does not contain a sentence that can shock the delicacy, or wound the feelings of any class' (p. vi).

   Uses tiny print; very detailed; shows the tendency to focus on history to be expected from the full title and the author's other books; displays a belief in progress which the 'vilest and wickedest of kings and governments cannot stop' (p. 63).

   Makes occasional use of direct quotations, which, if long, are likely to be introduced by a mention of the author's name.

TORQUEMADA, Antonio de **mSG tr**

- **1600 (01)**
- *The Spanish Mandeuile of miracles*. Or, the garden of curious flowers. Wherin are handled sundry points of humanity, philosophy, diunity, and geography, beautified with many strange and pleasant histories. First written in Spanish by Anthonio de Torquemada, and out of that tongue translated into English. It was dedicated by the author, to the right honourable and reuerent prelate, Don Diego Sarmento de Soto Maior, Bishop of Astroga. etc. It is diuided into sixe treatises, composed in a manner of dialogue, as in the next page shall appeare
- LONDON; printed by I.R. for Edmund Matts, and are to be solde at this shop, at the signe of the hand [sic] and Plow in Fleet-streete
- 19 cm.; [10] pp., 158 numbered leaves

- ***Libraries*** : CSmH, CtY, DFo, DLC, ICN, **MH**, NNUT
- ***Notes***

1. Only one of the six parts into which the book is divided is devoted to special geography, and in that part it is limited to a treatment of northern lands; there it is noticeably concerned with marvels and wonders. Sources are cited; they are chiefly Classical, such as Strabo and Pliny.

- **[1618] (02)**
- *The Spanish Mandevile of myracles* ... or the garden ...
- LONDON; imprinted by Bernard Alsop, by the asssigne of Richard Hawkins, and are to be solde at his house by Saint Annes church neere Aldergate
- 19 cm.; pp. [6], 325, [3]
- ***Libraries*** : CSmH, CtY, DFo, DLC, MB, **MH**, NIC, NNH, NNUT, OCl, PU

TOUSSAINT, François Xavier **Y SG Can**

- **{1871} (01)**
- *Abridgement of modern geography for the use of elementary schools* by ... professor at Laval Normal School[.] Translated by the Ursuline ladies
- QUEBEC; C. Darveau, printer and publisher
- 17 cm.; pp. [5], 6-91
- ***Libraries*** : : CaQQS
- ***Source*** CIHM
- ***Notes***

1. 'Our object in this elementary geography for the use of beginners, has been, to present in a simple, analytical, and clear manner, the most useful and correct geographical information' ([3]). Information is presented in numbered paragraphs, and with lists of questions intended 'as a means of engraving the subject more deeply on the mind of the pupil' ([4]). Special geography, beginning with the Americas, and Canada within them, occupies pp. 14-87.
2. A microform version is available from CIHM.

TRAVELLER **Y mSG**

- **{[1820]} (01)**
- *The traveller*; or, an entertaining journey round the habitable globe; being a novel and easy method of studying geography. Illustrated with forty-two plates, consisting of views of the principal cities of the world, and the costume of its various inhabitants
- LONDON; printed for J. Harris and son
- 17 cm.; pp. viii, 204
- ***Libraries*** : : CaOTP
- ***Source*** St. John (1975, 1:192), which notes that the plates are dated 1820
- ***Notes***

1. NUC records further eds. (all London, J. Harris): [1820], 2nd, 204 pp. (CU, ICU); [182–], 3rd, 204 pp. (DLC, MH : CaBVaU); [1820], 4th, 219 pp. (MH, NN).

TREATISE **A**

- **1718 (01)**
- *A treatise of the description and use of both globes*. To which is annexed, a geographical description of our earth
- LONDON; printed for John Senex at the Globe in Salisbury Court, near Fleetstreet, and W. Taylor and the Ship in Pater-noster Row
- 12mo, 15 cm.; pp. [12], 114, [6]
- ***Notes***

1. The description of the world occupies pp. 69-114. No indication of intended audience, but consists of text, not lists and tables. Apart from some indication of location, the contents of the brief miscellany of information provided about each country is unpredictable.

TROUTBECK, Anne **Y SG?**

- **{[1869]} (01)**
- *Great steps for little scholars; or, geography simplified*, etc.
- LONDON, Manchester [printed]; Simpkin [Heywood, the printer]
- 16mo
- ***Libraries*** L
- ***Source*** BL and ECB, Index (1856-76:131)

TRUSLER, John **A n.c.**

- **1788-97 (01)**
- *The habitable world described*, or the present state of the people in all parts of the globe, from north to south; shewing the situation, extent, climate, productions, animals, etc. of the different kingdoms and states; including all the new discoveries; together with the genius, manners, customs, trade, religion, forms of government, etc. of the inhabitants, and every thing respecting them, that can be entertaining or informing to the reader, collected from the earliest and latest accounts of historians and travellers of all nations; with some that have never been published in this kingdom; and, nothing advanced but on the best authorities. With a great variety of maps and copper-plates, engraved in a capital stile, the subjects of which are mostly new, and such as have never yet been given in any English work
- LONDON; printed for the author
- 8vo, 22 cm., 20 vols.
- ***Libraries*** E, **L** : DLC, FU, MoU, MWA, NjNbS, ViU, WaS
- ***Notes***

1. Bibliographic information for the individual volumes is given below; it includes the countries described. Despite the claim made in the series title, not all the world is described. There is nothing on Latin America, or on Africa south of the Sahara.

   Vol. I, (1788). Printed ... at the Literary Press, No. 62, Wardour Street, Soho; and sold by all booksellers. Pp., [2], iv, 344. Describes Greenland, Labrador, Iceland, Lappland, Norway.

   Vol. II, (1788). Printed ... Press, No. 14, Red-Lion Street, Clerkenwell, and sold .... Pp., [2], iv, 324. Contain extracts from Pallas's *Tour through Siberia and Tartary*, with additions from other sources.

Vol. III, (1788). Printed ... Pp., [2], 327. Pallas's tour continued.

Vol. IV, (1788). Printed ... Press, No. 62 Wardour Street, Soho; and sold ... Pp., [2], 331, [1]. Pallas's tour continued.

Vol. V, (1789 [tp. 1788]). Printed ... Press, No. 62 Wardour Street, Soho; and sold ... Pp., [2], 352. Thos parts of the Russian empire not described by Pallas; China.

Vol. VI, (1790 [tp. 1788]). Printed ... Pp., [2], 354. China cont.; NW coast of America; Sweden.

Vol. VII, (1790 [tp. 1788]). Printed ... Pp., [4], 367. Sweden cont.; Denmark; northern North America; California; Persia.

Vol. VIII, (1790 [tp. 1788]). Printed ... Pp., [2], 327. Persia cont.; Poland; Prussia; Bohemia; Hungary.

Vol. IX, (1791 [tp. 1788]). Printed ... Pp., [2], 352. Hungary cont.; Germany.

Vol. X, (1791 [tp. 1788]). Printed ... Pp., [2], 337, [3]. Germany cont.; Burgundy (i.e., the Austrian Netherlands); United Netherlands.

Vol. XI, (1792 [tp. 1788]). Printed ... Pp., [2], 348. United Netherlands cont.; Turkey in Europe.

Vol. XII, (1792 [tp. 1790]). Printed ... Pp., [2], 368. Turkey in Europe cont.

Vol. XIII, (1793 [tp. 1790]). Printed ... Pp., [2], 366. Turkey in Europe cont.

Vol. XIV, (1793 [tp. 1790]). Printed ... Pp., [2], 376. Turkey in Asia; Egypt.

Vol. XV, (1794 [tp. 1790]). Printed ... Pp., [2], 380, [4]. Egypt cont.; Arabia; Italy.

Vol. XVI, (1794 [tp. 1795]). Printed for the author, and sold by J. Parsons, Paternoster-Row, and all booksellers. Pp., [2], 399. Italy cont.

Vol. XVII, (1795). Printed ... Pp., [2], 376. Italy cont.

Vol. XVIII, (1795). Printed ... Pp., [2], 240. Italy cont.; Switzerland.

Vol. XIX, (1796 [tp. 1795]). Printed ... sold by L. Legoux, No. 52, Poland Street, Oxford Road, and all booksellers. Pp., [2], 392. Switzerland cont.; Corsica; Spain.

Vol. XX, (1797 [no tp.]). Printed ... Pp. 408. Spain cont.; Portugal.

TURNER, Richard (Sr) **Y mSG**

- **{1762} (01)**
- *A view of the earth: being a short but comprehensive system of modern geography* ... To which is added, a description of the terrestrial globe ...
- LONDON; S. Crowder & Co. Worcester; S. Gamidge
- Folio; pp. [6], 35, [1]
- ***Libraries*** : DFo
- ***Source*** Bowen (1981:318) and NUC (NT 0397218)

- **1766, 2nd, with additions and improvements (02)**
- *A view of the earth ... geography*. Exhibiting I. A description of the figure, size, motion, etc. of the earth; with the uses and height of the atmosphere, or air surrounding it. II. Such geographical definitions, schemes, and descriptions, as form a necessary introduction to this branch of learning. III. The

situation and extent of the several kingdoms, and nations in each quarter; their chief cities; with the distance, direction, and difference in time from London. IV. An account of the several islands, trade, commodities, religion, number of inhabitants, principal mountains, and rivers in the world; also some observations on the less known parts, the trade winds, and monsoons. V. The explanation and use of a new set of maps, annex'd to the several quarters, drawn according to the most approved modern projections, and regulated by observations. VI. A description of Commodore Anson's voyage round the world; shewing the several islands and countries he touch'd at; and the places where he took any prize, particularly the rich Manilla ship. VII. A new and curious geographical clock, which points out the difference of time, with the hour, in the different nations upon earth, at one view. To which is added, a description of the terrestrial globe: with its application to a great variety of useful problems. Concluding with some curious phaenomena exhibited upon the globe in a darkened room; and a few select paradoxes, intended to excite the attention of the learner. The whole laid down in a manner so easy and natural, as to be understood in a few days. Addressed to the young gentlemen and ladies of Great Britain and Ireland

- ▸ LONDON; printed for S. Crowder, in Pater-noster-Row; and S. Gamidge, in Worcester
- ▸ Folio, 29 cm.; pp. [5], 45, [3]
- ▸ ***Libraries*** **L** (lacks the movable parts of the clock) : CtY, MH
- ▸ ***Notes***

1. Concerned with the systematic rather than the regional aspects of geography: the 9 pp. of special geography present information at the continental level. Most of the paradoxes presented depend upon the sphericity of the earth for their resolution; many also depend upon a misleading form of presentation. Solutions to the paradoxes are provided.
2. NUC records a further ed.: 1771, 3rd, London, 48 pp. (DLC, NNC, NSyU).

- ▸ **1787, 4th, with ... (see below) (03)**
- ▸ *A view of the earth* ... paradoxes and theorems, intended ... The fourth, with many additions and improvements; particularly an account of the lately discovered countries and islands in the great South Sea; and also, a large table of longitude and latitude of all the remarkable cities and towns in the world; being a necessary appendage to the use of the globe
- ▸ LONDON; printed ...
- ▸ Folio, 30 cm.; pp. [5], 52
- ▸ ***Libraries*** **L**, ShP : TxU
- ▸ ***Notes***

1. Compared with 1766, 2 pp. of information on the British Isles have been added; presumably this is the 'addition' referred to in the title of the 3rd ed. The geographical theorems that have been added since 1766 are concerned with latitude, longitude, and time.
2. ESTC and NUC record a further ed.: 1798, 5th, London, 52 pp. (SwnU : CtY, ICJ).

TURNER, Richard (Jr) **Y**

- ▸ **1780 (01)**

- *A new and easy introduction to universal geography.* In a series of letters to a youth at school; describing the figure, motions, and dimensions of the earth; the different seasons of the year; the situation and extent of the several empires, kingdoms, states, and provinces; their government, customs, religion, and manners, etc. To which is added six useful and necessary tables, etc. Illustrated with copper-plates, and a new set of pocket maps, drawn and engraved by the best authors on purpose for the work
- LONDON; printed for S. Crowder, No. 12, Pater-Noster-Row
- 16mo, 16 cm.; pp. [9], 230, [2]
- ***Libraries*** GU, **L** : CtY, NN
- ***Notes***

1. An elementary special geography; half the book is devoted to Europe.
2. NUC records further eds.: 1783 (IaU); 1787, 4th, Dublin, 255 pp. (DLC).

- **1789, 4th, improved ... (see below) (02)**
- *A new and easy introduction to universal geography*; in ... school: describing ... religion, manners and the characters of the sovereigns ... Illustrated ... set of maps ... artists, on purpose for this work. The fourth edition, improved and considerably enlarged by the author. To this edition is now first added, a large map of the world, on which are delineated the different tracts [sic] of Captain Cook's ships in his three voyages round the world; also a plate of the terrestrial globe; and a plate of a new and curious geographical clock
- LONDON; printed for S. Crowder, Pater-Noster-Row
- 12mo, 16 cm.; pp. xi, [1], 228
- ***Libraries*** BmUE, **L** : : CaOTP
- ***Notes***

1. An advertisement to this ed. identifies the additions made to it (a letter on the solar system, one on the 'geographical clock,' and three new illustrations). As this claim is made in identical words on behalf of the 5th ed., in the advertisement to it, it may have been made first in the 2nd ed.

- **{1791}, 5th, considerably enlarged (03)**
- *A new and easy introduction to universal geography.* In ... school; describing ... Illustrated ... work
- DUBLIN; printed by P. Byrne, No. 108, Grafton-street
- 12mo, 17 cm.; pp. [9], 255, [2]
- ***Libraries*** **L** : AU, NIC, PU
- ***Notes***

1. This Dublin ed. is much closer to the 1st ed. than it is to the 4th London ed. of 1789. The first 212 pp. seem unchanged; 22 of the additional 25 pp. are devoted to the terrestrial globe.

- **1792, 5th, improved ... (04)**
- *A new and easy introduction to universal geography* ... [as 1789]
- LONDON; printed ...
- 12mo, 16 cm.; pp. xi, [1], 228
- ***Libraries*** **L**
- ***Notes***

1. The pagination suggests that no changes have been made since 1789.

- **1797, 8th, improved ... (05)**
- *A new and easy introduction to universal geography* ... work. To which is added ... tracks ... world. Also ... globe, and ...
- LONDON; printed for C. Dilly, J. Johnson, G.G. and J. Robinson, F. and C. Rivington, G. Wilkie, J. Scatcherd, T.N. Longman, and C. Law
- 12mo, 17 cm.; pp. viii, 232
- ***Libraries*** **L**, SothnU
- ***Notes***

1. An advertisement to this ed. mentions that a small addition had been made to the 6th, but none since.
2. NUC records further eds. (both Dublin): 1799, 9th (MH); 1801, 10th, 208 pp. (NIC).

- **1802, 10th, improved ... (06)**
- *A new and easy introduction to universal geography ...*
- LONDON; printed for J. Johnson, G. and J. Robinson, F. and C. Rivington, G. Wilkie, J. Scatcherd, Longman and Rees, C. Law, and J. Mawman
- 12mo, 14 cm.; pp. xii, 250, [10]
- ***Libraries*** L : ViW
- ***Notes***

1. An advertisement to this ed. claims that the work has been updated; the pagination suggests that most of the changes have been made to the section on Europe.
2. BL records, but has mislaid, 11th ed., 1803, London.

- **1805, 12th, improved (07)**
- *A new and easy introduction to universal geography ...*
- LONDON; printed for J. Johnson, G. and J. Robinson ...
- 12mo, 14 cm.; pp. xii, 286
- ***Libraries*** L
- ***Notes***

1. The advertisement to this ed. states that about 400 questions were added to the 11th ed., 'which have been found very useful in examining the progress of pupils.' Compared with 1802, some note is taken of changes in the political map of Europe.

- **1808, 13th, improved ... (08)**
- *A new and easy introduction to universal geography ...*
- LONDON; printed for J. Johnson, F.C. and J. Rivington, G. Wilkie and J. Robinson, Scatcherd and Letterman, Longman, Hurst, Rees, and Orme, C. Law, and J. Mawman
- 12mo, 14 cm.; pp. xii, 295, [1]
- ***Libraries*** L : DLC
- ***Notes***

1. The advertisement to this ed. states that five letters on Great Britain and Ireland have been added.
2. St. John (1975, 1:192-3) records an ed.: 1819, London, 312 pp. (: : CaOTP).
3. NUC also records an undated ed. published in Salisbury (ViW), and one

published in 1741 [sic] in Dublin (PU). The latter is a ghost; PU reports that the book it holds is the 5th ed. of 1791; see (03), above.

TURNER, Richard **OG**

- **{1779} (01)**
- *A view of the earth, as far as it was known to the ancients ...*
- LONDON
- 21 cm.
- ***Libraries*** : CtY
- ***Source*** NUC (NT 0397217)
- ***Notes***

1. Included only because the short title could lead to confusion with the previous work.

TUTHILL, Louisa Caroline (Huggins), ed. **Y OG Am**

- **{1847} (01)**
- *My little geography*
- PHILADELPHIA; Lindsay & Blakiston
- 15 cm.; pp. [3], iv, xv-xvi, 17-138
- ***Libraries*** : AU, DLC, MH, **NN**, PU : CaAEU
- ***Notes***

1. Although the tp. states that T. is the editor, the whole text seems to be from the same hand. 'It is intended for the child's first book of scientific study' (Preface). It introduces physical geography; not a special geography.
2. NUC (NT 0401780). NUC also records another version of this ed., distinguished by the use of revised spelling: 'Mi lytl diografi, in Komstok's purfekt alfabet ... Filadelfia, Lindzi & Blekistun' (DLC).
3. There were later eds. in 1848 (Allibone 1854) and 1850 (NjP).
4. Discussed in Brigham and Dodge (1933:18-19), and Nietz (1961:210-11).

[TYTLER, J.] **A**

- **1788 (01)**
- *A new and concise system of geography*; containing a particular account of the empires, kingdoms, states, provinces, and islands, in the known world. In which is included a comprehensive history of remarkable and interesting events. With an introduction, exhibiting the principles of astronomy, as they are connected with the knowledge of geography. By the author of the continuation of Salmon's Geographical Grammar. Embellished with a set of accurate maps engraved purposely for the work
- EDINBURGH; printed for Peter Hill, at Thomson's Head, Parliament Square
- 8vo, 21 cm.; pp. [6], lxvi, 67-397, [2]
- ***Libraries*** **L** : CtY, CU
- ***Notes***

1. Catalogued by BL under System; it is attributed to Tytler by NUC (NT 0414263).

   If the author of this work was responsible for the 66 pp. on the solar system and physical geography, it is hard to believe that he was also responsible for the equivalent sections of SALMON and TYTLER (1777). In that book the approach is dominated by a concern for the use of the celestial and terrestrial globes, placing it in a tradition that goes back at least as far as MOXON (1659); in it are found also the 'problems and theorems of geography.' In this (1788) book, the section on physical geography contains passages that could almost pass unnoticed in a twentieth-century text (e.g., that on land and sea breezes), though it is true that a short section on geographical problems is included. The special geography is also very different; not only are countries described in a different sequence, but the information provided about them and their inhabitants is strikingly dissimilar, as is the organization of the information. It is not impossible that Tytler did write this book and was also responsible for editing a late series of SALMON (1749), which is the assumption embodied in the treatment of SALMON and TYTLER (1777) in this guide, but in that case his work as editor and as author were distinct. The style is noticeably less wordy than the contemporary eds. of Guthrie and Salmon.

TYTLER, James **NG**

- ▸ **1802 (01)**
- ▸ *A new system of geography, ancient and modern*
- ▸ SALEM; (no publisher or printer)
- ▸ 8vo, 22 cm.; pp. 15, [1]
- ▸ ***Libraries*** L
- ▸ ***Notes***

1. Not a book, but a proposal to publish one. The proposal was for 3 vols., 8vo, plus an atlas, for $9.00, c.o.d. Printing was to start when 500 subscriptions had been obtained. The justification is the abundance of errors in existing works, especially that of Guthrie. It would seem that insufficient response was forthcoming, for no work such as that proposed is known.

U

UNWIN, William Jordan **OG**

- ▸ **1862 (01)**
- ▸ *Modern geography: descriptive, political, and physical.* By ... Principal of Homerton College
- ▸ LONDON; Longman, Green, Longman, and Robeck
- ▸ 12mo, 16.5 cm.; pp. [3], iv, 124
- ▸ ***Libraries*** **L**
- ▸ ***Notes***

1. The information is presented in numbered paragraphs, organized on a continental and subcontinental basis; the features whose location are identified and about which information is given are those of physical geography; thus the book belongs to the tradition of *Reine Geographie* (Tatham 1957) rather than special geography.

UPSDALE, T. **?**
- **{1780} (01)**
- *A new description of the terrestrial globe*: or an abridged description of the earth
- LONDON
- 12mo
- ***Libraries*** n.r.
- ***Source*** Watt (1824, 2), who identifies the author as a French teacher

V

VAN WATERS, George **Y Am**
- **{1864} (01)**
- *The illustrated poetical geography*, to which is added the rules of arithmetic in rhyme
- NEW YORK
- 23 cm.; pp. vi, 7-96
- ***Libraries*** : DLC
- ***Notes***

1. Nietz (1961:197-8) may refer to this work.
2. A microform version is available from CIHM.

VAN WATERS, George **Y Am**
- **{1841} (01)**
- *The poetical geography*, being a classification of the principal rivers, towns, islands, mountains and lakes of the world, woven into verse: the object of which is to aid the memory in acquiring and retaining a knowledge of the present branch, with an appendix of interesting and useful knowledge, designed as a class book, for schools and private learners; made to accompany any of the common school atlases
- OGDENSBURGH; J.M. Tillotson
- 18 cm.; pp. 48
- ***Libraries*** : DLC

- **{1847} (02)**
- *The poetical geography*, to which is added the rules of arithmetic, and the outlines of the history of England, from Edward the First to Victoria, in verse, designed as an aid to memory, for the use of all ... made to accompany any of the atlases and arithmetics ...
- MILWAUKEE
- ***Libraries*** : CSmH, ICN
- ***Notes***

1. NUC records further eds.: 1848, Milwaukee, 96 pp. (CSmH, CTY, IU, MiEM, OClWHi, OO, RPB); 1848, Cincinnati, 72 pp. (CSmH, DHEW, DLC, MiU, OC, OO, RPB); [CIHM records 1849], Cincinnati (: DLC); and 1850, Cincinnati (: : CaOKU)]; 1850, Louisville, 80 pp. (CtY, ICU, IU,

KyU, NjP, OClWHi); 1851, Cincinnati, 80 pp. (OC, RPB); 1851, Louisville, 80 pp. (DAU, MH, MiD, MnHi); 1852, Cincinnati (MH, RPB); 1853, Cincinnati, 80 pp. (DLC, MH, NcD, OClWHi, RPB); 1854, Louisville, 96 pp. (IC [imperfect: p. 7-18 missing]); 1856, New York, 80 pp. (RPB); 1859, Cincinnati (CtY, MH); 1860, New York, 96 pp. (MnHi); and 1863, New York, 96 pp. (IU) [According to Culler (1945:311) this is the 2nd ed.]; [CIHM records 1864, New York].
2. Culler (1945, *passim*) and Nietz (1961:197-8) discuss the ed. of 1863.
3. Microform versions of the eds. of 1849, 1850, Cincinnati, and 1864 are available from CIHM.

VAREN, Bernhard (Varenius) **OG tr**

- **1733 (01)**
- *A compleat system of general geography*: explaining the nature and properties of the earth; viz. it's [sic] figure, magnitude, motions, situation, contents, and division into land and water, mountains, woods, desarts, lakes, rivers, etc. With particular accounts of the different appearances of the heavens in different countries; the seasons of the year over all the globe; the tides of the sea; bays, capes, islands, rocks, sand-banks, and shelves. The state of the atmosphere; the nature of exhalations; winds, storms, tornados, etc. The origin of springs, mineral-waters, burning mountains, mines, etc. The uses of making of maps, globes, and sea-charts. The foundation of dialling; the art of measuring heights and distance; the art of ship-building, navigation, and the ways of finding the longitude at sea. Originally written in Latin by Bernhard Varenius, M.D. Since improved and illustrated by Sir Isaac Newton and Dr. Jurin; and now translated into English; with additional notes, copper-plates, an alphabetical index, and other improvements. Particularly useful to students in the universities; travellers, sailors, and all those who desire to be acquainted with mix'd mathematics, geography, astronomy, and navigation. By Mr Dugdale. The whole revised and corrected by Peter Shaw, M.D.
- LONDON; printed for Stephen Austen, at the Angel and Bible, in St. Paul's Church-Yard
- 8vo, 20 cm., 2 vols.; Vol. I, pp. [1], xxiv, 520; Vol. II, pp. xvi, 521-790, [16]
- ***Libraries*** GU, **L** : IU, MH, MWA, NNC
- ***Notes***

1. The original Latin ed. was published in Amsterdam in 1650. Systematic physical geography, as it was understood at the time, together with some related topics. Emphasis is given to matters that are related to the sphericity of the earth and its relations with the sun and moon, hence the inclusion of the section on navigation, which was affected by tides, as well as the problems of identifying location at sea. The improvements and illustrations are apparently given in footnotes.
2. This is not the first translation into English. For that see BLOME (1680-2).

- **1734, 2nd, with large additions (02)**
- *A compleat system of general geography ...*
- LONDON; printed for Stephen ...

- 8vo, 20 cm., 2 vols.; Vol. I, pp. [1], xxiv, 528; Vol. II, pp. xvi, 529-898, [16]
- ***Libraries*** BrP, CoP, EU, **L**, LvU, RdU, SanU : DLC, ICJ, ICU, MH, MnU, OClW, PU
- ***Notes***

1. The changes are inserted in yet further footnotes.

- **1736, 3rd, with large additions (03)**
- *A compleat system of general geography ...*
- LONDON; printed ...
- 8vo, 20 cm., 2 vols.; Vol. I, pp. [1], xxiv, 528; Vol. II, pp. xvi, 529-898, [14]
- ***Libraries*** **L**, LSM, LU, RdU : CSt, MdBP, NbU, NN, OClJC, OkU, PMA, PPL, PU
- ***Notes***

1. The pagination suggests that no changes have been made since 1734.

- **1765, 4th, with large additions (04)**
- *A compleat system of general geography ...*
- LONDON; printed for L. Hawkes, W. Clarke, and R. Collins, at the Red-Lion, in Pater-Noster-Row
- 8vo, 20 cm., 2 vols.; Vol. I, pp. [1], xxiv, 528; Vol. II, pp. xvi, 529-898, [14]
- ***Libraries*** **L**, LRGS, LU : MH
- ***Notes***

1. Pagination and spot-checking suggest no changes since 1734.
2. Every history of geography that deals with the seventeenth century devotes at least some space to Varen/Varenius.
3. Major studies: Baker (1955); Bowen (1981:77-90, 276-83; the latter set of pages provides a translation of the dedication and the introduction to the original Latin ed. of Varen's *General Geography*); May (1973); Warntz (1964:109-18; 1989).

[VENNING, Mary Anne] **mSG**

- **1817 (01)**
- *A geographical present*; being descriptions of the principal countries of the world; with representations of the various inhabitants in their respective costumes, beautifully coloured
- LONDON; printed for Darton, Harvey and Darton, Gracechurch-Street
- 12mo, 16 cm.; pp. [8], 120
- ***Libraries*** L : CtY : CaOTP
- ***Notes***

1. NUC lists under Geographical (NG 0126984). British Museum (1881-1900) includes the following entry: 'A., M., author of the *Geographical Present*.'
2. NUC and St. John (1975, 1:193) record a further ed.: 1818, 2nd, London, 144 pp. (ICN, MnU : CaOTP).

- **1820, 3rd (02)**
- *A geographical present ...*

- LONDON; printed for Harvey and Darton, Gracechurch-street
- 14.5 cm.; pp. [1], 142
- ***Libraries*** L : **NN**, NRU, OClWHi
- ***Notes***

1. The text is slight, the emphasis being on the illustrations. It is not so much a school book as a miniature coffee-table book. Countries are dealt with in haphazard order. Attention is focused on the people who live in countries, rather than on the countries; however, the full range of topics dealt with in most special geographies is surveyed, though not consistently.

   The changes from the 1st ed. that were responsible for the increase in size not identified.
2. NUC records a further ed.: 1829, 1st American from 3rd London, New York, 200 pp. (DLC). There were later eds., but they dealt only with Europe (1830) or Africa (1831).

VINCENT, L. **Y SG?**

- **{1808} (01)**
- *Geographical exercises*
- Place/publisher n.r.
- ***Libraries*** n.r.
- ***Source*** Peddie and Waddington (1914:226)

VINSON, T. and H. MANN **Y Am**

- **{1818} (01)**
- *Universal geography improved*: or, a present view of the terraqueous globe ... Principally designed for schools, academies, etc.
- [D]EDHAM, MA; printed for the proprietor
- 18.5 cm.; xii, [13]-348 p.
- ***Libraries*** : NNC
- ***Source*** NUC (NV 0178717)
- ***Notes***

1. NUC notes: 'The same work published, 1818, under title: The material creation ... By Herman Mann.' Elson (1964:397) also notes the duplication, and adds: 'Vinson claims his to be the original.'

VISIBLE WORLD ... **Y mSG tr**

- **1791 (01)**
- *The visible world*; or, the chief things therein: drawn in pictures. Originally written in Latin and High Dutch; now rendered easy to the capacities of children
- LONDON; printed and sold by Darton and Harvey, 55, Gracechurch Street
- 12mo, 14 cm.; pp. ix, [3], 132
- ***Libraries*** L
- ***Notes***

1. Contains, among many other topics, some of the classes of information found in special geographies, including a section on Europe, but overall it falls outside the category.

VYSE, Charles **?**
- **{1774} (01)**
- *A new and complete geographical grammar ...*
- LONDON
- 18 cm.; pp. xvi, lxii, [63]-479, [6]
- ***Libraries*** : DLC
- ***Source*** NUC (NV 0253277)

W

W., C. **?**
- **{1867}, 2nd (01)**
- *Simple geography*
- LONDON; Preston [printed]
- 8vo
- ***Libraries*** L

WAKEFIELD, Priscilla {born Bell} **Y? mSG?**
- **{1807} (01)**
- *Sketches of human manners ...*
- LONDON?
- ***Libraries*** n.r.
- ***Source*** St. John (1975, 1:192)

- **{1811}, 3rd (02)**
- *Sketches of human manners*, delineated in stories, intended to illustrate the characters, religion, and singular customs, of the inhabitants of different parts of the world. The third edition
- LONDON; printed for Darton, Harvey, and Darton
- 14 cm.; pp. 243
- ***Libraries*** : : CaOTP
- ***Source*** St. John (1975, 2:816)
- ***Notes***

1. NUC records further eds.: 1811, 1st American, Philadelphia, 252 pp. (CLU, DLC, MH, MWA, NcU, NN, OClWHi, PHC, PPL); 1814, 4th, 243 pp. (CLU : CaOTP).

WALKER, A. **?**
- **{1808} (01)**
- *Geographical, historical and commercial grammar*
- Place/publisher n.r.
- ***Libraries*** n.r.
- ***Source*** Peddie and Waddington (1914:226)

WALKER, Adam ?
- **{1798}, 3rd (01)**
- *An easy introduction to geography and the use of globes*
- LONDON; C. Buckton
- 18mo; pp. 64
- ***Libraries*** : DLC

WALKER, John **A**
- **1788 (01)**
- *Elements of geography, with the principles of natural philosophy,* and sketches of general history. Containing I. The figure of the earth, and the elements of mechanics and astronomy. II. The oeconomy of the sublunary works of creation, living and inanimate. Cohesion, gravity, magnetism, electricity, optics, phonics, pneumatics, meteors, hydrostatics, etc. the structure of fossils, anatomy of plants and animals. III. Picturesque and general sketches of the different parts of the earth, and the varied appearances and manners of its inhabitants, both man and brute. With an account of J. Cooke's last voyage, which, in accounting for the peopling of the remote parts of the world, may serve the most incredulous as a cord to bind together all the nations of the earth in one great family, descended from one common stock. Also, the history of slavery, ancient and modern. IV. The rise, revolutions and fall of the principal empires of the world. In which the Jewish history is, as the most important, most fully entered into; with a particular account of the siege and final destruction of Jerusalem. V. Changes through different ages in the manners of mankind. In which the idolatry of the Ancients, the testimonies of the primitive Christians, and the Gothic and feudal manners, with the gradual refinement of Europe therefrom, are particularly described. VI. VII. VIII. IX. Descriptions of the different quarters of the world, Europe, Asia, Africa, and America. Their division into countries, provinces, etc. their climates, soils, animals, plants, minerals, mountains, rivers, lakes, canals, commerce, manufactures, curiosities, schools, learning, literati, religious profession, language, government, history, etc. Illustrated with ten copperplates. By ... teacher of the classics and mathematics, Usher's Island, Dublin. Being principally the substance of a course of lectures addressed to his pupils
- DUBLIN; printed for the author and sold by Robert Jackson, Meath-street, also by James Phillips, George-Yard, Lombard-street, London, and Joseph Crukshank, Philadelphia
- 8vo, 20.5 cm., 2 bks.; pp. [9], viii, 216, 167
- ***Libraries*** **L** : DLC : CaBViP, CaBViPA
- ***Notes***

1. Book I contains the first five sections, presenting the scientific principles underlying physical geography (ranging from physics to astronomy), systematic human geography or anthropology, behavioral zoology, political history, and social history. The tone is prolix and Newtonian, i.e., rational and deistic, even agnostic. Begins with ideas of Greek philosophers and follows with a discursive presentation of the most advanced views. The attitude is detached, philosophic, and with no reference to the then popular concepts of

a deity or providence; yet, it is not hostile to established religion. For example, missionary societies receive favourable mention as agents of civilization. Makes occasional use of long extracts from the works of others, but the authors quoted are not readily identified.

Book II contains the special geography of the four continents. This, too, is somewhat unconventional. In some countries the character of the people is presented first, the approach being empirical and sociopolitical (e.g., England). In other countries the conventional start is made, with location coming first. The usual range of topics is covered, generally in a discursive way, though some guidance is provided by the use of a conventional set of headings.

- **{1795}, 2nd (02)**
- *Elements of geography*, and of natural and civil history ...
- LONDON; printed and sold by Darton and Harvey
- 21.5 cm.; 8 p. l., 516 [i.e., 616] p.
- ***Libraries*** : CU, IU, NBuG, OU, PPL
- ***Source*** NUC (NW 0036828)
- ***Notes***

1. MWA report that they no longer have the copy listed in NUC.

- **{1796?} (03)**
- *Elements of geography*, and a natural and civil history. Containing I. The order of the spheres. II. The economy of the sublunary works of creation, inanimate and living. III. Picturesque and general sketches of the different parts of the earth; and of the varied appearances and manners of it's [sic] inhabitants. IV. The rise, revolution, and fall, of the principal empires of the world. V. Changes through different ages in the manners of mankind. VI. VII. VIII. IX. Descriptions of the different quarters of the world; Europe, Asia, and Africa, and America. Their divisions into countries, provinces, etc. Their climates, soils, animals, plants, minerals, mountains, rivers, lakes, canals, commerce, manufactures, curiosities, schools, learning, literati, religions, professions, language, government, history, etc.
- LONDON; Darton & Harvey, and others
- 8vo, 22 cm.; pp. [4], 620
- ***Libraries*** : MH
- ***Notes***

1. The title has been shortened for the English ed.; the wording of the title suggests that a Ptolemaic cosmology is presented, but the appearance is misleading. The general character of the text is unchanged; it would require a careful comparative reading to establish the extent of any changes in detail.

- **{1797}, 3rd (04)**
- *Elements of geography* ...
- DUBLIN; printed and sold by T.M. Bates
- Pp. 624
- ***Libraries*** : NjP, PU
- ***Notes***

1. BL and NUC record further eds. (both London): 1800, 3rd, 620 pp. (L :

INU, MB, N, NNC, NWM, PPL, PU, ViW); 1805, 4th, carefully rev. and enl. by Th. Smith, 671 pp. (: OO).
2. A microform version of (01) is available from CIHM.

WALKER, John **Dy**
- **1795 (01)**
- *The universal gazetteer*; being a concise description, alphabetically arranged, of the nations, kingdoms, states, towns, empires, provinces, cities, oceans, seas, harbours, rivers, lakes, canals, mountains, capes, etc. in the known world; the government, manners, and religion of the inhabitants, with the extent, boundaries, and natural productions, manufactures and curiosities of the different countries
- LONDON; Darton and Harvey
- 21.5 cm.; pp. [3], iv-xv, [736] (from NUC NW 0036846)
- ***Libraries*** **L** : CSt, CtY, DLC, MWA, PPAN, PPL, PSC, RPJCB, ViU
- ***Notes***
1. A conventional geographical dictionary.
2. BL and NUC record further eds. (all London): 1798, [805] pp. (L : CtY, DLC, NjP, OGK, OOxM); 1801, 3rd, Rev., considerably enl., and improved, by Arthur Kershaw, [929] pp. (: DLC, IU, KyLxT, KyU, MWA, NcD, NcU, NN, PPL); 1807, 4th, revised ... 1 v. [unpaged] (L : ICU, IU, MB, WaU); 1810, 5th (: MH).

- **1815, 6th, revised ... (see below) (02)**
- *The universal gazetteer*; being a concise description, alphabetically arranged, of all the nations ... towns, empires, provinces, cities, oceans, seas, harbours, rivers, lakes, canals, mountains, capes, etc. ... inhabitants, with the extent, boundaries, and natural productions, manufactures, and curiosities of the different countries. Containing several thousand places not to be met with in any similar gazetteer. Illustrated with fourteen maps ... Revised, considerably enlarged, and improved by B.P. Capper. The sixth edition
- LONDON; printed for F.C. and J. Rivington; J. Cuthell; J. Nunn; Scatcherd and Letterman; G. Wilkie; Darton, Harvey, and Darton; Longman, Hurst, Rees, Orme and co.; Lackington, Allen and co.; Otridge and son; C. Law; J. and A. Arch; B. and R. Crosby and co.; W. Baynes; J. Richardson; J. Mawman; J. Harris; R. Baldwin; Sherwood, Neeley and Jones; Gale, Curtis and Fenner; A.K. Newman and co.; Craddock and Joy; J. Walker and co.; G. Cowie and co.; Wilson and sons, York; and Doig and Sterling, Edinburgh
- 21.5 cm.; pp. [4], iv-viii, [800]
- ***Libraries*** **EU**, L : CLU, NjR
- ***Notes***
1. Physically similar to 16th ed. of BROOKES (1762) (06), and with the same principal publishers as well as several of the associates, but a cursory examination of the text suggests that the two are independent; many small places appear in only one or the other.
2. BL and NUC record a further ed.: 1822, 7th, London, [929] pp. (L : : CaBVaU).

WALLACE, Robert ?
- **{1856} (01)**
- *Outlines of descriptive geography*
- Place/publisher n.r.
- ***Libraries*** n.r.
- ***Source*** Allibone (1854)

WALSH, H. ?
- **{1871} (01)**
- *Geography made easy*
- LONDON
- ***Libraries*** n.r.
- ***Source*** Allford (1964:248)

WANTHIER, J.M. ?
- **{1814-15} (01)**
- *Geographical institutions*
- Place/publisher n.r.
- ***Libraries*** n.r.
- ***Source*** Peddie and Waddington (1914:226)

[WARREN, David M.] **Y SG? Am**
- **{1874} (01)**
- *A brief course in geography*
- PHILADELPHIA; Cowperthwait & co.
- 30 cm.; pp. 92
- ***Libraries*** : DLC, MH
- ***Source*** NUC (NW 0084546), which notes: '"The text and questions have been written by Mr Arthur Sumner"– Pref.'
- ***Notes***

1. NUC records further eds. (all Philadelphia, Cowperthwait, 92 pp.): 1875 (MH, PP); 1876, c.1874 (PP); 1878 (DNAL); 1881 (ViU) [NW 0084550 notes: (1) '"The text and questions have been written by Mr Arthur Sumner"– Pref.' (2) 'With this is bound his A special geography of the New England states. Philadelphia, c1879']; and 1886 (ViU) [NW 0084551 contains the same two notes as 0084550].

WARREN, D[avid] M., ed. **Y Am**
- **{1857} (01)**
- *The common-school geography*: an elementary treatise on mathematical, physical, and political geography. [For the use of schools. Illustrated by many copper-plate and electrotyped maps, and embellished with numerous fine engravings.] Prepared under the direction of D.M. Warren. The maps by James H. Young
- PHILADELPHIA; H. Cowperthwait & co.
- 30 cm.; pp. 100

- ▸ ***Libraries*** : DLC, OO, PPL
- ▸ ***Source*** NUC (NW 0084552), which notes: '"Most of the text has been written by Mr Arthur Sumner" - Pref.'
- ▸ ***Notes***

1. The words in the title set in [ ] are taken from the NUC entry for the ed. of 1863.
2. NUC records further eds. (all Philadelphia; H. Cowperthwait; where noted *New England ed.*, a special geography of New England is added at the end): 1858, (MH, MiU); 1859, Rev., 100 pp. (DLC); 1861, Rev. (MH); 1863, Rev. (MH, OkU); 1864, 100 pp. (AU, DLC, MH, PPeSchw); 1866, New England ed. (MH); 1867 (OU); 1868, New England ed. (MH); 1869, 108 pp. (CU) [NW 0084561 notes: 'Cover title: Warren's intermediate geography']; 1870, New England ed. (MH); 1871, New England ed. (MH); 1872 (PP); 1875, New England ed. (MH); 1876, New England ed. (MH); 1877, New England ed., 114 pp. (DHEW, DLC); 1878, 114 pp. (MH, ViU); 1879 (MH, PPF); 1880, New England ed. (MH); 1881, New England ed. (MH); 1883, New England ed., 114 pp. (NN); 1884, New England ed. (MH); 1886, New England ed., 34 pp. at end (MH); [1887], 133 pp. (DLC, MB, PP, PPD); [c1887], 133, 34 pp. (VtU) {NW 0084576 notes: (1) '(Warren's series of geographies, 2)'; (2) '"A special geography of the New England States" (34 p. at end)'}. NUC also records eds. in [1892?, c1887], [1897], and [c1905].
2. To the entries for the eds. of 1876, 1878, and 1887, NUC provides notes stating that 'most of the text was written by Mr Arthur Sumner.'
3. Carpenter (1963:266-7) provides a general discussion; Culler (1945, *passim*), and Nietz (1961:213) probably contain references to the ed. of 1872. Hauptman (1978:432) includes a reference to the ed. of 1887 in a general discussion; Elson (1964:403-4) cites the eds. of 1869 and 1887.

WARREN, D[avid] M. **Y SG? Am**

- ▸ **{1860}a** (01)
- ▸ *A first book of geography*; illustrated by numerous maps and engravings
- ▸ SAN FRANCISCO; H.H. Bancroft & co. Sacramento; J.J. Murphy
- ▸ 21.5 cm.; pp. 88
- ▸ ***Libraries*** : DLC

WARREN, D[avid] M., ed. **Y SG? Am**

- ▸ **{1860}b**, rev. (01)
- ▸ *The intermediate geography: an elementary treatise on* mathematical, physical, and political geography [for the use of schools. Illustrated by many copper-plate and electrotyped maps drawn expressly for the purpose work by James H. Young, and embellished with numerous fine engravings]. Rev. ed.
- ▸ SAN FRANCISCO; H.H. Bancroft & co.
- ▸ 30 cm.; pp. 100
- ▸ ***Libraries*** : DLC
- ▸ ***Source*** NUC (NW 0084594), which notes: 'Most of the text was written by Mr Arthur Sumner'

▸ *Notes*
1. The words in the title set in [ ] are taken from the NUC entry for the ed. of 1864.
2. NUC records a further ed.: 1864, Philadelphia, 100 pp. (WaPS).

[WARREN, David M.] **Y SG? Am**
▸ **{1886} (01)**
▸ *A new brief course in geography*
▸ PHILADELPHIA; Cowperthwait & co.
▸ 30.5 cm.; pp. 96
▸ ***Libraries*** : DLC, NBuG, PP, PPD
▸ ***Source*** NUC (NW 0084596), which notes: 'The text and the questions have been written by Mr. Arthur Sumner'
▸ *Notes*
1. NUC also records an ed. in [1896?, c1886].

WARREN, D[avid] M. **Y Am**
▸ **{1868} (01)**
▸ *A new primary geography*[; illustrated by numerous maps and engravings ...]
▸ PHILADELPHIA; Cowperthwait & co.
▸ 22.5 cm.; pp. 88
▸ ***Libraries*** : DLC
▸ ***Notes***
1. The words set in the title set in [ ] are taken from the NUC entry for the ed. of 1874.
2. NUC records further eds. (all Philadelphia; Cowperthwait): 1869 (MH); 1872, 88 pp. (MB); 1874 [c1872], 88 pp. (MB); 1875, 88 pp. (DHEW, MH); 1876 (MH); 1878 [c1875], 88, 8 pp. (MH, PPT); 1879 (MH); 1880 (MH, OO); 1882 (MH); 1884 (DHEW, MH); [c1886], 88 pp. (MH, NN, OO, PPD) [NW 0084610 notes that the author is identified only by the cover title, which is 'Warren's primary geography']; 1886, 88 pp. (DHEW, DLC) [NW 0084611 supplies the author's name]. NUC also records two separate eds. in [1890?, c1886].
3. According to Culler (1945:311), the ed. of 1874 is the 2nd ed. Culler (1945, *passim*) discusses that ed.

WARREN, D[avid] M. **Y SG? Am**
▸ **{1858} (01)**
▸ *A primary geography*: illustrated by numerous maps and engravings
▸ Philadelphia; H. Cowperthwait & co.
▸ 21.5 cm.; pp. 23
▸ ***Libraries*** : DLC
▸ *Notes*
1. NUC records further eds. (Philadelphia, Cowperthwait, unless stated otherwise): 1859, Boston (MH); 1863, 88 pp. (OO); 1864, 88 pp. (DLC); 1867 (MH, PU).

WARREN, William **Y Am**

- **{1842} (01)**
- *A systematic view of geography*, [containing a general, particular, and miscellaneous view of the world] ...
- BANGOR; E.F. Durren
- 18.5 cm.; vi p., 1 l., [13]-172 p.
- ***Libraries*** : DLC, PU
- ***Source*** NUC (NW 0087371)
- ***Notes***

1. The words in the title set in [ ] are taken from the NUC entry for the first of the two eds. of 1846 listed in the next note.
2. BL and NUC record further eds. (all Portland, ME): 1843, 3rd, 180 pp. (L : DLC, MH); 1846, [44] pp. (: DLC, MH, OO); and 1846, 7th (MH).
3. Brigham and Dodge (1933:19) discuss (01). Culler (1945, *passim*) discusses the ed. of 1843.

WATSON, Chas. P.
See (04) of [PINNOCK, William] (1821).

WATSON, Frederic **Dy**

- **{1773} (01)**
- *A new and complete geographical dictionary*. Containing a full and accurate description of the several parts of the known world ... The situation, extent, and boundaries, of all the empires, kingdoms, provinces, states, etc. in Europe, Asia, Africa and America ... By ... and several other gentlemen
- LONDON; printed for the authors, and sold by G. Kearsly
- 37 cm.; pp. [600]
- ***Libraries*** : CU, ICU, MWA, NIC, PHi, TxU, WaPS
- ***Source*** NUC (NW 0113649)

- **{1783} (02)**
- *Geographical dictionary ...*
- LONDON
- ***Libraries*** : PPL

WATSON, John **Dy**

- **{1794} (01)**
- *Universal gazetteer; or, modern geographical index, etc.*
- LONDON
- 8vo
- ***Libraries*** n.r.
- ***Source*** Allibone (1854)

WATT, Alexander **?**

- **{1819}, new (01)**
- *Denholm's synopsis of geography*. New ed.

- GLASGOW
- *Libraries* n.r.
- *Source* Allibone (1854)

WATTS, Isaac **Y mSG**

- **1726 (01)**
- *The knowledge of the heavens and the earth made easy*: or, the first principles of astronomy and geography explain'd by the use of globes and maps: with a solution of the common problems by a plain scale and compasses as well as by the globe. Written several years since for the use of learners
- LONDON; printed for J. Clark and R. Hett, at the Bible and Crown in the Poultry near Cheapside; E. Matthews at the Bible in Pater-noster Row, and R. Ford at the Angel in the Poultry near Stocks-Market
- 8vo, 20 cm.; pp. [2], xii, [2], 219 (86 and 87 misnumbered 102 and 103), [13]
- *Libraries* **L** : CLU, CSt, CtY, IU, MH, MiU, NN, NNC, OrU
- *Notes*

1. Concerned primarily with astronomy and the consequences of the sphericity of the earth for establishing location and time. Pp. 71-88 are devoted to special geography, but they contain little more than a list of countries and their locations, plus an irregular identification of principal cities, rivers, islands, capes, bays, etc. For further information of this type, Watts refers his reader to Gordon's *Geographical grammar*.

   Asserts in the dedicatory pages: 'There is not a son or daughter of Adam but has some concern in both of them [i.e., astronomy and geography].' This statement was used, with an attribution to Watts, as an epigram by more than one author of special geography in the decades that followed. Presumably they saw in this book a recognized authority in geography.
2. Discussed in Warntz (1964:119-23).

- **1728, 2nd, corrected (02)**
- *The knowledge of the heavens and the earth ...*
- LONDON; printed ...
- 8vo, 19 cm.; pp. [2], xi, [1], 222
- *Libraries* AbU, **L** : CLU, ICU, MH, MWA, NIC, NjP, OkU : CaBVaU
- *Notes*

1. The special geography is still on pp. 71-88, and spot-checking suggests that it is unchanged, apart from the occasional rewording of a phrase.

- **1736, 3rd, corrected (03)**
- *The knowledge of the heavens and the earth ...*
- LONDON; printed for Richard Ford, at the Angel; and Richard Hett, at the Bible and Crown, both in the Poultry
- 8vo, 20 cm.; pp. xiii, [1], 222, [12]
- *Libraries* **L** : CLU, MB, MdBJ, MeB, MH, MWA, N, NjR, NN
- *Notes*

1. Spot-checking and the pagination suggest no change.

- **{1745}, 4th, corrected ... (04)**
- *The knowledge of the heavens and the earth* ... maps. With ...
- LONDON; printed for T. Longman and T. Shewell, at the Ship in Pater-noster Row; and J. Brackstone, at the Globe in Cornhill
- 8vo, 21 cm.; pp. xiii, [1], 222, [12]
- ***Libraries*** : CoU, IU, KMK, MB
- ***Source*** NUC (NW 0120340)

- **1752, 5th, corrected (05)**
- *The knowledge of the heavens and the earth ...*
- LONDON; printed ... Longman at the Ship, and J. Buckland at the Buck, in Pater-noster-Row; J. Oswald at the Rose and Crown in the Poultry; J. Waugh, at the Turk's-Head in Lombard-Street, and J. Ward, at the King's Arms in Cornhill
- 8vo, 21 cm.; pp. xiii, [1], 222, [12]
- ***Libraries* L** : IU, MB, MH, PMA, PP
- ***Notes***

1. Spot-checking and the pagination suggest no change.

- **1760, 6th, corrected (06)**
- *The knowledge of the heavens and the earth ...*
- LONDON; printed for T. Longman, J. Buckland, and W. Fenner, in Pater-noster-row; J. Waugh in Lombard-street; J. Ward in Cornhill; and E. Dilly in the Poultry
- 8vo, 20 cm.; pp. xiii, [11], 222
- ***Libraries* L** : IaU, ICU, MB, MH, MU, NIC, PPL, TxU
- ***Notes***

1. Apart from the transfer of the table of contents from the back to the front of the book, still no sign of change.

- **{1765}, 7th, corr. (07)**
- *The knowledge of the heavens and the earth ...*
- LONDON; T. Longman and J. Buckland (etc.)
- 21 cm.; pp. [xiii], 222
- ***Libraries*** : MH, NIC, NNUT
- ***Source*** NUC (NW 0120344)

- **1772, 8th, corrected (08)**
- *The knowledge of the heavens and the earth ...*
- LONDON; printed for T. Longman, and J. Buckland, in Pater-Noster-Row; T. Field, in Leadenhall Street; E. and C. Dilly, in the Poultry; and G. Pearch, in Cheapside
- 8vo, 21 cm.; pp. ii-xxv, [1], 222
- ***Libraries* L** : DN-Ob, MB
- ***Notes***

1. Still no sign of change.

WEBB, Maria, *Mrs* ?

- **{1851} (01)**

- *Modern geography simplified*
- LONDON
- ***Libraries*** n.r.
- ***Source*** Allibone (1854)

WELLS, Edward **Y mSG**

- **1701 (01)**
- *A treatise of antient and present geography*, together with a sett of maps, designed for the use of young students in the universities
- OXFORD; printed at the Theatre
- 8vo, 22 cm.; pp. [13], 183, [1]
- ***Libraries*** **L**, O : CtY, MoU
- ***Notes***

1. Intended primarily for undergraduates. It provides a reference for those who wish to learn the current names of places referred to in Classical literature, and vice versa. As such it is the earliest example identified in this guide of what came to be known as *systems of ancient and modern geography*. The phrase ancient and modern occurs on the tps. of books earlier than this (e.g., CLUVER 1657), but this is the first English book to provide lists relating Classical and contemporary toponyms systematically. The maps referred to in the title were published in a separate volume.

   The text is largely devoted to tables in which the subdivisions of countries are given, together with their chief towns. The modern subdivisions are listed first; then those used by Classical authors. The maps were intended to make comparisons between 'ancient' and 'modern' geography easy through the use of maps at the same scale, for each period, in the case of any given country. Information is not limited to the world known to Classical authors, however, but covers the whole world known to Wells. It thus qualifies as special geography, though of the driest kind.

   The Preface contains statements that provide insight into the contemporary understanding of geography as a 'science,' and also a few comments on the quality of other geographies of the period.

- **1706, 2nd (02)**
- *A treatise of antient and present geography* ... Together ... maps, both of antient and present geography, designed ...
- LONDON; printed for A. and J. Churchill, at the Black Swan in Pater-Noster-Row
- 8vo, 20 cm.; pp. [15], 175 (i.e., 179 because of [4] between 42 and 43), [29]
- ***Libraries*** LcU, **O** : CU, DeU, IU
- ***Notes***

1. The Preface is longer, a consequence of both a fuller detailing of the contents of the book, and of the principal additions; the latter consist chiefly of an index.

- **1717, 3rd, with ... (see below) (03)**
- *A treatise of antient and present geography* ... universities. The third edition, with such alterations and additions, as have been chiefly occasioned by the

change of affairs as to the government of countries, etc. since the second edition
- LONDON; printed for William Churchill, at the Black Swan ...
- 8vo, 20 cm.; pp. [15], 179, [29]
- ***Libraries*** O : DLC, ICN, ICU
- ***Notes***

1. Apart from a few minor reorganizations of the text, only three additions seem to have been made. However, the 4 pp. that were left unnumbered in the previous ed. have been included in the pagination this time. The index was not adjusted.

- **1726, 4th, with such alterations ... (04)**
- *A treatise of antient and present geography*. Together ... maps, both of antient and present geography, shewing the difference between each, by bare inspection of the maps, and design'd ... universities ... The fourth edition ... since the third edition
- LONDON; printed for W. Bowyer, for R. and J. Bonwicke, J. Walthoe, R. Wilkin, and T. Ward
- 8vo, 18 cm.; pp. [17], 181, [15]
- ***Libraries*** L, O : DCL, OClWHi

- **1738, 5th, with such alterations ... (05)**
- *A treatise of antient and present geography* ... The fifth ... countries
- LONDON; printed ... for R. Wilkin, A. Bettesworth and C. Hitch, J. and J. Bonwicke, J. and P. Knapton and T. Longman, S. Birt, T. Osborn and E. Wicksted
- 8vo, 20 cm.; pp. [17], 181, [33]
- ***Libraries*** L, O
- ***Notes***

1. NUC (NW 0180376) records WaT as holding a 4th ed. published in 1738; the latter reports, however, that it cannot find any record of such a book in its collection.

WHITE, Emerson Elbridge **Y SG? Am**

- **{1854} (01)**
- *Class-book of geography*. [in two parts] ... To which are added special topics on the descriptive geography of each political division ...
- CLEVELAND, O[hio]; Medill, Cowles & co's steam press
- 17 cm.; pp. 59, [1]
- ***Libraries*** : OCl, OClWHi
- ***Source*** NUC (NW 0249190)

- **{[1863]} (02)**
- *Class-book of geography*, containing a complete syllabus of oral instruction on the method of object teaching; also, map exercises, systematically arranged for class drill ...
- CINCINNATI; W.B. Smith & co.
- 16mo, 17 cm.; pp. iv, 5-64

▸ *Libraries* L : DLC, MiU, OC, OCl, OClWHi, OO
▸ *Source* BL and NUC (NW 0249192)

WHITE, Emerson Elbridge **Y Am**
▸ **{1856} (01)**
▸ *Class-book of local geography*, in which all the important geographical names [of the globe] are systematically arranged [for the use of classes] and pronounced according to the best authority
▸ CLEVELAND; J.B. Cobb & co.
▸ 16 cm.; pp. 46
▸ *Libraries* : DLC, OClWHi, OO
▸ *Source* NUC (NW 0249193, and 0249194)
▸ *Notes*
1. The words in the title set in [ ] are taken from the NUC entry for the ed. of 1860.
2. NUC records a further ed.: 1860, Cleveland, 46 pp. (OClWHi, OKentU, MiU). Culler (1945:311) records a further ed.: 1863, Cincinnati, library n.r.
3. Culler (1945, *passim*) discusses the ed. of 1863.

WHITE, George **Y SG? n.c.**
▸ **{1859} (01)**
▸ *Constructive geography*; being a series of exercises, by which a child may effectually learn geography, England. (Scotland)
▸ LONDON
▸ 4to, 2 pts.
▸ *Libraries* L
▸ *Notes*
1. BL reports: 'No more published.'

WHITE, John **?**
▸ **{1832}, 2nd (01)**
▸ *An abstract of general geography*. Second edition
▸ EDINBURGH
▸ 12mo
▸ *Libraries* L

▸ **{1836}, 5th (02)**
▸ [*Abstract of general geography*]
▸ EDINBURGH
▸ 12mo
▸ *Libraries* L
▸ *Notes*
1. BCB, Index (1837-57:110) records 5th ed. in 1844, (?), Simpkin.
2. BL records further eds. (all Edinburgh): 1857, 169th; 1858, 170th; 1859, 171st; 1861, 173rd, rev.; 1864, 177th; 1865, 179th, rev.; 1876, 190th, enlarged; 1878, 191st, rev.; 1879, 192nd, rev., 96 pp.; 1884, 194th, rev.,

96 pp. BL also records the 196th ed. in 1888, the 197th in 1890, and the 199th in 1897.

The failure of the British Museum, as it then was, to secure a copy of every ed. of this title caused no distress in some quarters.

WHITE, John **Y**

- **{1822} (01)**
- *A system of modern geography, with an outline of astronomy*[: comprehending an account of the principal towns, remarks on the climate, soil, productions ... with a complete system of sacred geography, and numerous problems on the globes, compiled from a new and improved plan, from the best and most recent authorities, for the use of schools and private students]
- EDINBURGH
- 12mo
- ***Libraries*** L
- ***Notes***

1. The words in [ ] are derived from CIHM, which records [1842?], 2nd (: : CaOLU).
2. BCB, Index (1837-57:110) records an ed. in 1844(?), Simpkin. BL records further eds. as follows (all Edinburgh): [1850?], 13th; 1857, 22nd, rev., 1859, 23rd, rev.; 1862, 24th, enlarged, 298 pp.; 1863, 24th, enlarged, 309 pp. [BL notes: 'This second 24th edition is a duplicate of the preceding with a new titlepage and 11 additional pages of text']; 1866, 25th, rev.; 1870, 28th, rev.; 1871, 29th; 1872, 31st; 1881, 33rd, 309 pp.; and 1887, 34th, 309 pp.

WIFFEN, Jeremiah Holmes **Y SG?**

- **{[1812]} (01)**
- *The geographical primer*; designed for the younger classes of learners ... with an appendix, containing fourteen hundred questions on the principal maps ...
- LONDON; printed by William Darton
- 14.5 cm.; pp. xix, 196
- ***Libraries*** : PSC
- ***Source*** NUC (NW 0290577), which notes: 'imperfect: p. 3-12, 17-20 wanting; p. 67 wrongly numbered p. 6'

- **{1813} (02)**
- *Geographical primer*
- Place/publisher n.r.
- ***Libraries*** n.r.
- ***Source*** Peddie and Waddington (1914:226)

WILLARD, Emma **OG Am**

- **{1822} (01)**
- *Ancient geography*, as connected with chronology, and preparatory to the study of ancient history. Compiled chiefly from D'Anville, Adam, Lavoisne,

and other standard works. To which are added, problems on the globes, and rules for the construction of maps. To accompany the Modern Geography by William C. Woodbridge
- ▸ HARTFORD, [CT]; Oliver D. Cooke
- ▸ 19 cm.; pp. 88
- ▸ ***Libraries*** : CtY, NN, PU
- ▸ ***Source*** NUC (NW 0313758). See WOODBRIDGE [and WILLARD] (1824).

WILLARD, Emma **Y Am**
- ▸ **{1826} (01)**
- ▸ *Geography for beginners*, or, the instructer's [sic] assistant ...
- ▸ HARTFORD; Oliver D.Cooke & Co.
- ▸ 15 cm.; pp. 110
- ▸ ***Libraries*** : CtY, DLC, KyBgW, NN
- ▸ ***Source*** Sahli (1941:383) and NUC (NW 0313789)
- ▸ ***Notes***

1. According to Brigham and Dodge (1933:9), also published jointly with W.C. Woodbridge. They also refer to a review of 'this little text' in 'The American Journal of Science and Arts ... [that] comments on collections of facts imperfectly grouped, as common in geography, "so little connected by an associating principle as to overload the memory and fatigue the mind"' (Brigham and Dodge 1933:10).
2. NUC records: 2nd, rev. and corr., 1829, Hartford; O.D. Cooke, 123 pp. (AU, MH, NcD).
3. Nietz (1961:201, 222-3) and Sahli (1941, *passim*) discuss (01).

WILLETTS, Jacob **mSG Am**
- ▸ **1819 (01)**
- ▸ *A compendious system of geography*: being a description of the earth, and of the manners, customs, religion, government, manufactures, commerce, and curiosities of the various nations who inhabit it. To which is added plain directions for constructing maps, illustrated by plates. With an atlas. Intended as a sequel to Willetts; 'Easy grammar of geography'
- ▸ POUGHKEEPSIE, [NY]; printed and published by P. Potter, for himself, and for S. Potter & co., no. 55, Chesnut St., Philadelphia
- ▸ 19.5 cm.; pp. [5], iv-x, 466, [2]
- ▸ ***Libraries*** : CSmH, **MWA**, N, NIC
- ▸ ***Notes***

1. The Preface refers to an earlier small treatise, 'An Easy Grammar of Geography' (see WILLETTS 1814, below), to which this work, though 'intended to be complete in itself ... still [the author] consider[s] them as part of the same system' (p. [5]).

   It is a highly condensed special geography, being little more than a series of lists of places of different types.
2. NUC records: 2nd, rev., 1822, Poughkeepsie, [NY]; P. Potter, 323 pp. (ICRL, NNU-W, OC).

WILLETTS, Jacob **Y Am**

- **{1814} (01)**
- *An easy grammar of geography*, for the use of schools, upon ... J. Goldsmith's much approved plan
- POUGHKEEPSIE; P. & S. Potter
- ***Libraries*** : MH, MWA
- ***Source*** NUC (NW 0317432)
- ***Notes***

1. According to the Preface of WILLETTS (1819) (see above), more than 30,000 copies of this work had been sold in the five years following its appearance.
2. According to Brigham and Dodge, based on the ed. of 1819, Willets [sic] 'approves of the review exercises [used in PHILLIPS 1804], puts astronomy at the end, and has a separate atlas, thinking it useless to communicate knowledge of the earth without maps' (Brigham and Dodge 1933:9). They also comment, seemingly in an ironic fashion, on the 'orderly' character of the review questions.
3. BL and NUC record further eds. (all Poughkeepsie, [NY], and Potter, if a publisher specified): 1815, 2nd., corr. and enl., 197 pp. (L : CtY, NNC, PSC); 1815, 3rd, 200 pp. (: CtY, MWA, NjP, NN, NP, NSyU, PSC, ViU); 1817, 4th (: MH, MWA, N, NNC); 1818, 210 pp. (: CtY, MH, MWA, NRAB); 1819, 6th, 215 pp. (: MWA, NcA, NNC, NRCR, PHi); 1820, 7th (: CtY, NP); 1822, 8th, 215 pp. (: CtY, DLC, MiU, NNC, OClWHi); 1823, 9th, 215 pp. (: PU); 1823, 10th, 212 pp. (: CtY, NBuU); 1826, 9th (: MH [imperfect, lacks after p. 214]); 1826, 10th, 215 pp. (: NN); and 1826, 12th, 208 pp. (: CtY, PHi).

- **{1826}, 13th (02)**
- *Willetts' geography*. A geography for the use of schools ... Thirteenth edition ...
- POUGHKEEPSIE; P. Potter
- 14.5 cm.; pp. xii, 14-197
- ***Libraries*** : MH, NBu, NBuG, NN, PSC
- ***Source*** NUC (NW 0317445)
- ***Notes***

1. Can only be the 13th ed. if it is still the *Grammar* despite the change of title.
2. NUC records further eds. (both Poughkeepsie, [NY], P. Potter): 1831, 13th (MH); and 1832, 19th, 197 pp. (DLC).
3. Brigham and Dodge (1933:9) discuss the 6th ed. of 1819; Nietz (1961:204) and Sahli (1941, *passim*) discuss the 10th ed. of 1826; Elson (1964:396) cites the eds. of 1815 and 1822.

WILLETTS, Jacob **Y Am**

- **{1838} (01)**
- *Willetts' new and improved school geography*; accompanied by a new and correct atlas, drawn and engraved on steel, expressly for this work
- POUGHKEEPSIE, [NY]; Potter and Wilson ... [stereotyped by Francis F. Ripley, New York]
- 18 cm.; pp. 173

- *Libraries* : CSmH, CtY, PSC
- *Source* NUC (NW 0317452)
- *Notes*

1. NUC records further eds. (all Poughkeepsie, [NY]; W. Wilson): 1841, 173 pp. (AU, N, NN); 1842, 5th (MH, NWM); 1843, 173 pp. (DLC).
2. Culler (1945, *passim*) discusses the 5th ed. of 1842.

WILLIAMS, Charles ?

- **{n.d.} (01)**
- *Visible world*
- Place/publisher n.r.
- *Libraries* n.r.
- *Source* Allibone (1854)
- *Notes*

1. Williams might be the author of VISIBLE WORLD (1791).

WILLIAMS, David ?

- **{1817} (01)**
- *The geographical mirror*. Containing an accurate ... description of the known world ... a comparative view of ancient and modern geography, etc.
- LONDON
- 12mo
- *Libraries* L

WILLOX, John ?

- **1852 (01)**
- *Lizar's illustrated and descriptive atlas of modern geography*, a series of maps of all the countries in the wold [sic], with pictorial designs, illustrative of the habits, customs and natural productions, together with brief descriptions of the various countries
- EDINBURGH
- 4to
- *Libraries* n.r.
- *Source* James Thin, Edinburgh; Catalogue 332, *General Second-hand and Antiquarian*, item 1255

WILMHURST, *Miss* **Y SG?**

- **{1825} (01)**
- *An introduction to geography to accompany children's maps*
- LONDON
- *Libraries* n.r.
- *Source* Allford (1964:244)

WILSON, Horace Hayman ?

- **{n.d.} (01)**

▸ *Compendium of history and geography*
▸ Place/publisher n.r.
▸ ***Libraries*** n.r.
▸ ***Source*** Allibone (1854)

WILSON, H.H. **?**
▸ **{1847} (01)**
▸ *Geography and history*
▸ ? Hamilton
▸ 12mo
▸ ***Libraries*** n.r.
▸ ***Source*** BCB, Index (1837-57:109). The relationship of this title to the previous entry is unclear.

WILSON, R. **?**
▸ **{1849} (01)**
▸ *Geography simplified*
▸ ? Simpkin
▸ 18mo
▸ ***Libraries*** n.r.
▸ ***Source*** BCB, Index (1837-57:110)
▸ ***Notes***
1. See WILSON, S. (1849).

WILSON, S. **?**
▸ **{[1849]}, 3rd (01)**
▸ *Geography simplified*: being a brief summary of the principal features of the great divisions of the earth; with a more detailed account of the British empire, and the chief towns in other parts of the world. Third edition
▸ LONDON
▸ 24mo
▸ ***Libraries*** L
▸ ***Notes***
1. See WILSON, R. (1849).

WILSON, Thomas **Y SG?**
▸ **{[1843]} (01)**
▸ *The first catechism of geography*
▸ Place/publisher n.r.
▸ 12mo
▸ ***Libraries*** L
▸ ***Notes***
1. ECB, Index (1856-76:130-1) records two further eds.: 1862 and 1873.

WILSON, T.
See GOODRICH, Samuel Griswold (Parley, Peter) (18–?).

WITS ... **NG**

- **1599 (01)**
- *Wits theatre of the little world*
- LONDON; printed by I.R. for N.L. and are to be sold at the west door of Paules
- 8vo, 12 cm.; pp. [9], 270, [13]
- ***Libraries*** **L** : CSmH, CtY, DFo, DLC, ICN, MB, MiU, MWiW, PU, TxU
- ***Notes***

1. Not a special geography. The 'little world' of the title is the human individual, as opposed to the great world of the earth or the universe.
   NUC (NA 0194477) notes: 'Published anonymously. Dedicated to "I.B." i.e., John Bodenham under whose patronage the work was produced and to whom it has been often attributed. In a copy preserved in the British Museum the dedication is signed "Robert Allott" and Bodenham's name is printed in full.'

WOODBRIDGE, William Channing **Y mSG Am**

- **{1844} (01)**
- *Modern school geography, on the plan of comparison and classification*; with an atlas[, exhibiting, on a new plan, the physical and political characteristics of countries, and the comparative size of countries, towns, rivers, and mountains]
- HARTFORD; Belknap and Hamersley
- 18 cm.; pp. 352
- ***Libraries*** : DLC, MB, OCl, OO
- ***Source*** NUC (NW 0432215)
- ***Notes***

1. The words in the title set in [ ] are taken from the NUC entry for the ed. of 1846.
2. Intended to replace WOODBRIDGE (1821)*.
3. BL and NUC records further eds. (both Hartford, Belknap): 1845, 2nd, with improvements and additions, 350 pp. (: CtY, DLC, MH, NNC); 1846, 3rd, 350 pp. (**L** : MB, MH).

- **1846, 4th (02)**
- *Modern school geography* ... By ... member of the Geographical Societies of Paris, Frankfort, and Berlin. With improvements and additions
- HARTFORD; Belknap ...
- 17.5 cm.; pp. [2], [iii], iv-xxiv, [1], 26-350
- ***Libraries*** : MHi : **CaBVaU**
- ***Notes***

1. Seeks to provide a book suitable for students of various ages and abilities by the use of type of different sizes. In the Preface, (pp. v-viii) inveighs against the practice of requiring students to memorize the answers to set questions. Summarizes the history of his textbooks in geography and of their relation-

ship to those of Mrs. Willard (see notes to [01] and [02] of WOODBRIDGE 1824, below). Advises teachers to make their students draw maps. Pp. ix-44 are devoted to maps and related topics. Physical geography occupies pp. 45-138, and systematic human geography pp. 139-54. Information about countries is provided, under the heading of *Statistical Geography* on pp. 155-344; of this, pp. 155-216 are devoted to North America.
2. Brigham and Dodge are struck by the 'new emphasis on physical geography' (Brigham and Dodge 1933:12), and observe that 'in this volume the glacial studies of Agassiz, of about 1840, were bearing early fruit in the schools.'
3. NUC records further eds. (Hartford, Belknap until 1849, Hamersley thereafter): 1847, 5th, 350 pp. (CtY, MH, NNC : CaBVaU); 1849, 7th (MH); 1850, 8th (MH); 1851, 9th (MH); 1852, 10th (MH); and 1854, 11th (MH).
4. Brigham and Dodge (1933:12) discuss (01); Hauptman (1978:435) draws on it, and (1978:431) refers to it in a general discussion; Elson (1964:402) cites (02) and the 5th ed. of 1847.

WOODBRIDGE, William Channing **Y Am**
- **{[1831]} (01)**
- *Preparatory lessons for beginners: or, first steps in geography*
- Place/publisher n.r.
- 14.5 cm.; pp. 36
- ***Libraries*** : DLC
- ***Source*** NUC (NW 0432227)

WOODBRIDGE, William C. [and? Emma WILLARD] **Y mSG Am**
- **{[1821]}, [3rd?] (01)**
- *Rudiments of geography, on a new plan*, designed to assist the memory by comparison and classification; with numerous engravings of manners, customs, and curiosities. Accompanied with an atlas, exhibiting the prevailing religions, forms of government, degrees of civilization, and the comparative size of towns, rivers, and mountains
- [HARTFORD, CT?]
- 15.5 cm.; pp. [vii]-xiv, [15]-194
- ***Libraries*** : NBuG
- ***Source*** NUC (NW 0432229)
- ***Notes***
1. The joint publication with Emma Willard is asserted by Brigham and Dodge (1933:9). See also notes to (04) and (06), below.

- **{1822}, 2nd, rev. ... (see below) (02)**
- *Rudiments of geography ... plan*; designed ... memory of comparison ... 2nd ed., rev. and adapted to the use of schools
- HARTFORD; Samuel G. Goodrich
- 15 cm.; pp. 208
- ***Libraries*** : CtY, MoU, ViU
- ***Source*** NUC (NW 0432230)

- **{1823}, 3rd, with comparative ... (see below) (03)**
- *Rudiments of geography ... plan*, designed ... memory by comparison ... rivers and mountains. Third ed. With comparative views of cities, and other important additions
- HARTFORD; O.D. Cooke & Sons
- 15 cm.; 1 p. l., [vii]-xiv, [15]-208 (i.e., 218) p.
- ***Libraries*** : DNLM, MH, NNC, NSyU, TNJ
- ***Source*** NUC (NW 0432232), which notes: 'Includes extra pages 103b-103e, 127b-127c, 155b-155c, 171b-171c'
- ***Notes***

1. NUC records further eds. (both Hartford, [CT], Cooke, 208 pp.): 1823, 4th (OO); 1825, 5th from 3rd improved (CtY, IU, MiD, NN, OU).

- **1826, 6th from 3rd improved (04)**
- *Rudiments of geography ...*
- HARTFORD; O.D. Cooke & co.
- 15.5 cm.; pp. [5], viii-xiv, [15]-208
- ***Libraries*** : CtY, CU, ICN, IU, MH, **NN**, NNC, OCl, OClWHi
- ***Notes***

1. A short version of WOODBRIDGE (1824) designed for elementary students. In the Preface, W. asserts that the 'plan of teaching geography from the maps, and the "interrogative system", usually ascribed to *Guy* and *Goldsmith*, were used in this country before their introductions of their works, by the *Rev. William Woodbridge*, formerly of New-Jersey, and adopted in books prepared for his pupils' (p. x). He also states that he began work on this, his own work, in 1813.
2. NUC records an ed.: 1828, 8th from 3rd improved, Hartford (MH).

- **{1828}, 9th from 3rd improved, with corrections (05)**
- *Rudiments of geography ...*
- HARTFORD; O.D. Cooke & co.
- 15 cm.; 1 p. l., [vii]-xiv, [15]-208 [i.e., 218] p.
- ***Libraries*** : DSI, IU, MH, NNC, ViU
- ***Source*** NUC (NW 0432238), which notes: (1) 'Includes extra pages 103b-103e, 127b-127c, 155b-155c, 171b-177[sic]c'; (2) 'Date of imprint on cover: 1827'; (3) 'First ed., 1821'

- **1828 (06)**
- *Rudiments of geography ... plan*. Designed ...
- LONDON; printed for Geo. B. Whittacker, Ave-Maria Lane
- 12mo, 16 cm.; pp. xxxii, 214, [4]
- ***Libraries*** **L**
- ***Notes***

1. Contains both general (pp. 1-44) and special (pp. 44-186) geography. The sequence of the continents is: Europe, Asia (including islands in the Pacific), Africa, Polar regions, North America, West India islands, South America; each receives roughly equivalent attention. The book ends with: 'General views of the regions, climates, animals, vegetables, and minerals of the earth, and of the arts, commerce, literature, and customs of its inhabitants' (pp. 188-214), a compendium of fact and theory, opinion and prejudice.

2. BL and NUC record further eds. (all Hartford, and derived from the 3rd improved ed., except as noted): 1829 10th, from ... with corrections, 208 [i.e., 218] pp. (: CtHT, IU, MH, NNC, OU); 1830, 11th, from ... 208 [i.e., 218] pp. (: CtY, MB, MH, NNC, OCl, OClWHi); 1830, 12th, from ... 208 [i.e., 218] pp. (: MiU, NN, ViU); 1830 13th, from ... (: MH); 1831, 14th, from ... 208 [i.e., 220] pp. (: MH, NNC, PPPrHi); 1831, 15th rev. and improved, 208 [i.e., 220] pp. (: CU, MH, NNC); 1832, new, enl., cor., and improved, 208 pp. (: DLC, NcRS, OClWHI); 1832, 15th, with preparatory lessons for beginners (: MH); 1833, 17th (: MH, NBu, OrU); 1833, 3rd (L); 1835, 18th (: DLC); 1838, 19th, with preparatory lessons, a series of questions, 208 pp. (L : CU-I, OU, UU). BCB, Index (1837-57:110) records an ed. in 1843.
3. Nietz (1961:223) is probably referring to this work. Nietz (1961:208, 213, 218-19, 221) and Sahli (1932, *passim*) discuss the 18th ed. of 1835; Hauptman (1978:429) discusses the 5th from 3rd ed. of 1825; Elson (1969:399-400) cites eds. of 1828, 1829, and 1831.

WOODBRIDGE, William C. **Y mSG Am**

- **1824 (01)**
- *System of modern geography, on the principles of comparison and classification ...*
- *Notes*

1. See next entry, especially the note.

WOODBRIDGE, William C. [and Emma WILLARD] **Y mSG Am**

- **{1824}* (01)**
- *A system of universal geography, on the principles of comparison and classification* ... Illustrated with maps and engravings; and accompanied by an atlas ...
- HARTFORD; O.D. Cooke & sons
- 18 cm.; 2 p. l., vii-xxx, 336 [2] p.
- ***Libraries*** : CtY, DLC, MH, MnU, OClWHi
- ***Source*** NUC (NW 0432262)
- *Notes*

1. In addition to being issued separately, as here, this book was also issued bound with WILLARD (1822); see note to (02) below. The story can be traced in NUC, where NW 0432275 records what, at first sight, appears to be a different work, as follows: 'Universal geography, ancient and modern; on the principles of comparison and classification. Modern geography by ... Ancient geography by Emma Willard, 19 cm, xxx, 336, 88 p. [Hartford, O.D. Cooke, 1824].' However, a note to NUC [NW 0432264; see (02), which follows] states: 'First edition [of A system of universal geography] published 1821 under title: Universal geography, ancient and modern.' NW 0432264 also notes: 'Bound with Willard, Emma (Hart); Ancient Geography, Hartford.' As a consequence, the two titles (i.e., *A system of universal geography* and *Universal geography, ancient and modern*) will be treated as a single series in the sequence that follows.

- **1827, 2nd, improved (02)**
- *A system of universal geography* ... engravings; and accompanied by an atlas, exhibiting, in connexion with the outlines of countries, their climate and productions; the prevailing religions, forms of government, and degrees of civilization; and the comparative size of towns, rivers, and mountains. Published with: Ancient geography, as connected with chronology, and preparatory to the study of ancient history: accompanied with an atlas ... Compiled chiefly from D'Anville, Lavoisine, Malte Brun, and other standard works. To which are added, problems on the globes, and rules for the construction of maps. To accompany the Modern Geography by William C. Woodbridge. Second edition, improved
- HARTFORD, [CT]; Oliver D. Cooke & co. J. & J. Harper, printers
- 18 cm., 2 pts.; Pt. 1, pp. [8], ix-xxxi, 336, 4, [2]; Pt. 2, pp. [5], vi-viii, 9-96
- *Libraries* : DLC, MH, MnU, NIC, **NN**, ScU, ViU : **CaAEU**
- *Notes*

1. Two books animated by a single spirit and published jointly. The two authors had, independently, arrived at a similar method of teaching geography. WOODBRIDGE (1821) represents one version of this method; when it was published, he learned that Mrs. Willard was preparing a similar book. 'It was thought desirable that both should unite in the support of one work, composed of the Modern Geography then prepared by the author, and a system of Ancient Geography which had been used by Mrs Willard in the instruction of her pupils' (WOODBRIDGE, 3rd ed. of 1821, x). When this work (i.e., WOODBRIDGE & WILLARD 1824) was published, both authors held jointly the copyright to each of the two books that, according to their tps., they had written separately. Each also wrote a preface to their separate contributions explaining the history of publication and justifying their contributions.

   The prefaces suggest what the text confirms, namely, that the authors take a systematic approach to their subject, doing so with respect to both physical and human geography. The result is that their books are not special geographies, but rather consist of sets of facts, each of which is presented first systematically, then regionally. The sets of facts are classed together in three major units: physical, civil, and statistical geography. These sets are subdivided, sometimes in ways that overlap to a considerable degree. Regional summaries are provided at a variety of levels.

   No doubt the order in which this material is presented seemed logical to the authors, but the rules that govern the order are not easily discerned, either in the very detailed tables of contents or in the text.

   As the tp. states, the book was accompanied by two atlases, and it was the intention of the authors that students should learn by examining the maps and relating to them the facts presented in the text. Moreover, they were to do so by drawing maps displaying the relevant information. The text provides supplementary information of types not easily mapped.

   The text of *Modern geography* opens with a section on mathematical geography (pp. 1-10). Physical geography occupies pp. 10-164. Civil geography takes up pp. 165-257. It is composed of: races of men, languages, civilization, government, religion, learning, literary institutions (including universities), national character, agriculture, roads, buildings and cities, arts and manufactures, and commerce. Statistical geography (pp. 258-336) deals with the 'sources of national power,' and could be viewed as a condensed but conventional special geography. Woodbridge's contribution ends with a four-page section on geology; it seems very out of place.

Willard's contribution includes summary descriptions of the countries, and, separately the cities known to Classical authors. In addition, she provides surveys of their mythology and history, with the latter being an amalgam of Classical and Biblical traditions.

The text is preceded by two pages of 'recommendations' drawn from book reviews and solicited letters.

2. See also note to (03), below.

- **1829, 3rd (03)**
- *A system of universal geography* ... atlas, exhibiting, in connexion with the outlines of countries, their climate and productions; the prevailing religions, form of government, and degrees of civilization; and the comparative size of towns, rivers, and mountains
- HARTFORD; published by Oliver D. Cooke & Co., W.W. Dean, printer
- 12mo, 19 cm.; pp. xxxi, [1], 336, [2], 4
- ***Libraries*** L : CtHT, CtY, IU, MB, MH, MiU, MsSM, NBuHi, OClWHi
- ***Notes***

1. Intended for university students. It was 'constructed upon the plan of the School Geography, by the same author' (p. ix). This must refer to WOODBRIDGE, 1821*. The revisions for this 3rd ed. were undertaken in Paris with the aid of von Humboldt.

- **{1830} New ... (see below) (04)**
- *A system of universal geography* ... classification. By ... member of the Geographical Society of Paris. A new edition, illustrated with maps and engravings; and accompanied by an atlas, exhibiting in connexion with the outlines of countries, their climate and productions; the prevailing religions, forms of government, and degrees of civilization; and the comparative size of towns, rivers, and mountains. Published for the British provinces in North America
- HARTFORD; Oliver D. Cooke & Co.
- 18.5 cm; pp. [1]-2, [iii]-[xxxii], [1]-338, [1]-4
- ***Libraries*** : : CaOTP
- ***Source*** BoC 7245

- **{1833}, 5th (05)**
- *A system of universal geography* ...
- HARTFORD; O.D. Cooke & co.
- 18 cm.; 2 p. l., ix-xxxi, 24-336, [2], 4, [2] p.
- ***Libraries*** : ICU, MH, NNC, OClWHi, ViU
- ***Source*** NUC (NW 0432266 and 0432278)

- **{1835}, 6th (06)**
- *A system of universal geography* ...
- HARTFORD; Beach & Beckwith
- 20 cm.; pp. xxiv, 336, 2

- ***Libraries*** : GASU, MiD, NBuHi, NN, ViU
- ***Source*** NUC (NW 0432268 and 0432279)

- **{1836}, 7th (07)**
- *A system of universal geography ...*
- HARTFORD; J. Beach
- 12mo, 19 cm.; pp. 336
- ***Libraries*** : MH, MiD, NIC, NN : CaBVaU
- ***Source*** NUC (NW 0432269 and 0432280)

- **{1838}, 8th (08)**
- *A system of universal geography ...*
- HARTFORD; J. Beach
- 19 cm.; 2 p. l., [vii]-xxiv, 336, 2 p.
- ***Libraries*** : CtHT, MH, NN, NNC, ViU
- ***Source*** NUC (NW 0432270 and 0432281)

- **1839, 8th (09)**
- *A system of universal geography* ... Woodbridge. Illustrated ... connection ... mountains
- HARTFORD, [CT]; John Beach; and for sale by the principal booksellers in the United States
- 18 cm.; 2 pts.; Pt. 1, pp. [13], viii-xxiv, 24, 336, 2; Pt. 2, [5], vi-viii, 9-96
- ***Libraries*** : NN
- ***Notes***

1. The 24 pp. that follow the introductory material, and that have been added since 1827, contain 'Questions on maps,' and belong to the tradition of special geography.
2. NUC records further eds. (all Hartford, Belknap and Hammersley, 24, 336, 22 pp.): 1840, 9th (CtHT, TxHU); 1841, 9th (MH); 1841, 10th (MH, ViU); and 1842, 10th (DLC, ViU).

- **1844, new, rev., and enlarged (10)**
- *Woodbridge and Willard's universal geography ...*
- HARTFORD; Belknap and Hamersley
- 19.5 cm.; pp. viii, [11]-38, [49]-474
- ***Libraries*** : DLC, MiD, NBuU, OO : CaBVaU
- ***Source*** NUC (NW 0432259 and 0432287)
- ***Notes***

1. NUC (NW 0432259) records the title as 'System of modern geography.' This slight change in the short title seems unlikely to mean a new work. It is taken, rather, to be a consequence of the major revision that took the book from 336 to 474 pp. in length.
2. NUC records further eds. (Hartford, Belknap and Hamersley, 474 pp., except as noted): 1845 (MH, NN, NNC, ViU : CaBVaU); 1847 (CtY, CU, MH, T); 1849 (ViU); [1850] (CtY); [1858] (CtY); [1861], W.J. Hamersley, pp. n.r. (NN); and [1866], W.J. Hamersley (NNC).
3. Nietz (1961:221) and Sahli (1932, *passim*) are probably discussions of the 7th ed. of 1836; that ed. is also discussed by Hauptman (1978, n. 62); Elson (1964:401, 403) cites the eds. of 1845 and 1866.

WOODBRIDGE, W.C. and E. WILLARD **Gt**

- **1824**
- *Universal geography, ancient and modern ...*
- Certainly a ghost; see note 1 to WOODBRIDGE (1824), above.

WOODBURN, James A[lbert] and J.H. McMILLAN **Y mSG? Am**

- **{1881} (01)**
- *The geographical handbook.* Fifteen hundred questions in geography, with answers
- XENIA, O.[sic]; Aldine
- 16 cm.; 116 p.
- ***Libraries*** : DLC

WOOLEY, J. **?**

- **{1850} (01)**
- *Geography, descriptive*
- ? J.W. Parker
- 8vo
- ***Libraries*** n.r.
- ***Source*** BCB, Index (1837-57:110)

WORCESTER, Joseph Emerson **Y Am**

- **{1819} (01)**
- *Elements of geography, ancient and modern.* With an atlas
- BOSTON; T. Swan
- 17.5 cm.; pp. viii, [9]-322, [2]
- ***Libraries*** : CSt, CtY, DLC, ICRL, IU, MB, NjR, NSyU, OClW, PMA
- ***Source*** NUC (NW 0445659)
- ***Notes***

1. For information about the contents of this book, see notes to (03) and (04), below.

- **{1822}, 2nd (02)**
- *Elements of geography ...*
- BOSTON; Cummings and Hilliard
- 18 cm.; pp. xi, [1], 332, [2]
- ***Libraries*** : AU, DLC, FU, MB, MH, NN
- ***Source*** NUC (NW 0445660)
- ***Notes***

1. NUC records further eds. (Boston, Hilliard, Gray, Little & Wilkins, except as noted): 1824, Cummings, Hilliard, 293 pp. (DLC); 1825, Cummings, Hilliard, 293 pp. (DGU, KyU, MB, MH, NNC); 1827, 293 pp. (DGU, DLC, MB, PP); 1828, 293 pp. (WU); 1829, 271 pp. (DLC, MB, MH, ViU); 1829, 293 pp. (DLC, MH, PHi); and 1830, New, 271 pp. (GU).

- **1831, new (03)**
- *Elements of geography ...*

- BOSTON; Hilliard, Gray, Little and Wilkins
- 12mo, 19 cm.; pp. xi, [1], 324
- ***Libraries*** **L** : DeU, MH
- ***Notes***

1. The preface opens: 'This work, in its original form, was first published in 1819; after passing through two editions, it was stereotyped, and in that state it was printed a number of times; it was then written entirely anew ... After the publication of the second edition, in this new form, it has now been a second time stereotyped; and it may be expected to remain substantially as it now is, till a considerable change shall become desirable' (p. v).

   The preface of WORCESTER (1826, republished in 1838, p. iii) states that this book (*Elements of geography*) was intended for 'academies' and 'higher schools,' and that it had been 'adopted by several colleges among the books which are required to be studied before entering on a collegiate course.'

   It is primarily a special geography, with the United States receiving most attention, but the disproportion is not very great. There are separate sections on physical, Ancient-Classical, and Scripture geography, and on the use of globes. The statistical material is included in a separate section at the end; in some eds. it was transferred to the atlas, which was issued separately.
2. NUC records further eds. (Boston, Hilliard [usually with others], except as noted): 1832, New, 324 pp. (DLC, MB); 1833 (MH, OOxM); 1835, New, 324 pp. (MH, NcU, NNC : CaQQS); 1836, New (KAS, MB, MH); 1838, New, 324 pp. (MB, MH, PU, ViU); and 1839, New, D.H. Williams, 322+ pp. (DLC, MCE).

- **{1839}, improved (04)**
- *Elements of geography, modern and ancient*, with a modern and an ancient atlas
- BOSTON; D.H. Williams
- 18 cm.; vi, 257 p., 1 l., 74 p.
- ***Libraries*** : DLC
- ***Source*** NUC (NW 0445675)
- ***Notes***

1. The Preface still opens with the statement: 'This work, in its original form, was first published in 1819.' It thus seems that the reversing of the order of the terms *ancient* and *modern* in the title does not reflect a change in the book great enough to justify treating it as a new title. All the same, the Preface does state that changes have been made since the book was originally issued, but it does so in terms that are so general as to have little meaning.

   Pp. 11-211 are devoted to special geography, with the U.S. receiving most attention. There follow sections on physical geography, statistical tables, the use of globes, and the construction of maps.

   The second series of numbered pages contains the Ancient-Classical and the Scripture geographies. Throughout, the text consists of numbered paragraphs, with questions on the text keyed by way of numbers.
2. The Preface also states (p. iv): 'The *Modern* and *Ancient Geography* have heretofore been published together, but neither of them separately; but it has been thought advisable, in order to accommodate different wants, to publish the whole together as heretofore, and both the Modern and Ancient part in a separate form; and they are now offered to the public under the following

titles; viz. "*Elements of Modern and Ancient Geography, with a Modern and an Ancient Atlas;*" "*Elements of Modern Geography, with an Atlas;*" and "*Elements of Ancient Classical and Scipture Geography, with an Atlas*".' Given the existence of WORCESTER (1831; see below), where NUC (NW 0445681) points to the separate publication of a book with the title *Elements of geography, with an atlas*, it is possible that this undated preface may actually have first appeared in 1831.
3. BL and NUC record further eds. (all Boston, 257, 74 pp.): 1840, D.H. Williams, improved (: NcD); 1842, D.H. Williams, Rev. (: MH); and 1844, Lewis and Sampson, rev. (L : DLC, MB, MH, OClWHi, PV, ViU); 1844, Lewis and Sampson, revised and improved (: NN); 1846, Lewis & Sampson, rev. and improved (AkU, MWA) [Culler (1945:311) reports *Elements of geography, modern and ancient*, 2nd ed. for this date. According to him it was published by Phillips & Sampson, in Boston]; 1847, Rev., Boston, Lewis & Sampson (MH).
4. Carpenter (1963:254-5) provides a general discussion; Nietz (1961:219) and Sahli (1932, *passim*) discuss (02), and Elson (1964:397) cites it as well as other eds.
5. Microform versions of (01) and the ed. of 1835 are available from CIHM.

WORCESTER, J[oseph] E[merson] **?**

- **{1831}* (01)**
- *Elements of geography*, with an atlas. A new ed.
- BOSTON; Hilliard, Gray, Little and Wilkins
- 19 cm.; pp. xi, 324
- ***Libraries*** : DLC
- ***Source*** NUC (NW 0445681)
- ***Notes***

1. See note 2 to (04) of WORCESTER (1819); see above.

WORCESTER, J[oseph] E[merson] **Y Am**

- **{1826} (01)**
- *An epitome of geography*, with an atlas
- BOSTON; Hilliard, Gray, Little and Wilkins
- 15.5 cm.; vi p., 1 l., 165 p.
- ***Libraries*** : DLC, ICJ, MH, NIC
- ***Source*** NUC (NW 0445731)

- **{1828} (02)**
- ... *An epitome of geography*, with an atlas.
- BOSTON; Hilliard ...
- 20.5 cm.; vii, 165 p.
- ***Libraries*** : MB, MH, NNC
- ***Source*** NUC (NW 0445732), which notes: 'Later published under title: A geography for common schools'
- ***Notes***

1. NUC record further eds. (Boston except as noted): 1830, 162 pp. (DLC); 1834, Philadelphia (MH, NjR); and 1837, Boston, 165 pp. (MH).

▸ **1838 (03)**
▸ *An epitome of geography ...*
▸ BOSTON; Hilliard, Gray & co.
▸ 12mo, 16 cm.; pp. viii, 165
▸ ***Libraries* L**
▸ ***Notes***
1. Approximately equivalent amounts of information are given for each continent. There are short sections on ancient and scripture geography at the end. Intended as a counterpart to WORCESTER (1819) for young children or for others whose education is limited.
2. NUC records an ed.: 1839, Boston; Hilliard, Gray, 176 pp. (NNC, PP) [NW 0445743 notes: 'Originally published, 1826, under title: An epitome of geography'].
3. Elson (1964:399) cites (02) and the ed. of 1839.

WORCESTER, Joseph Emerson **Y Am**
▸ **{1820} (01)**
▸ *An epitome of modern geography*, with maps. For the use of common schools
▸ BOSTON; Cummings and Hilliard
▸ 1 p. l., [iii]-iv, [5]-156 p.
▸ ***Libraries*** : IU, MH, N, Nh, PMA
▸ ***Source*** NUC (NW 0445739)

WORCESTER, Joseph Emerson **Y Am**
▸ **(1839) (01)**
▸ *A geography for common schools ...*
▸ ***Notes***
1. See note 2 in (03) of WORCESTER, J.E. (1826), above.

WORCESTER, Joseph Emerson **Dy Am**
▸ **{1817} (01)**
▸ *A geographical dictionary*, or universal gazetteer; ancient and modern ...
▸ ANDOVER; printed by Flagg and Gould for the author. Published and sold by Henry Whipple, Salem
▸ 22.5 cm., 3 vols.
▸ ***Libraries*** : CSt, DGU, DI-GS, DLC, KyU, MdBP, MeB, MiU, MWA, NIC, NWM, OCl, OClW, OO, PHi, PPL, PPPrHi, PU, PWcS, TxU, ViU
▸ ***Source*** NUC (NW 0445741)

▸ **1817 (02)**
▸ *A geographical dictionary ...* modern
▸ ANDOVER, [MA]; printed by Flagg and Gould for the author. Published and sold by Henry Whipple, Salem
▸ 22.5 cm., 2 vols.; unpaged
▸ ***Libraries*** : **NN**
▸ ***Notes***
1. A special geography arranged in alphabetical order of places, rather than a

gazetteer in the modern sense. Acknowledges drawing on the 'gazetteer of Cruttwell' (CRUTTWELL 1798) and, for ancient geography, 'the learned D'Anville' (p. iv).
2. There was another ed.: 1823, Boston; Cummings and Hilliard, No. 1 Cornhill; 22.5 cm., 2 vols.; Vol. 1; pp. [3], iv-viii, 972; Vol. 2; pp. [2], 960. (DI-GS, DLC, DN, MH, MWA, NcD, Nh, NN, NWM, OO, OOxM, OU, PHi, PP, PV).

WORCESTER, Samuel **Y Am**
- **{1830} (01)**
- *A first book of geography*
- BOSTON; Crocker & Brewster, etc.
- ***Libraries*** : MH

- **{1831}, 2nd, with improvements (02)**
- *A first book of geography*
- BOSTON; Crocker & Brewster. New York; J. Leavitt
- 14.5 cm.; pp. 80
- ***Libraries*** : DLC
- ***Notes***
1. Discussed in Sahli (1941, *passim*), and Brigham and Dodge (1933:15-16); Elson (1964:400) cites (02).

WORKMAN, Benjamin **Y Am**
- **{1790}, 3rd (01)**
- *Elements of geography, designed for young students in that science.* In seven sections - Sect. I. Of the solar system. Sect. II. Of the earth in particular. Sect. III. Of maps and globes. The three foregoing sections contain the scientific or astronomical part of geography, digested in a clear and comprehensive manner. Sect. IV. Of the different religions, governments, and languages of nations. Sect. V. Of the political divisions of the earth, into empires, kingdoms, etc. or the historical part of geography. Sect. VI. Of natural philosophy; or the properties of matter, etc. Sect. VII. Of chronology ... 3rd ed.
- PHILADELPHIA; Printed and sold by John M'Culloch, in Third-street, no. 1, above Market-street
- 14 cm.; pp. iv, [5]-148
- ***Libraries*** : CtY, DLC, MWA, NN, PHi, PU, RPB, RPJCB
- ***Source*** NUC (NW 0451309)
- ***Notes***
1. NUC records further eds. (all Philadelphia, 180 pp., except as noted): 1793, 4th (NNC, PHi, PU); [Elson (1964:393) records 1794, 4th, Boston, I. Thomas]; 1795, 5th (IU, MH, NN, PHi); 1796, 6th (CtY, DLC, MWA, PHi, PU, RPJCB); 1799, 7th (CtY, PHi); 1803, 9th (NNC); 1804, 10th (PHC, PHi); and 1805, 11th, much improved and enlarged, 196 pp. (MB, PHi, PPL); Elson (1964:395) records 1807, 11th.

- **{1807}, 12th (02)**
- *Elements of geography ...*

- PHILADELPHIA; Ebenezer M'Culloch, no. 1, North Third street
- 15 cm.; pp. [5], iv, 5-191
- ***Libraries*** : DLC, **NN**
- ***Notes***

1. The first three sections occupy pp. 5-61; the special geography, dominated by the U.S., pp. 97-161. Asia and Africa are allotted 8 pp. between them.
2. NUC records further eds. (all Philadelphia): 1809, 13th, 191 pp. (CtY, DLC, PBm, PHi, PU); 1811, 14th, 208 pp. (DLC, PHi); 1814, 15th, 216 pp. (CtY, PHi, PPAmP, PPL, PPWa, PSC); 1816, 16th, 216 pp. (DLC, MiU, PPL).
3. Discussed in Brigham and Dodge (1933:8), and cited by Elson (1964:393-5).

WORKMAN, Benjamin **Y Am**

- **{1816} (01)**
- *Epitome of Workman's Geography*, containing such parts only as are necessary to be committed to memory. Published at the desire of Sunday teachers
- PHILADELPHIA; published by Wm. M'Carty; printed by M'Carty & Davis
- 15 cm.; pp. 106
- ***Libraries*** : NjNbS, NNC, PHC, PHi, PP, PPRF
- ***Source*** NUC (NW 0451327)

WORLD ... **Y SG?**

- **{[ca. 1845?]} (01)**
- *The world and its inhabitants*
- LONDON; Darton & Clark
- 20.5 cm.;pp. [18]
- ***Libraries*** : : CaOTP
- ***Source*** St. John (1975, 2:817), which notes: 'Similar in format and design to *Reuben Ramble's travels in the eastern counties of England* by Samuel Clark'

WORLD ... **A**

- **1853 (01)**
- *The world; or, general geography*
- LONDON; the National Society, and sold at the depository, Sanctuary, Westminster
- 12mo, 14 cm.; pp. [1], 28
- ***Libraries*** L
- ***Notes***

1. A very slight special geography.

WORLD ... **OG**

- **1862, new (01)**
- *The world: an introduction to general geography*. New ed.
- LONDON; the National Society's depository, Sanctuary, Westminster
- 16mo, 14 cm.; pp. 32

- ***Libraries*** L
- ***Notes***

1. Even though all the bibliographic information suggests that this is a new ed. of WORLD (1853), it is not a special geography, but indeed a general one as Varen used the word. The preface includes the claim that the objective of the book is 'to serve as a text-book for pupil teachers and the higher classes in National Schools, who are advancing to a scientific study of the subject; and as a help to masters in the arrangement of their lessons.'

WORLD DISPLAYED **mSG**

- **{1760-1} (01)**
- *The world displayed*; or, a curious collection of voyages and travels, selected from the writers of all nations. In which the conjectures and interpolations of several vain editors and translators are expunged, every relation is made concise and plain, and the divisions of countries and kingdoms are clearly and distinctly noted. Illustrated and embellished with variety of maps and prints by the best hands
- LONDON; printed for J. Newbery: and James Hooey, jun., Dublin
- 20 vols.
- ***Libraries*** L : CtY, InU, MH, NN : CaOTP
- ***Source*** NUC (NW 0453610)
- ***Notes***

1. NUC provides information about the contents of the individual volumes; this makes it clear that this is indeed a collection of voyages and not a special geography.
2. On this occasion, the NUC entry does not make it clear that this is the earliest ed.; that fact is confirmed by BL.

WRIGHT, Edward **OG**

- **{1613} (01)**
- *The description and vse of the sphaere deuided into three principal partes*: whereof the first intreateth especially of the circles of the vppermost moueable sphaere, and of the manifould vses of euery one of them seuerally: the second sheweth the plentifull vse of the vppermost sphaere, and of the circles therof ioyntly: the third conteyneth the description of the orbes whereof the sphaeres of the sunne and moone haue beene supposed to be made, with their motions and vses
- LONDON; J. Tap
- 8vo, 18 cm.; pp. 104
- ***Libraries*** EU, L, LU, O, SkPt : CtY, DFo, DLC, NN, NNE, PPL
- ***Source*** Pollard and Redgrave (1946; 1976) and NUC (NW 0465525)
- ***Notes***

1. Included here because the full title provides a glimpse of the Ptolemaic model of the universe, still all but universally accepted 70 years after the death of Copernicus.
2. Facsimile ed., 1696 (Johnson/TOT).
3. Pollard and Redgrave (1946; 1976) and NUC (NW 0465528) record another ed., in 1627 (C, CSyS, L [tp. defective], LU, LUU : CSmH).

WRIGHT, Edward **NG**

- **{1614} (01)**
- *A short treatise of dialing*: shewing, the making of all sorts of sun-dials, horizontal, erect, direct, declining, inclining, reclining; vpon any flat or plaine superficies, howsoeuer placed, with ruler and compasse onely, without any arithmaticall calculation ...
- LONDON; printed by Iohn Beale for William Welby
- 4to
- ***Libraries*** C, L, LWS, MC, O, OChch, OT, [IDM {imperfect}] : DFo
- ***Source*** Pollard and Redgrave (1946; 1976) and NUC (NW 0465535)
- ***Notes***

1. Included here because the full title gives a strong hint as to the meaning, in the seventeenth century, of what was meant by 'dialing' (a word that appears quite often in the full titles of special geographies). Pollard and Redgrave (1976) record another variant of this ed. (C, L, LUU, O, [IDT] : DFo, MeB, MH).

WRIGHT, George Newenham **Dy?**

- **{1834-7} (01)**
- *A new and comprehensive gazetteer*, being a delineation of the present state of the world ... constituting a systematic dictionary of geography ... Illustrated by a series of maps, forming a complete atlas; and a selection of appropriate views ...
- LONDON; Thomas Kelly
- 8vo, 23.5 cm., 5 vols. [4 according to NUC]
- ***Libraries*** L : DLC, IU, OC
- ***Source*** BL and NUC (NW 0466496)
- ***Notes***

1. NUC also records: 'Supplement. Illustrated by a series of appropriate views ... London, T. Kelly, 1838, 23.5 cm, vii, [1]-632 p.'

WYNNE, R. **OG tr**

- **1778 (01)**
- *An introduction to the study of geography; or, a general survey of Europe.* By A.F. Busching, Professor of Divinity and Philosophy at Gottingen. Translated from the second German edition, with improvements
- LONDON; printed for J. Bow, in Pater-noster Row
- 12mo, 16.5 cm.; pp. [2], 220
- ***Libraries*** **L** : MH
- ***Notes***

1. No preface, and hence no statement of the objective, nor of the way in which Büsching's six large 4to volumes have been so reduced in bulk. Not a special geography; instead the information has been organized in systematic form, beginning with political topics (e.g., forms of government), through maps and extent of countries, to climate and economic topics (chiefly types of product, both naturally occurring and subject to human modification), language, population, religion, and military topics. These are discussed first in

general terms, and then some remarks are made associating them with a specific place or places (e.g., porcelain earth with Meissen, p. 49).
2. MH catalogues under Büsching.

WYNNE, Richard, comp. **?**
- **{1787} (01)**
- *A short introduction to geography, to which is added an abridgement of astronomy*. Compiled by ... and translated into French and Italian by Catharine Wynne
- LONDON; printed for the author by J. Ryder
- Pp. xi, 271
- ***Libraries*** : CU, PPL
- ***Source*** NUC (NW 0487018)

Y

YOUNG, Francis **Y**
- **{1870} (01)**
- *Geography, class and home lesson book*
- ? Allman
- ***Libraries*** n.r.
- ***Source*** ECB, Index (1856-1876:130)

- **{1871} (02)**
- *The class and home-lesson book of geography*
- LONDON; Guildford [printed]
- 8vo
- ***Libraries*** L

YOUNG, Francis **Y**
- **{1859} (01)**
- *Elementary geography*. General geography of the world
- LONDON and New York; Routledge, Warnes & Routledge
- 8vo; pp. 79
- ***Libraries*** L : MH
- ***Source*** BL and NUC (NY 0026617). The latter gives the size as 24mo.

YOUNG, Francis **Y**
- **{[1869]} (01)**
- *An explanatory introduction to geography*, etc.
- LONDON; J. Cassell
- 8vo
- ***Libraries*** L

Z

ZOUCH, Richard **mSG**

- **1613 (01)**
- *The dove: or passages of cosmography*
- LONDON; printed for George Norton, and are to be sould at his shop vnder the blacke Bell, neere Temple-Barre
- 8vo, 15 cm.; pp. [71], of which 52 carry verse
- ***Libraries*** L, MRu, O : CSmH, DFo, NNPM
- ***Notes***

1. The one pre–nineteenth-century English representative of the Classical tradition of presenting in verse matters that we today consider unsuited for that literary form, because they are matters of fact. By two criteria it is a special geography; first, the continents are dealt with sequentially, although America is ignored; second, within the continents, the relative location of the countries is always established. The countries of Europe receive relatively great attention, with many references to rivers. Some cities have complete verses devoted to them. According to DNB the poem is written after the manner of the 'Periegesis' of DIONYSIUS (1572).

Sources

Part 1
General Catalogues, Bibliographies, and Related Works

Allibone, S. Austin, 1854. *A Critical Dictionary of English Literature, and British and American Authors, Living and Deceased, from the Earliest Accounts to the Middle of the Nineteenth Century: With forty indexes of subjects*. 3 vols. Philadelphia: Childs and Peterson

Alston, Sandra and Karen Evans, eds., 1985. *A Bibliography of Canadiana: Being items in the Metropolitan Toronto Library relating to the early history and development of Canada*. Second Supplement, vol. 2, 1801-49 (contains entries #6800-8040. The numbers 6685-799 were never allocated to items in the collection; this gap in the sequence is presumably associated with the sequence in which the volumes of the second supplement were published) Toronto: Metropolitan Library Board

- 1986. *A Bibliography of Canadiana: Being items in the Metropolitan Toronto Library relating to the early history and development of Canada*. Second Supplement, vol. 3, 1850-67 (contains entries #8041-9655) Toronto: Metropolitan Library Board

- 1989. *A Bibliography of Canadiana: Being items in the Metropolitan Toronto Library relating to the early history and development of Canada*. Second Supplement, vol. 1, 1512-1800 (contains entries #6287-684) Toronto: Metropolitan Library Board

Amtmann, B., 1971. *Contributions to a short title catalogue of Canadiana*, 4 vols. Montreal: published privately

- 1977. *A Bibliography of Canadian Children's Books and Books for Young People, 1841-1867*. Montreal: Bernard Amtmann

Averley, G., A. Flowers, F.J.G. Robinson, E.A. Thompson, R.V. and P.J. Wallis, assisted by J.G.B. Heal and B. Jones (computer services), and L.E. Menhennet and H. Robson, 1979. *Eighteenth-Century British Books: A subject catalogue, extracted from the British*

Museum General Catalogues of Printed Books, vol. 4, 'Geography and History.' London: Dawson, for Project for Historical Biography, University of Newcastle upon Tyne

Borba de Moraes, Rubens, 1958. *Bibliographia Brasiliana*. 2 vols. Amsterdam and Rio de Janeiro: Colibris Editora

Bowers, Fredson, 1949. *Principles of Bibliographical Description*. New York: Princeton University Press (reissued 1962, Russell and Russell)

Boyle, Gertrude M. and Marjorie Colbeck, eds., 1959. *A Bibliography of Canadiana, First Supplement: Being items in the Public Library of Toronto, Canada, relating to the early history and development of Canada* (contains entries #4647-6286). Toronto: Toronto Public Library

British Catalogue, 1858. *Index to the British Catalogue of Books, published during the years 1837 to 1857 inclusive. Compiled by Sampson Low*. London: Publishers Circular (reprinted 1963 and 1976, New York: Kraus Reprint)

British Museum, 1881-1900. *Catalogue of the Printed Books in the Library of the British Museum*. 58 vols. London: Trustees of the British Museum

– 1959-66. *British Museum, General Catalogue of Printed Books, to 1955*. 263 vols. London: Trustees of the British Museum

– 1968. *British Museum, General Catalogue of Printed Books, Ten-Year Supplement, 1956-1965*. 50 vols. London: Trustees of the British Museum

– 1971-2. *British Museum, General Catalogue of Printed Books, Five-Year Supplement, 1966-1971*. 26 vols. London: Trustees of the British Museum

Brydges, Samuel E., 1805-9. *Censura Literaria. Containing titles, Extracts, and Opinions of old English Books, especially those which are scarce. To which are added Necrographia Authorum, or memoirs of deceased authors; and the Ruminator, consisting of original, moral, and critical essays, with other literary disquisitions*. 10 vols. bound in 5. London: Longman, Hurst, Rees, and Orme. (Vol. 1 was issued anonymously. Only vols. 4-6 have the full title given here. Reprinted 1966, New York: AMS Press)

Catalogue générale, 1897-1981. *Catalogue générale des livres imprimés de la Bibliothèque Nationale*. 231 vols. Paris: l'Imprimerie Nationale

Cox, Edward G., 1935. 'General travels and descriptions,' ch. 3 of Cox, *A Reference Guide to the Literature of Travel: Including Voyages, Geographical Descriptions, Adventures, Shipwrecks and Expeditions*, 1:69-87. Seattle: University of Washington Press

– 1938. 'Geography,' ch. 13 of Cox, *A Reference Guide*, 2:332-57.

Cumulative Title Index to the Library of Congress Shelflist: A combined listing of the MARC and REMARC databases, through 1981. 1983. 158 vols. Arlington, VA: Carrollton Press

English Catalogue, 1877[?]. *Index to the English Catalogue of Books. Compiled by Sampson Low.* vol. 2, 1856 to Jan. 1876. London: Publishers Circular (reprinted 1963 and 1979, New York: Kraus Reprint)

- 1881[?]. *Index to the English Catalogue of Books. Compiled by Sampson Low.* vol. 3, Jan. 1874 to Dec. 1880. London: Publishers Circular (reprinted 1979, New York: Kraus Reprint)

- 1890[?]. *Index to the English Catalogue of Books. Compiled by Sampson Low.* vol. 4, Jan. 1881 to Dec. 1889. London: Publishers Circular (reprinted 1979, New York: Kraus Reprint)

General Catalogue, 1786. *A general catalogue of books in all languages, arts, and sciences, printed in Great Britain, and published in London, from the year MDCC to MDCCLXXXVI. Classed under the several branches of literature, and alphabetically disposed under each head, with their sizes and prices.* London: printed for W. Bent, Paternoster-Row

London Catalogue, 1811. *The London Catalogue of Books, with their sizes and prices. Corrected to August MDCCCXI.* London: printed for W. Bent, Paternoster-Row

Low, Sampson. See (1) British Catalogue, and (2) English Catalogue.

Maggs, F.B., 1962. *Voyages and Travels in All Parts of the World: A Descriptive Catalogue.* London: Maggs Bros.

- 1964. *Voyages and Travels, Australia and the Pacific: A Descriptive Catalogue*, vol. 4. London: Maggs Bros.

National Union Catalog, 1968-81. *The National Union Catalog, pre-1956 Imprints: A cumulative author list representing Library of Congress printed cards and titles reported by other American libraries. Compiled and edited with the cooperation of the Library of Congress and National Union Catalog Subcommittee of the Resources Committee of the Resources and Technical Services Division, American Library Association.* 754 vols. London: Mansell

Nineteenth Century Short Title Catalogue: Series I, Phase I, 1801-1815; extracted from the catalogues of the Bodleian Library, the British Library, the Library of Trinity College, Dublin, the National Library of Scotland, and the University Libraries of Cambridge and Newcastle, 1984. 6 vols. Cambridge: Chadwyck-Healey for Avero

Peddie, Robert A. and Q. Waddington, 1914. *The English Catalogue of Books (including the original 'London catalogue'), giving in one alphabet, under author, title, and subject, the size, price, month and year of publication, and publisher of books issued in the United Kingdom of Great Britain and Ireland: 1801-1836.* London: Publishers' Circle (facsimile ed. 1963, New York: Kraus Reprint)

Peddie, Robert A., 1933. *Subject Index of Books Published up to and including 1880, A-Z.* London: Grafton

Pollard, A.W. and G.R. Redgrave (with the help of others), 1946. *A Short-Title Catalogue of Books Printed in England, Scotland, and*

Ireland, and of English Books Printed Abroad; 1475-1640. London: Bibliographical Society. 1976, 2nd ed.; revised and enlarged by W.A. Jackson, F.S. Ferguson, and K.F. Pantzer; vol. 2 (I-Z) only. 1986, 2nd ed.; revised and enlarged; begun by W.A. Jackson and F.S. Ferguson; completed by K.F. Pantzer; vol. 1 (A-H) only

Sabin, Joseph, 1868-1936. *A Dictionary of Books Relating to America, from Its Discovery to the Present Time*. 29 vols. New York (reprinted 1961-2, bound in 15 vols., Amsterdam: N. Israel)

Staton, Frances and Marie Tremaine, eds., 1934. *A Bibliography of Canadiana: Being items in the Public Library of Toronto, Canada, relating to the early history and development of Canada* (contains entries #1-4646). Toronto: Toronto Public Library

St. John, Judith, 1975. *The Osborne Collection of Early Children's Books, 1476-1910*. 2 vols. Toronto: Toronto Public Library (vol. 1 1959, reprinted with corrections 1975; vol. 2 prepared with the assistance of D. Tenny and H.I. MacTaggart)

Watson, G., ed., 1974. *The New Cambridge Bibliography of English Literature: 600-1600*, vol. 1. Cambridge: Cambridge University Press

Watt, Robert, 1824. *Bibliotheca Britannica; Or a general index of British and Foreign Literature. In two parts: - authors and subjects*. 4 vols. Edinburgh: A. Constable

Wing, Donald, 1945. *Short-Title Catalogue of Books Printed in England, Scotland, Ireland, Wales, and British America, and English Books Printed in Other Countries; 1641-1700*. 3 vols. New York: Index Society. 2nd ed., revised and enlarged, 1972 (vol. 1), 1982 (vol. 2). New York: Index Committee of the Modern Language Association of America

PART 2
BOOKS, PAPERS, AND OTHER PUBLICATIONS RELATED TO SPECIAL GEOGRAPHY AND TO THE HISTORY OF IDEAS

Adams, J.R.R., 1987. 'The mass distribution of geographical literature in Ulster, 1750-1850,' *Geographical Journal* 153(3):383-7

Allford, G.R., 1964. 'The development of geography textbooks used in England before 1902,' M.Ed. Thesis, University of Leeds

Anonymous, 1885. 'Scotland and geographical work,' *Scottish Geographical Magazine* 1(1):17-25

Andrews, H.F., 1986. 'A French view of geography teaching in Britain in 1871,' *Geographical Journal* 152(2):225-31

Anstey, R.L., 1958. 'Arnold Guyot, teacher of geography,' *Journal of Geography* 57(9):441-9

Bagrow, L., 1964. *History of Cartography*; revised and enlarged by R.A. Skelton. London: C.A. Watts

Baker, J.N.L., 1928. 'Nathanael Carpenter and English geography in the seventeenth century,' *Geographical Journal* 71:261-71 (reprinted in Baker 1963a, 1-13)

- 1935. 'Academic geography in the seventeenth and eighteenth centuries,' *Scottish Geographical Magazine* 51:129-44 (reprinted in Baker 1963a, 14-32)

- 1948. 'Mary Somerville and geography in England,' *Geographical Journal* 111:207-22 (reprinted in Baker 1963a, 51-71)

- 1955a. 'Geography and its history,' *Advancement of Science* 12:188-98 (reprinted in Baker 1963a, 84-104)

- 1955b. 'The geography of Bernhard Varenius,' *Transactions and Papers of the Institute of British Geographers* (21):51-60 (reprinted in Baker 1963a, 105-18)

- 1963a. *The History of Geography: Papers by J.N.L. Baker*. Oxford: Basil Blackwell

- 1963b. 'The history of geography in Oxford,' in Baker 1963a, 119-29

Beazley, C.R., 1901. 'Sebastian Munster,' *Geographical Journal* 17:423-5

Billinge, M., 1983. 'The Mandarin dialect: an essay on style in contemporary geographical writing,' *Transactions of the Institute of British Geographers*, New Series, 8(4):400-20

Blouet, Brian W., ed., with the assistance of T.L. Stitcher, 1981. *The Origins of Academic Geography in the United States*. Hamden, CT: Archon Books

Bockenhauer, M.H., 1990. 'Connections: geographic education and the National Geographic Society,' paper delivered at the annual meeting of the American Educational Research Association, Boston, 19 April 1990

Bowen, M., 1981. *Empiricism and Geographical Thought: From Francis Bacon to Alexander von Humboldt*. Cambridge: Cambridge University Press

Bricker, C., with the assistance of R.V. Tooley, 1976. *Landmarks of Mapmaking: An Illustrated Survey of Maps and Mapmaking*. Oxford: Phaidon Press

Brigham, Albert P. and R.E. Dodge, 1933. 'Nineteenth century textbooks in geography,' *Thirty-Second Yearbook of the National Society for the Study of Education*, 3-27

Broc, N., 1980. *La géographie de la renaissance (1420-1620)*. Paris: Bibliothèque Nationale, Comité des Travaux Historiques et Scientifiques, Section de Géographie

Brown, R.H., 1941. 'The American geographies of Jedidiah Morse,' *Annals of the Association of American Geographers* 31:145-217

- 1951. 'A letter to the Reverend Jedidiah Morse,' *Annals of the Association of American Geographers* 41:188-98

Butterfield, H., 1957. *The Origins of Modern Science: 1300-1800*. London: G. Bell (1st ed., 1949)

Büttner, M., 1973. *Die Geographie Generalis vor Varenius: geographisches Weltbild und Providentialehre*. Wiesbaden: Franz Steiner Verlag
- 1978. 'Bartholomäus Keckermann, 1572-1609,' in T.W. Freeman and P. Pinchemel, eds., *Geographers: Biobibliographical Studies* 2:73-9. London: Mansell
- 1979a. 'The significance of the Reformation for the reorientation of geography in Lutheran Germany,' *History of Science* 17(3):151-69
- 1979b. 'Philipp Melanchthon, 1497-1560,' in T.W. Freeman and P. Pinchemel, eds., *Geographers: Biobibliographical Studies* 3:93-7. London: Mansell
Büttner, M. and K.H. Burmeister, 1979. 'Sebastian Münster, 1488-1522,' in T.W. Freeman and P. Pinchemel, eds., *Geographers: Biobibliographical Studies* 3:99-106. London: Mansell
Campbell, E.M.J., 1950. 'The early development of the atlas,' *Geography* 35:187-95
Capel, H., 1981a. *Filosofia y ciencia en la Geografia contemporanea*. Barcelona: Barcanova
- 1981b. 'Institutionalization of geography and strategies of change,' in D.R. Stoddart, 37-69
Carpenter, C., 1963. *History of American Schoolbooks*. Philadelphia: University of Pennsylvania Press
Chomsky, N., 1968. *Language and Mind*. New York: Harcourt, Brace and World
Claval, Paul, 1972. *La pensée géographique: introduction à son histoire*. Paris: Société d'Édition d'Enseignement Supérieur
Cohen, Murray, 1977. *Sensible Words: Linguistic Practice in England, 1640-1785*. Baltimore: Johns Hopkins University Press
Crone, G.R., 1949. 'John Green: notes on a neglected eighteenth century geographer and cartographer,' *Imago Mundi* 6:85-91
- 1951, 'Further notes on Bradock Mead, alias John Green, an eighteenth century cartographer,' *Imago Mundi* 8:69-70
- 1978. *Maps and Their Makers: An Introduction to the History of Cartography*, 5th ed. London: Dawson
Culler, Ned, 1945. 'The development of American geography textbooks, 1840-1890,' Ph.D. Thesis, University of Pittsburgh
Dawson, J.A., 1969. 'Some early theories of settlement location and size,' *Journal of the Town Planning Institute* 55(10):444-8
Dickinson, R.E. and O.J.R. Howarth, 1933. *The Making of Geography*. Oxford: Clarendon Press (reprinted 1976, Westport, CT: Greenwood Press)
Dilke, O.A.W., 1987. 'The culmination of Greek cartography in Ptolemy,' in J.B. Harley and D. Woodward, eds., *The History of Cartography. Vol. I: Cartography in Prehistoric, Ancient, and Medieval Europe and the Mediterranean*. Chicago: University of Chicago Press

Downes, Alan, 1971. 'The bibliographic dinosaurs of Georgian geography (1714-1830),' *Geographical Journal* 137(3):379-87
Dryer, C.R., 1924. 'A century of geographic education in the United States,' *Annals of the Association of American Geographers* 14(3): 117-49
East, W.G., 1956. 'An eighteenth-century geographer: William Guthrie of Brechin,' *Scottish Geographical Magazine* 72(1):32-7
Elson, R.M., 1964. *Guardians of Tradition: American schoolbooks of the nineteenth century*. Lincoln, NE: University of Nebraska Press
Emery, F.V., 1958a. 'English regional studies from Aubrey to Defoe,' *Geographical Journal* 124:315-25
Fahy, G., 1981. 'Geography and geographic education in Ireland from early Christian times to 1960,' *Geographical Viewpoint* 10:5-30
Ferrell, E.H., 1981. 'Arnold Henry Guyot, 1807-1884,' in T.W. Freeman, ed., *Geographers: Biobibliographical Studies* 5:63-71. London: Mansell
Firth, C.H., 1918. *The Oxford School of Geography*. Oxford: B.H. Blackwell
Fox, H.S.A. and D.R. Stoddart, 1975. 'The original *Geographical Magazines*, 1790 and 1874,' *Geographical Magazine* 47(8):482-7
Gilbert, A.H., 1919. 'Pierre Davity and his "geography" and its use by Milton,' *Geographical Review* 8:323-38
Gilbert, E.W., 1951. 'Seven lamps of geography: an appreciation of the teaching of Sir Halford J. Mackinder,' *Geography* 36:21-43 (reprinted in Gilbert 1972, 128-36, as ch. 7, 'Victorian Methods of Teaching Geography')
– 1962. '"Geographie is better than divinitie",' *Geographical Journal* 128(4):494-7 (reprinted in Gilbert 1972, 44-58)
– 1972. *Some British pioneers in Geography*. Newton Abbot, UK: David & Charles
Glacken, C.J., 1967. *Traces on the Rhodian Shore*. Berkeley and Los Angeles: University of California Press
Glick, T.F., 1984. 'History and philosophy of geography,' *Progress in Human Geography* 13(2):275-83
Gordon, C., ed., 1980. *Power/Knowledge: Selected Interviews and Other Writings, 1972-1977, Michel Foucault*. New York: Pantheon Books
Granö, O., 1981. 'External influence and internal change in the development of geography,' in Stoddart, 1981:17-36
Hamelin, Louis-Edmond, 1960. 'Bibliogaphie annotée concernant la pénétration de la géographie dans le Québec,' *Cahiers de Géographie de Québec*, No. 8, 345-58
Hart, P.J., 1957. 'The teaching of geography in nineteenth century Britain,' M.A. Thesis, University of London
Hauptman, L.M., 1978. 'Westward the course of empire: geography

schoolbooks and manifest destiny, 1783-1893,' *The Historian* 40(3): 423-37
Hodgen, M.T., 1953. 'Johann Boemus (fl. 1500): an early anthropologist,' *American Anthropologist* 55:284-94
– 1954. 'Sebastian Muenster (1489-1552): a sixteenth-century ethnographer,' *Osiris* 11:504-29
Jeans, J., 1951. *The Growth of the Physical Sciences*, 2nd ed. Cambridge: Cambridge University Press
Keltie, J.S., 1885. 'Geographical education,' *Scottish Geographical Magazine* 1(10):497-505
Laughlin, C.D. and E.G. d'Aquili, 1974. *Biogenetic Structuralism*. New York: Columbia University Press
Leighly, J., 1977. 'Matthew Fontaine Maury, 1806-1873,' in T.W. Freeman, M. Oughton, and P. Pinchemel, eds., *Geographers: Biobibliographical Studies* 1:59-63. London: Mansell
Levine, P., 1986. *The Amateur and the Professional*. Cambridge: Cambridge University Press
Livingstone, D.N., 1988. 'Science, magic and religion: a contextual reassessment of geography in the sixteenth and seventeenth centuries,' *History of Science* 26(3):268-94
Lochhead, Elspeth N., 1980. 'The emergence of academic geography in Britain in its historical context,' Ph.D. Thesis, University of California, Berkeley.
Lovejoy, A.O., 1936. *The Great Chain of Being*. Cambridge, MA: Harvard University Press
Maclean, K., 1975. 'George G. Chisholm: his influence on university and school geography,' *Scottish Geographical Magazine* 41(2):70-8
– 1988. 'George Goudie Chisholm, 1850-1930,' in T.W. Freeman, ed., *Geographers, Biobibliographical Studies* 12:21-33. London: Mansell
Mayo, W.L., 1964. 'The development of secondary school geography as an independent subject in the United States and Canada,' Ph.D. Thesis, University of Michigan
Mikesell, M., 1981. 'Continuity and change,' in B.W. Blouet, 1-15
Mill, Hugh R., 1930. *The Record of the Royal Geographical Society: 1830-1930*. London: Royal Geographical Society
Newton, R.R., 1977. *The Crime of Claudius Ptolemy*. Baltimore: Johns Hopkins University Press
Nietz, J.A., 1961. *Old Textbooks: Spelling, Grammar, Reading, Arithmetic, Geography, American History, Civil Government, Physiology, Penmanship, Art, Music, as Taught in the Common Schools from Colonial Days to 1900*. Pittsburgh: University of Pittsburgh Press
Ogden, C.K. and I.A. Richards, 1923. *The Meaning of Meaning: A study of the influence of language upon thought and of the science of symbolism*. London: International Library of Psychology, Philosophy, and Scientific Method
Parks, George B., 1928. *Richard Hakluyt and the English Voyages*.

New York: American Geographical Society (2nd ed., 1961. New York: Frederick Ungar)
Piaget, Jean, 1959. *The Language and Thought of the Child*, 3rd ed. London: Routledge and Kegan Paul (original French edition, 1923)
Popper, K.R., 1969. *Conjectures and Refutations: The Growth of Scientific Knowledge*, 3rd ed. London: Routledge and Kegan Paul
Robertson, C.J., 1973. 'Scottish geographers: the first hundred years,' *Scottish Geographical Magazine* 89(1):5-18
Robinson, Brian S., 1971. 'A note on geographical description in the age of discoveries,' *Professional Geographer* 23(3):208-11
– 1973. 'Elizabethan society and its named places,' *Geographical Review* 63(3):322-33
Robinson, H., 1951. 'Geography in the dissenting academies,' *Geography* 36:179-86
Rumble, H.E., 1943. 'Morse's school geographies,' *Journal of Geography* 42:174-80
Russell, B., 1945. *A History of Western Philosophy*. New York: Simon and Schuster
Sahli, J.R., 1941. 'An analysis of early American geography textbooks from 1784 to 1840,' Ph.D. Thesis, University of Pittsburgh
Savard, Pierre, 1962a. 'Les débuts de l'enseignement de l'histoire et de la géographie au Petit Séminaire de Québec (1765-1830), I,' *Revue d'histoire de l'Amérique française* 15(4):509-25
– 1962b. 'Les débuts de l'enseignement de l'histoire et de la géographie au Petit Séminaire de Québec (1765-1830), II,' *Revue d'histoire de l'Amérique française* 16(1):43-62
– 1962c. 'Les débuts de l'enseignement de l'histoire et de la géographie au Petit Séminaire de Québec (1765-1830), III,' *Revue d'histoire de l'Amérique française* 16(2):188-212
A Seventh Centurie of Rare Adventures and Painful Peregrinations in divers parts of the World, 1941. London: Maggs Bros. (being Catalogue No. 706, *Voyages and Travels*, Part VII, issued by Maggs Bros [Booksellers by Appointment to His late Majesty])
Singer, C., 1929. 'Francis Bacon,' *Encyclopaedia Britannica*, 14th ed., 2:883-7. London and New York: Encyclopaedia Britannica
– 1959. *A Short History of Scientific Ideas to 1900*. Oxford: Clarendon Press
Sitwell, O.F.G., 1972. 'John Pinkerton: an armchair geographer of the early nineteenth century,' *Geographical Journal* 138:470-9
– 1980. 'Where to begin? A problem for the writers of special geography in the seventeenth and eighteenth centuries,' *Canadian Geographer* 24:294-9
Stoddart, D.R., 1967. 'Growth and structure of geography,' *Transactions and Papers of the Institute of British Geographers* (41):1-19
– 1981. *Geography, Ideology and Social Change*. Oxford: Basil Blackwell

Tatham, G. 1957. 'Geography in the Nineteenth century,' in T.G. Taylor, 28-69

Taylor, E.G.R., 1930. *Tudor Geography, 1485-1583*. London; Methuen (reprinted 1968, New York: Octagon Books)

- 1934. *Late Tudor and Early Stuart Geography, 1583-1650*. London: Methuen (reprinted 1968, New York: Octagon Books)

- 1937. 'Robert Hooke and the cartographical projects of the late seventeenth century (1666-1696),' *Geographical Journal* 40(6):529-40

- 1940. 'The "English Atlas" of Moses Pitt, 1660-83,' *Geographical Journal* 95:292-9

- 1948. 'The English worldmakers of the Seventeenth century and their influence on the earth sciences,' *Geographical Review* 38:104-12

- 1949. 'The measure of the degree: 300 B.C. - A.D. 1700,' *Geography* 34:121-31

Taylor, T.G., 1957. *Geography in the Twentieth Century*, 3rd ed., enlarged. London: Methuen

Thomson, J. Oliver, 1948. *History of Ancient Geography*. Cambridge: Cambridge University Press

Tooley, R.V., 1952. *Maps and Mapmakers*, 2nd ed., revised. London: Batsford

Tozer, H.F., 1935. *A History of Ancient Geography*, 2nd ed., with additional notes by M. Cary. Cambridge: Cambridge University Press (reprinted 1971, New York: Biblo and Tannen)

Vaughan, J.E., 1972. 'Aspects of teaching geography in England in the early Nineteenth century,' *Pedagogica Historica* 12(1):128-47

- 1985. 'William Hughes, 1818-1876,' in T.W. Freeman, ed., *Geographers: Biobibliographical Studies* 9:47-53. London: Mansell

Wade, Mason, 1954. 'The contribution of Abbé John Holmes to education in the Province of Québec,' *Culture* 15(1):3-16

Warntz, W., 1964. *Geography Now and Then: Some Notes on the History of Academic Geography in the United States*. New York: American Geographical Society, Research Series, No. 25

- 1981. '*Geographia Generalis* and the earliest development of American academic geography,' in B.W. Blouet, 245-63

- 1989. 'Newton, the newtonians, and the *Geographia Generalis Varenii*,' *Annals of the Association of American Geographers* 79(2):165-91

Whitehead, A.N., 1927. *Science and the Modern World*. Cambridge: Cambridge University Press

Wilcock, A.A., 1974. '"The English Strabo": the geographical publications of John Pinkerton,' *Transactions of the Institute of British Geographers* (61):35-45 (March)

- 1975. 'From cosmography to geography,' *Geographical Magazine* 47(8):487-9

- 1986. 'Büsching translated, 1762,' *Transactions of the Institute of British Geographers*, New Series, 11(4):490-5

Williams, R., 1973. *The Country and the City*. New York: Oxford University Press
Wise, J.H., 1969. 'The nature and development of geography in the secondary, grammar, and comprehensive schools of England and Wales,' 3 vols., Ph.D. Thesis, University of Iowa
Wise, M.J., 1975. 'A university teacher of geography,' *Transactions of the Institute of British Geographers* (66):1-16 (November)
Wright, J.K., 1925. *The Geographical Lore of the Time of the Crusades*. New York: American Geographical Society (republished 1965, with an introduction by C.J. Glacken, New York: Dover Publications)
– 1959. 'Some British "grandfathers" of American geography,' in Miller, R. and J.W. Watson, eds., 1959. *Essays in Memory of Alan G. Ogilvie*. London: T. Nelson, 144-65

Index

The Chronology of Publication

All the books that appear in the main entries, except for ghosts (whether certain or merely probable), are listed below in chronological order of publication, by short title preceded by their author. In the case of anonymous books, only a short title is given. The fact that some entry is not the first edition, when that is known, is stated only when it distinguishes between books that would otherwise seem to have been first published in the same year.

1481, *Here begynneth ... this presente volume named the Mirrour of the World*
1482, Higden, *Polychronicon*
1502/1503, Arnold, *The copy of a carete cumposynge the circuit of the worlde*
1511 (or 1521 or 1522), *Of the newe landes*
1520, Rastell? *A new interlude and a mery, of the nature of the iiij elements*
1530, 1532, 1535 Ptolemy, *Here begynneth the compost of Ptholomeus*
1550, Proclus, *The descripcion of the sphere or frame of the worlde*
1553, Münster, *A treatyse of the newe India*
1559, Cuningham, *The cosmographical glasse*
1566, Pliny (the Younger), *A summarie of the antiquities, and wonders of the worlde*
1572, Dionysius, *The Surueye of the world*
1572, Münster, *A brief collection and compendious extract of strau[n]ge and memorable thinges*
1573, Powell, *Certaine brief and necessarie rules of geographie*
1578, Bourne, *A booke called the treasure for traueilers*
1585, Mela, *The worke of Pomponius Mela*
1587, Solinus, *The excellent and pleasant worke of ... Solinus Polyhistor*
1589, Blundeville, *A briefe description of vniuersal mappes and cardes*
1591, Henisch, *The principles of geometrie, astronomie, and geographie*
1594, Blundeville, *M. Blundevill his exercises, containing six treatises*
1599, Abbot, *A briefe description of the whole worlde*
1599, Thorie, *The theatre of the earth*
1599, *Wits theatre of the little world*
1600, Torquemada, *The Spanish Mandeuile of miracles*

1601, Botero, *The travellers breviat*
1601, Galvåo, *The discoveries of the world from their first originall*
1602, Ortelius, *An epitome of Ortelivs his theatre of the world*
1606, Ortelius, *Theatrvm orbis terrarvm*
1607, Stafforde, *A geographicall and anthologicall description of all the ... kingdomes*
1608, *A geographical description of all the empires and kingdoms*
1611, Boemus, *The manners, lawes, and cvstomes of all nations*
1613, Wright, E., *The description and vse of the sphaere*
1613, Zouch, *The dove: or passages of cosmography*
1614, Wright, E., *A short treatise of dialing*
1615, Avity, *The estates, empires, and principallities of the world*
1621, Heylyn, *Microcosmos, or a little description of the great world*
1621, Scribonius, *Naturall philosophy: or a description of the world*
1625, Carpenter, *Geography delineated forth in two bookes*
1627, Speed, *A prospect of the most famovs parts of the world*
1628, Earle, *Microcosmography*
1630, Pemble, *A briefe introdvction to geography*
1631, Barclay, *The mirrour of mindes*
1633, Devereux, R. and others, *Profitable instructions*
1635, Mercator, *Historia mundi: or Mercator's atlas*
1635, Swan, *Speculum mundi*
1636, Mercator, *Atlas or a geographicke description of the ... countries ... of the world*
1636, Raleigh, *Tvbvs historicvs ... discovering all the ... kingdomes of the world*
1637, Taylor, John, *The carriers cosmographie*
1646, Speed, *A prospect of the most famovs parts of the world*
1649a, Gerbier, *The first lectvre, of an introduction to cosmographie*
1649b, Gerbier, *The second lecture being an introduction to cosmographie*
1649c, Gerbier, *The first lectvre, of geographie*
1652, Heylyn, *Cosmographie in four bookes*
1654, Blaeu, *A tutor to astronomy and geography*
1654, Campanella, *A discourse touching the Spanish monarchy*
1657, Clarke, S., *A geographicall description of all the countries*
1657, Cluver, *An introduction into geography, both ancient and moderne*
1658, Fage, *A description of the world*
1659, Moxon, *A tutor to astronomie and geographie*
1659, P., R., *A geographical description of the world*
1659, Porter, *A compendious view, or cosmographical, and geographical description of the whole world*
1660, Le Blanc, *The world surveyed*
1662, *A geographical dictionary*
1665, Moxon, *A tutor to astronomy and geography*
1670, Blome, *A geographical description of the four parts of the world*
1670, Hussey, *Memorabilia mundi: or, choice memories, of the history and description of the world*
1671, Meriton, *A geographical description of the world*
1675, Leybourn, *An introduction to astronomy and geography: being a plain and easie treatise of the globes*

1848b, Hiley, *Elementary geography*
1848, Taylor, Jefferys, *Glance at the globe*
1848, Krishnamohana, *Geography*
1848, Lyon, *The musical geography*
1848, Pelton, *Key to Pelton's new and improved series of outline maps*
1848, Smith, R.M., *Modern geography, for the use of schools, academies, etc*
1848, Taylor, J., *A glance at the globe, and at the world around us*
1849, Fisher, R.S., *The book of the world*
1849, Fowle, *An elementary geography*
1849, Hutchinson, *Geography, easy lessons*
1849, Mather, *Manual of geography*
1849, Mitchell, S.A., *Mitchell's intermediate or secondary geography*
1849, Mortimer, *Near home; or, the countries of Europe*
1849, Nicolay, *'On history and geography,' in: Introductory lectures delivered at Queen's College*
1849, Olney, *Olney's quarto geography*
1849, Reid, A., *A first book of geography*
1849, Reid, H., *Geography, first book of*
1849, Smith, R.M., *The child's first book in geography*
1849, Wilson, R., *Geography simplified*
1849, Wilson, S., *Geography simplified: being a brief summary of the ... great divisions of the earth*
1850, *The college elementary geography*
1850, Farr, *Manual of geography for schools*
1850-7, *A gazetteer of the world*
1850a, Goodrich, S.G., *A comprehensive geography and history, ancient and modern*
1850b, Goodrich, S.G., *A primer of geography*
1850, Hughes, W., *A child's first book of geography*
1850, Milner, T., *A universal geography, in four parts: historical, mathematical, physical, and political*
1850, Putz, *Handbook of modern geography and history*
1850, Sargeant, *Papa and mamma's easy lessons in geography*
1850, Sterne, *A physical and political school geography*
1850, Wooley, J., *Geography, descriptive*
1851, Anderson, J., *A practical system of modern geography for exercises on maps*
1851, Giles, *Geography in question and answer, for the use of little children*
1851, Hart, *A popular system of practical geography, for the use of schools*
1851, Hughes, E., *Geography for elementary schools*
1851a, Hughes, W., *Geography for beginners*
1851b, Hughes, W., *A manual of European geography*
1851, Johnston, A.K., *Dictionary of geography, descriptive ... forming ... [a] gazetteer of the world*
1851, Martin, R.M., *The illustrated atlas and ... history of the world, geographical ... and statistical*
1851, Milner, T.? *Milner's elements of geography for the use of schools*
1851, Pelton, *Key to Pelton's hemispheres*
1851, Webb, *Modern geography simplified*
1852, Allison, *Geography*

Other Titles

* Known not to be the first edition.

Short Titles

In using this short-title index, note the following points:

1. The order of listing is alphabetical, but note that punctuation affects the order.
2. Titles are given in italics.
3. In the case of an anonymous book, its location in the main entries is given, as a rule, by the opening words of the short title; e.g., *An abridgement of geography* follows ABBOT and precedes ADAM, A. In those cases where the rule does not apply, the phrase in the title that is used as the criterion for locating the book in the main entries is given, in the list that follows, in lower case, as though it were the author.
4. Cross references are indicated by the direction to see another book. If the reference is to an author, look for it in the main entries (a number in parentheses indicates a subentry subsequent to the earliest known edition). If the key words are in italics, the reference is to a title in this list.